Lecture Notes in Artificial Intelligence 13515

Subseries of Lecture Notes in Computer Science

More information about this subseries at https://link.springer.com/bookseries/1244

Michelangelo Ceci · Sergio Flesca ·
Elio Masciari · Giuseppe Manco ·
Zbigniew W. Raś (Eds.)

Foundations of Intelligent Systems

26th International Symposium, ISMIS 2022
Cosenza, Italy, October 3–5, 2022
Proceedings

Springer

Editors
Michelangelo Ceci 🆔
Università degli Studi di Bari Aldo Moro
Bari, Italy

Sergio Flesca 🆔
Università della Calabria
Rende, Italy

Elio Masciari 🆔
Università Federico II di Napoli
Naples, Italy

Giuseppe Manco 🆔
ICAR-CNR
Rende, Italy

Zbigniew W. Raś 🆔
University of North Carolina
Charlotte, NC, USA

Polish-Japanese Academy of Information
Technology
Warsaw, Poland

ISSN 0302-9743 ISSN 1611-3349 (electronic)
Lecture Notes in Artificial Intelligence
ISBN 978-3-031-16563-4 ISBN 978-3-031-16564-1 (eBook)
https://doi.org/10.1007/978-3-031-16564-1

LNCS Sublibrary: SL7 – Artificial Intelligence

This Springer imprint is published by the registered company Springer Nature Switzerland AG
The registered company address is: Gewerbestrasse 11, 6330 Cham, Switzerland

Preface

This volume contains the papers selected for presentation at the 26th International Symposium on Methodologies for Intelligent Systems (ISMIS 2022), which was held in Rende, Italy, during 3–5 October, 2022. The Symposium was organized by ICAR-CNR, the Department of Computer Engineering, Modeling, Electronics and Systems Engineering (DIMES) at the University of Calabria, Computer Science Department at University of Bari "Aldo Moro" and DIETI Department at University of Naples "Federico II". ISMIS is a conference series that started in 1986. Held twice every three years, it provides an international forum for exchanging scientific, research, and technological achievements in building intelligent systems. In particular, major areas selected for ISMIS 2022 include health informatics, social media analysis, recommendation systems, natural language processing, explainable AI, intelligent systems, classification and clustering, and complex data analysis. This year, we also organized a special session on "Industrial Applications".

After the pandemic, this is the first edition that the Symposium is held in person, although the hybrid mode is still kept for obvious precautionary reasons. We decided promote in-person participation, in order to renew the usual level of interactions and fruitful exchange of ideas that has always characterized the Symposium in the past.

ISMIS 2022 received 71 submissions that were carefully reviewed by three or more Program Committee members or external reviewers. Papers submitted to the special session on "Industrial Applications" were subject to the same reviewing procedure as those submitted to the regular session. After a rigorous reviewing process, 31 regular papers, 11 short papers and 4 industrial papers were accepted for presentation at the conference and publication in the ISMIS 2022 proceedings volume.

It is truly a pleasure to thank all people who helped this volume come into being and made ISMIS 2022 a successful and exciting event. In particular, we would like to express our appreciation for the work of the ISMIS 2022 Program Committee members and external reviewers who helped assure the high standard of accepted papers. We would like to thank all authors of ISMIS 2022, without whose high-quality contributions it would not have been possible to organize the conference. We are grateful to the organizers of the special session on "Industrial Applications" at ISMIS 2022: Luciano Caroprese and Fabio Fumarola.

The invited talks for ISMIS 2022 were "Adaptive Machine Learning for Data Streams" by Albert Bifet from the University of Waikato and the Institut Polytechnique de Paris, and "Interactive Machine Learning" by Andrea Passerini from the University of Trento. Abstracts of the two invited talks are included in these proceedings. We wish to express our thanks to the invited speakers for accepting our invitation to give the talks.

We are thankful to Ronan Nugent and Anna Kramer of Springer Nature for their work on the proceedings and for supporting the ISMIS 2022 Best Paper and Best Student Paper awards. Finally, we would like to thank our platinum sponsor Relatech S.p.A. (https://www.relatech.com), and gold sponsors Revelis S.r.l. (https://www.reveli

s.eu), Polo ICT (https://poloinnovazioneict.org), Pixeltek (https://www.pxltk.com) and DevSkill (https://www.devskill.school) for their valuable support.

We believe that the proceedings of ISMIS 2022 will become a valuable source of reference for your ongoing and future research activities.

August 2022

Michelangelo Ceci
Sergio Flesca
Elio Masciari
Giuseppe Manco
Zbigniew W. Raś

Organization

Symposium Chair

Giuseppe Manco   ICAR-CNR, Italy

Program Committee Co-chairs

Michelangelo Ceci   Università degli Studi di Bari, Italy
Sergio Flesca   Università della Calabria, Italy
Elio Masciari   Università Federico II di Napoli, Italy

Industry Session Chairs

Luciano Caroprese   ICAR-CNR, Italy
Fabio Fumarola   Prometeia SPA, Italy

Steering Committee Chair

Zbigniew Ras   UNC-Charlotte, USA & Polish-Japanese
     Academy of IT, Poland

Steering Committee Members

Annalisa Appice   Università degli Studi di Bari, Italy
Michelangelo Ceci   Università degli Studi di Bari, Italy
Alexander Felfernig   Graz University of Technology, Austria
Mohand-Said Hacid   Université Claude Bernard Lyon 1, France
Denis Helic   Graz University of Technology, Austria
Nathalie Japkowicz   American University, USA
Marzena Kryszkiewicz   Warsaw University of Technology, Poland
Gerhard Leitner   University of Klagenfurt, Austria
Jiming Liu   Hong Kong Baptist University, Hong Kong
Henryk Rybinski   Warsaw University of Technology, Poland
Andrzej Skowron   Polish Academy of Sciences, Warsaw,
     Poland
Dominik Slezak   University of Warsaw, Poland
Martin Stettinger   Graz University of Technology, Austria

Proceedings Chair

Paolo Mignone   Università degli Studi di Bari, Italy

Program Committee

Esra Akbas   Oklahoma State University, USA
Marharyta Aleksandrova   University of Luxembourg
Troels Andreasen   Roskilde University, Denmark
Giuseppina Andresini   Università degli Studi di Bari, Italy
Annalisa Appice   Università degli Studi di Bari, Italy
Arunkumar Bagavathi   Oklahoma State University, USA
Robert Bembenik   Warsaw University of Technology, Poland
Petr Berka   University of Economics, Prague, Czech
                                               Republic
Gloria Bordogna   CNR, Italy
Jose Borges   University of Porto, Portugal
François Bry   Ludwig Maximilian University of Munich,
                                               Germany
Jerzy Błaszczyński   Poznań University of Technology, Poland
Jianhua Chen   Louisiana State University, USA
Konrad Ciecierski   Research and Academic Computer
                                               Network, Warsaw, Poland
Roberto Corizzo   American University, Washington DC,
                                               USA
Agnieszka Dardzinska   Bialystok University of Technology,
                                               Poland
Stephan Doerfel   Kiel University of Applied Sciences,
                                               Germany
Saso Dzeroski   Jozef Stefan Institute, Slovenia
Christoph F. Eick   University of Houston, USA
Tapio Elomaa   Tampere University, Finland
Alexander Felfernig   TU Graz, Austria
Stefano Ferilli   Università degli Studi di Bari, Italy
Naoki Fukuta   Shizuoka University, Japan
Paolo Garza   Politecnico di Torino, Italy
Martin Gebser   University of Klagenfurt, Austria
Jacek Grekow   Bialystok Technical University, Poland
Mohad-Saïd Hacid   Université Lyon 1, France
Denis Helic   Graz University of Technology
Lothar Hotz   Hamburger Informatik
                                               Technologie-Center, Germany
Michele Ianni   Università della Calabria, Italy

Adam Jatowt	University of Innsbruck, Austria
Roman Kern	Know-Center GmbH, Germany
Dragi Kocev	Jozef Stefan Institute, Slovenia
Roxane Koitz-Hristov	Graz University of Technology, Austria
Grazina Korvel	Vilnius University, Lithuania
Bożena Kostek	Gdansk University of Technology, Poland
Dominique Laurent	Université Cergy-Pontoise, France
Gerhard Leitner	University of Klagenfurt, Austria
Marie-Jeanne Lesot	LIP6 - UPMC, France
Rory Lewis	University of Colorado at Colorado Springs, USA
Antoni Ligeza	AGH University of Science and Technology, Poland
Yang Liu	Hong Kong Baptist University, Hong Kong
Corrado Loglisci	Università degli Studi di Bari, Italy
Henrique Lopes Cardoso	University of Porto, Portugal
Donato Malerba	Università degli Studi di Bari, Italy
Domenico Mandaglio	Università della Calabria, Italy
Yannis Manolopoulos	Open University of Cyprus, Cyprus
Małgorzata Marciniak	Polish Academy of Sciences, Poland
Mamoun Mardini	University of Florida, USA
João Mendes Moreira	INESCTEC, Portugal
Radosław Michalski	Wrocław University of Science and Technology, Poland
Paolo Mignone	Università degli Studi di Bari, Italy
Mikhail Moshkov	KAUST, Saudi Arabia
Agnieszka Mykowiecka	IPI PAN, Poland
Mirco Nanni	ISTI-CNR Pisa, Italy
Amedeo Napoli	LORIA Nancy, France
Vincenzo Pasquadibisceglie	Università degli Studi di Bari, Italy
Ruggero G. Pensa	University of Torino, Italy
Jean-Marc Petit	Université de Lyon & INSA Lyon, France
Ingo Pill	Silicon Austria Labs, Austria
Gianvito Pio	Università degli Studi di Bari, Italy
Luca Piovesan	DISIT, Università del Piemonte Orientale, Italy
Olivier Pivert	IRISA-ENSSAT, France
Lubos Popelinsky	Masaryk University, Czech Republic
Anita Raja	City University of New York, USA
Jan Rauch	Prague University of Economics and Business, Czech Republic
Henryk Rybiński	Warsaw University of Technology, Poland

Hiroshi Sakai	Kyushu Institute of Technology, Japan
Giovanni Semeraro	Università degli Studi di Bari, Italy
Urszula Stanczyk	Silesian University of Technology, Poland
Oscar Tamburis	University of Naples Federico II, Italy
Herna Viktor	University of Ottawa, Canada
Eugenio Vocaturo	Università degli Studi della Calabria, Italy
Alicja Wieczorkowska	Polish-Japanese Academy of Information Technology, Poland
David Wilson	UNC Charlotte, USA
Yiyu Yao	University of Regina, Canada
Beata Zielosko	University of Silesia, Poland

Additional Reviewers

Altieri, Massimiliano
Guarascio, Massimo
Hussain, Hussain
Iovine, Andrea
Islam, Muhammad Ifte Khairul
Kovács, Benjamin
Leite, Bernardo
Lo Scudo, Fabrizio
Minici, Marco
Pellicani, Antonio
Pisani, Francesco S.
Saifuddin, Khaled Mohammed
Sousa, Afonso
Tanvir, Farhan
Tarzariol, Alice
Tassel, Pierre

Invited Talks

Adaptive Machine Learning for Data Streams

Albert Bifet

University of Waikato, and Institut Polytechnique de Paris

Abstract. Advanced analysis of big data streams from sensors and devices is bound to become a key area of data mining research as the number of applications requiring such processing increases. Dealing with the evolution over time of such data streams, i.e., with concepts that drift or change completely, is one of the core issues in stream mining. In this talk, I will present an overview of data stream mining, and I will introduce some popular open source tools for data stream mining.

Interactive Machine Learning

Andrea Passerini

University of Trento

Abstract. With artificial intelligence and machine learning becoming increasingly more pervasive in our societies and everyday lives, there is a growing need for interactive approaches bringing the human in the loop and allowing systems to adapt to the needs and specificities of each user. In this talk I will present some promising frameworks for interactive machine learning, highlighting their pros and cons, and discussing open challenges towards truly human-centric and personalized learning systems.

Contents

Explainability

Intelligent Systems

Classification and Clustering

Complex Data

Medical Applications

Industrial Applications

Social Media and Recommendation

Granular Emotion Detection in Social Media Using Multi-Discipline Ensembles

Robert H. Frye(✉) ⓘ and David C. Wilson

University of North Carolina at Charlotte, Charlotte, NC 28223, USA
{rfrye13,davils}@uncc.edu
https://www.charlotte.edu/

Abstract. A variety of applications across industry and society have started to adopt emotion detection in short written text as a key enabling component. However, the task of detecting fine-grained emotions (e.g. love, hate, sadness, happiness, etc.) in short texts such as social media remains both challenging and complex. Particularly for high-stakes applications such as health and public safety, there is a need for improved performance. To address the need for more accurate emotion detection in social media (EMDISM), we investigated the performance of ensemble classification approaches, which combine baseline models from machine learning, deep learning, and transformer learning. We evaluated a variety of ensemble approaches in comparison to the best individual component model using an EMDISM Twitter dataset with more than 1.2M samples. Results showed that the most accurate ensemble approaches performed significantly better than the best individual model.

Keywords: Emotion detection · Sentiment analysis · Social media · Ensemble · Transformer learning · Machine learning · Deep learning

1 Introduction

Understanding a person's emotional context by way of sentiment analysis or finer-grained emotion detection from written text can play a significant role in intelligent systems and modern applications, such as in commercial, political, or security areas [50]. Sentiment analysis (SA) is an application of Natural Language Processing (NLP) focused on determining the polarity of emotions in a textual or spoken sample (i.e., positive, negative, neutral). On a finer-grained level, emotion detection refines the task of sentiment analysis into classifying a sample as representative of specific emotions (e.g., happy, sad, angry, etc.). Illustrative commercial applications include identifying angry customers based on email content [23] as well as proper routing and escalation of messages to appropriate customer representatives [28].

Correctly identifying specific emotions in written text is challenging, even with richer data where texts are longer and well-written stylistically. However, texts in modern communication are more often aligned in structure with social

© The Author(s), under exclusive license to Springer Nature Switzerland AG 2022
M. Ceci et al. (Eds.): ISMIS 2022, LNAI 13515, pp. 3–12, 2022.
https://doi.org/10.1007/978-3-031-16564-1_1

media interactions (shorter, less formally written), which present even greater challenges. *Emotion detection in social media* (EMDISM) must consider the less formal nature of the communication medium, with little regulation of writing styles and generally smaller sample sizes for analysis [21]. EMDISM is important for a variety of application contexts. For example, marketers and airlines apply sentiment analysis or EMDISM to assess emotional responses to advertising and understand overall customer satisfaction with travel experiences based on social media posts [24,43,47]. Beyond commercial applications, mental health providers monitor social media to identify indicators of depression [14], and security researchers are working to identify emerging threats from extremists [3] and other violent actions [34] from social media posts. Developing improved EMDISM approaches is broadly important for industry and society, and improving accuracy is a key open research question.

Our research is focused on the potential for improving accuracy in EMDISM applications by investigating ensemble approaches. In this paper, we present an in-depth evaluation of ensemble EMDISM approaches combining 15 common classifiers from 3 classification disciplines in 21 unique combinations across 4 categories of ensembles. We discuss key design decisions and experimental results indicating which ensembles were more effective than singleton classifiers and present significance testing demonstrating ensembles are often more accurate than singleton classifiers.

2   Related Work

In previous related research, we characterize three primary types of approach for sentiment analysis and emotion detection: machine learning (ML), deep learning (DL), and transformer learning (TL). Our research focuses on creating ensembles comprised of ML, DL, and TL classifiers, which have been previously applied to the tasks of text-based sentiment analysis or EMDISM. We present background research on individual component DL, ML, and TL classifiers, as well as ensemble approaches for leveraging combinations of component classifier outcomes.

2.1   Classifiers

Traditional **machine learning** (ML) classifiers generally apply logic or statistical analysis for text classification, and were among the earliest text classification algorithms. *Decision trees* have been applied to numerous classification problems, including EMDISM [36], and are a type of supervised learning algorithm, which builds classification structures based on partitioning data into subsets of samples with similar characteristics. Decision trees are one of the easiest classification methods for humans to understand, as they can be presented as graphs resembling trees, where each branch is a decision point and each leaf is a classification node. Ranganathan [36] applied decision trees to Twitter EMDISM of five emotions with reported accuracies between 88% to 96%. *Support vector machine* (SVM) [41] classifiers attempt to define a theoretical hyperplane used to segregate large vectors of sparsely populated data into discrete clusters with maximized distances between clusters, and given the sparse vector representations generated through tokenization of text. SVM has been widely applied to SA and EMDISM [8,32]. *Support vector classification* (SVC) [16] is used for processing high dimensional sparse vectors by "...reducing the number of objects in the training set that are used for defining the classifier." *LinearSVC* [46] is a variant of SVC designed to better scale to larger datasets. *Logistic regression* [18] uses independent variables to predict between binary classes, and has been applied in a one versus rest approach for SA and EMDISM [35].

Deep learning (DL) classifiers utilize layered neural networks and backwards propagation of error correction to create class predictions from tokenized embedding layers. DL classifiers for text classification generally consist of an embedding layer of tokenized text data, one or more hidden layers of decision neurons, and an output layer for predicting sample classes [52]. Complex neural networks have been developed, including *convolutional neural networks* (CNN) [30], which establish progressively smaller filters on samples to retain data about the context of one token to other tokens around it, and *recurrent neural networks* (RNN) which use an internal memory of previous steps to preserve contextual information about the relationships between tokens. *Bidirectional RNNs* (B-RNN) [38] and *long short-term memory* (LSTM) [26] neural networks were adapted versions of RNNs designed to address the vanishing or exploding gradient problem. B-RNNs use stacked RNNs to capture the context before and after a token, by training one RNN with tokens in the original order and the other RNN with tokens in reverse order. LSTM uses a combined forget gate, input gate, hidden memory layer, and output gate at each time step in the training process, and several variations of LSTM have been created including *gated recurrent units* (GRU) [11], *bidirectional GRU* (BiGRU) [10], *bidirectional LSTM* (BiLSTM) [39], and *convolutional LSTM* (C-LSTM) [22]. GRU combines LSTM's input and forget gates and merges the hidden memory layer and cell states, and BiGRU and BiLSTM add a bidirectional layer to GRU and LSTM respectively. C-LSTM adds memory of the class label to each gate in the LSTM layer.

Transformer learning (TL) classifiers, first proposed by Vaswani et al. [42], are a specific type of neural network which replace the convolutions and recurrence of DL classifiers with a paired encoder and decoder and a self-attention mechanism, which combine to effectively capture the context of each token in relation to other tokens in each sample. As TL classifiers avoid the need for recurrence or convolution, they generally require fewer epochs to fine-tune their base models and are more accurate than DL classifiers. *BERT* (Bidirectional Encoder Representations from Transformers) was developed by Devlin et al. [15] and used a masked language model approach to train their base model. BERT achieved an SST-2 accuracy score for the GLUE benchmarks [44] of 91.6% for binary SA. *RoBERTa* [31] attempted to improve upon BERT by training by training with larger batch sizes, more training epochs, and a larger vocabulary. RoBERTa achieved an SST-2 accuracy of 92.9%. *XLNet* [49] avoids the introduction of noise caused by inserting masking and separator tokens during BERT pre-training, and also considers permutations of factorization orders to

Table 1. Ensembles applied to text-based sentiment analysis or emotion detection.

Approach	Ensemble components	Type	Metric	Score
Kang et al. (2018) [27]	Hidden Markov Models	SA	Acc.	86.10%
Xia et al. (2011) [48]	NB, SVM, Maximum Entropy	SA	Acc.	80%–88%
Da Silva et al. (2014) [13]	NB, SVM, Random Forest, Logistic Regression, Lexicon	SA	Acc.	70%–79%
Araque et al. (2017) [2]	NB, ME, SVM, RNN, Lexicon	SA	Acc.	85%–94%
Perikos et al. (2016) [33]	NB, ME, knowledge-based	SA	Acc.	89%
Baziotis et al. (2018) [4]	Bi-directional LSTM ensemble	SA-Irony	Acc.	78.50%
Cao and Zukerman (2012) [9]	Lexicon-based, NB, ensemble SVM	ED-5 star	Acc.	70%–75%
Duppada et al. (2018) [17]	XG Boost and Random Forest	ED-4 class	Acc.	83.60%
Bickerstaffe et al. (2010) [5]	SVM, Decision Trees	ED-4 star	Acc.	49%–76%
Al-Omari et al. (2019) [1]	Fully connected NN, LSTM	ED-4 class	F1	0.67
Yue et al. (2018) [51]	CNN, RCNN, LSTM	ED-5 class	F1	0.468

capture the bidirectional context of tokens and maximize the probability that a token sequence would be present in each permutation. XLNet was 94.4% accurate in the SST-2 task. Lample and Conneau [29] developed the cross lingual model, *XLM*, to extend the concepts of BERT to additional languages, using 7500 training samples from 15 languages. *XLM-RoBERTa* (XLM-R) integrated concepts from XLM and BERT by applying MLM training with a larger vocabulary consisting of 250K tokens from 100 different languages compared to the 30K vocabulary used for BERT. XLM-R reported 95.0% accuracy in the SST-2 task. Clark et al. presented *ELECTRA* (Efficiently Learning an Encoder that Classifies Token Replacements Accurately) [12], which was designed to offset an imbalance caused by introducing masked tokens during pre-training BERT base models but not during fine-tuning. ELECTRA delivered SST-2 accuracy between 89.1% and 96.7% depending on training duration and which dataset was used for fine-tuning.

2.2 Ensembles

Ensemble classifiers are designed to offset the weaknesses of one or more classifiers with the strengths of other classifiers. Hansen and Salamon [25] suggested ensembles can be more accurate than singleton classifiers and that the correct first step for creating ensembles was to assess individual classifiers for accuracy to determine their suitability for inclusion in an ensemble. Boosting [37] is a process whereby iterative training and adjusting of weights is used to turn weak classifiers into strong classifiers, and AdaBoost [20] uses a weighted voting ensemble which is still in popular use. Bootstrap aggregating (bagging) [6] concepts included simple voting among base learners trained on different replicas of data, and this ensemble voting approach is still in use for SA and emotion detection today [4,32,48]. Burke [7] described numerous architectures for creating hybrids (ensembles) for recommender systems, including weighted voting, cascading, and switching approaches, among others. We adopt Burke's characterizations in discussing our ensemble approaches. Several research teams have created and applied ensembles combining various classifiers for sentiment analysis or emotion detection. Table 1 provides a list of ensemble researchers, the ensemble components they assessed, and the metrics reported for each approach [1,2,4,5,9,13,17,27,33,48,51]. Previous ensemble research has generally focused on binary sentiment analysis or classifying a more limited sampling of emotions with one of a few classifiers, whereas we have developed and assessed ensembles to classify a larger number of emotions (7) developed from a broader, cross-disciplinary selection of ML, DL, and TL classifiers.

3 Ensemble Approach and Evaluation

The specific challenge our research addresses focuses on potential performance improvements in finer-grained emotion detection in social media text. To address this challenge, we investigated the potential of ensemble approaches to improve performance in EMDISM. We conducted an in-depth evaluation of ensemble EMDISM approaches combining 15 common classifiers from 3 classification disciplines in 21 unique combinations across 4 categories of ensembles.

3.1 Experimental Setup

Our experiments were completed on a Micro-star International Z390 Gaming Infinite X Plus 9 desktop computer, with 48 GB of RAM, an Intel(R) Core(TM) i7-9700K CPU, and one NVIDIA GeForce RTX 2080 GPU. Our experimental platform was created in Python—using the Scikit-learn library for ML models, partitioning training/testing data, and analyzing results; Keras Tensorflow for DL model creation; the HuggingFace's Transformers and Simple Transformer libraries for TL model fine-tuning; Pandas and Numpy for dataframe and array processing; and NLTK for preprocessing text. We selected the EMDISM dataset developed by Wang et al. [45], hereafter referenced as the *HT* dataset. The HT dataset originally consisted of 2.5M Twitter tweets labeled with seven emotions—**joy, sadness, anger, love, thankfulness, fear**, and **surprise**—which are closely aligned with Ekman's six basic emotions [19]. At the time of our experimentation, the text detail of only 1.2M HT tweets remained available for hydration from Twitter with 349,419 samples of joy, 299,412 of sadness, 261,806 of anger, 153,017 of love, 72,505 of thankfulness, 65,010 of fear, and 11,978 of surprise. We followed common pre-processing steps [39,40] to de-noise the dataset. Specifically, we removed URLs, usernames, hashtags, and numbers, cast all text to lowercase, un-escaped html escape strings, replaced duplicate punctuation with singles (e.g. !!! became !), stripped extra whitespace, and lemmatized verbs. For experimentation, we performed 10-fold cross-validation testing and compared validation loss and accuracy curves to avoid overfitting.

3.2 Analysis of Individual Component Approaches

To create our ensembles, we followed the recommendations of Hansen and Salamon [25] in that we assembled and assessed a cross-discipline list of candidate ML, DL, and TL classifiers, focusing specifically on classifiers which had been applied to the task of sentiment analysis or emotion classification. In assessing

Emotion	Machine Learning Classifiers				Deep Learning Classifiers					Transformer Learning Classifiers				
	Decision Trees	Linear SVC	Logistic Regression	SVM	BiGRU	BiLSTM	C-LSTM	GRU	LSTM	BERT	ELECTRA	RoBERTa	XLM-RoBERTa	XLNet
joy	82.45%	79.35%	78.60%	82.01%	84.89%	85.15%	86.26%	79.33%	83.44%	90.86%	90.57%	90.33%	88.24%	89.92%
sadness	79.87%	64.48%	62.92%	41.76%	76.20%	78.99%	79.98%	71.70%	73.92%	89.50%	87.89%	87.86%	85.80%	87.42%
anger	83.71%	72.14%	68.71%	72.33%	82.45%	79.44%	83.20%	74.62%	82.00%	90.36%	89.89%	88.38%	87.20%	89.26%
love	77.79%	41.31%	35.98%	12.06%	66.24%	63.70%	68.11%	55.03%	60.96%	82.33%	80.53%	78.22%	76.13%	77.97%
thankfulness	81.64%	50.96%	46.46%	44.01%	69.06%	69.74%	71.30%	59.35%	64.47%	84.03%	82.39%	79.13%	77.64%	80.60%
fear	76.45%	28.38%	22.57%	8.45%	57.17%	55.25%	57.17%	42.84%	45.92%	76.94%	74.20%	70.40%	66.89%	73.24%
surprise	81.87%	9.66%	3.06%	1.78%	39.91%	42.61%	43.16%	18.38%	37.43%	64.49%	58.54%	55.22%	46.25%	55.28%

Fig. 1. Heatmap of classification accuracy by emotion for each classifier - greater than 80% - green, 50–80% - yellow, below 50% - red. (Color figure online)

individual models, we focused on base models and common implementations of each approach, including ML classifiers (decision trees, linear SVC, logistic regression, Naïve Bayes, SVM), DL classifiers (GRU, BiGRU, LSTM, C-LSTM, BiLSTM), and TL classifiers (BERT, ELECTRA, RoBERTA, XLM-R, XLNet). For additional detail on hyperparameter selection see [21]. We followed the same basic outline in assessing each model, in that we pre-processed our dataset and saved a clean version for reuse across all models compared. Next we trained or fine-tuned each model, performed 10-fold cross-validation to compute average accuracy, and created a heatmap (see Fig. 1) to assess how each performed in classifying specific emotions. This served to help identify strengths and weaknesses among individual component models, and informed the creation of the ensemble approaches we explored. We selected BERT, the most accurate singleton classifier, as a baseline for comparing ensemble performance.

3.3 Analysis of Ensemble Approaches

Based on the analysis of individual component approaches, we created 21 ensembles, including simple voting, weighted voting, cascading, and cascading/switching ensembles. Simple voting ensembles were created by pooling predictions from selected classifiers, as described by the names of their approaches (e.g. TL(all) is an ensemble including BERT, ELECTRA, RoBERTa, XLM-R, and XLNet), with each component receiving one vote per sample. Weighted voting ensembles were designed to leverage the greater accuracy of decision trees for the least represented classes in the HT dataset, adding votes from decision trees only when fear or surprise were predicted. The weighted voting ensembles are identified with abbreviations, where B is BERT, E is ELECTRA, R is RoBERTa, D is decision trees, F is fear, S is surprise, and 2 (when present) indicates that 2 votes were added whenever decision trees predicted fear (**BER+DS2**) or fear and suprise (**BER+DFS2**) instead of 1 vote. The cascading and cascading/switching ensembles were designed to append new super-class labels to the HT dataset to segment the data into subsets for training individual super-class and sub-class models. For example the cascading ensemble named **BERT 5, Dectree 2** indicates the super-classes were segmented to include the 5 most represented classes (joy, sadness, anger, love, and thankfulness) in one class and the 2 least represented (fear and surprise) in another class. A BERT model was trained to classify each sample as belonging to one of these super-classes and

#	Simple Voting	Accuracy
1	All Models	83.17%
2	BRT+XLMR+XLNET	87.90%
3	CLSTM+TL(all)	81.00%
4	DTree+TL(all)	89.37%
5	DTree+BRT+ELECTRA	88.11%
6	DL(all)	77.83%
7	DL(all)+TL(all)	85.28%
8	ML(all)	65.26%
9	TL(all)	88.46%

#	Weighted Voting	Accuracy
10	BE+DS	88.11%
11	BE+DS2	83.80%
12	BE+DFS	88.11%
13	BE+DFS2	83.49%
14	BER+DS	89.42%
15	BER+DS2	88.06%
16	BER+DFS	89.42%
17	BER+DFS2	88.04%

#	Cascading	Accuracy
18	BERT 4,3	88.23%
19	BERT 2,2,3	87.54%

#	Cascading/Switch	Accuracy
20	BERT 3, Dectree 4	86.37%
21	BERT 5, Dectree 2	88.06%

Fig. 2. Ensemble average accuracy.

#	Ensemble	Average Accuracy	Weighted Precision	Weighted Recall	Weighted F-Measure
16	BER_DFS	89.42%	0.89535	0.89423	0.89441
14	BER_DS	89.42%	0.89535	0.89423	0.89441
4	Dectree_All_TLs	89.37%	0.89332	0.8937	0.89311
9	All_TLs	88.46%	0.88416	0.88456	0.88375
18	BERT 4,3	88.23%	0.88184	0.88225	0.88175
	BERT (baseline)	87.85%	0.87796	0.87851	0.87804

Fig. 3. Comparing 5 most accurate ensembles with the BERT baseline.

this result was passed to one of two other models (a BERT model for the 5 most represented and a decision tree model trained to predict the 2 least represented classes) trained to predict from either the top 5 classes or bottom 2 classes respectively. The cascading hybrid **BERT 4,3** leverages one BERT model fine-tuned for the initial super-class prediction and two additional BERT models fine-tuned to predict within the sub-classes. The entire set of predictions was then reassembled and assessed for accuracy, with significance testing via ANOVA between the 5 most accurate models and the BERT baseline, as well as average accuracy, weighted precision, weighted recall, and weighted f-measure for each.

4   Results and Discussion

Of the individual classifiers we evaluated, the most accurate were the TL algorithms (in order from most to least accurate - BERT, ELECTRA, RoBERTa, XLNet, XLM-R), followed by decision trees, then all DL algorithms (C-LSTM, BiGRU, BiLSTM, LSTM, GRU in descending order), and finally the remaining ML algorithms (Linear SVC, Logistic regression, Naïve Bayes, SVM).

12 of 21 ensembles created were more accurate than the BERT baseline accuracy of 87.851%, including 4 of 9 simple voting ensembles, 6 of 8 weighted voting ensembles, 1 of 2 cascading ensembles, and 1 of 2 cascading/switching ensembles. The most accurate ensembles were weighted voting ensembles BER_DFS and BER_DS, with 89.423% average accuracy. Figure 2 shows accuracy across all tested ensembles and Fig. 3 shows a detail comparison of the accuracy, precision, recall, and f-measure for the top 5 ensembles and the BERT baseline. We also performed a single factor analysis of variance between BERT and the 5 most accurate ensembles and found that the variance was statistically significant, with

a p-value of 9.92e−59. The addition of weighted votes for fear appeared to have little affect on the accuracy of our ensembles, with no difference in accuracy scores for BER_DFS and BER_DS. The ensembles which were less accurate than the BERT baseline consisted primarily of reference models created to assess novel approaches rather than realistically expected to outperform the baseline.

5   Conclusions and Future Work

Results show that ensembles can provide more accurate results than the most accurate single classifier, with at least 5 ensembles providing significantly more accurate results than BERT (89.423% for our best ensemble compared to 87.851% for the baseline). These also showed performance improvement compared to the BERT baseline in precision, recall, and f-measure. Results also showed that simple voting, weighted voting, cascading, and cascading/switching ensembles may all provide measurably more accurate results, when designed to offset the weaknesses of one approach with the strengths of another approach.

Future work includes testing further ensemble variations, including dictionary classifiers, to understand tradeoffs in ensemble architectures, evaluation with additional EMDISM datasets under development, and extending our research to identify imbalance thresholds wherein voting and switching ensembles are most effective. Overall, results demonstrate the potential of ensemble approaches for performance improvement in EMDISM, with the potential to benefit a wide variety of applications that rely on accurate understanding of emotion contexts.

References

1. Al-Omari, H., et al.: EmoDet at SemEval-2019 task 3: emotion detection in text using deep learning. In: Proceedings of the 13th International Workshop on Semantic Evaluation (2019)
2. Araque, O., et al.: Enhancing deep learning sentiment analysis with ensemble techniques in social applications. Expert Syst. Appl. **77**, 236–246 (2017)
3. Asif, M., et al.: Sentiment analysis of extremism in social media from textual information. Telematics Inform. **48**, 101345 (2020)
4. Baziotis, C., et al.: Ntua-slp at semeval-2018 task 3: tracking ironic tweets using ensembles of word and character level attentive RNNs. arXiv:1804.06659 (2018)
5. Bickerstaffe, A., Zukerman, I.: A hierarchical classifier applied to multi-way sentiment detection. In: Proceedings of the 23rd International Conference on Computational Linguistics, pp. 62–70. Association for Computational Linguistics (2010)
6. Breiman, L.: Bagging predictors. Mach. Learn. **24**(2), 123–140 (1996)
7. Burke, R.: Hybrid recommender systems: survey and experiments. User Model. User-Adap. Inter. **12**(4), 331–370 (2002)
8. Burnap, P., et al.: Multi-class machine classification of suicide-related communication on twitter. Online Soc. Networks Media **2**, 32–44 (2017)
9. Cao, M.D., Zukerman, I.: Experimental evaluation of a lexicon-and corpus-based ensemble for multi-way sentiment analysis. In: Proceedings of the Australasian Language Technology Association Workshop 2012, pp. 52–60 (2012)
10. Cho, K., et al.: Learning phrase representations using RNN encoder-decoder for statistical machine translation. arXiv preprint arXiv:1406.1078 (2014)

11. Chung, J., Gulcehre, C., Cho, K., Bengio, Y.: Empirical evaluation of gated recurrent neural networks on sequence modeling. arXiv preprint arXiv:1412.3555 (2014)
12. Clark, K., Luong, M.T., Le, Q.V., Manning, C.D.: ELECTRA: pre-training text encoders as discriminators rather than generators (2020)
13. Da Silva, N.F., Hruschka, E.R., Hruschka, E.R., Jr.: Tweet sentiment analysis with classifier ensembles. Decis. Support Syst. **66**, 170–179 (2014)
14. De Choudhury, M., et al.: Predicting depression via social media. In: Seventh international AAAI conference on weblogs and social media (2013)
15. Devlin, J., Chang, M.W., Lee, K., Toutanova, K.: Bert: pre-training of deep bidirectional transformers for language understanding (2019)
16. Duin, R.P.: Classifiers in almost empty spaces. In: Proceedings 15th International Conference on Pattern Recognition. ICPR-2000, vol. 2, pp. 1–7. IEEE (2000)
17. Duppada, V., Jain, R., Hiray, S.: Seernet at semeval-2018 task 1: domain adaptation for affect in tweets. arXiv preprint arXiv:1804.06137 (2018)
18. Efron, B.: The efficiency of logistic regression compared to normal discriminant analysis. J. Am. Stat. Assoc. **70**(352), 892–898 (1975)
19. Ekman, P.: Basic emotions. In: Handbook of Cognition and Emotion, pp. 45–60 (1999)
20. Freund, Y., Schapire, R.E.: A decision-theoretic generalization of on-line learning and an application to boosting. J. Comput. Syst. Sci. **55**(1), 119–139 (1997)
21. Frye, R.H., Wilson, D.C.: Comparative analysis of transformers to support fine-grained emotion detection in short-text data. In: The Thirty-Fifth International Flairs Conference (2022)
22. Ghosh, S., Vinyals, O., Strope, B., Roy, S., Dean, T., Heck, L.: Contextual LSTM (CLSTM) models for large scale NLP tasks. arXiv preprint arXiv:1602.06291 (2016)
23. Gupta, N., Gilbert, M., Fabbrizio, G.D.: Emotion detection in email customer care. Comput. Intell. **29**(3), 489–505 (2013)
24. Gupta, S.: Applications of sentiment analysis in business. Towards Data Science. https://towardsdatascience.com/applications-of-sentiment-analysis-in-business-b7e660e3de69
25. Hansen, L.K., Salamon, P.: Neural network ensembles. IEEE Trans. Pattern Anal. Mach. Intell. **10**, 993–1001 (1990)
26. Hochreiter, S., Schmidhuber, J.: Long short-term memory. Neural Comput. **9**(8), 1735–1780 (1997)
27. Kang, M., Ahn, J., Lee, K.: Opinion mining using ensemble text hidden Markov models for text classification. Expert Syst. Appl. **94**, 218–227 (2018)
28. Khan, J.: Sentiment analysis : Key to empathetic customer service. Ameyo. https://www.ameyo.com/blog/sentiment-analysis-key-to-empathetic-customer-service
29. Lample, G., Conneau, A.: Cross-lingual language model pretraining (2019)
30. LeCun, Y., Haffner, P., Bottou, L., Bengio, Y.: Object recognition with gradient-based learning. In: Shape, Contour and Grouping in Computer Vision. LNCS, vol. 1681, pp. 319–345. Springer, Heidelberg (1999). https://doi.org/10.1007/3-540-46805-6_19
31. Liu, Y., et al.: Roberta: a robustly optimized Bert pretraining approach (2019)
32. Oussous, A., Lahcen, A.A., Belfkih, S.: Impact of text pre-processing and ensemble learning on Arabic sentiment analysis. In: Proceedings of the 2nd International Conference on Networking, Information Systems & Security, p. 65. ACM (2019)
33. Perikos, I., Hatzilygeroudis, I.: Recognizing emotions in text using ensemble of classifiers. Eng. Appl. Artif. Intell. **51**, 191–201 (2016)

34. Pujol, F.A., Mora, H., Pertegal, M.L.: A soft computing approach to violence detection in social media for smart cities. Soft. Comput. **24**(15), 11007–11017 (2019). https://doi.org/10.1007/s00500-019-04310-x
35. Ramadhan, W., Novianty, S.A., Setianingsih, S.C.: Sentiment analysis using multinomial logistic regression. In: 2017 International Conference on Control, Electronics, Renewable Energy and Communications (ICCREC), pp. 46–49. IEEE (2017)
36. Ranganathan, J., Hedge, N., Irudayaraj, A., Tzacheva, A.: Automatic detection of emotions in twitter data-a scalable decision tree classification method. In: Proceedings of the RevOpID 2018 Workshop on Opinion Mining, Summarization and Diversification in 29th ACM Conference on Hypertext and Social Media (2018)
37. Schapire, R.E.: The strength of weak learnability. Mach. Learn. **5**(2), 197–227 (1990)
38. Schuster, M., Paliwal, K.K.: Bidirectional recurrent neural networks. IEEE Trans. Signal Process. **45**(11), 2673–2681 (1997)
39. Smetanin, S.: Emosense at semeval-2019 task 3: Bidirectional LSTM network for contextual emotion detection in textual conversations. In: Proceedings of the 13th International Workshop on Semantic Evaluation, pp. 210–214 (2019)
40. Symeonidis, S., et al.: A comparative evaluation of pre-processing techniques and their interactions for twitter sentiment analysis. Expert Syst. Appl. **110**, 298–310 (2018)
41. Vapnik, V.: The nature of statistical learning theory. Springer, New York (2000). https://doi.org/10.1007/978-1-4757-3264-1
42. Vaswani, A., et al.: Attention is all you need. In: Advances in Neural Information Processing Systems 30 (2017)
43. Walther, C.: Sentiment analysis in marketing: What are you waiting for? CMS Wire. https://www.cmswire.com/digital-marketing/sentiment-analysis-in-marketing-what-are-you-waiting-for/
44. Wang, A., Singh, A., Michael, J., Hill, F., Levy, O., Bowman, S.R.: Glue: a multitask benchmark and analysis platform for natural language understanding (2019)
45. Wang, W., et al.: Harnessing twitter "big data" for automatic emotion identification. In: 2012 International Conference on Privacy, Security, Risk and Trust and 2012 International Conference on Social Computing, pp. 587–592. IEEE (2012)
46. Wang, X., et al.: A novel hybrid mobile malware detection system integrating anomaly detection with misuse detection. In: Proceedings of the 6th International Workshop on Mobile Cloud Computing and Services, pp. 15–22. ACM (2015)
47. Wolfe, J.: Want faster airline customer service? try tweeting. The New York Times. https://www.nytimes.com/2018/11/20/travel/airline-customer-service-twitter.html
48. Xia, R., Zong, C., Li, S.: Ensemble of feature sets and classification algorithms for sentiment classification. Inf. Sci. **181**(6), 1138–1152 (2011)
49. Yang, Z., Dai, Z., Yang, Y., Carbonell, J., Salakhutdinov, R., Le, Q.V.: Xlnet: generalized autoregressive pretraining for language understanding (2020)
50. Yue, L., Chen, W., Li, X., Zuo, W., Yin, M.: A survey of sentiment analysis in social media. Knowl. Inf. Syst. **60**(2), 617–663 (2018). https://doi.org/10.1007/s10115-018-1236-4
51. Yue, T., Chen, C., Zhang, S., Lin, H., Yang, L.: Ensemble of neural networks with sentiment words translation for code-switching emotion detection. In: Zhang, M., Ng, V., Zhao, D., Li, S., Zan, H. (eds.) NLPCC 2018. LNCS (LNAI), vol. 11109, pp. 411–419. Springer, Cham (2018). https://doi.org/10.1007/978-3-319-99501-4_37
52. Zhang, L., et al.: Deep learning for sentiment analysis: a survey. Wiley Interdisc. Rev. Data Min. Knowl. Discov. **8**(4), e1253 (2018)

Sentiment Polarity and Emotion Detection from Tweets Using Distant Supervision and Deep Learning Models

Muhamet Kastrati[1(✉)], Marenglen Biba[1], Ali Shariq Imran[2],
and Zenun Kastrati[3]

[1] Department of Computer Science, University of New York Tirana, Tirana, Albania
muhamet.kastrati@gmail.com, marenglenbiba@unyt.edu
[2] Department of Computer Science, Norwegian University of Science
and Technology (NTNU), 2815 Gjøvik, Norway
ali.imran@ntnu.no
[3] Department of Informatics, Linnaeus University, 351 95 Växjö, Sweden
zenun.kastrati@lnu.se

Abstract. Automatic text-based sentiment analysis and emotion detection on social media platforms has gained tremendous popularity recently due to its widespread application reach, despite the unavailability of a massive amount of labeled datasets. With social media platforms in the limelight in recent years, it's easier for people to express their opinions and reach a larger target audience via Twitter and Facebook. Large tweet postings provide researchers with much data to train deep learning models for analysis and predictions for various applications. However, deep learning-based supervised learning is data-hungry and relies heavily on abundant labeled data, which remains a challenge. To address this issue, we have created a large-scale labeled emotion dataset of 1.83 million tweets by harnessing emotion-indicative emojis available in tweets. We conducted a set of experiments on our distant-supervised labeled dataset using conventional machine learning and deep learning models for estimating sentiment polarity and multi-class emotion detection. Our experimental results revealed that deep neural networks such as BiLSTM and CNN-BiLSTM outperform other models in both sentiment polarity and multi-class emotion classification tasks achieving an F1 score of 62.21% and 39.46%, respectively, an average performance improvement of nearly 2–3 percentage points on the baseline results.

Keywords: Sentiment polarity · Emotion detection · Distant supervision · Emoji · Deep learning · Twitter · Classification

1 Introduction

Nowadays, microblogging and social networks are highly influential in a wide range of settings, from daily communication, sharing ideas, opinions, emotions,

M. Ceci et al. (Eds.): ISMIS 2022, LNAI 13515, pp. 13–23, 2022.
https://doi.org/10.1007/978-3-031-16564-1_2

and reactions with others, shopping behavior, political issues, and reacting to crises, just to name a few [12]. Over the past few years, researchers have shown a growing interest in text-based emotion detection on online social networks, notably Twitter and Facebook. The huge amount of text generated by Twitter users is a rich source to obtain people's emotions, which are an integral part of human life and have a strong influence on people's behaviors and actions [27].

Emotion detection from text is a sub-field of sentiment analysis that aims to extract and analyse emotions that can be explicit or implicit in the sentence [20]. While sentiment analysis is concerned with classifying sentiments as positive, negative, or neutral, emotion detection on the other hand deals with extracting fine-grained emotions such as anger, disgust, fear, joy, sadness, and surprise.

There are various learning approaches used to detect emotions in text, including the lexicon- and rule-based approach [17,23], the machine learning-based approach [21,22,28,29], and the deep learning-based approach [1,5,10,11,24].

Machine learning and deep learning models are widely employed to build sentiment analysis and emotion recognition systems [7,10]. More recently, deep neural networks such as CNN and RNN (including its variants LSTM and GRU) have gained popularity due to the state-of-the-art performance obtained on various natural language processing (NLP) tasks [13]. Supervised learning is the most widely used approach in machine learning, including deep and shallow learning [19]. However, training supervised learning models require a large amount of human-labeled data, which is not the case for several real-world applications, and text emotion detection is no exception [28].

To address this issue, we have collected a large-scale emotion dataset of tweets from Twitter. Inspired by the research study conducted in [3], emotion-indicative emojis are used for automatic labeling of the dataset. Then, several supervised conventional machine learning algorithms and deep learning models are tested on the newly collected dataset to establish the baseline results and examine an approach on sentiment polarity and emotion detection that better suits the dataset in order to improve the performance of the classifier models.

The core contributions of this work are:

– Collecting and curating a real-world large-scale dataset of tweets that are automatically labelled with categorical emotions based on Ekman's model [8] employing distant supervision using emotion indicative emojis.
– New knowledge with regard to performance comparison of supervised conventional machine learning algorithms and deep neural networks for sentiment polarity classification and emotion detection on our created dataset.

The rest of the paper is organized as follows: Sect. 2 presents related work on emotion analysis and approaches used for dataset creation. Section 3 presents the research method followed by an overview of the experimental settings provided in Sect. 4. Section 5 presents the results and analysis, while conclusions and directions for future work are given in Sect. 6.

2   Related Work

During the past decade, several studies have been conducted with regard to the sentiment analysis tasks in Twitter posts. Most of these studies can generally be grouped into two main research directions based on their core contributions: i) data curation/labeling techniques for sentiment analysis tasks, ii) polarity/emotion classification. The first group entails studies concerning data collection and (semi)automatic labeling techniques. For instance, the research work conducted in [9], introduced for the first time distant supervision labels (emoticons) for classifying the sentiment polarity of tweets. The study presents one of the most widely used Twitter sentiment datasets for sentiment analysis tasks known as Sentiment140. Another similar study that uses a distant supervision strategy for automatic labeling is presented in [6]. In particular, hashtags and text emoticons for sentiment annotation are applied in both studies to generate labels. A similar study that applies emojis as distantly supervised labels to detect Plutchik's emotions is conducted in [26].

There is another strand of research that focuses on creating datasets for the emotion detection task. For example, the research study in [22] presents Twitter Emotion Corpus annotated using distant supervision with emotion-specific hashtags for emotion annotation. An extended dataset called Tweet Emotion Intensity dataset is presented later in [21] where the authors created the first dataset of tweets annotated for anger, fear, joy, and sadness intensities using best-worst scaling technique. The researchers in [16] present the first emoji sentiment lexicon, known as the Emoji Sentiment Ranking as well as a sentiment map that consists of 751 most frequently used emojis. The sentiment of the emojis is computed from the sentiment of the tweets in which they occur. Similar work is conducted in [3] where a large-scale dataset of tweets in Urdu language for sentiment and emotion analysis is presented. The dataset is automatically annotated with distant supervision using emojis. A list of 751 most frequently used emojis are applied for annotation.

The second group of research works focuses on polarity and emotion classification using conventional machine learning algorithms and deep neural networks. Such a study is conducted in [24], where the authors proposed a classification approach for emotion detection from text using deep neural networks including Bi-LSTM, and CNN, with self-attention and three pre-trained word-embeddings for words encoding. Another similar example where LSTM models are used for estimating the sentiment polarity and emotions from Covid-19 related tweets is proposed in [10] and in [2]. The later study also introduced a new approach employing emoticons as a unique and novel way to validate deep learning models on tweets extracted from Twitter. Another study focusing on emotion recognition using both emoticon and text with LSTM is conducted in [11].

3   Design and Research Methodology

This study uses a quantitative research approach composed of five major phases. The first phase entails the collection of emoji tweets on Twitter, belonging to the time period from 01 January until 31 December 2021. To be able to collect enough tweets to meet our needs, we selected 41 emojis indicative of the emotion used in [3] and collected tweets that contained at least one of the selected emojis, and only those tweets that were tagged by Twitter as English (retweets excluded). In the second phase of the study, a text pre-processing is performed to remove extra attributes related to tweets (author id, date of creation, language, source, etc.), duplicate tweets, extract emojis from tweets, remove hashtags/mentions, URLs, emails, phone number, non-ASCII characters and tweets with length less or equal to five characters. Additionally, all tweets were converted to lowercase. In the third phase, automatic labeling of collected tweets is carried out through distant supervision using emotion-indicative emojis. Table 1 shows an example of mapping emojis to emotion and sentiment, respectively. The emoji description and sentiment are used to do the mapping from emoji into one of Ekman's six basic categorical emotions. Consequently, the sentiment polarity class is derived from emotions, including positive sentiment polarity from joy and surprise, and negative sentiment polarity from anger, disgust, fear and sadness. In the fourth phase, a representation model to prepare and transform the tweets to an appropriate numerical format to be fed into the emotion classifiers is performed. A bag of word representation model with its implementation, term frequency inverse document frequency ($tf - idf$) is employed.

The final phase of the study involves the sentiment analyser (binary classification) and the emotion analyser for the multi-class classification of tweets along the six basic categorical emotions, namely anger, disgust, fear, joy, sadness, or surprise. The analyser involves several classifiers including conventional machine learning algorithms and deep neural networks for emotion detection. A high-level pipeline of the proposed sentiment and emotion analyser depicting all the five phases elaborated above is illustrated in Fig. 1.

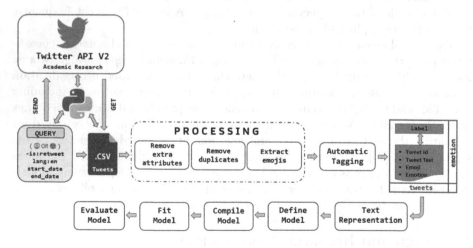

Fig. 1. High-level pipeline of the proposed solution.

Table 1. Example of mapping emojis to emotion labels.

Tweet	Emoji Description	Emotion	Polarity
He's slowly taking over my poor, confused heart 😽	[Angry face, -0.299]	Anger	Negative
Those Comrade memes are shitty but I can't lie some of them manage to Crack me up 😂	[Face with tears of joy, 0.221]	Joy	Positive
I take internet beef too personally cannot get into any sort of argument here bc it will ruin my month bc it'll be all I think about 😭	[Loudly crying face, -0.093]	Sadness	Negative
OMG that's one of the rare cases where cartoons are REAL! 😱	[Face screaming in fear, 0.190]	Fear	Negative

4   Experimental Settings

This section briefly describes the dataset (emoji tweets) as well as the classifier models used to perform the sentiment and emotion classification task.

4.1   Dataset

The dataset consists of 1,832,279 tweets posted between January 1 and December 31, 2021, with the same distribution of tweets every day. The whole data collection process was conducted through Twitter API v2 for academic research product track using Python 3. The dataset is balanced for sentiment (51% for positive and 49% for negative), but is imbalanced for emotion, and its statistics are given in Table 2.

Table 2. Dataset statistics

Sentiment polarity	# Of instances	%	Emotion	# Of instances	%
Positive	934,435	51	Joy	547,047	30
			Surprise	387,388	21
Negative	897,844	49	Sadness	298,742	16
			Disgust	207,838	11
			Anger	207,514	11
			Fear	183,750	10
Total	1,832,279	100	Total	1,832,279	100

4.2   Conventional Machine Learning Models

The conventional machine learning models employed in this study for sentiment and emotion classification include Naive Bayes (NB), Logistic Regression (LR), Support Vector Machine (SVM), Decision Tree (DT), and AdaBoost, as they are known for their good performance [15] and efficiency even for handling millions of tweets [18]. All the algorithms are trained in scikit-learn library in Jupyter Notebook in Anaconda, with default values for all parameters for all classifiers.

4.3   Deep Neural Networks

We selected DNN, BiLSTM, CNN (Conv-1D), GRU and CNN-BiLSTM combined, as these models are well known for their state-of-the-art performance in almost all NLP tasks, including sentiment and emotion analysis [4,13,14,24,25]. All these models are trained and tested in google colab using Keras Python library for deep learning using the TensorFlow backend. Table 3 presents various deep neural networks along with their model configurations as well as their accuracy obtained on the test set (on 10% test data) for each of the models.

Table 3. Configuration and accuracy of the deep learning models.

Classifier	Model configuration/Parameters	Sentiment Polarity	Emotion detection
DNN	Embedding Layer with 100 Dimension, GlobalMaxPooling1D, Layers with 128, 64, 32 with ReLU, (Dense 2 with Sigmoid)/Dense 6 with Softmax	61.61%	38.40%
CNN (1D)	Embedding Layer with 100 Dimension, Layers with 64, 32 with ReLU, GlobalMaxPooling1D, Dense 32 with ReLU, (Dense 2 with Sigmoid)/Dense 6 with Softmax	60.29%	38.20%
BiLSTM	Embedding Layer with 100 Dimension, BiLSTM Layers with 32, 32 with ReLU, GlobalMaxPooling1D, Dense 10 with ReLU, (Dense 2 with Sigmoid)/Dense 6 with Softmax	62.06%	39.69%
GRU	Embedding Layer with 100 Dimension, GRU Layers with 32, 32 with ReLU, GlobalMaxPooling1D, Dense 10 with ReLU, (Dense 2 with Sigmoid)/Dense 6 with Softmax	62.11%	39.38%
CNN-BiLSTM	Embedding Layer 100, SpatialDropout1D(0.3), Conv1D with 32 with ReLU, BiLSTM with 32 with ReLU, Flatten layer, Dense 64 with ReLU, (Dense 2 with Sigmoid)/Dense 6 with Softmax	62.20%	39.27%

5   Results and Analysis

We conducted a set of experiments to investigate the performance of both conventional machine learning and deep learning models on the classification of sentiment polarity and emotions task. The following parameter settings are used to conduct experiments. Dataset is divided in two sets: training and test sets, with 10% of samples used for testing the model. Model training was set to 50 epochs and the *'EarlyStopping'* criteria with its arguments: *monitor='val_loss'* and *patience* = 3, is used to stop classifiers. The batch size of 2048 gave us the best result.

The findings illustrated in Fig. 2 show that the best performance with regard to F1 score is achieved by deep learning models on both sentiment polarity and emotion classification tasks.

Class-wise performance with respect to F1 score for the task of sentiment polarity classification is shown in Fig. 2. For the sake of space, we present results obtained from only two best performing models, including one from conventional machine learning (Logistic Regression) and one from deep learning (CNN-BiLSTM). The results show that CNN-BiLSTM generally outperforms the Logistic Regression model in sentiment polarity classification achieving an F1 score of 62.21%. It is interesting to note that BiLSTM slightly performs better than Logistic Regression, achieving an F1 score of 59.85% for the negative class and 64.28% for the positive class. This slight improvement is accounted to the network architecture and it might be higher if more complex architectures would have been used to train the BiLSTM.

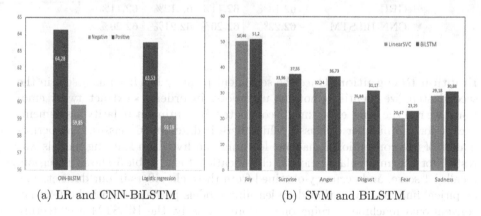

(a) LR and CNN-BiLSTM   (b) SVM and BiLSTM

Fig. 2. F1 score of best performing algorithms on (a) sentiment polarity and (b) emotion detection tasks

Next, we examined the class-wise performance of classifiers on the task of emotion classification. The obtained results from two best performing models, one from conventional machine learning (SVM) and one from deep learning (BiL-STM) are illustrated in Fig. 2. The result show that BiLSTM generally outperforms the SVM (LinearSVC) model in multi-class emotion classification achieving an F1 score of 39.46%. It is worth pointing out that a better performance is achieved by BiLSTM at all classes of emotions.

Sentiment Assessment. The next round of experiments is conducted to investigate the performance of various classifiers on the task of sentiment polarity classification. The results summarized in Table 4 show that a better performance is achieved by deep learning classifiers. In particular, the combined CNN-BiLSTM architecture slightly outperforms the other deep learning models achieving an F1 score of 62.21%.

Table 4. Performance of ML and DL models for sentiment polarity assessment

Classifier	Precision	Recall	F1 score	Accuracy
Naïve Bayes	61.34%	61.33%	61.33%	61.33%
Logistic Regression	61.49%	61.48%	61.48%	61.48%
SVM	61.42%	61.42%	61.42%	61.42%
Decision Tree	64.11%	51.63%	57.20%	51.63%
AdaBoost	57.10%	55.42%	56.25%	55.42%
DNN	61.60%	61.61%	61.60%	61.61%
CNN	60.39%	60.29%	60.34%	60.29%
BiLSTM	62.09%	62.06%	62.07%	62.06%
GRU	62.13%	62.11%	62.12%	62.11%
CNN-BiLSTM	62.22%	62.20%	62.21%	62.20%

Emotion Recognition. Once the sentiment polarity has been assessed, in the second step, we identify emotions in tweets. In order to extract tweet emotions, we run the same experiments conducted for sentiment polarity assessment, except for the number of classes which here is different, 6 classes. The performance of five conventional machine learning and five deep learning models was tested for the multi-class emotion classification task. Table 5 shows precision, recall, F1 score, and accuracy obtained from these classifiers in our dataset. The empirical findings reveal that deep learning models perform slightly better than conventional machine learning ones. More precisely, the BiLSTM architecture slightly outperforms the other deep learning models achieving an F1 score of 39.46%, compared to the best performing conventional machine learning algorithm (NB) which achieved an F1 score of 38.06% on the same task.

Table 5. Performance of conventional ML and DL models for emotion detection

Classifier	Precision	Recall	F1 score	Accuracy
Naïve Bayes	38.84%	37.32%	38.06%	37.32%
Logistic Regression	37.81%	37.93%	37.87%	37.93%
SVM	37.16%	37.97%	37.56%	37.97%
Decision Tree	26.00%	30.48%	28.06%	30.48%
AdaBoost	34.52%	32.07%	33.25%	32.07%
DNN	37.43%	38.40%	37.91%	38.40%
CNN	37.26%	38.20%	37.72%	38.20%
BiLSTM	39.23%	39.69%	39.46%	39.69%
GRU	38.52%	39.38%	38.95%	39.38%
CNN-BiLSTM	38.85%	39.27%	39.06%	39.27%

6   Conclusion and Future Work

This article presented and evaluated the use of emotion-indicative emojis to automatically label a large corpus of tweets with basic categorical emotions they express using Ekman's model. Supervised conventional machine learning and deep learning models are used for both sentiment polarity and detection of emotions from users' tweets on the created dataset. The experimental results showed that the BiLSTM and the combined CNN-BiLSTM architecture outperform the other models with a slight difference in accuracy and F1 score. As future work, we will focus on further increasing the size of the dataset as the deep neural networks benefit from the presence of a huge amount of samples. We will also focus on addressing the class imbalance in the dataset and experiment with filter options to further clean the dataset from problematic instances/tweets. Additionally, experimenting with larger deep learning architectures, pre-trained word embedding models, and attention mechanism, is interesting to be further investigated in the future.

References

1. Abdul-Mageed, M., Ungar, L.: Emonet: fine-grained emotion detection with gated recurrent neural networks. In: Proceedings of the 55th Annual Meeting of the Association for Computational Linguistics, pp. 718–728 (2017)
2. Batra, R., Imran, A.S., Kastrati, Z., Ghafoor, A., Daudpota, S.M., Shaikh, S.: Evaluating polarity trend amidst the coronavirus crisis in peoples' attitudes toward the vaccination drive. Sustainability **13**(10), 5344 (2021)
3. Batra, R., Kastrati, Z., Imran, A.S., Daudpota, S.M., Ghafoor, A.: A large-scale tweet dataset for Urdu text sentiment analysis. arXiv e-prints, p. 2021030572 (2021)
4. Chandio, B.A., Imran, A.S., Bakhtyar, M., Daudpota, S.M., Baber, J.: Attention-based RU-BiLSTM sentiment analysis model for roman Urdu. Appl. Sci. **12**(7), 3641 (2022)

5. Colnerič, N., Demšar, J.: Emotion recognition on twitter: comparative study and training a unison model. IEEE Trans. Affect. Comput. **11**(3), 433–446 (2018)
6. Davidov, D., Tsur, O., Rappoport, A.: Enhanced sentiment learning using twitter hashtags and smileys. In: Coling 2010: posters, pp. 241–249 (2010)
7. Edalati, M., Imran, A.S., Kastrati, Z., Daudpota, S.M.: The potential of machine learning algorithms for sentiment classification of students' feedback on MOOC. In: Arai, K. (ed.) IntelliSys 2021. LNNS, vol. 296, pp. 11–22. Springer, Cham (2022). https://doi.org/10.1007/978-3-030-82199-9_2
8. Ekman, P.: Facial expression and emotion. Am. PSV **48**(4), 384 (1993)
9. Go, A., Bhayani, R., Huang, L.: Twitter sentiment classification using distant supervision. CS224N project report. Stanford **1**(12), 2009 (2009)
10. Imran, A.S., Daudpota, S.M., Kastrati, Z., Batra, R.: Cross-cultural polarity and emotion detection using sentiment analysis and deep learning on covid-19 related tweets. IEEE Access **8**, 181074–181090 (2020)
11. Islam, J., Ahmed, S., Akhand, M., Siddique, N.: Improved emotion recognition from microblog focusing on both emoticon and text. In: 2020 IEEE Region 10 Symposium (TENSYMP), pp. 778–782. IEEE (2020)
12. Kapoor, K.K., Tamilmani, K., Rana, N.P., Patil, P., Dwivedi, Y.K., Nerur, S.: Advances in social media research: past, present and future. Inf. Syst. Front. **20**(3), 531–558 (2018)
13. Kastrati, M., Biba, M.: A state-of-the-art survey on deep learning methods and applications. Int. J. Comput. Sci. Inf. Secur. (IJCSIS) **19**(7), 53–63 (2021)
14. Kastrati, Z., Ahmedi, L., Kurti, A., Kadriu, F., Murtezaj, D., Gashi, F.: A deep learning sentiment analyser for social media comments in low-resource languages. Electronics **10**(10), 1–19 (2021)
15. Kastrati, Z., Imran, A.S.: Performance analysis of machine learning classifiers on improved concept vector space models. Future Gener. Comput. Syst. **96**, 552–562 (2019)
16. Kralj Novak, P., Smailović, J., Sluban, B., Mozetič, I.: Sentiment of emojis. PLoS ONE **10**(12), e0144296 (2015)
17. Krommyda, M., Rigos, A., Bouklas, K., Amditis, A.: Emotion detection in twitter posts: a rule-based algorithm for annotated data acquisition. In: 2020 International Conference on Computational Science and Computational Intelligence (CSCI), pp. 257–262. IEEE (2020)
18. Kumar, H.H., Gowramma, Y., Manjula, S., Anil, D., Smitha, N.: Comparison of various ml and dl models for emotion recognition using twitter. In: 2021 Third International Conference on Intelligent Communication Technologies and Virtual Mobile Networks (ICICV), pp. 1332–1337. IEEE (2021)
19. LeCun, Y., Bengio, Y., Hinton, G.: Deep learning. Nature **521**(7553), 436–444 (2015)
20. Medhat, W., Hassan, A., Korashy, H.: Sentiment analysis algorithms and applications: a survey. Ain Shams Eng. J. **5**(4), 1093–1113 (2014)
21. Mohammad, S.M., Bravo-Marquez, F.: Wassa-2017 shared task on emotion intensity. arXiv preprint arXiv:1708.03700 (2017)
22. Mohammad, S.M., Kiritchenko, S.: Using hashtags to capture fine emotion categories from tweets. Comput. Intell. **31**(2), 301–326 (2015)
23. Mohammad, S.M., Turney, P.D.: Crowdsourcing a word-emotion association lexicon. Comput. Intell. **29**(3), 436–465 (2013)

24. Polignano, M., Basile, P., de Gemmis, M., Semeraro, G.: A comparison of word-embeddings in emotion detection from text using bilstm, CNN and self-attention. In: Adjunct Publication of the 27th Conference on User Modeling, Adaptation and Personalization, pp. 63–68 (2019)
25. Skenduli, M.P., Biba, M., Loglisci, C., Ceci, M., Malerba, D.: User-emotion detection through sentence-based classification using deep learning: a case-study with microblogs in albanian. In: Ceci, M., Japkowicz, N., Liu, J., Papadopoulos, G.A., Raś, Z.W. (eds.) ISMIS 2018. LNCS (LNAI), vol. 11177, pp. 258–267. Springer, Cham (2018). https://doi.org/10.1007/978-3-030-01851-1_25
26. Suttles, J., Ide, N.: Distant supervision for emotion classification with discrete binary values. In: Gelbukh, A. (ed.) CICLing 2013. LNCS, vol. 7817, pp. 121–136. Springer, Heidelberg (2013). https://doi.org/10.1007/978-3-642-37256-8_11
27. Wang, W., Chen, L., Thirunarayan, K., Sheth, A.P.: Harnessing twitter "big data" for automatic emotion identification. In: 2012 International Conference on Privacy, Security, Risk and Trust and 2012 International Conference on Social Computing, pp. 587–592. IEEE (2012)
28. Wood, I., Ruder, S.: Emoji as emotion tags for tweets. In: Proceedings of the Emotion and Sentiment Analysis Workshop, Portorož, pp. 76–79 (2016)
29. Yousaf, A., Umer, M., Sadiq, S., Ullah, S., Mirjalili, S., Rupapara, V., Nappi, M.: Emotion recognition by textual tweets classification using voting classifier (LR-SGD). IEEE Access 9, 6286–6295 (2020)

Disruptive Event Identification in Online Social Network

Aditi Seetha, Satyendra Singh Chouhan[✉], Sanskar Soni, Dev Milan Mehta, and Vinush Vishwanath

MNIT, Jaipur 302017, India
{2021rcp9548,sschouhan.cse,2018ucp1265,2018ucp1382, 2018ucp1491}@mnit.ac.in

Abstract. This paper presents a disruptive event identification framework on the social network platform Twitter. There are numerous data-driven event prediction methods in the literature that employ supervised, unsupervised and semi-supervised techniques. Most of these works use explicit feature extraction techniques, such as TF-IDF, which fails to capture the semantic encoding between the words; and uses a machine learning-based classifier for the event and non-event prediction. The proposed framework is based on classification then clustering approach. First, it uses a sentence transformer, BERT, to encode the sentences and a Deep Neural Network (DNN) classifier to classify event and non-event tweets. Next, the community detection technique is used for detecting optimized event communities (clusters), and these event communities are annotated using various keyword extraction techniques. The effectiveness of the proposed framework is validated using two real-time datasets, showcasing that the proposed framework can successfully identify disruptive events.

Keywords: Disruptive events · Twitter social network · Event classification

1 Introduction

A disruptive event is an event that obstructs a routine process to fulfill its own goals. Nowadays, social media has become a primary source for discussing what is happening around real-world events [1,2]. People tend to share emotions about every incident and event on social media. In recent years, utilizing social network data for identifying disruptive events has increased, and it helps to create situational awareness throughout the crisis [3]. Twitter is a micro-blogging platform where more than 500 million tweets come in a day. For example, Twitter played an essential role in the Farmer Protest in India. The people scattered around the world have amplified the protest using social media. Any misleading information can quickly escalate the peaceful protest into violence. Therefore, a key challenge is identifying events threatening social safety and security. The main goal of developing this framework is to analyze Twitter data to identify disruptive events.

M. Ceci et al. (Eds.): ISMIS 2022, LNAI 13515, pp. 24–34, 2022.
https://doi.org/10.1007/978-3-031-16564-1_3

There are three ways for event detection using Twitter streams: supervised, semi-supervised, and unsupervised [4]. Semi-supervised techniques employ two stages for event detection, i.e., classification and then clustering. In [5], authors proposed a method for event identification in social media streams by employing word embeddings and hierarchical agglomerative clustering. Some researchers used semi-supervised event identification techniques to identify events in which tweets are represented using the TF-IDF feature vector [6]. Next, they used a Naive Bayes classifier to segregate the event-related tweets from the irrelevant tweets and an online clustering algorithm to form clusters. In [7], researcher(s) employed an unsupervised technique to detect newsworthy events by extracting bursty segments and performing clustering based on the probability distribution of segments, retweet count, user diversity, and user popularity.

In the above works, some of the researchers used machine learning classifiers like Naive Bayes for the event and non-event classification. In [6], authors used TF-IDF sentence encoding, which fails to capture the semantic similarity between the words and also computes similarity directly in the word-count space, which is slow for large vocabularies. Some researchers captured semantics and structural information using sentence transformers or multi-weight-view-graph and used the HDBSCAN algorithm for clustering, which requires the number of clusters to be known beforehand [5,8]. The researchers employed online clustering, which only scans data once and fails to detect sub-communities among the communities [6].

In light of the above works, we present a novel framework for disruptive event identification on the Twitter social network. It identifies the events using classification and then clustering approach. First, it encodes the collected tweets using the BERT sentence transformer [9]. Next, deep neural network model is used to separate out event tweets from all tweets. Thereafter, we use a community detection approach for detecting the optimized communities (disruptive event clusters) from the event tweets. Finally, the community representatives are extracted using the keyword extraction techniques.

We have validated the effectiveness of the proposed framework on two datasets collected from Twitter's Tweepy API. The results show that it can successfully identify disruptive events and non-event tweets with an accuracy of 98.9% and effectively forms clusters using of Community Detection method with a Completeness score of 79.2%. Also, it effectively identifies the disruptive event annotation using various keyword extraction techniques with a maximum accuracy of 98%.

2   Disruptive Event Identification Framework

The proposed framework predicts disruptive events from the tweets from the Twitter stream within the current time window t. The proposed method consists of two modules, as shown in Fig. 1: Data Acquisition and Preprocessing module and Clustering and Event Identification module. A detailed discussion of each module is given below.

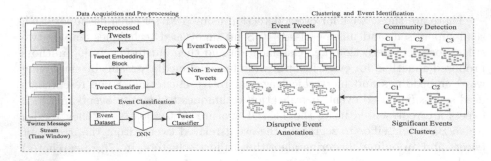

Fig. 1. Proposed framework

2.1 Data Acquisition and Preprocessing

We developed a data acquisition system based on the Twitter social media platform's Application Programming Interface (API). 'Tweepy' API is used for tweet extraction. Tweets are collected by keeping the bounding box fixed to India and the language set to English for a time window of 1 day. We perform basic preprocessing of the tweets as tweets are highly unorganized and involve redundant information. Preprocessed tweets are fed to the tweet embedding block for generating sentence embeddings. In the tweet embedding block, we used a BERT transformer to generate the encoding of preprocessed tweets. Given n preprocessed tweets say, $m_1, m_2, \ldots, m_n$. The preprocessed tweet m_i fed into BERT sentence Encoder and computed token level hidden representation (shown below).

$$[B_{i,0}; \ldots; B_{i,k}; \ldots; B_{i,l}] = Sentence\_encoder(m_i) \tag{1}$$

where,

$$B_{i,k} \in \mathbb{R}^{len(m_i) \times d} \tag{2}$$

where $0 \leq k \leq l$, l is the number of hidden layer, d is the size of hidden representation and $len(m_i)$ is the length of tokenized sentence. Then, pooling function p is applied to $B_{i,k}$ to derive diverse sentence level views $b_{i,k} \in \mathbb{R}^d$ from all layers, i.e., $b_{i,k} = p(B_{i,k})$. In last, we apply the sampling function σ:

$$R^l = \{b_{i,k} \,|0 \leqslant k \leq l|\} \tag{3}$$

BERT employs mean pooling to compute the mean of all the output layers (by default, mean pooling is used). The output R^l is fed as an input to the tweet classifier.

We have used DNN (Dense Neural Network) consisting of two hidden layers with 64 and 32 neurons each followed by a classification layer, which was trained on the Event and Non Event dataset (Sect. 3.1). The preprocessed tweets are given as input to the Tweet classifier, classifying them as event and non-event tweets.

Algorithm 1: Algorithm for predicting Disruptive Events

Input: Twitter message stream for time window t=0,...,n;
Output: Set of Disruptive Events;
Tweet Classifier = DNN($EventDataset$);
for $t = 0$ to n **do**
    for m_i in $TweetStream$ **do**
        $m_i' \leftarrow Preprocess(m_i)$;
        if $TweetClassifier(m_i') == 1)$ **then**
            $D_t = D_t \cup m_i'$
        end
    end
    $C_t \leftarrow CommunityDetection(D_t)$ (Algorithm 2) **for** c in C_t **do**
        if $|c| < threshold$ **then**
            $C_t = C_t - c$
        end
    end
    for c in C_t **do**
        $l_e \leftarrow EventSummarization(c)$;
        $L_e = L_e \cup l_e$;
    end
end
End

2.2 Clustering and Event Identification

Now, these event-related tweets are given to the clustering algorithm to find clusters corresponding to the same disruptive event. We have used the *Louvain* and *Pycombo* community detection approach to create the communities (disruptive event clusters). We have also applied other methods of community detection to compare their clustering efficiency. Only those clusters with a number of tweets greater than the threshold are considered significant clusters. The reasons for using community detection are i) It helps detect the optimized communities for social media networks. ii) It can detect sub-communities within a community. iii) It does not require the number of clusters before applying clustering like the k-means algorithm. Once the clusters are formed, keyword extraction techniques are used to annotate disruptive events. The goal is to summarize the information being discussed in each cluster. Each cluster may contain hundreds of tweets, and extracting the top keywords among the cluster is crucial. We have used various keyword extraction techniques for extracting the most relevant five words within each cluster.

The pseudo-code of the proposed methodology is given in Algorithms 1 and 2. The community detection (Algorithm 2) is explained below. It takes embeddings of event tweets generated using the BERT sentence transformer and a list of threshold values for the[1]neighborhood blending(Algorithm 2). We apply neighborhood search to find the embeddings that resemble other embeddings (Algorithm 2, Step 1). It takes embeddings and threshold values and gives an output match index list and similarity list. The neighborhood search compares each pair of embeddings using the cosine similarity metric.

---

[1] https://pypi.org/project/NeighborBlend/.

Algorithm 2: Community Detection Algorithm

Input: Event Tweets embeddings generated using BERT sentence transformer
and threshold values for neighborhood search ;
Output: Set of communities (clusters) ;
1. **for** *thres in thres_vals* **do**
   | match_index_list, similarities_list = *neighborhood_search*(emb, thres)
   | emb = blend_neighborhood(emb, match_index_lst, similarities_lst)
end
2. graph = *create_graph*(match_index_list);
3. membership = *community_louvain.best_partition*(graph);
4. predicted_communities = |(*memberships*)|;
5. final_communities = *refine_clustering*(emb)

Embedding whose cosine similarity matches with threshold value is kept as neighbor embedding. *Similarity_list* contains similarity score of all the embeddings along with their neighbor's embeddings. The *Match_index_list* contains the index of the embeddings with their neighbor's embedding index. In the next step, the embeddings are modified and updated using the Neighborhood blending strategy (Algorithm 2, line 3). The neighborhood blending takes embedding, *match_index_list*, and *similarity_list* as input and results in the updated embedding. The updated embedding is calculated by adding the weighted sum of its neighborhood embedding with similarity scores as their weights to the original embedding. For example, A embedding v1 matches with three of its neighbors embedding like v2, v3, and v4. The resulted embedding after neighborhood blending is shown in Eq. 4.

$$v1 = N(v1 + v2 * cos(v1, v2) + v3 * cos(v1, v3) + v4 * cos(v1, v4)) \quad (4)$$

where N() is a normalization function.

We have applied the neighborhood search and neighborhood blending strategy for the two different threshold values for improving the tweet embedding (Algorithm 2, Step 1). The next step creates the graph by taking the match_index_list as an input. Louvain community detection is applied to the graph generated by the create_graph function. Community detection assigns different communities to each node of the graph. We calculate the modularity gain for each node by removing the current node from the graph and placing it in the neighbor's community. The node will be placed in the community if the modularity gain is positive; otherwise, the node will remain in the same community. This step is repeated until there is no change in the network and maximum modularity is achieved. Once the communities are created, we take one community as an instance and apply all the above steps until community detection using Louvain and check if the number of communities after refine_clustering is less than the number of communities before it, which means some of the communities have been blended. Otherwise, the number of communities will remain the same as before the *refine_clustering* (Algorithm 2, Step 5). Final communities have been obtained by applying to *refine_clustering*. Now, to find out the significant communities, we check if the number of members in the community is greater than the threshold. Only then we keep them as significant communities. In the next step, event summarization techniques are used for each community to extract the keyword representing the community (Algorithm 1).

3   Result and Discussion

This section discusses experimental details, used datasets, and performance metrics. Moreover, we formulate some research queries for evaluations of the proposed framework. Proposed framework implemented in Python 3.0. The tweet classifier keeps the threshold as 0.6 for the sigmoid function. For each embedding, the closest neighbors were taken with a similarity threshold more significant than 0.5 for the first iteration and 0.6 for the second iteration. The communities were refined using thresholds of 0.4 in the first iteration and 0.5 in the second for refine_clustering function. The number of minimum members present in a community is set to 100. All the parameters were selected using a random search strategy. All the source code and dataset are available at GitHub repository[2,3].

3.1   Dataset Used

In this work, we used two datasets for the evaluation of our proposed framework.

- **Event and Non Event Dataset**
  The dataset contains tweets of past and ongoing events that may cause disruption and tweets related to the non-event category. We label the disruptive event tweets as '1' and non-event tweets as '0'. It consists of 7 attributes (creation time of a tweet, retweet count, follower count, location, username, statuses count, and label(event and non-event tweet) with 263,561 records, out of which 168,706 records are of non-event class and 94,855 records of the event class.
- **Twitter Dataset.** The Twitter dataset consists of tweets gathered from 1st Jan 2022 using Twitter streaming Tweepy API by keeping the time window of 1 day. The total number of tweets extracted per day is almost 25 thousand to 30 thousand from the bounding box fixed to India and the language set to English. For the training and testing phases of our research, we use human annotators to manually label the clusters to evaluate the performance of the clustering algorithm. Please refer to [10] annotation guidelines and agreement measures for more information.

3.2   Performance Metrics Used

We have used standard classification measures: Precision, Recall, F1-score. Apart from these measures, some other performance measures used for clustering's performance evaluation are given in Table 1.

Table 1. Performance metrics used

Performance metrics	Formula
FMS (Fowlkes-Mallows Score)	$FMS = \dfrac{T_p}{\sqrt{(T_p+F_p)*(T_p+F_n)}}$
Normalized Mutual Information (NMI)	$NMI = \dfrac{\sum_{c,k} n_{c,k} \log \frac{n_{c,k} \cdot N}{n_c \cdot n_k}}{\sqrt{\left(\sum_c n_c \log \frac{n_c}{N}\right)\left(\sum_c n_c \log \frac{n_c}{N}\right)}}$
Homogeneity (H)	$H = 1 - \dfrac{\sum_{c,k} n_{c,k} \log \frac{n_{c,k}}{n_k}}{\sum_c n_c \log \frac{n_c}{N}}$
Completeness (C)	$C = 1 - \dfrac{\sum_{c,k} n_{c,k} \log \frac{n_{c,k}}{n_c}}{\sum_c n_c \log \frac{n_k}{N}}$ here, n_k is the number of samples in cluster k, $n_{c,k}$ is the number of samples in class c as well as cluster k, N is the number of total samples in the dataset
V measure	$V = \dfrac{2*H*C}{H+C}$
Adjusted Rand Score (ARS)	$ARI = \dfrac{(RI - expected\_RI)}{(max(RI) - expected\_RI)}$, The value of ARS ranges from 0 to 1.
Precision@k	$Precision@k = \dfrac{No.\ of\ correctly\ identified\ events}{No.\ of\ top\ k\ of\ identified\ events}$
(Normalized Discounted Cumulative Gain)	$NDCG = \dfrac{\sum_{i=1}^{n} \frac{r_i}{\log_2(i+1)}}{\sum_{i=1}^{|R|} \frac{r_i}{\log_2(i+1)}}$ The NDCG value varies from 0 to 1

Table 2. Results of different classifiers for disruptive event classification

Measures	Classification algorithm				
	DNN	SVM	Decision tree	Naive Bayes	Random Forest
Precision	**0.984**	0.888	0.749	0.854	0.749
Recall	**0.992**	0.873	0.748	0.829	0.748
F1-score	**0.988**	0.88	0.748	0.841	0.748
Accuracy	**0.989**	0.893	0.773	0.859	0.773

We formulated four research queries and attempted to find the answers of the same from the experimental analysis. Research queries are as follows:

RQ1: How well does the tweet classifier in the proposed framework perform the event and non-event tweet classification?

RQ2: How well does the clustering algorithm in the proposed framework performs clustering?

RQ3: What is the comparative performance of the different keyword extraction techniques for disruptive event annotation task?

RQ4: How does the proposed framework perform in the prediction of disruptive events?

---

[2] https://github.com/seethaaditi/Disruptive-Event-Identification.
[3] https://github.com/devmehta01/DiPD.

Table 3. Results of community detection techniques and clustering algorithms

Measures	Types of community detection							
	Louvain	Pycombo	Girvan-Newman	Leidenaly	Connected components	K-means +TF-IDf	K-means+ CountVectorizer	K-means +LDA
Homogieniety	0.754	**0.788**	0.264	0.196	0.244	0.624	0.618	0.286
Completeness	**0.792**	0.766	0.041	0.004	0.032	0.758	0.748	0.417
NMI	**0.772**	**0.777**	0.071	0.007	0.057	0.685	0.677	0.34
V measure	**0.772**	**0.777**	0.071	0.007	0.057	0.685	0.677	0.34

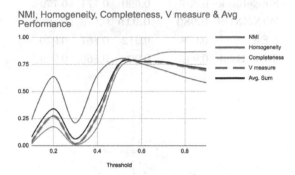

Fig. 2. Clustering performance of louvain community detection

Result Analysis for RQ1. The result of the tweet classifier based on the various classification techniques on the Event and Non-Event dataset has been shown in Table 2. The table contains the result of five classification techniques for the Event and Non-Event dataset in various performance metrics. From Table 2, we can observe that DNN achieved the highest precision, recall, F-1 score, and accuracy for the Event and Non-Event dataset. The tweet classifier built using DNN outperforms SVM, Decision Tree, Naive Bayes, and Random Forest.

Result Analysis for RQ2. For this research question, we have built and compared different types of community detection methods with K-means using TF-IDF, LDA, and count vectorizer. Table 3 includes the result of five different community detection variants and K-means using LDA and TF-IDF in terms of various performance measures. All the above clustering techniques have been applied to Twitter Dataset. The Louvain and Pycombo community detection methods have improved performance for most cases. The K-means using TF-IDF and count vectorizer follows it. The Girwan Newman, Leidenaly, and Connected Components methods did not perform well. K-means using LDA worked better than three of the above methods.

Optimum Threshold Selection. All the performance metrics discussed above were given equal importance. Hence, the maximal point of average should give us the optimum threshold value. Therefore, at a threshold value of 0.5, the value of all performance measures is optimal (Fig. 2).

Table 4. Results of the different keyword extraction techniques

Methods	Accuracy	ARS	NMI	FMS
KeyBERT	0.937	0.911	**0.930**	0.930
TF - IDF	0.978	0.911	0.865	0.929
YAKE	0.970	0.917	0.87	0.933
TextRank	0.845	0.690	0.774	0.779
SingleRank	0.857	0.690	0.771	0.779
WINGNUS	**0.980**	**0.919**	0.876	**0.935**
KEA	0.978	0.912	0.867	0.930
Topic Rank	0.934	0.908	0.866	0.926

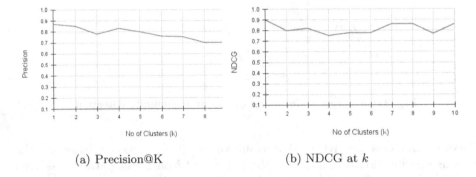

(a) Precision@K   (b) NDCG at k

Fig. 3. Precision@K and NDCG at k of our proposed framework.

Result Analysis for RQ3. We have compared and implemented eight keyword extraction methods for answering the research question. The table includes Accuracy, ARS, NMI, and FMS performance measures for eight keyword extraction methods. All eight methods have been applied to the Twitter dataset. Table 4 shows that the WINGNUS and KeyBERT produce better performance for all of the measures. TextRank and SingleRank yield lower performance than all of the eight techniques.

Result Analysis for RQ4. To answer this research question, NDCG and Precision@k were calculated. The data was gathered over a single day. The total number of events generated by our suggested framework daily ranges from 1 to 10. So, for k values ranging from 1 to 10, we compute Precision@k and NDCG. For Precision@k and NDCG, three annotators have been assigned to annotate the event clusters to calculate the relevancy score. We computed the precision value for the different number of clusters (precision@k). As seen in Fig. 2, the precision@5 score is 0.8, indicating that four of the top five events were accurately identified. Similarly, the precision value has been calculated for various values of k, as shown in Fig. 3(a).

NDCG score at different k values is calculated by dividing the Discounted Cumulative Gain (DCG) for the event ranking of k events obtained from the proposed framework by the ideal ranking of k events. The optimum ranking is determined by sorting the k events list by relevancy score. In this case, the relevance score is calculated by dividing the total number of annotators by the number of annotators who identified the acquired event as relevant. Figure 3(b) displays the NDCG score for various k values. Our proposed framework, as illustrated in Fig. 3, is effective and performs well in both the NDCG and Precision@K evaluation metrics.

4   Conclusion

In this paper, we have proposed a novel framework to identify disruptive real-world events. Disruptive Event identification was performed in two stages: Data acquisition & Preprocessing and Clustering & Event Identification. Experiments were conducted to evaluate and compare the state-of-art methods with the proposed framework on two real-time datasets. Our framework achieved effective results in the identification of disruptive events. In future works, we will explore and evaluate the annotated events using other techniques to provide sound output to decision-makers. Further, we can apply continual learning to train a neural network model for event prediction to learn previously identified events while retraining on new events.

Acknowledgment. This work is supported by a Research Grant under National Supercomputing Mission (India), Grant number: *DST/NSM/R&D_HPC_Applications/2021/24*.

References

1. Karimi, S., Shakery, A., Verma, R.M.: Enhancement of twitter event detection using news streams. Nat. Lang. Eng., 1–20 (2022)
2. George, Y., Karunasekera, S., Harwood, A., Lim, K.H.: Real-time spatio-temporal event detection on geotagged social media. J. Big Data **8**(1), 1–28 (2021). https://doi.org/10.1186/s40537-021-00482-2
3. Alsaedi, N., Burnap, P.: Feature extraction and analysis for identifying disruptive events from social media. In: Proceedings of the International Conference on Advances in Social Networks Analysis and Mining, pp. 1495–1502 (2015)
4. Xiao, K., Qian, Z., Qin, B.: A survey of data representation for multi-modality event detection and evolution. Appl. Sci. **12**(4), 2204 (2022)
5. Hettiarachchi, H., Adedoyin-Olowe, M., Bhogal, J., Gaber, M.M.: Embed2detect: temporally clustered embedded words for event detection in social media. Mach. Learn. **111**(1), 49–87 (2022)
6. Alsaedi, N., Burnap, P., Rana, O.: Can we predict a riot? disruptive event detection using twitter. ACM Trans. Internet Technol. (TOIT) **17**(2), 1–26 (2017)
7. Morabia, K., Murthy, N.L.B., Malapati, A., Samant, S.: Sedtwik: segmentation-based event detection from tweets using Wikipedia. In: Proceedings of the Conference of the North American Chapter of the Association for Computational Linguistics, pp. 77–85 (2019)

8. Wang, Y., Wang, J., Lin, H., Zhang, Y., Yang, Z.: Dependency multi-weight-view graphs for event detection with label co-occurrence. Inf. Sci. **606**, 423–439 (2022)
9. Reimers, N., et al.: Sentence-bert: sentence embeddings using siamese bert-networks. In: Proceedings of the 2019 Conference on Empirical Methods in Natural Language Processing, pp. 671–688 (2019)
10. Becker, H., Naaman, M., Gravano, L.: Beyond trending topics: real-world event identification on Twitter. In: Proceedings of the International AAAI Conference on Web and Social Media **5**, 438–441 (2011)

Modeling Polarization on Social Media Posts: A Heuristic Approach Using Media Bias

Sadia Kamal, Jade Gullic(✉), and Arunkumar Bagavathi(✉)

Oklahoma State University, Stillwater, USA
{sadia.kamal,jade.gullic,abagava}@okstate.edu

Abstract. Developing machine learning models to characterize political polarization on online social media is challenging due to the lack of annotated data, noise in social media datasets, and large volume of datasets. The common research practice is to analyze the biased structure of online user communities for a given topic or to qualitatively measure the effects of polarized topics on social media. However, there is a very limited work to analyze polarization at the ground-level like the social media posts itself which are heavily dependent on annotated data. Understanding the level of political leaning in social media posts is important to quantify the bias of online user communities. In this work, we show that current machine learning models can give better performance in predicting political leaning of social media posts. We also propose two heuristics based on news media bias and post content to collect the labeled data for supervised machine learning algorithms. We experiment the proposed heuristics and machine learning approaches to study political leaning on posts collected from two ideologically diverse social media forums: *Gab* and *Twitter* without the availability of human-annotated data.

Keywords: Media bias · Social network analysis · Predictive analytics

1 Introduction

Online news media forums and social networks are the primary information system to rapidly disseminate the current news to the global population. Several factors like media bias of news media houses [3], and cognitive bias of social media users [12] contribute in setting the political ideology, ranging from *far-left* to *far-right* [14], of social media users. Exposure to such polarization can not only develop online disagreements and ideology segregation but may also lead to offline extreme or even violent activities. Thus, it is essential to develop quantitative methods to characterize social media polarization with the advancements in data science and machine learning. The existing computational approaches study online political polarization in multiple aspects using machine learning with both supervised [12] and unsupervised approaches. However, the existing literature focuses on analyzing political polarization at the user-level, topic-level

© The Author(s), under exclusive license to Springer Nature Switzerland AG 2022
M. Ceci et al. (Eds.): ISMIS 2022, LNAI 13515, pp. 35–43, 2022.
https://doi.org/10.1007/978-3-031-16564-1_4

or news-level. Characterizing the political leaning of social media posts has the potential to become a pre-cursor for hate speech detection, fake news detection, and influence prediction on social media [14]. In this work, we present methods to identify political leaning ($\mathcal{P} \in \{$left, center, right$\}$) at a fine-grained social media posts using supervised machine learning. Since there are no existing labeled data for supervised political polarization prediction at the posts-level, we present heuristics to assign labels for social media posts in this paper. News media houses take a political stand with their interpretation of topics in news articles. We utilize the political leaning of such news media houses as surrogate data to heuristically label social media posts.

We experiment with multiple text representation learning frameworks and supervised machine learning models for the political leaning prediction task. We use existing social media posts from Twitter [4] and Gab [6] and compare the performance of the proposed methods between two datasets. In this work, our contributions are two-fold:

1. We present two heuristic methods to identify the political leaning of social media posts on Twitter and Gab using news media bias
2. We measure the performance of traditional machine learning algorithms to predict the political bias of posts using state-of-the-art text representation learning methods

2   Related Work

Recent years have seen growing concern that social media forums such as Twitter and Gab may cause political bias in people, affecting presidential elections [2] and news consumption [7]. Machine learning approaches have become increasingly popular for detecting political bias in text content using traditional lexicon-based classifiers based on "bag-of-words" techniques [8]. There are several problems with these approaches, including the overreliance on primary-level lexical information and the neglect of semantic structure. Some researchers used attention-based multi-view model [10] to identify the political leanings of topics. To compare different document representation choices, deep learning approaches, semantically meaningful word embeddings and attention mechanisms are analyzed in [5]. Studying bias is quite popular in the data science research. Examples include statistical measures for news media bias [13], fair models for political leaning prediction on news articles [1], and examining an online community behavior with multi-modal data [9]

3   Datasets

In this work, we use web resources to capture the political leaning of online news domains along with existing datasets collected from social media forums. We describe the datasets used in our experiments and methods in this section.

3.1   News Domains Data

We use the political leaning of news domains to associate the political lean-ing of social media posts. Thus first, we collect the political leaning ($\mathcal{P} \in$ {left, center, right}) of news domains. Given a set of n news domains $\mathcal{D} \in$ {$d_1, d_2, \ldots, d_n$}, we associate a political leaning of each news domain $d_i \in \mathcal{D}$ with one of the political leaning labels using $d_i \rightarrow \mathbb{1}_{\mathcal{P}}$. With the help of web scrapping tools, we created a media bias dataset from allsides.com[1]. The dataset collected from allsides.com contains online news domains like *CNN, Fox News, CNBC*, and *nytimes* along with their corresponding political leaning. Our media bias dataset contains a total of 422 news domains, out of which *158* news domains are labeled *left*, *166* are labeled *center*, and *98* are labeled *right*.

3.2   Social Media Posts

In our work, we use publicly available datasets from social media sites Twitter [4] and Gab [6] for all experiments.

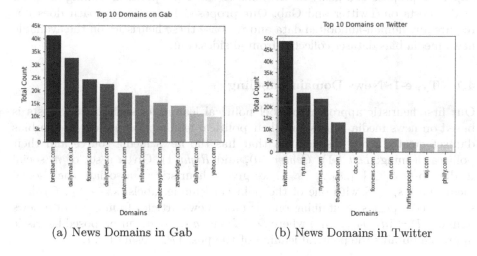

(a) News Domains in Gab   (b) News Domains in Twitter

Fig. 1. Top 10 news domains in Gab and Twitter datasets

Twitter. Tweets with news article URLs that discuss political topics from mainstream news media sources are included in the Twitter dataset [4]. The timeline of the Twitter dataset is from January 2018 to September 2018. The dataset comprises *289,738* tweets with URLs from *60* handpicked mainstream news domains covering a diverse range of political views.

---

[1] http://www.allsides.com/media-bias/media-bias-rating-methods.

Gab. The Gab dataset [6] contains a total of 40 million posts including posts, replies, and re-posts collected in 2018 which mostly support far-right ideologies. We further select only the posts with more than five words along with news URLs to get post context on news articles. In total, our dataset contains *1,368,028* posts with URLs from *355* news domains.

Figures 1a and 1b show the top ten news domains appearing in both Twitter and Gab datasets. We can see that the *Gab* dataset contains large number of news articles from right-aligned news media outlets like *breitbert*, *dailymail*, and *foxnews*. On the other hand, *Twitter* posts are leaning more towards left aligned mainstream news media forums such as *nytimes*, and *theguardian*.

4   Methodology

To quantify the characteristics of political polarization on online social media with machine learning approaches, we require annotations of social media posts. Due to the near impossibility of hand curating millions of posts, the existing approaches analyze polarization at the user-level [9] and news-level [1]. In this paper we propose two heuristic methods to label the political leaning of social media posts on Twitter and Gab. Our proposed heuristic approach does not require any human-annotated data and we base these heuristics on the available news media bias dataset collected from allsides.com.

4.1   Type-1: News Domain Labeling

Our first heuristic approach to label political leaning of social media posts is based on news media houses and their political leaning. We use our media bias dataset created from allslides.com that has the news media source and their political leaning: *Left* (-1), *Center* (0), and *Right*(1). Given a set of m social media posts $\mathcal{S} \in \{s_1, s_2, \ldots, s_m\}$, we give a heuristic to associate each social media post $s_i \in \mathcal{S}$ with one of the political leaning labels: $s_i \to \mathbb{1}_{\mathcal{P}}$. Given a social media post s_i containing one or more news articles from a set of q news domains $\widehat{\mathcal{D}} = \{\hat{d}_1, \hat{d}_2, \ldots, \hat{d}_q\}$ where $\widehat{\mathcal{D}} \subset \mathcal{D}$ and $q <<< n$, our proposed heuristic approach obtains the political leaning of the post $(\mathcal{P}_{s_i})$ using Eq. 1.

$$\mathcal{P}_{s_i} = \begin{cases} -1 \ if \ \widehat{\mathcal{P}}_{s_i} < 0.1 \\ 0 \ if \ -0.1 < \widehat{\mathcal{P}}_{s_i} < 0.1 \\ +1 \ if \ \widehat{\mathcal{P}}_{s_i} > 0 \end{cases} \tag{1}$$

where $\widehat{\mathcal{P}}_{s_i} = \frac{\sum_{j=1}^{m} \mathcal{P}_{d_j}}{m}$; $-\infty < \widehat{\mathcal{P}}_{s_i} < +\infty$ is an unscaled political leaning of the social media post s_i and $\mathcal{P}_{d_j}$ is the political leaning of news domain $d_j \in \widehat{\mathcal{D}}$. Since we use these labels for supervised machine learning algorithms, we use Eq. 1 to assign final political leaning label of the social media post $\mathcal{P}_{s_i}$. We define our heuristics for social media posts with news domain URLs, it will not assign any labels if there are no news domain URLs in social media posts.

4.2   Type-2: Sentiment Labeling

Our second heuristic approach to labelling political leaning of social media posts is based on both news media bias and social media post content. The proposed method is an extension of our first approach and it intends to improve the accuracy of the labels by including the post's sentiment in news articles. We measure the sentiment score of social media posts by combining scores collected from three sentiment analyzer tools.

Textblob: Textblob is a popular sentiment analysis tool which returns polarity and subjectivity score. In this paper we use only the polarity score from TextBlob and we represent it as $\alpha \in [-1, +1]$. $\alpha = -1$ refers to negative sentiment and $\alpha = +1$ refers to positive sentiment.

Vader: Vader is another popular sentiment analysis tool which gives the probability of a sentence being positive, negative or neutral. We represent the computed score from Vader as $\beta \in \{-1, 0, +1\}$, where -1 is negative, 0 is neutral and $+1$ is positive.

Afinn: Afinn is constructed with a large corpus of labeled lexicons where each lexicon has a polarity score associated with it. Similar to TextBlob, we represent the polarity score of Afinn as $\gamma \in [-1, +1]$.

Given a set of social media posts $\mathcal{S}$, we compute the sentiment score (τ) of the post $s_i \in \mathcal{S}$ by combining sentiment values from three sentiment analyzer tools as $\tau_{s_i} = \frac{\alpha+\beta+\gamma}{3}$. With a sentiment value of a post τ_{s_i}, we compute the unscaled political leaning ($\widehat{\mathcal{P}}$) as $\widehat{\mathcal{P}}_{s_i} = \widehat{\mathcal{P}}_{s_i} \times \tau_{s_i}$. We further update the political leaning of the given post using Eq. 1. In other words, we switch the $\mathcal{P}_{s_i}$ only if the overall sentiment of the post (τ_{s_i}) is negative.

5   Results

In this section, we present the results of our labeling heuristics. Also, we experiment with existing machine learning approaches to predict the political leaning of social media posts.

5.1   Political Leaning Labeling

In Fig 2a and 2b we show the political leaning labeling summary for both the datasets with our Heuristic 1 and Heuristic 2 labeling methods. Labels collected using our heuristic 1 labeling method correlate with our qualitative analysis given in Fig. 1. As depicted in Fig. 2a, it is evident that Gab posts tend to focus on right-leaning news outlets whereas Twitter posts primarily share articles from left-leaning news sources. However, this scenario changes when we also consider post content using our heuristic 2 as shown in Fig. 2b. More than 50% of Gab posts support left and center-aligned posts, while most of the posts on Twitter still support left-leaning news.

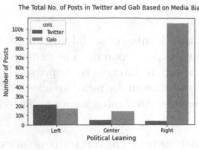

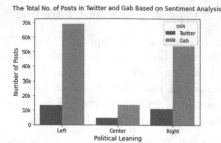

(a) News Domain Labeling   (b) Sentiment analysis labeling

Fig. 2. Political bias labeling summary

5.2 Methods for Learning Text Representations

We experiment with three text representation learning methods that are based on both word frequency and contextual features to extract features of our labeled social media posts.

Bag-of-Words: Count vectorizer or Bag-of-Words model generates vectors of social media posts based on only the occurrence frequency of each word.

TFIDF Vectorizer: TFIDF model measures the importance of a word based on the occurrence frequency of words in all social media posts along with the relevance of words to a social media post.

Word2vec: We use the skip-gram version of Word2Vec [11] model that tries to predict the probability of occurrence of context words, given a keyword w from T words. The context is assigned using a sliding window of size c.

Table 1. ML models performance (%) with news domain labeling accuracy

Twitter Data				
—	SVM	LR	Naive bayes	NN
W2V	0.44	0.50	0.40	0.55
Count	0.62	0.62	0.58	**0.70**
TFIDF	0.63	0.63	0.58	**0.71**
Gab Data				
W2V	0.51	0.50	0.44	0.62
Count	0.95	0.95	0.91	**0.97**
TFIDF	0.95	0.95	0.91	**0.96**

Table 2. ML models performance (%) with sentiment analysis labeling accuracy

Twitter Data				
—	SVM	LR	Naive bayes	NN
W2V	0.42	0.41	0.39	0.51
Count	0.60	0.60	0.56	**0.68**
TFIDF	0.60	0.60	0.57	**0.68**
Gab Data				
W2V	0.51	0.50	0.44	0.62
Count	**0.85**	0.77	0.70	**0.85**
TFIDF	0.83	0.76	0.72	**0.84**

5.3 Political Leaning Prediction

We use traditional machine learning models to predict political leaning $\mathcal{P}$ of social media posts using text features extracted in our previous step. We give the hyperparameters used in our models below:

SVM: We use SVC class from sklearn library with $C=1.0$ and RBF kernel.

Logistic Regression: From sklearn library we use logistic regression module with the default hypermeters $C=1.0$ and solver as *lbfgs*.

Naive Bayes: We use *GaussianNB* in our experiments.

Neural Network: We use a sequential neural network architecture with one input layer, three Dense layers and one classifier head which is fine-tuned with *Categorical Crossentropy* loss function and *Adam* optimizer.

In Tables 1 and 2 we present the accuracy results of machine learning models using each text representation model to predict the political leaning of posts collected from Twitter and Gab. The results in Table 1 show the accuracy of each classifier for the data labeled using our first heuristic approach. In general, we notice that the TFIDF feature set yields the best results for the Twitter dataset, and the BoW feature set yields the best results for the Gab dataset. Based on the results, we can see that Gab has a significantly higher accuracy score than Twitter due to the availability of a large quantity of diverse data. In addition, we observe that of all the models, our neural network model gives the best classification performance.

Table 2 shows the results of each classifier's accuracy using our second heuristic approach. Although the accuracy of ML models is relatively low for these datasets, we notice that the ML algorithms can identify political leaning of Gab posts better than that of Twitter posts.

It is evident from Tables 1 and 2 that a neural network classifier outperforms other machine learning models to predict the political leaning of Twitter and Gab posts. We also note that machine learning models trained with features extracted using simple methods like BoW and TFIDF outperform models trained with state-of-the-art text representation learning methods like Word2Vec. While we note that models on Gab perform better the accuracy of all models decreases by about 14% for the Gab dataset when we use the Heuristics-2 labels.

6 Conclusion and Future Work

In this paper, we provided two methods to analyze political leaning in social media posts. In particular, we presented a methodology to obtain political leanings of social media posts from the large-scale Twitter and Gab datasets using our created news media bias dataset. Finally, we explore state-of-the-art text representation techniques to extract the features to train our ML models to predict the political leaning of any given post. However, despite the effectiveness of our methods, more comprehensive studies with some ground-truth data can help update the methods for more robust social media political bias modeling. In the future, we can extend this study by including methods for learning graph representations for multimodal bias detection.

References

1. Baly, R., Martino, G.D.S., Glass, J., Nakov, P.: We can detect your bias: Predicting the political ideology of news articles. arXiv preprint arXiv:2010.05338 (2020)
2. Belcastro, L., Cantini, R., Marozzo, F., Talia, D., Trunfio, P.: Learning political polarization on social media using neural networks. IEEE Access **8**, 47177–47187 (2020)
3. Bernhardt, D., Krasa, S., Polborn, M.: Political polarization and the electoral effects of media bias. J. Public Econ. **92**(5–6), 1092–1104 (2008)
4. Brena, G., Brambilla, M., Ceri, S., Di Giovanni, M., Pierri, F., Ramponi, G.: News sharing user behaviour on twitter: a comprehensive data collection of news articles and social interactions. In: AAAI ICWSM, vol. 13, pp. 592–597 (2019)
5. Cruz, A.F., Rocha, G., Cardoso, H.L.: On document representations for detection of biased news articles. In: Proceedings of the 35th Annual ACM Symposium on Applied Computing, pp. 892–899 (2020)
6. Fair, G., Wesslen, R.: Shouting into the void: a database of the alternative social media platform gab. In: AAAI ICWSM, vol. 13, pp. 608–610 (2019)
7. Garimella, V., Smith, T., Weiss, R., West, R.: Political polarization in online news consumption. In: AAAI ICWSM, pp. 152–162 (2021)
8. Gerrish, S.M., Blei, D.M.: Predicting legislative roll calls from text. In: ICML, pp. 489–496 (2011)
9. Hosseinmardi, H., Ghasemian, A., Clauset, A., Mobius, M., Rothschild, D.M., Watts, D.J.: Examining the consumption of radical content on youtube. In: Proceedings of the National Academy of Sciences, vol. 118, no. 32 (2021)
10. Kulkarni, V., Ye, J., Skiena, S., Wang, W.Y.: Multi-view models for political ideology detection of news articles. In: EMNLP, pp. 3518–3527 (2018)
11. Mikolov, T., Chen, K., Corrado, G., Dean, J.: Efficient estimation of word representations in vector space. In: Workshop Proceedings of ICLR (2013)
12. Nair, S., Ng, K.W., Iamnitchi, A., Skvoretz, J.: Diffusion of social conventions across polarized communities: an empirical study. Soc. Netw. Anal. Min. **11**(1), 1–17 (2021). https://doi.org/10.1007/s13278-021-00726-2

13. Sales, A., Balby, L., Veloso, A.: Media bias characterization in Brazilian presidential elections. In: ACM Conference on Hypertext and Social Media, pp. 231–240 (2019)
14. Vicario, M.D., Quattrociocchi, W., Scala, A., Zollo, F.: Polarization and fake news: early warning of potential misinformation targets. ACM TWEB **13**(2), 1–22 (2019)

Sarcasm Detection in Tunisian Social Media Comments: Case of COVID-19

Asma Mekki[✉], Inès Zribi, Mariem Ellouze,
and Lamia Hadrich Belguith

ANLP Research Group, MIRACL, University of Sfax, Sfax, Tunisia
asma.elmekki.ec@gmail.com

Abstract. The goal of this study is to learn more about how the public sees COVID-19 pandemic behaviors and to identify important themes of concern expressed by Tunisian social media users during the epidemic. Around 23K comments were collected, written in both Arabic and Latin characters in the Tunisian dialect. Native language experts manually tagged these comments for sarcasm identification (sarcastic and non-sarcastic). In addition to health, our dataset contains comments on entertainment, social, sports, religion, and politics, all of which are impacted by COVID-19. This research examines the sarcasm expressed in Tunisian social media comments regarding the novel COVID-19 from its appearance in the first half of 2020. We also provide benchmarking findings applying machine learning and deep learning algorithms for sarcasm detection. We obtained an accuracy of above 80%.

Keywords: COVID-19 · Sarcasm detection · Deep learning · Tunisian dialect

1 Introduction

The World Health Organization[1] (WHO) declared COVID-19, also known as COrona VIrus Disease of 2019, as a pandemic on March 11, 2020. Governments imposed stringent requirements for population control by analyzing cases and effectively deploying available resources. Citizens experienced worry, dread, and anxiety as a result of the exponentially growing number of cases throughout the world. The worldwide population's mental and physical health has been proven to be directly proportionate to this pandemic. As a result, several measures must be implemented through demystifying relevant facts and information. According to the WHO, millions of people were tested positive throughout the world, with nearly one million confirmed COVID-19 cases registered in Tunisia between January 3, 2020 and February 2022.

The purpose of this study is to examine social media comments encompassing all handles associated with COVID-19. Various studies have used Twitter, Reddit, and Facebook for sarcasm detection, however, Facebook remains Tunisia's

---

[1] https://www.who.int.

M. Ceci et al. (Eds.): ISMIS 2022, LNAI 13515, pp. 44–51, 2022.
https://doi.org/10.1007/978-3-031-16564-1_5

most popular media. Therefore, this study examines over 23,000 Tunisian comments that were scraped from Facebook between February 1, 2020 and May 31, 2020, as well as the same period in 2021. Furthermore, native speakers manually labeled the sarcasm of the comments. The most frequently used words in the texts were also calculated mathematically to show how COVID-19 affects Tunisians' responses. Moreover, state-of-the-art machine learning and deep learning classifiers were applied and evaluated on the word vectors to determine the model with the best accuracy.

The rest of this paper is organized as follows. Section 2 includes a background on the provided work, including the basics of social media, related works and a brief presentation of Tunisian dialect. Section 3 details the data collection, data preprocessing and data annotation steps. Section 4 discusses the results of the experiments.

2   Background

The COVID-19 condition can cause panic and make people feel afraid, helpless and overwhelmed. They may experience anxiety as a result of their social isolation and daily pandemic updates. As a result of this crisis, more people are turning to the internet to keep in touch with family, friends, and colleagues. In this context, Facebook reported that overall messages on its platform has increased by more than 50%. This would include Facebook, Messenger, Instagram and WhatsApp combined. With this ongoing growth and the increase in the number of Internet users around the world, the processing of social networks comments and messages became a necessity. For example, Tunisia is increasingly connected. Until January 2022, the number of Facebook user community in Tunisia is around 70% (8,602,900) of the population[2]. In May 2020, at the start of the pandemic, the number of Tunisian Facebook members was roughly 7,870,000. In comparison to May 2019, we notice an increase of more than 1.2 million users (10% of the population in Tunisia). The number of Tunisian Facebook community increased by about 500,000 compared to February 2020, before the quarantine in Tunisia.

2.1   Related Work

Since people share their emotions and opinions more freely than ever, the use of sarcasm detection is becoming more popular for identifying and analyzing these feelings. Regarding COVID-19, sarcasm detection has been of considerable interest to the research community to advance processing techniques to understand the behavioral awareness of the public for different languages around the world. In this study, we focus on Arabic language and its dialects.

In the study published by Alhajji et al. [1], sentiment analysis of Arabic tweets was carried out by applying Naïve Bayes to run Arabic sentiment analysis using a

---

[2] https://napoleoncat.com.

Python library (Natural Language Toolkit (NLTK)). They used a labeled dataset of Arabic tweets in which its sentiment labels are based on emoji lexicons. The dataset was split into 50% positive and 50% negative. A total of 47K tweets were utilized for training, with 11K being used in the testing process. The evaluation results obtained with an unigram Naïve Bayes achieved 0.89 accuracy and 0.89 F-score.

Althagafi et al. [2] examined opinion mining and sentiment analysis in relation to online learning in Saudi Arabia during COVID-19. They collected about 10K tweets and divided them into three categories (positive, negative and neutral). They found that most tweets expressed a neutral feeling when they analyzed sentimental characteristics such as polarity and subjectivity. Authors developed three machine learning models (Naïve Bayes, Random Forest and KNN), with Random Forest having the best classification accuracy (84%).

Madani et al. [12] collected tweets expressed in the most used languages by Moroccan users (Spanish, English, French, and Arabic). The dataset collected from March 2020 until October 2020 proves that the majority of COVID-19-related tweets are negative. They proposed a dictionary based method applying the SenticNet dictionary[3] and four features using the TextBlob python library. They compared their method to four machine learning algorithms (SVM, Naïve Bayes, Random Forest and Decision Tree). The four algorithms yielded results ranging from 60% to 65%.

AraCOVID19-SSD [3] is a manually annotated COVID-19 sarcasm detection and sentiment analysis dataset containing 5 162 Arabic tweets. The annotated dataset has been evaluated using Logistic Regression, Random Forest, SVM and transformers models. SVM achieved the best F-score for sarcasm detection with 95.97%, while AraBERT attained the highest F-score for sentiment analysis at 92.26%.

Habbat et al. [9] created a dataset of 37K Moroccan tweets, gathered between March 1st and June 28th, 2020, during the COVID-19 confinement period. They applied six machine learning algorithms to classify the collected dataset. The results showed that logistic regression classifier yielded the best sentiment prediction, with a performance of 68.28% of the F-measure metric.

In summary, sarcasm detection has shown to be a helpful source of information mining, particularly in cases when a huge quantity of data about the public is required, such as investigating public behavior to the COVID-19 crisis and its impact on public life. However, to the best of researcher's knowledge, no such study has been conducted on sarcasm detection for Tunisian Dialect (TD).

2.2   Tunisian Dialect

Tunisian Dialect (TD) is a North African dialect of Arabic that is spoken by almost 12 million people in Tunisia [15]. It differs from the Modern Standard Arabic (MSA) in different levels (morphology, syntax, pronunciation and vocabulary). Its vocabulary includes terms from a variety of languages, including Mal-

---

[3] https://sentic.net.

tese, Berber, French and English. Although it is mainly spoken, it is written in social media platforms, blogs, and certain novels, as well as comic books, advertisements, newspapers, and popular music.

3   Dataset

During the COVID-19's lockdown, people used social media to express their feelings and get information. Furthermore, humor and sarcasm may be employed to improve well-being and as coping strategies during challenging situations. As a result, social media helps in analysing public sarcasm use and its dynamics during the pandemic, offering insights about prevalent sentiment and its network impacts. Therefore, in this section we present the TunCorS (Tunisian Corpus for Sarcasm detection) collected dataset.

3.1   Data Collection and Preprocessing

Since it has been statistically proven, Facebook is the most popular social network in Tunisia when compared to other internet and social media platforms. Therefore, the data was gathered from public Facebook pages such as Covid-19 Tunisia, Mosaïque FM, Shems FM, Elhiwar Ettounsi, Attessia TV, etc. We retrieved the comments using an online website[4]. This study's data was collected between February and May of 2020, as well as the same period in 2021. We obtained over 60K comments in total, written in TD in Arabic and Latin letters. After crawling the data, the initial step is to filter any comments that are not related to COVID-19, remaining only 26K comments. The file is then cleaned of duplicate comments. Moreover, the "re" python module is used to clean the Facebook comments, removing unnecessary symbols such as stop words, URLs, numeric values, and punctuation marks. We also used COTA Orthography system [4] to automatically normalize the errors.

3.2   Data Annotation

It is important to remember that fitting a model for classification requires the discovery of relevant features first. As a result, we used Stanford-TUN [13] to add a Part-Of-Speech (POS) tag for each word of the dataset. We applied the segmentor provided by [14] since the input for POS tagging needed to be segmented. We also added the month and year of the comment. Experts manually annotated all comments for sarcastic annotation. Table 1 presents examples for each label (Table 2).

---

[4] https://exportcomments.com.

Table 1. TunCorS examples.

Label	Script	Example
Not sarcastic	Arabic	سافا ماشية وتطيح
		sAfA mA$yp wtTyH
		Well, it is on its way down
		ربي يسترنا
		rby ystrnA
		God protects us
	Latin	nchalah jey khir
		God willing, the next is better
		9adech min tahlill 3mlou
		How many tests were done
Sarcastic	Arabic	كان قالو.الوباء زاد منصدقوش .وكان قالو نقص.زادا منصدقوش؟
		kAn qAlw..AlwbA' zAd mnSdqw$ ..wkAn qAlw nqs..zAdA mnSdqw$?
		If they said... that the epidemic was increasing, we didn't believe them... and if they said that it was decreasing we don't believe them either?
		ياخي شكون قاعد يعمل في تحليل ههه
		yAxy $kwn qAEd yEml fy tHlyl hhh
		Who is doing the analysis hhh
	Latin	dima el 9odem 3morna ma nwa5rou hmdlh hhh
		Always forward never backward God willing hhh
		Makther esbou3 hhh 777
		too many sevens hhh 777

Table 2. Dataset statistics.

Year	Month	Comments	Words	Longest comment	Sarcasm %
2020	February	343	4 538	303	21.05
	March	2 357	41 344	270	14.43
	April	3 941	60 206	558	16.86
	May	4 874	56 974	284	15.11
2021	February	3 140	46 931	557	10.29
	March	2 290	33 175	411	27.14
	April	2 914	39 310	600	24.75
	May	3 491	40 621	428	26.34

4   Experimental Results

Sarcasm detection may be considered as text classification issues (in our case, binary classification for sarcasm detection). In this section, we run several experiments to find the best machine learning and deep learning algorithms for our task. The data was divided into 80% as training set, 5% as validation set and remaining 15% as testing set.

In this section, we experimented several algorithms using the collected corpus. Researchers can use the results as a baseline. Indeed, we tested four machine learning algorithms (K-Nearest Neighbors (KNN) [7], Decision Tree [16] , Random Forest [10] , AdaBoost (Adaptive Boosting) [8] and Support Vector Classifier (SVC) [17]) and two deep learning algorithms (Simple Convolutional Neural Network (CNN) [5] and Long Short-Term Memory (LSTM) [11]) . Two experiments were carried out for each classifier depending on the vectorizer (see Table 3). It helps us convert text data to computer understandable numeric data.

All the evaluation results detailed in Table 3 are generated using default parameters. For deep learning based models, we applied a 128 batch size, 100 epochs, Adam optimizer and Categorical Cross-entropy loss function.

Of all the machine learning models examined, AdaBoost shows the highest accuracy of 0.816 for sarcasm detection. However, deep learning models outperformed machine learning models. CNN model extracts higher-level features using convolutional layers and maximum pooling layers, whereas LSTM model can capture long-term dependencies between word sequences and is thus more suited to text classification.

We conducted non-parametric tests on the dataset to see whether there were any significant results in this study. The p-value for the non-parametric independent Wilcoxon test [6] is 0.023. Since the p-value is below the threshold of 0.05, we can conclude that the results of the sarcasm models are significantly different.

Table 3. Results of sarcasm detection experiments using various classifiers in Tunisian dialect.

Classifier	Vectorizer	Precision	Recall	Accuracy
KNN	TF-IDF vectorizer	0.793	0.785	0.786
	CBOW	0.799	0.802	0.800
Decision tree	TF-IDF vectorizer	0.795	0.799	0.796
	CBOW	0.810	0.805	0.805
Random vorest	TF-IDF vectorizer	0.781	0.775	0.777
	CBOW	0.786	0.779	0.780
AdaBoost	TF-IDF vectorizer	0.794	0.789	0.793
	CBOW	**0.818**	**0.809**	**0.816**
SVC	TF-IDF vectorizer	0.700	0.702	0.700
	CBOW	0.703	0.708	0.704
Simpler CNN	TF-IDF vectorizer	0.838	0.844	0.840
	CBOW	0.840	0.845	0.841
LSTM	**TF-IDF vectorizer**	**0.849**	**0.840**	**0.841**
	CBOW	0.835	0.840	0.835

The fact that Tunisians first expressed a sarcastic attitude toward the pandemic is one of the study's outcomes. It should also be noted that, despite the disease's massive spread over time, Tunisians had more or less neutral opinions about the whole pandemic period, with the exception of April (i.e. after the first wave). According to the collected dataset, people in Tunisia exhibited more sarcastic attitudes throughout the lockdown, which began in March 2020. More possible conclusion is that individuals are unsure about how to prevent this disease, as seen by the large amount of sarcastic comments collected from the dataset, particularly in 2021.

5 Conclusion

Sarcasm detection can provide important information regarding trends in the topic of the COVID-19 crisis on social media as well as alternate opinions on the COVID-19 pandemic, which has sparked widespread public concern. This study shows that Facebook comments are an effective communication source for gaining a better understanding of Tunisians' public concern and awareness regarding COVID-19. In this paper, we provided TunCorS, an open-source dataset of 23K Tunisian dialect comments labelled for sarcasm detection. The created dataset was thoroughly tested using multiple classification algorithms to validate its practical applicability. As future work, we plan to continue enhancing the annotated dataset with additional comments to keep it up-to-date with the latest events and discussions about the COVID-19 pandemic.

References

1. Alhajji, M., Al Khalifah, A., Aljubran, M., Alkhalifah, M.: Sentiment analysis of tweets in Saudi Arabia regarding governmental preventive measures to contain covid-19. In: MDPI AG (2020)
2. Althagafi, A., Althobaiti, G., Alhakami, H., Alsubait, T.: Arabic tweets sentiment analysis about online learning during covid-19 in Saudi Arabia. Int. J. Adv. Comput. Sci. Appl **12**, 620–625 (2021)
3. Ameur, M.S.H., Aliane, H.: Aracovid19-ssd: Arabic covid-19 sentiment and sarcasm detection dataset. arXiv preprint arXiv:2110.01948 (2021)
4. Besdouri, F.Z., Mekki, A., Zribi, I., Ellouze, M.: Improvement of the cotaorthography system through language modeling. In: IEEE/ACS 18th International Conference on Computer Systems and Applications, pp. 1–7. IEEE (2021)
5. Collobert, R., Weston, J.: A unified architecture for natural language processing: Deep neural networks with multitask learning. In: Proceedings of the 25th International Conference on Machine Learning, pp. 160–167 (2008)
6. Demšar, J.: Statistical comparisons of classifiers over multiple data sets. J. Mach. Learn. Res. **7**, 1–30 (2006)
7. Fix, E., Hodges, J.L.: Discriminatory analysis. nonparametric discrimination: Consistency properties. Int. Stat. Rev./Rev. Internationale de Statistique **57**(3), 238–247 (1989)
8. Freund, Y., Schapire, R.E.: A decision-theoretic generalization of on-line learning and an application to boosting. J. Comput. Syst. Sci. **55**(1), 119–139 (1997)

9. Habbat, N., Anoun, H., Hassouni, L.: Sentiment analysis and topic modeling on Arabic twitter data during covid-19 pandemic. Indonesian J. Innov. Appl. Sci. (IJIAS) **2**(1), 60–67 (2022)
10. Ho, T.K.: Random decision forests. In: Proceedings of 3rd International Conference on Document Analysis and Recognition, vol. 1, pp. 278–282. IEEE (1995)
11. Hochreiter, S., Schmidhuber, J.: Long short-term memory. Neural Comput. **9**(8), 1735–1780 (1997)
12. Madani, Y., Erritali, M., Bouikhalene, B.: Analyzing Moroccan tweets to extract sentiments related to the coronavirus pandemic: a new classification approach. In: Fakir, M., Baslam, M., El Ayachi, R. (eds.) CBI 2021. LNBIP, vol. 416, pp. 33–42. Springer, Cham (2021). https://doi.org/10.1007/978-3-030-76508-8_3
13. Mekki, A., Zribi, I., Ellouze, M., Belguith, L.H.: Treebank creation and parser generation for Tunisian social media text. In: IEEE/ACS 17th International Conference on Computer Systems and Applications, pp. 1–8. IEEE (2020)
14. Mekki, A., Zribi, I., Ellouze, M., Belguith, L.H.: Sentence boundary detection of various forms of Tunisian Arabic. Lang. Resour. Eval. **56**(1), 357–385 (2022)
15. Mekki, A., Zribi, I., Ellouze Khmekhem, M., Hadrich Belguith, L.: Critical description of TA linguistic resources. In: The 4th International Conference on Arabic Computational Linguistics (ACLing 2018) & Procedia Computer Science, November 17–19 2018. Dubai, United Arab Emirates (2018)
16. Salzberg, S.L.: C4. 5: Programs for Machine Learning by j. Ross Quinlan. Morgan Kaufmann Publishers, Inc. Burlington (1993)
17. Vapnik, V.N.: The Nature of Statistical Learning Theory. Springer-Verlag, New York (1995). https://doi.org/10.1007/978-1-4757-2440-0

Multimodal Deep Learning and Fast Retrieval for Recommendation

Daniele Ciarlo[1,3] and Luigi Portinale[1,2(✉)]

[1] Computer Science Institute, DiSIT, University of Piemonte Orientale,
Alessandria, Italy
luigi.portinale@uniupo.it
[2] Inferendo srl, Alessandria, Italy
[3] ORS Group, Roddi, Italy
daniele.ciarlo@ors.it

Abstract. We propose a retrieval architecture in the context of recommender systems for e-commerce applications, based on a multi-modal representation of the items of interest (textual description and images of the products), paired with a locality-sensitive hashing (LSH) indexing scheme for the fast retrieval of the potential recommendations. In particular, we learn a latent multimodal representation of the items through the use of CLIP architecture, combining text and images in a contrastive way. The item embeddings thus generated are then searched by means of different types of LSH. We report on the experiments we performed on two real-world datasets from e-commerce sites, containing both images and textual descriptions of the products.

Keywords: Multimodal embeddings · Recommender systems · Locality sensitive hashing

1 Introduction

Recommender Systems (RS) are software products based on machine learning having the goal of learning user preferences for specific items or services in very different contexts such as e-commerce, e-learning, e-tourism, e-health, and others. They can employ various methods such as collaborative filtering, content-based, hybrid, and knowledge-based approaches [15]. One promising direction in Recommender Systems (RS) research is the exploitation of different modalities during item search, such as textual descriptions and product images. However, since different modalities usually involve a massive amount of data which is also typically high-dimensional, multimodal retrieval requires huge storage space and a long retrieval time.

In the present paper, we propose a retrieval architecture in the context of e-commerce applications, based on a multi-modal representation of the items of interest (textual description and images of the products), paired with a locality-sensitive hashing (LSH) indexing scheme [5] for the fast retrieval of the potential

M. Ceci et al. (Eds.): ISMIS 2022, LNAI 13515, pp. 52–60, 2022.
https://doi.org/10.1007/978-3-031-16564-1_6

recommendations. In particular, we resorted to the use of CLIP (Contrastive Language-Image Pre-Training) [12], an approach proposed by Open-AI where a joint embedding of images and textual descriptions is learned in a contrastive way. In fact, CLIP learns a multi-modal embedding space by jointly training an image encoder and a text encoder in order to maximize the cosine similarity of the image and text embeddings of the correct pairs in a batch, while minimizing the cosine similarity of the embeddings of incorrect pairings. In a RS context, this allows one to exploit a single representation of both the item description and item image for searching the most suitable recommendation taking into account a multi-modal input. The unified embedding space can then be searched through a suitable indexing scheme; we propose to adopt LSH in order to trade-off precision of the results and retrieval time. The hypothesis we decided to test concerns the relevance of the retrieved results and the response time of the architecture, depending on the type and quantity of resources employed such as the type of LSH method, the number of hash tables, the number of LSH functions, and the type of visual encoder.

2 Multimodal Embedding of Text and Images

The first task we needed to address was the generation of multimodal embeddings. As reported above, we resorted to CLIP (Contrastive Language-Image Pre-Training), a task-agnostic model trained on a wide variety of images and texts. The model is trained in a contrastive way [17] to predict, given a batch of N pairs (image, text), which of the N^2 pairs is the correct one. To this end, the model learns a multimodal embedding space by simultaneously training an image and a text encoder with the goal of maximizing the cosine similarity of the correct pairs and of minimizing the similarity of incorrect ones.

Specifically, we considered different pre-trained versions of CLIP, and we finally selected two of them: RN50 and ViT-L/14. They differ in the image encoder employed: the first model uses ResNet50 [6], while the second uses a Vision Transformer (ViT) [4]. They represent a trade-off between the time needed to generate an embedding and its size. In particular, embeddings generated with RN50 are larger in size, but require less time to be generated compared with those generated by ViT-L/14 version. In particular, the RN50 architecture has been defined by applying a model tweak to ResNet50 called ResNet-D [7] which consists in a modification of the downsampling block of ResNet50. in addition, we also integrate a low-pass filtering to anti-alias to maintain shift-invariance [19]. Moreover, we substituted the global average pooling layer with an attention pooling implemented with a single attention layer as in the transformer architecture. Finally, a scaling strategy based on the approach described in [16] has been adopted, in order to obtain a better balance among depth, width and resolution of the image encoder.

Concerning the transformer-based architecture ViT-L/14, we implemented the version as described in [4]. In particular, the configuration includes the following parameters: a 14×14 image patch size, 24 layers, a latent embedding size of 1024, an MLP size of 4096 and 4 heads of attention, for a total number of

$307M$ learnable parameters. As we will see in Sect. 4, our experiments shows that the two types of embedding provided by the different architectures have similar performances. Both architectures use the same text encoder which is actually a Transformer with the architecture modifications proposed in [13]

3   Search and Retrieval Through LSH

Once the embeddings are generated, we need a retrieval mechanism, in order to select, from a given query, the embeddings most similar to it. A natural choice is to resort to Locality Sensitive Hashing (LSH) as an indexing scheme, because it allows the balance between precision of the retrieved results and retrieval time. This feature is crucial to our setting, since we want an architecture offering the possibility of prioritizing either precision or retrieval speed based on the configuration used. LSH [5] is a technique where similar data are hashed into the same "buckets" with high probability. This allows one to implement (approximate) nearest-neighbour queries as collision detection in a set of hash tables suitably designed [8].

LSH uses two main hyper-parameters: the number k of hash functions and the number L of hash tables; we also adopted a *multi-probe* LSH approach [11] with the goal of limiting the number of hash tables. In fact, one major limitation of standard LSH methods is that L must be large enough in order to achieve a good quality of the retrieval. In the multi-probe approach, multiple buckets that are likely to contain the query results are "probed" in a given hash table. A main feature of LSH algorithms concerns the proximity of buckets containing similar objects: if an object is close to a query but not hashed to the same bucket, it is likely to be in a bucket that is "close by", since the hash values of the two buckets only differ slightly. The idea of multi-probe is then to build a probing sequence that allows to probe, for each hash table, a set of buckets that are close to the one in which the query is indexed. The number of probes (i.e., number of extra hash buckets to check) to be used can be chosen dynamically, by setting the desired retrieval precision level [11].

Concerning the implementation, we decided to use FALCONN: FAst Lookups of Cosine and Other Nearest Neighbors [14], a widely tested and efficient library that implements LSH based algorithms. FALCONN supports two main hash families: hyperplane LSH [3] and cross polytope LSH [1]. They have theoretical guarantees for cosine similarity; this is particularly relevant in our case, since CLIP generates multimodal embeddings that maximize the cosine similarity for similar inputs. Both LSH families are implemented using multi-probe LSH [11] to minimize memory usage.

4   Experimental Analysis

We performed an experimental analysis by considering two different datasets:

- **AF**: a dataset of about 190K Amazon products related to fashion containing reviews, item metadata and images; we consider only the product image and the product name. Since several products have several associated images, we elaborated more than 280k images.
- **FC**: a dataset of photo related products from a medium e-commerce Italian site provided by INFERENDO (www.inferendo.ai); also in this case, we consider only the item image and name. In this dataset, several items do not have an associated image, so for those product we have produced embeddings based only on text. We elaborated 11.5K images and 342K names.

As previously described, the reference architecture we propose is based on a neural network pre-trained with a contrastive language-image approach (CLIP), and on an LSH indexing scheme, in order to speed-up the retrieval of the generated multi-modal embeddings (the FALCONN library).

We tested the architecture with the two datasets described above, and by considering 3 different phases:

1. embedding generation through CLIP by considering two underlying visual encoder architectures: RN50 and ViT-L/14;
2. LSH indexing by considering different choices concerning: the LSH algorithm used: cross polytope (CP) or random hyperplane (RH)); the number of hash functions (k); the number of tables (L); the number of probes (chosen dynamically to maintain a retrieval precision greater than 0.9);
3. evaluation of a set of multi-modal queries on the selected configuration

In particular, for each dataset we considered 16 different experimental configurations concerning: the LSH algorithm (hyperplane or cross polytope), the number of tables (30 or 50), the number of hash functions (16 or 17) and the image encoder net (RN50 or ViT-L/14).

We set up, for each dataset, $N = 75$ queries as follows: 25 queries representing actual items in the corresponding dataset, 25 textual queries manually generated from item descriptions, and 25 graphical queries from the web. We focused on a top-5 recommendation problem, where the 5 most similar items to the query were considered as relevant. For each query we computed Precison@k ($P@k$) and Recall@k ($R@k$) for $k = 1 \ldots 5$ from which we computed the Average Precision (AP) given by the formula

$$AP = \sum_{k=1}^{5} P@k(R@[k-1] - R@k) \ \ (R@0 = 0)$$

The relevant items for a query are obtained through a linear scan of the item embeddings, and by selecting the 5 items corresponding to embeddings having the largest *cosine similarity* with respect to the query.

We finally compute the Mean Average Precision (mAP) over all the queries as

$$mAP = \frac{1}{N} \sum_{i=1}^{N} AP_i$$

where AP_i is the average precision of i-th query.

Table 1. Mean average precision achieved and number of probes used by the various configurations on **AF**.

Hash families	n. tables	n. hash	RN50		ViT	
			mAP	n. Probes	mAP	n. Probes
Hyperplane	30	16	0.981	43	0.983	118
		17	0.986	119	0.974	156
	50	16	0.984	137	0.997	84
		17	0.987	88	0.986	88
Cross polytope	30	16	0.998	30	0.980	37
		17	1.000	78	0.970	30
	50	16	0.998	50	0.985	50
		17	0.987	88	0.986	105

Table 2. Mean average precision achieved and number of probes used by the various configurations on **FC**.

Hash families	n. tables	n. hash	RN50		ViT	
			mAP	n. Probes	mAP	n. Probes
Hyperplane	30	16	0.919	82	0.913	207
		17	0.945	64	0.914	178
	50	16	0.951	50	0.932	203
		17	0.947	73	0.933	50
Cross polytope	30	16	0.930	34	0.930	46
		17	0.960	54	0.919	30
	50	16	0.963	50	0.943	50
		17	0.932	114	0.930	1197

Table 1 reports the results for dataset **AF** and Table 2 the results for dataset **FC**. For each configuration we also reported the number of probes that are dinamically computed in the specific situation.

We can notice that, independently on the tested configuration, the performance in terms of mAP of the LSH retrieval on the generated multimodal embedding is really good, with no clear indication of a given configuration as definitely better than the others. The qualitative performance is slightly better in case of **AF** dataset, which is the largest one in size (more item images and descriptions provides a better final multimodal representation).

We also report in Table 3 the average number of unique candidates that must be checked by LSH retrieval, together with the corresponding time for answering the query, in case of dataset **AF**[1]. In each configuration, only a few thousands

---

[1] Similar results are obtained for dataset **FC**.

candidates are selected, and the answer is given in a few milliseconds on average. It is worth noting that the cross polytope version of LSH is usually selecting a larger number of candidates with respect to the hyperplane based algorithm; since qualitative results on mAP reported above are actually comparable for the two approaches, this suggests that the use of a simple hyperplane LSH algorithm can be really effective in the tested situations.

Table 3. Average number of unique candidates per query and average query time for the various configurations on **AF**.

Hash families	n. tables	n. hash	RN50		ViT	
			n. cand	Avg. query time	n. cand	Avg. query time
Hyperplane	30	16	3727.476	0.00328	4983.236	0.00419
		17	4525.53	0.00432	3466.555	0.00334
	50	16	6413.391	0.00693	5451.542	0.00472
		17	4717.643	0.00447	4318.597	0.00425
Cross polytope	30	16	20098.753	0.01329	2938.648	0.00352
		17	5601.449	0.00652	7878.953	0.00532
	50	16	32077.866	0.02304	19237.653	0.01217
		17	4717.641	0.0045	4318.589	0.00423

Finally, by way of example Fig. 1 shows the top-1 results (i.e. the most similar item) obtained for 3 sample queries on the **AF** dataset.

When dealing with data expressed in different modalities as in recommender systems, the information carried by each modality should be exploited in a way that is in principle both complete (no part of the information is ignored) and consistent (different expressions of the same information must be coherently fused). The combination of deep learning and hashing methods applied to multimodal retrieval can significantly improve the retrieval efficiency, since deep features extracted from the model contain richer semantic information and have a stronger ability to express the original data. However, two possible ways can be devised: data-dependent and data-independent approaches [2].

In data-dependent approaches, one tries to learn both a representation and an indexing scheme from the original multimodal data, usually in a supervised fashion. The strict integration of feature learning and hash learning is proposed in [9] with the DCMH (Deep Cross Modal Hashing) framework and in [18] with the PRDH (Pairwise Relationship Deep Hashing) framework, where the objective functions directly takes into account the similarity between modalities (text and images), and the preservation of such similarities when producing the hash code. In [10], an adversarial strategy called SSAH (Self-Supervised Adversarial Hashing) is adopted, where different discriminators (one for each modality) are trained in such a way to make the hash codes of different modalities more closely related to the original data. This idea is furtherly elaborated in [20], where an

(a) Round charm necklace

(b) Retrieved

(c) Flower dress

(d) Retrieved

(e) Striped T-shirt

(f) Retrieved

Fig. 1. Some results on **AF** dataset

attention mechanism is also introduced for this task, resulting in the so called ADAH (Attention-aware Deep Adversarial Hashing) framework.

Our proposal concerns the simpler data-independent approach; instead of training a deep model with the goal of getting both the multimodal representation and the hashing, we separately deal with the problem of learning the latent representation and exploiting a suitable data-independent indexing for retrieval. By implementing this architecture, we verified the feasibility of building a multimodal retrieval system that, employing LSH, achieves fast retrieval times and high quality of the retrieved elements. Even if the number of hash functions and tables should be data dependent, we showed that with a constant and relatively small number of resources of this kind we can achieve results of very good quality (measured in terms of mAP) and in reasonable time. The only data dependent aspects is restricted to the number of probes used, which is dynamically set on the basis of the required retrieval precision. The experiments showed that using high quality embeddings, specifically generated to correctly represent cosine similarity among inputs, we can get these results with a variety of LSH configurations.

Future work will focus on further testing the architecture against larger datasets, and on evaluating the robustness of the approach when data expressed in only one modality are mixed with data expressed in both textual and visual modality in different relative percentages.

References

1. Andoni, A., Indyk, P., Laarhoven, T., Razenshteyn, I.P., Schmidt, L.: Practical and optimal LSH for angular distance. CoRR abs/1509.02897 (2015). http://arxiv.org/abs/1509.02897
2. Cao, W., W.Fen, Lin, Q., Cao, G., He, Z.: A review of hashing methods for multimodal retrieval. IEEE Access **8**, 15377–15391 (2020)
3. Charikar, M.: Similarity estimation techniques from rounding algorithms. In: Proceedings of 34th Annual ACM Symposium on Theory of Computing, pp. 380–388 (2002)
4. Dosovitskiy, A., et al.: An image is worth 16×16 words: Transformers for image recognition at scale. CoRR abs/2010.11929 (2020). https://arxiv.org/abs/2010.11929
5. Gionis, A., Indyk, P., Motwani, R.: Similarity search in high dimensions via hashing. In: Proc. 25th VLDB 99. pp. 518–529 (1999)
6. He, K., Zhang, X., Ren, S., Sun, J.: Deep residual learning for image recognition. CoRR abs/1512.03385 (2015), http://arxiv.org/abs/1512.03385
7. He, T., Zhang, Z., Zhang, H., Zhang, Z., Xie, J., Li, M.: Bag of tricks for image classification with convolutional neural networks. In: Proc.CVPR2019 (2019)
8. Indyk, P., Motwani, R.: Approximate nearest neighbors: towards removing the curse of dimensionality. In: Proc. 13th ACM STOC1998. pp. 604–613 (1998)
9. Jiang, Q., Li, W.: Deep cross-modal hasing. In: Proc. IEEE Conf. Comput. Vis. Pattern Recognit. (CVPR). pp. 3232–3240 (2017)
10. Li, C., Deng, C., Li, N., Liu, W., Gao, X., Tao, D.: Selfsupervised adversarial hashing networks for cross-modal retrieval. In: Proc. CVPR2018. pp. 4242–4251 (2018)
11. Lv, Q., Josephson, W., Wang, Z., Charikar, M., Li, K.: Multi-probe lsh: Efficient indexing for high-dimensional similarity search. In: Proc. 33rd VLDB2007. p. 950–961 (2007)
12. Radford, A., Kim, J., Hallacy, C., Ramesh, A., Goh, G., Agarwal, S., Sastry, G., Askell, A., Mishkin, P., Clark, J., Krueger, G., Sutskever, I.: Learning transferable visual models from natural language supervision. In: Proc. of the 38th International Conference on Machine Learning (ICML 21) (2021)
13. Radford, A., Wu, J., Child, R., Luan, D., Amodei, D., Sutskever, I.: Language models are unsupervised multitask learners (2019), http://www.persagen.com/files/misc/radford2019language.pdf
14. Razenshteyn, I., Schmidt, L.: Falconn - fast lookups of cosine and other nearest neighbors. https://github.com/FALCONN-LIB/FALCONN
15. Ricci, F., Rokach, L., Shapira, B.: Recommender Systems Handbook: 2nd edition. Springer (2015)
16. Tan, M., Le, Q.: EfficientNet: Rethinking model scaling for convolutional neural networks. In: Proc. 36th PMLR2019 (2019)
17. Tian, Y., Krishnan, D.: Contrastive multiview coding. In: Proc. 16th European Conference on Computer Vision (ECCV) (2020)

18. Yang, E., Deng, C., Liu, W., Tao, D., Gao, X.: Pairwise relationship guided deep hashing for cross-modal retrieval. In: Proc. 31st AAAI2017. pp. 1618–1625 (2017)
19. Zhang, R.: Making convolutional networks shift-invariant again. In: Proc. 36th PMLR2019 (2019)
20. Zhang, X., Lai, H., Feng, J.: Attention-aware deep adversarial hashing for cross-modal retrieval. In: Proc. Eur. Conf. Comput. Vis. (ECCV). pp. 591–606 (2018)

Natural Language Processing

Mining News Articles Dealing with Food Security

Hugo Deléglise[1,3], Agnès Bégué[1,3], Roberto Interdonato[1,3(✉)],
Elodie Maître d'Hôtel[2,4], Mathieu Roche[1,3], and Maguelonne Teisseire[1,5]

[1] TETIS, Univ Montpellier, AgroParisTech, CIRAD, CNRS, INRAE,
Montpellier, France
roberto.interdonato@cirad.fr
[2] MOISA, Univ Montpellier, CIHEAM-IAMM, CIRAD, INRAE, Institut Agro,
Montpellier, France
[3] CIRAD, UMR TETIS, 34398 Montpellier, France
[4] CIRAD, UMR MOISA, 34398 Montpellier, France
[5] INRAE, Montpellier, France

Abstract. Food security is a major concern in West Africa, particularly in Burkina Faso, which has been the epicenter of a humanitarian crisis since the beginning of this century. Early warning systems for food insecurity and famines rely mainly on numerical data for their analyses, whereas textual data, which are more complex to process, are rarely used. To this end, we propose an original and dedicated pipeline that combines different textual analysis approaches (e.g., word embedding, sentiment analysis, and discrimination calculation) to obtain an explanatory model evaluated on real-world and large-scale data. The results of our analyses have proven how our approach provides significant results that offer distinct and complementary qualitative information on the food security theme and its spatial and temporal characteristics.

Keywords: Food security · Sentiment analysis · Spatiotemporal analysis · Term discrimination · Text mining · Word embedding

1 Introduction

Hunger remains a major problem in many parts of the world. Although a large scale and permanent solution to this situation is far from being achieved, steady progress was made in the first 15 years of this century. Among West African countries, Burkina Faso is in one of the most severe situations, with an undernourishment prevalence of 21.3% from 2015–2017 [7]. Burkina Faso is also one of the countries most affected by the phenomenon commonly known as the "triple burden of malnutrition", characterized by the coexistence of overnutrition, undernutrition and micronutrient deficiencies in the population.

Following several food crises in the 1970s and 1980s in different regions of the world, several food security alert and monitoring systems (FSMSs) were created

© The Author(s), under exclusive license to Springer Nature Switzerland AG 2022
M. Ceci et al. (Eds.): ISMIS 2022, LNAI 13515, pp. 63–73, 2022.
https://doi.org/10.1007/978-3-031-16564-1_7

by governmental organizations and NGOs. The objective of these systems, which are still very active today, is to prevent food crises and to help countries plan food aid programs to optimize their food production and distribution channels. In this study, we examine the ability of text mining methods to extract and analyze the qualitative information used as proxies for the national and regional food situation and its evolution over the last ten years in Burkina Faso from a corpus of newspapers from the country. The aim is to provide explainable indicators complementary to the automatic predictions of food security scores, i.e., as the ones obtained in our previous works based on the application of machine learning approaches on heterogeneous [3] and textual [2] data.

However, the difficulties associated with the implementation of text-mining approaches are linked to the structural complexity of textual data and are the subject of a large number of studies. We now detail the approaches proposed in the scientific literature to address these difficulties. In the domain of agriculture, which is closely related to food security, information extraction from textual data is a topic that has been attracting increasing interest [6]. In this field, several studies have focused on sentiment analysis [16], named entity extraction (i.e., places, dates or individuals related to agriculture) [12], etc.

The originality and methodological contributions of this work are presented at 3 levels: (1) its multidisciplinary aspect involving the combination of approaches based on text mining (e.g., word embedding and sentiment analysis) for the analysis of food security, which has been little studied from this perspective; (2) spatiotemporal analysis based on the content of French texts; and (3) extension of discrimination measures to address spatiotemporal data. The usefulness of this approach is to propose an explanatory framework complementary to the outputs of the predictive models usually applied to other types of data (e.g., digital data and satellite images). While we focus on the study case of Burkina Faso, the proposed method is generic and can be applied to any other area in the world. Section 2 presents the proposed approach. Section 3 outlines and discusses the information extracted from a dedicated corpus in French.

2   Proposed Approach

2.1   Text Mining Approaches

Studies use text mining methods to extract information on food security-related events from newspapers which proposes a framework for automatic detection of food crises [19]. Their method consists of extracting the most characteristic vocabulary (keywords) by tf-idf (term frequency-inverse document frequency) for each article [15], which is a method of weighting characteristic terms of texts, and then extracting the named entities with a Bi-LSTM-CNN-CRF framework [20]. A weight is associated with each keyword according to its semantic similarity (by Word2vec) with the terms of the article title. Each article, through a set of weighted keywords and associated named entities, is classified by single-pass clustering [14]. The ability of tf-idf to extract relevant and specific vocabulary from newspaper articles has also been demonstrated [1]. In this context, some text-mining methods are integrated in our pipeline:

Word2vec. Word2vec (w2v) [13] is a family of automatic language processing models for word embedding, i.e., the transformation of terms and texts into vectors. W2v is based on two-layer neural networks and aims at learning vector representations of terms in texts so that terms that share similar contexts (i.e., are often surrounded by the same terms) are represented by close numerical vectors. In our study, a CBOW (continuous bag of words) architecture is used (preferred to the skip-gram architecture, which requires more execution time while sometimes offering less satisfactory performance for processing newspaper articles [10]). CBOW aims at predicting the appearance of a term by using as proxies the terms that are close to it in the text. The model is trained on a large training corpus (French Wikipedia, in our study) by traversing each term and its neighbors and obtaining a set of feature vectors that represent each term in the text as the output.

Term Polarity. The polarity of a term is a criterion that indicates whether it is positive, negative or neutral [9, 17]. In our context, the average polarity of texts dealing with food security can give us relevant information about their worrisome or even alarming character. There are currently few methods for performing sentiment analysis on French texts. To evaluate the negativity of a term, we use the French version of the sentiment analysis model VADER (Valence Aware Dictionary and Sentiment Reasoner) implemented by the Python package vaderSentiment-fr[1]. This model is based on a lexicon of 7500 terms classified as positive or negative and on contextual rules that can modify the valence of the terms (e.g., the use of negation, punctuation, capitalization, and adverbs). This model was chosen because it has a good compromise between its simplicity of implementation and execution time and its classification performance, performing better than many existing methods, some of which are based on the use of machine learning [8].

tf-idf. To evaluate the discrimination of the terms of an article, we use the concept of tf-idf (term frequency-inverse document frequency) [15], which measures to what extent a term is characteristic of a text by evaluating its relevance and its singularity. Its principle is based on a formula in which two values, tf (term frequency) and idf (inverse document frequency), are multiplied together. tf corresponds to the frequency of a term in a text, and it therefore increases when a term is frequent in the text. idf measures the importance of a term according to its distribution in all the texts studied rather than based on its frequency in a particular text.

2.2 Our Food Security Pipeline

The objective of our pipeline is to perform a spatiotemporal analysis of food security based on the terminology of this domain linked to the textual proxies we define. In this framework, we propose an original and dedicated pipeline that combines different textual analysis approaches (e.g., word embedding, sentiment

---

[1] https://pypi.org/project/vaderSentiment-fr/.

analysis, and discrimination calculation). To this end, we present in this section the methodology deployed to obtain a spatial and temporal explanatory context of the Burkinabe food situation from the corpus of newspapers studied. Figure 1 summarizes the analysis plan. For this a first general lexicon on food security is therefore used to detect articles of interest, we name this lexicon "*GLEX*" (Generalist LEXicon). Then, two other more detailed lexicons are used to detect the expressions of "food security" and "crisis" themes used and thus obtain a more qualitative view of the content of the articles. We call these two detailed lexicons on food security and on crises "*FLEX*" (Food LEXicon) and "*CLEX*" (Crises LEXicon), respectively. These lexicons are freely available [5].

First, we present step (1) of selecting relevant articles. For this, we compute by w2v the semantic similarity between each article and the generalist lexicon *GLEX*, used as a basis to identify articles on the theme "food security". The principle is to consider an article as dealing with food security if its semantic similarity with *GLEX* by w2v is higher than a threshold x (chosen and validated in the Appendix document [4]). This aims to detect the articles of interest to focus the analyses.

Second, we establish in step (2) the textual proxies of food security on the selected articles. To this end, we perform the following operations:

- We keep for the selected articles their w2v score calculated during step (1), which quantifies their degree of connection with the food security theme and constitutes a proxy of this domain.
- We compute the negativity rate of the articles we propose as a proxy, i.e., the frequency of the negative terms in each article (Formula 1), to obtain information on the alarming nature of the articles' content.

$$Neg(art) = \frac{nb_{terms\_neg}(art)}{nb_{terms}(art)} \qquad (1)$$

where Neg is the negativity rate of an art article, and nb_{terms_neg} and nb_{terms} represent the number of negative terms and the number of terms of an art article, respectively, based on the French version of the VADER model (Valence Aware Dictionary and Sentiment Reasoner).
The hypothesis assumes that articles published during periods and in areas of food insecurity are associated with more negative valences than in a context of food sufficiency. An article is considered to be negative if its negativity rate is greater than 0.1 (the threshold validation methodology is detailed in the Appendix document [4]).
- We study the most used vocabulary in articles related to food security to detect whether the vocabulary adopted is consistent with the trends and crises that have affected food security in the country and thus to have a more explanatory perspective of the data. To accomplish this, we calculate for each article the frequency of 119 expressions from the two detailed lexicons *FLEX* and *CLEX*.

Third, we describe step (3) of global, regional and annual analysis of the proxies defined in step (2). To take into account the spatiotemporal aspect of food security, the proxies presented are then aggregated at different granularities to perform targeted analyses at the global, regional and annual levels and thus be able to visualize the trends and food crises that have affected the country over the last decade. The proxies are aggregated at three levels:

Global level: this level of analysis provides a general view of the characteristics of the country's food situation between 2009 and 2018 and can be used as a comparison for targeted analyses (regional and annual). The proportions of articles dealing with food security and negative articles are calculated over the entire corpus. We consider the average frequency of occurrence of each term in the detailed $FLEX$ and $CLEX$ lexicons across all articles in the corpus.

Regional level: this level of analysis aims to provide a representation of the food situation and its characteristics at the regional level. We illustrate our analyses with three regions: the Centre, Hauts-Bassins and Sahel regions. These three regions were chosen because they are among the most frequently cited in the articles in the corpus and are associated with distinct health situations [18]. Our approach consists of considering an article as associated with a region if a locality of the region is mentioned at the beginning of the article (i.e., in the title or in the first sentence of the article). The proportions of articles dealing with food security and negative articles were calculated for each of the 3 regions. To extract the characteristic regional vocabulary, we compute the tf-idf of each term of the $FLEX$ and $CLEX$ lexicons on the articles of each considered region. In our context, tf-idf allows us to highlight the expressions of food security and crises that are frequent in the articles related to a certain region and that are more specifically used in the articles of the region (i.e., more than for the other articles).

Annual level: this level of analysis provides annual characteristics of the food situation in Burkina Faso and tracks its evolution from 2009 to 2018. Each article is associated with its year of publication, which is extracted in the metadata linked to the article. This proposal, called the TIR (Tf-Idf ratio), is based on the concept of tf-idf and proves to be more suitable in our context, allowing us to distinguish rare and year-specific expressions more than tf-idf. More precisely, we first compute for each expression of the lexicons $FLEX$ and $CLEX$ the tf-idf of the expression on average on the articles of the year (Formula 2); then, in a second step, we compute the ratio of this tf-idf by the tf-idf of the expression on average on the articles of other years (TIR ratio, (Formula 3)).

$$TF - IDF_{moy}(t, A_y) = \frac{\sum_{art \in A_y} TF - IDF(t, art)}{N_y} \tag{2}$$

where $TF - IDF_{moy}$ is the average tf-idf of the term "t" on the articles "art" belonging to the set A_y of the articles of year y; we note N_y is the cardinality of this set.

$$TIR(t, A_y) = \frac{TF - IDF_{moy}(t, A_y)}{TF - IDF_{moy}(t, A_z)} \tag{3}$$

where TIR is the ratio of the tf-idf of the term "t" averaged over the articles belonging to the set A_y of articles in year y to the tf-idf of the term "t" averaged over the articles belonging to the set A_z of articles in different years of year y.

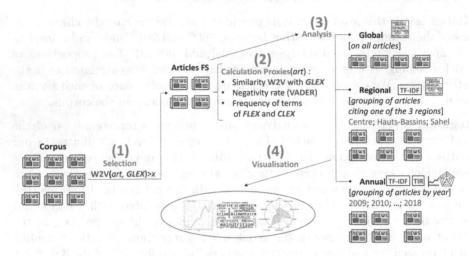

Fig. 1. General illustration of the analysis plan. The main steps are numbered: (1) selection of articles of interest; (2) computation of textual proxies on these articles; (3) global, spatial and temporal analysis; and (4) visualization.

3 Experiments

3.1 Corpus of Newspapers

Currently, the main Burkinabe newspapers have their own news website on which they publish their articles. For the creation of our newspaper corpus, we turned to two Burkinabe newspapers whose websites allow for good data accessibility: Burkina24 and LeFaso. These newspapers are among the most read newspapers in the country and have a large number of articles on various topics online. We extracted a total of 22856 articles between 2009 and 2018 (5595 for Burkina24 and 17261 for LeFaso), a period during which food security has undergone significant variations and several crises. The articles were filtered and then lemmatized with the Python package $Spacy^2$.

3.2 Results

Regional Analysis. We focus here on certain regions and observe whether the food security proxies aggregated over these regions are associated with the known regional food situation. The three regions studied are the Centre, Hauts-Bassins and Sahel regions.

---

[2] https://spacy.io/api/lemmatizer.

In Table 1, we see that the Hauts-Bassins region, which is the least food insecure of the regions presented, is associated with the lowest proportions of "food security" theme articles and negative articles. Conversely, the Sahel region, which is experiencing the most severe situation, has the highest proportions of "food security" themed articles and negative articles, which are both significantly higher than those at the national level. These data are consistent with expectations: the more an area is plagued by food insecurity and/or crises, the more articles will mention these topics and be negative.

In Fig. 2, we see that for the Hauts-Bassins region, the least poor, the most important expressions of the lexical field of food security are neutral (e.g., "riz" (*rice*), "agriculture" (*agriculture*), and "campagne agricole" (*agricultural campaign*)), whereas in the Sahel region, the expressions of food security are more negative (e.g., "malnutrition" (*malnutrition*) and "crise alimentaire" (*food crisis*)). Regarding the expressions of the theme of crises, the word clouds highlight concerns that are characteristic of each region. For example, the expression "inondation" (*flood*) is the most important in the Centre region, which is plagued by this problem, whereas the expression "foncier" (*land*) is the most important in the Hauts-Bassins region, where land management is a major problem.

Table 1. Comparison of the percentage of articles on the theme of "food security" (FS) and the percentage of negative articles for the three regions of the Centre, Hauts-Bassins, and Sahel and for Burkina Faso (BF).

	Centre	Hauts-Bassins	Sahel	BF
% of FS theme articles	6.5	4.9	10.7	7.3
% of negative articles	4.8	1.3	12.1	6.4

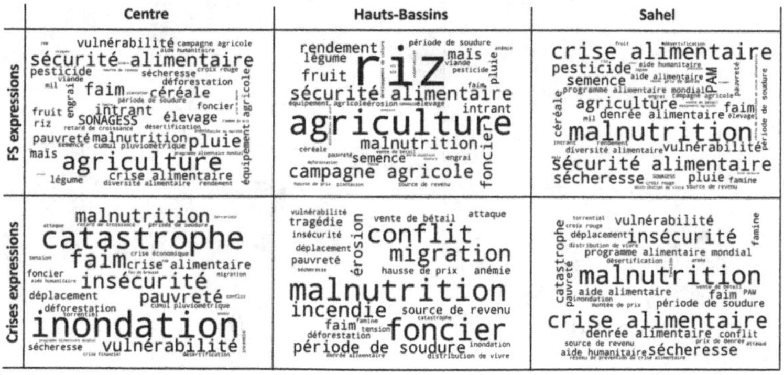

Fig. 2. Word clouds of expressions from the *FLEX* (FS expressions) and *CLEX* (Crises expressions) lexicons, based on "food security" theme articles related to three regions (Centre, Hauts-Bassins and Sahel). The size of the terms is proportional to their average tf-idf.

Annual Analysis. In this section, we examine whether food security proxies point over time to elements that are consistent, nuanced, or even contradictory to observations and events that have taken place over the past decade that may have affected food security. Namely, there has been a decline in food security since 2013 as well as events negatively impacting food security (e.g., flooding, drought, and conflict). The objective here is to determine the annual characteristics and evolution of the Burkinabe food situation through the food security proxies considered. We compute for each expression in the $FLEX$ and $CLEX$ lexicons the tf-idf of the expression averaged over the articles of the year, as well as the ratio TIR that we proposed.

Finally we analyze the evolution of the food security and crisis vocabularies used in the articles as a function of time (see Fig. 3a). In Fig. 4 (a), which represents the evolution of the proportion of negative articles by year from 2009 to 2018, we see a trend for negative articles to decrease in proportion. This may seem counterintuitive, and it may be explained by a certain freedom of the press that has tended to decline over the last decade (see Fig. 4 (b)). Figure 3b shows the evolution of the tf-idf of 5 expressions on average on the articles of each year between 2009 and 2018. We can see an upward trend in the tf-idf of the terms "sécurité alimentaire" (*food security*) and "malnutrition" (*malnutrition*), which have been increasingly used over the last decade. Moreover, some peaks correspond to the year of occurrence of events that took place over the period: the tf-idf of the expression "sécheresse" (*drought*) was the highest in 2012, which experienced a severe drought. The tf-idf for "conflits" (*conflict*) and "déplacement" (*displacement*) peaks in 2013, when conflicts in the Sahel led to the displacement of people from Sahelian countries bordering Burkina Faso.

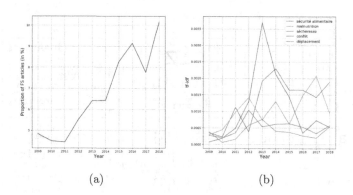

(a)   (b)

Fig. 3. (a) Change in the proportion (in percentage) of "food security" (FS) theme articles from 2009 to 2018 on the corpus studied. (b) Evolution of the average tf-idf of 5 expressions from the two detailed lexicons $FLEX$ and $CLEX$ between 2009 and 2018.

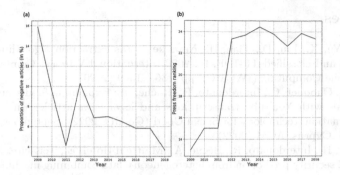

Fig. 4. Changes from 2009 to 2018 in the proportion (in percent) of negative articles per year among articles on the theme of "food security" (a) and Burkina Faso in the press freedom index (Reporters Without Borders(https://rsf.org/en/index-methodologie-2022)) (b).

4   Conclusion and Future Work

In this study, we examined the ability of text mining methods to extract spatial and temporal thematic information on food security from newspaper articles by examining the context of Burkina Faso.

We proposed, combined and extended, with adapted text mining methods (the Word2vec lexical embedding model, the VADER sentiment analysis model and the tf-idf term importance weighting method) three types of proxies defined on a set of articles, allowing us to obtain distinct and complementary information on the food security theme. This type of approach and the associated results can be exploited as complementary information to the outputs of predictive models (i.e., based on machine and deep learning). Indeed, machine and deep learning models applied to other types of data (e.g., digital data and satellite images) have strong predictive power but often lack explicability and interpretability. These models can then be validated, nuanced or explained by qualitative information from textual data that could make sense to domain experts and advance their understanding of complex food security phenomena.

To improve the thematic search in a finer way than with the word embedding applied in this work with w2v, technologies based on BERT (bidirectional encoder representations from transformers) and trained models for French, such as CamemBERT or FlauBERT, could also be integrated [11].

Acknowledgments. This work was supported by the French National Research Agency under the Investments for the Future Program #DigitAg, referred to as ANR-16-CONV-0004.

References

1. Ao, X., Yu, X., Liu, D., Tian, H.: News keywords extraction algorithm based on textrank and classified TF-IDF. In: 2020 International Wireless Communications and Mobile Computing (IWCMC), pp. 1364–1369 (2020)
2. Ba, C.T., Choquet, C., Interdonato, R., Roche, M.: Explaining food security warning signals with YouTube transcriptions and local news articles. In: Conference on Information Technology for Social Good (GoodIT 2022), September 7–9, 2022, Limassol, Cyprus. ACM (2022)
3. Deléglise, H., Interdonato, R., Bégué, A., Maître d'Hôtel, E., Teisseire, M., Roche, M.: Food security prediction from heterogeneous data combining machine and deep learning methods. Expert Syst. Appl. **190**, 116189 (2022). https://doi.org/10.1016/j.eswa.2021.116189
4. Deléglise, H., Roche, M., Interdonato, R., Teisseire, M., Bégué, A., Maître d'Hôtel, E.: Automatic extraction of food security knowledge from newspaper articles - Appendix. Working paper, Agritrop (2022). https://agritrop.cirad.fr/600423/
5. Deléglise, H., Schaeffer, C., Maître d'Hôtel, E., Bégué, A.: Lexiques en français sur la sécurité alimentaire et les crises (2021). https://doi.org/10.18167/DVN1/C5PU01, dataverse CIRAD
6. Drury, B., Roche, M.: A survey of the applications of text mining for agriculture. Comput. Electron. Agric. **163**, 104864 (2019). https://doi.org/10.1016/j.compag.2019.104864
7. FAO, ECA: Addressing the threat from climate variability and extremes for food security and nutrition. FAO (2018)
8. Gilbert, C.H.E.: Vader: A parsimonious rule-based model for sentiment analysis of social media text. In: Eighth International Conference on Weblogs and Social Media (ICWSM-14) (2014)
9. Hemmatian, F., Sohrabi, M.K.: A survey on classification techniques for opinion mining and sentiment analysis. Artif. Intell. Rev. **52**(3), 1495–1545 (2017). https://doi.org/10.1007/s10462-017-9599-6
10. Jang, B., Kim, I., Kim, J.W.: Word2vec convolutional neural networks for classification of news articles and tweets. PLoS ONE **14**(8), e0220976 (2019)
11. Le, H., et al.: Flaubert: unsupervised language model pre-training for French (2020)
12. Malarkodi, C., Lex, E., Sobha, L.: Named entity recognition for the agricultural domain. In: 17th International Conference on Intelligent Text Processing and Computational Linguistics (CICLING 2016); Research in Computing Science (2016)
13. Mikolov, T., Chen, K., Corrado, G., Dean, J.: Efficient estimation of word representations in vector space. In: Proceedings of Workshop at ICLR 2013, January 2013
14. Papka, R., Allan, J., et al.: On-line new event detection using single pass clustering. Univ. Massachusetts. Amherst **10**(290941.290954) (1998)
15. Salton, G., Buckley, C.: Term-weighting approaches in automatic text retrieval. Inf. Process. Manag. **24**(5), 513–523 (1988)
16. Surjandari, I., Naffisah, M., Prawiradinata, M.: Text mining of twitter data for public sentiment analysis of staple foods price changes. J. Ind. Intell. Inf. **3** (2014). https://doi.org/10.12720/jiii.3.3.253-257
17. Szabolcsi, A.: Positive polarity - negative polarity. Nat. Lang. Linguist. Theor. **22**(2), 409–452 (2004)
18. WFP: Burkina Faso : Analyse Globale de la Vulnérabilité, de la Sécurité Alimentaire et de la Nutrition. WFP (2014)

19. Xiao, K., Wang, C., Zhang, Q., Qian, Z.: Food safety event detection based on multi-feature fusion. Symmetry **11**(10) (2019). https://doi.org/10.3390/sym11101222

20. Yu, H.: Named Entity Recognition with Deep Learning. Ph.D. thesis, Auckland University of Technology (2019)

Identification of Paragraph Regularities in Legal Judgements Through Clustering and Textual Embedding

Graziella De Martino[1] and Gianvito Pio[1,2](✉)

[1] Department of Computer Science, University of Bari Aldo Moro, Via Orabona, 4, 70125 Bari, Italy
gianvito.pio@uniba.it
[2] Big Data Laboratory, National Interuniversity Consortium for Informatics, Via Ariosto, 25, 00185 Rome, Italy

Abstract. In an era characterized by fast technological progresses, working in the law field is very difficult if not supported by the right tools. In this paper, we present a novel method, called JPReg, that identifies paragraph regularities in legal case judgments to support legal experts during the preparation of new legal documents (i.e., paragraphs of existing documents that are similar to those of a document under preparation). JPReg adopts a two-step approach that first clusters similar documents, according to their semantic content, and then identifies regularities in the paragraphs for each cluster. Text embedding methods are adopted to represent documents and paragraphs into a numerical feature space, and an Approximated Nearest Neighbor Search method is adopted to efficiently retrieve the most similar paragraphs with respect to those of a target document. Our extensive experimental evaluation, performed on a real-world dataset, shows the effectiveness and the computational efficiency of the proposed method even in presence of noise in the data.

Keywords: Legal information retrieval · Embedding · Clustering · Approximate nearest neighbor search

1 Introduction

The legal sector is generally characterized by a slow response to new scenarios that appear every day in the modern society. In this context, the adoption of Artificial Intelligence (AI) methods can be helpful to improve the efficiency of the processes in this field. Among the several attempts that we can find in the literature in this direction, we can mention the work presented in [13], where the authors applied AI techniques to measure the similarity among legal case documents, that can be useful to speed up the identification and analysis of judicial precedents. Another relevant example is the work in [14], where the authors considered the semi-automation of some legal tasks, such as the prediction of judicial decisions of the European Court of Human Rights.

Following this line of research, in this paper, we present a novel method, called JPReg (Judgement Paragraph REGularities), that identifies paragraph regularities in legal case judgements, to support legal experts during the preparation of new legal documents. Methodologically, JPReg is based on a two-step approach that first groups documents into clusters, according to their semantic content, and then identifies regularities in the paragraphs for each cluster. Documents and paragraphs are represented into a semantic numerical feature space through text embedding methods, while the retrieval of similar paragraphs relies on an Approximated Nearest Neighbor Search (ANNS) approach, that provides a significantly higher computational efficiency with respect to classical similarity/distance-based methods. Therefore, given a (possibly incomplete or under preparation) document, henceforth called *target* document, JPReg supports the retrieval of similar paragraphs appearing in a set of *reference* documents related to previous transcribed legal case judgments.

Document clustering has received a lot of attention by the research community, but together with the design of advanced algorithms (e.g., for distributed computation, or co-clustering) [3,7,8,18], the most critical aspect is in the design of a proper representation of the objects/items at hand [11,16], as well as of similarity measures [10,13,17].

In this context, JPReg has the main advantage of properly combining embedding methods, to capture the textual semantics, with a two-step approach, that consists in learning a different numerical representation for each group of documents, rather than one single model for the whole collection of documents. This aspect allows JPReg to capture specific peculiarities of paragraphs according to the topic represented by the cluster they fall into.

Our extensive experimental evaluation, performed on a real-world dataset, proves the effectiveness and the computational efficiency of the proposed method. In particular, its ability of modeling different topics of legal documents, as well as of capturing the semantics of the textual content, appear very beneficial for the considered task, and make JPReg very robust to the possible presence of noise in the data and capable to significantly outperform state-of-the-art competitors.

2 The Proposed Method JPReg

Before describing JPReg, in the following, we provide some useful definitions:

- **Training set D_T**: a collection of legal judgments, represented as textual documents, adopted to train our models;
- **Reference set D_R**: a collection of legal judgments, represented as textual documents, from which we are interested to identify paragraph regularities;
- **Target document d**: a legal judgment (possibly under preparation) about which we are interested to identify paragraph regularities from D_R.

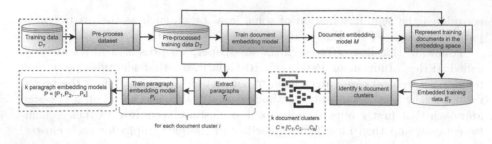

Fig. 1. Workflow followed by JPReg in the training phase. Green- and red-dotted rectangles represent inputs and outputs, respectively. (Color figure online)

The training set and the reference set may fully (or partially) overlap i.e., $D_T = D_R$ (or $D_T \cap D_R \neq \emptyset$), namely, the set of documents adopted to train our models may be the same as (or overlap with) the collection from which we want to identify paragraph regularities with respect to the target document. Note that JPReg is fully unsupervised and the target document d is never contained in either the training set or in the reference set (i.e., $d \notin (D_T \cup D_R)$).

JPReg consists of the three main phases, which are detailed in the following.

2.1 Training of Document and Paragraph Embedding Models

In Fig. 1, we show the workflow followed by JPReg in the first phase. JPReg first applies some pre-processing steps to the documents in D_t, namely: *i)* lowercasing of the text, *ii)* removal of the punctuation and digits, *iii)* lemmatization, and *iv)* removal of rare words. The pre-processed documents are then used to train a document embedding model M, that is subsequently exploited to represent each document of the training set D_T in the latent feature space, obtaining the set of embedded training documents E_T. Such documents are then partitioned into k clusters $[C_1, C_2, ..., C_k]$ by adopting the k-means clustering algorithm. Each cluster of documents becomes the input for a further learning step at the paragraph level: documents falling in the same cluster will contribute to the learning of a specific paragraph embedding model. Algorithmically, for each document cluster $C_i, 1 \leq i \leq k$, we extract the paragraphs from the documents falling into C_i and train a paragraph embedding model P_i. This approach allows us to learn more specific paragraph embedding models, according to the topic possibly represented by the identified clusters.

The embedding models, both at the document level and at the paragraph level, are learned by JPReg through neural network architectures based on Word2Vec Continuous-Bag-of-Words (CBOW) [16] or Doc2Vec [11] distributed memory (PV-DM). Their adoption is motivated by the fact that previous works demonstrated the superiority of Word2Vec and Doc2Vec over classical counting-based approaches, such as TF-IDF, since they take into account both the syntax and the semantics of the text [6,13]. In addition, their ability to capture the semantics and the context of single words and paragraphs allow them to prop-

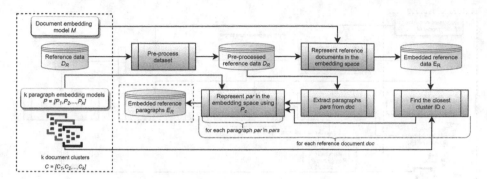

Fig. 2. Workflow followed by JPReg for the paragraph embedding of the reference set. Green- and red-dotted rectangles represent inputs and outputs, respectively. (Color figure online)

erly represent new (previously unseen) documents which features have not been explicitly observed during the training phase.

2.2 Embedding of the Paragraph of the Reference Set

In Fig. 2, we show the workflow followed by JPReg to represent the paragraphs of the documents belonging to the reference set into a latent feature space. Analogously to the training phase, we pre-process the documents of the reference set D_R. Then, each document is embedded using the previously learned document embedding model M. The embedded representation of the document is used to identify the closest document cluster that corresponds to the optimal paragraph embedding model (i.e., P_c), that can be adopted for its paragraphs. We stress the fact that this step performed by JPReg allows it to identify the most proper paragraph embedding model that was learned from a subset of the paragraphs possibly related to a similar topic.

The set of all the embedded paragraphs E_R is finally returned by this phase. Paragraph regularities for a given target document d under preparation will be identified from such set E_R.

2.3 Identification of Paragraph Regularities

The final phase, which workflow is represented in Fig. 3, starts by following the same steps mentioned in Sect. 2.2 to represent each paragraph of the target document d in the paragraph embedding space. Specifically, the most proper paragraph embedding model is adopted to embed its paragraphs, selected by identifying the closest document cluster with respect to d. For each embedded paragraph, we finally identify the top-n most similar paragraphs from the set of embedded paragraphs E_R belonging to the reference set.

As mentioned in Sect. 1, their identification could straightforwardly be based on the computation of vector-based similarity/distance measures (e.g., cosine

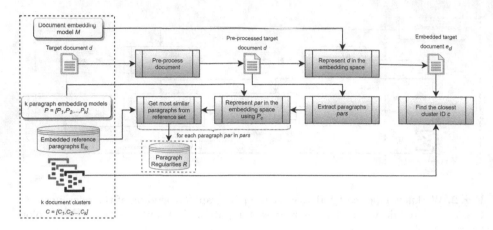

Fig. 3. Workflow followed by JPReg in the identification of paragraph regularities. Green- and red-dotted rectangles represent inputs and outputs, respectively. (Color figure online)

similarity, Euclidean distance, etc.) between the identified numerical representation of the paragraphs of the target document d and that of all the embedded paragraphs of the reference set E_r. However, in a real-world scenario, such a pairwise comparison would be computational intensive and would lead to inefficiencies and delays. To overcome this issue, we adopt an approximated and highly efficient approach for the identification of the top-n most similar paragraphs, that is based on random projections. In particular, we propose an approach based on Annoy [1], where the idea is to perform an approximated nearest neighbor search (ANNS), which consists of two phases: *index construction* on the paragraphs of the reference set, and *search*, that occurs when we actually need to identify the top-n most similar paragraphs with respect to a paragraph of the target document. During the index construction, we build T binary trees, where each tree is built by partitioning the input set of vectors recursively, by randomly selecting two vectors and defining a hyperplane that is equidistant from them. It is noteworthy that even if based on a random partitioning, vectors that are close to each other in the feature space are more likely to appear close to each other in the tree. During this process, a priority queue is exploited, and each tree is recursively traversed, where the priority of each split node is defined according to the distance to the query vector (here, a paragraph of the target document). This process leads to the identification of T leaves, where the query vector falls into. The distance between the query vector and the vectors falling into the such leaves is then exploited to return the top-n most similar paragraphs [12].

The adoption of this approach reduces the time complexity of the search phase for each paragraph of the target document from $O(|E_r|)$, in the case of the adoption of classical NNS approaches, to $O(log_2(|E_r|))$, that is the average length of a path in the trees from the root to a leaf node.

3   Experiments

All the experiments were performed using a real-world dataset consisting of 4,181 official public EU legal documents, provided by EUR-Lex[1], in a 10-fold cross-validation setting. All the documents of the testing set were considered as target documents, while the reference set was built by constructing 20 replicas of each paragraph of the documents in the testing set, perturbed by introducing a controlled amount of noise, as done in [4]. In particular, the noise was introduced by replacing a given percentage of words of each paragraph by random words selected from the Oxford dictionary[2]. In our experiments, we considered different levels of noise, namely, 10%, 20%, 30%, 40%, 50% and 60%, in order to assess the robustness of the proposed approach to different amounts of noise.

In order to quantify the specific contribution of the adopted embedding strategies, we compared the results obtained through Word2Vec and Doc2Vec with those achieved using a baseline approach, i.e., the classical TF-IDF. In all the cases, we adopted a 50-dimensional feature vector. Note that we use 50 features, since it is a commonly used dimensionality in other pre-trained embedding models. For TF-IDF, we selected the top-50 words showing the highest frequency across the set of legal judgments.

We evaluated the contribution of the JPReg two-step model with different numbers of clusters, i.e., with $k \in \{\sqrt{|D_T|}/2, \sqrt{|D_T|}, \sqrt{|D_T|} \cdot 2\}$, and compared the observed performance with that obtained without grouping training documents into clusters (henceforth denoted as *one-step model*).

We also performed an additional comparison with state-of-the-art methods. Specifically, we compared JPReg with **LEGAL-BERT-EURLEX**, that is the LEGAL-BERT model[3] fine-tuned by [2] using the EUR-LEX dataset, and **BERT-PLI**[4] which is based on BERT, fine-tuned with a small set of legal documents, proposed by [21] in the Competition On Legal Information Extraction/Entailment (COLIEE). Note that these competitors are embedding models, which are able to represent paragraphs as numerical feature vectors taking into account the semantics and the context of the textual content. Specifically, both LEGAL-BERT-EURLEX and BERT-PLI represent paragraphs in a 768-dimensional feature space. The embedding of each paragraph was computed as the mean of the embedding of its tokens.

Finally, we evaluated the effectiveness and the computational efficiency of the ANNS approach implemented in JPReg for the identification of the *top*-n most similar paragraphs. Specifically, we performed an additional comparative analysis against a non-approximated solution based on the cosine similarity, on a subset of 100 documents randomly selected from the dataset.

As evaluation measures, we collected precision@n, recall@n and F1-score@n, averaged over the paragraphs of target documents and over the 10 folds, with $n \in$

---

[1] https://eur-lex.europa.eu/homepage.html.
[2] raw.githubusercontent.com/cduica/Oxford-Dictionary-Json/master/dicts.json.
[3] https://huggingface.co/nlpaueb/legal-bert-base-uncased.
[4] https://github.com/sophiaalthammer/bert-pli.

$\{5, 10, 15, 20, 50, 100\}$. Specifically, for each paragraph of a target document in the testing set, we considered as True Positives the number of correctly retrieved (perturbed) replicas from the reference set. Note that, in this paper, for space constraints we only show the results in terms of F1-score@20.

3.1 Results

In Table 1, we report the F1-score@20 for the baseline one-step model and for the two-step model with different values of k, considering different embedding strategies and different levels of noise. From the table, we can observe that, although the TF-IDF led to acceptable results, the adoption of the embedding methods implemented in JPReg is significantly beneficial. Moreover, although Doc2Vec is natively able to work with word sequences, Word2Vec always obtains better results. This is possibly due to the fact that several paragraphs of different legal documents may share a similar topic, and the adoption of the unique sequence ID to associate the context with the document, as done by Doc2Vec (see [11] for details), may lead to overfitting issues.

From Table 1 and from Fig. 4, it is possible to clearly observe the contribution of the two-step process we propose. Indeed, the results show that the proposed two-step model outperforms the one-step model, in all the situations. In particular, the two-step model is much more robust to the presence of noise: although we can still observe a lower F1-score when the noise amount increases, its impact is much less evident. We can also observe that in general, the number of clusters k seems to not significantly affect the results, even if the best results are observed with $k = \sqrt{|D_T|} \cdot 2$. This means that the documents are distributed

Table 1. F1-score@20 results obtained with different embedding strategies (T = TF-IDF; D = Doc2Vec; W = Word2Vec) and different levels of noise. The upper-left sub-table shows the results obtained with the one-step model, while the other subtables show the results obtained by JPReg with different numbers of clusters. The best result in a given subtable is shown in boldface, while the absolute best result is underlined.

One-step model

	Noise %					
	10%	20%	30%	40%	50%	60%
T	0.696	0.587	0.456	0.305	0.154	0.047
D	0.885	0.815	0.687	0.510	0.327	0.180
W	**0.927**	**0.904**	**0.861**	**0.780**	**0.648**	**0.475**

Two-step model - $k = \sqrt{|D_T|}/2$

	Noise %					
	10%	20%	30%	40%	50%	60%
T	0.835	0.764	0.678	0.573	0.444	0.291
D	0.918	0.889	0.847	0.779	0.675	0.533
W	**0.944**	**0.930**	**0.905**	**0.859**	**0.776**	**0.652**

Two-step model - $k = \sqrt{|D_T|}$

	Noise %					
	10%	20%	30%	40%	50%	60%
T	0.854	0.789	0.709	0.611	0.492	0.349
D	0.923	0.896	0.856	0.795	0.703	0.575
W	**0.948**	**0.935**	**0.913**	**0.872**	**0.797**	**0.681**

Two-step model - $k = \sqrt{|D_T|} \cdot 2$

	Noise %					
	10%	20%	30%	40%	50%	60%
T	0.868	0.808	0.732	0.641	0.530	0.399
D	0.928	0.901	0.862	0.804	0.718	0.601
W	<u>**0.952**</u>	<u>**0.941**</u>	<u>**0.922**</u>	<u>**0.885**</u>	<u>**0.817**</u>	<u>**0.710**</u>

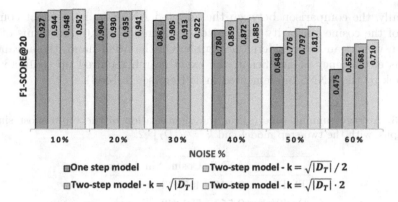

Fig. 4. F1-score@20 results obtained with the two-step model (with different values of k) and with the one-step model. As embedding strategy, we considered Word2Vec.

Table 2. F1-score@20 results obtained by JPReg (two-step model, $k = \sqrt{|D_T|} \cdot 2$, Word2Vec) and by the competitors, with different levels of noise. The best result for a given noise amount is shown in boldface.

	Noise %					
	10%	20%	30%	40%	50%	60%
JPReg	**0.952**	**0.941**	**0.922**	**0.885**	**0.817**	**0.710**
LEGAL-BERT-EURLEX	0.820	0.446	0.172	0.063	0.025	0.010
BERT-PLI	0.432	0.078	0.024	0.011	0.007	0.004

among several topics and that learning *specialized* paragraph embedding models is helpful to retrieve significant paragraph regularities.

Focusing on the comparison with state-of-the-art systems, in Table 2 we report the F1-score@20 results obtained by JPReg (two-step model, $k = \sqrt{|D_T|} \cdot 2$, Word2Vec) and by the considered competitors, with different levels of noise. From the results, we can easily observe that JPReg always outperforms both LEGAL-BERT-EURLEX and BERT-PLI, independently on the amount of noise in the data. Specifically, while the impact of noise is very strong on competitors, JPReg appears very robust and, thus, adoptable in real contexts even when the amount of noise in the data is high. The significantly lower F1-score@20 results achieved by the competitors, when documents are affected by high levels of noise, can be mainly due to the higher dimensionality of their embedding space (768), with respect to that adopted in JPReg (50). Indeed, although BERT-based models exhibit very interesting results in several NLP tasks [5], their high-dimensional feature space makes them more susceptible to the curse of dimensionality on tasks based on the computation of distances/similarities [9], like in the task at hand.

Finally, the comparison between the adopted ANNS and the exact computation of the cosine similarity emphasized a difference of 0.6% in terms of F1-score@n, which can be considered negligible. On the other hand, the advantage in terms of efficiency is significant: the exact search required up to 1000x the time took by the ANNS implemented in JPReg (see Table 3).

Table 3. Average running time (s) for the identification of the top-n most similar paragraphs, with the two-step model and $k = \sqrt{|D_T|} \cdot 2$.

	ANNS	Cosine Similarity
TF-IDF	**0.513**	407.612
Doc2Vec	**0.551**	580.842
Word2Vec	**0.610**	668.040

4 Conclusions

In this paper, we presented JPReg, a method to identify paragraph regularities in legal judgments. JPReg represents the documents and their paragraphs in a numerical feature space by exploiting embedding methods able to capture the context and the semantics. JPReg is based on a two-step approach, that groups similar documents into clusters and learns multiple paragraph embedding models able to represent specific topics. Finally, JPReg demonstrated to be able to identify paragraph regularities efficiently, thanks to the adopted ANNS strategy.

The accuracy and efficiency exhibited by the developed approach on real data make JPReg a useful tool in real-world scenarios, also when large collections of legal documents, possibly affected by noise, have to be analyzed.

For future work, we will exploit JPReg to provide suggestions during the preparation of new legal documents, by exploiting process mining methods. Moreover, we will evaluate the possibility to adopt transfer learning methods [15, 19] to exploit the models learned for a document cluster for the paragraphs of other, similar, clusters. Finally, we will investigate the possibility to implement JPReg as a service [20], to make it easily available to legal practitioners.

Acknowledgments. The authors acknowledge the support of the University of Bari Aldo Moro through the project "Horizon Europe Seeds - S34 - Hub Universitario Multidisciplinare. Processi innovativi per la digitalizzazione e la valorizzazione del patrimonio culturale scritto" (H99J21017690006).

References

1. Bernhardsson, E.: Annoy at github (2015). https://github.com/spotify/annoy
2. Chalkidis, I., Fergadiotis, M., Malakasiotis, P., Aletras, N., Androutsopoulos, I.: LEGAL-BERT: the muppets straight out of law school. In: Findings of the Association for Computational Linguistics: EMNLP 2020, pp. 2898–2904 (2020)

3. Corizzo, R., Pio, G., Ceci, M., Malerba, D.: DENCAST: distributed density-based clustering for multi-target regression. J. Big Data **6**(1), 1–27 (2019). https://doi.org/10.1186/s40537-019-0207-2
4. De Martino, G., Pio, G., Ceci, M.: PRILJ: an efficient two-step method based on embedding and clustering for the identification of regularities in legal case judgments. Artif. Intell. Law (2021). https://doi.org/10.1007/s10506-021-09297-1
5. Devlin, J., Chang, M.W., Lee, K., Toutanova, K.: BERT: Pre-training of deep bidirectional transformers for language understanding. In: Proceedings of Conference of the North American Chapter of the Association for Computational Linguistics: Human Language Technologies, Volume 1 (Long and Short Papers), pp. 4171–4186. Association for Computational Linguistics, Minneapolis, Minnesota, June 2019
6. Donghwa, K., Seo, D., Cho, S., Kang, P.: Multi-co-training for document classification using various document representations: TF-IDF, LDA, and Doc2Vec. Inf. Sci. **477**, 15–29 (2018)
7. Ester, M., Kriegel, H.P., Sander, J., Xu, X.: A density-based algorithm for discovering clusters in large spatial databases with noise. In: Proceedings of the 2nd International Conference on Knowledge Discovery and Data Mining, KDD 1996, pp. 226–231 (1996)
8. Hess, S., Pio, G., Hochstenbach, M., Ceci, M.: Broccoli: overlapping and outlier-robust biclustering through proximal stochastic gradient descent. Data Min. Knowl. Disc. **35**(6), 2542–2576 (2021)
9. Kumar, A., Makhija, P., Gupta, A.: Noisy text data: Achilles' heel of bert. In: Proceedings of the Sixth Workshop on Noisy User-generated Text (W-NUT 2020), pp. 16–21, January 2020
10. Kumar, S., Reddy, P.K., Reddy, V.B., Suri, M.: Finding similar legal judgements under common law system. In: Madaan, A., Kikuchi, S., Bhalla, S. (eds.) DNIS 2013. LNCS, vol. 7813, pp. 103–116. Springer, Heidelberg (2013). https://doi.org/10.1007/978-3-642-37134-9_9
11. Le, Q., Mikolov, T.: Distributed representations of sentences and documents. In: 31st International Conference on Machine Learning, ICML 2014 4, May 2014
12. Li, W., Zhang, Y., Sun, Y., Wang, W., Zhang, W., Lin, X.: Approximate nearest neighbor search on high dimensional data - experiments, analyses, and improvement (v1.0). CoRR (2016)
13. Mandal, A., Chaki, R., Saha, S., Ghosh, K., Pal, A., Ghosh, S.: Measuring similarity among legal court case documents. In: Proceedings of the 10th Annual ACM India Compute Conference, pp. 1–9. Association for Computing Machinery (2017)
14. Medvedeva, M., Vols, M., Wieling, M.: Using machine learning to predict decisions of the European court of human rights. Artificial Intelligence and Law 28(2) (2020)
15. Mignone, P., Pio, G.: Positive unlabeled link prediction via transfer learning for gene network reconstruction. In: Ceci, M., Japkowicz, N., Liu, J., Papadopoulos, G.A., Raś, Z.W. (eds.) ISMIS 2018. LNCS (LNAI), vol. 11177, pp. 13–23. Springer, Cham (2018). https://doi.org/10.1007/978-3-030-01851-1_2
16. Mikolov, T., Sutskever, I., Chen, K., Corrado, G., Dean, J.: Distributed representations of words and phrases and their compositionality. Advances in Neural Information Processing Systems 26, October 2013
17. Minocha, A., Singh, N., Srivastava, A.: Finding relevant Indian judgments using dispersion of citation network. In: Proceedings of the 24th International Conference on World Wide Web, pp. 1085–1088. Association for Computing Machinery (2015)
18. Pio, G., Ceci, M., Loglisci, C., D'Elia, D., Malerba, D.: Hierarchical and Overlapping Co-Clustering of mRNA: miRNA Interactions. In: ECAI 2012. Frontiers in Artificial Intelligence and Applications, vol. 242, pp. 654–659. IOS Press (2012)

19. Pio, G., Mignone, P., Magazzú G., Zampieri, G., Ceci, M., Angione, C.: Integrating genome-scale metabolic modelling and transfer learning for human gene regulatory network reconstruction. Bioinformatics **38**(2), 487–493 (09 2021)
20. Redavid, D., Corizzo, R., Malerba, D.: An OWL ontology for supporting semantic services in big data platforms. In: Proceedings of 2018 IEEE International Congress on Big Data, BigData Congress 2018, pp. 228–231 (2018)
21. Shao, Y., Mao, J., Liu, Y., Ma, W., Satoh, K., Zhang, M., Ma, S.: Bert-pli: Modeling paragraph-level interactions for legal case retrieval. In: Proceedings of International Joint Conference on Artificial Intelligence, IJCAI-20, pp. 3501–3507 (2020)

Aspect Term Extraction Improvement Based on a Hybrid Method

Sarsabene Hammi(✉), Souha Mezghani Hammami, and Lamia Hadrich Belguith

ANLP Research Group, MIRACL Lab, FSEGS, University of Sfax, Sfax, Tunisia
hsarsabene@gmail.com, souha.hammami@ihecs.usf.tn,
lamia.belguith@fsegs.usf.tn

Abstract. The sentiment analysis task has been given great attention in the recent years, especially by enterprises and customers in commercial domain. In fact, companies tend to identify the customers' opinions regarding their services, industrialized products, etc. In this context, the Aspect Based Sentiment Analysis (ABSA) was introduced to determine the clients' viewpoints and extract the different aspects (e.g., price, quality, etc.) of entity (e.g., laptops) and assign them a sentiment polarity. In the present work, we are interested only in the aspect extraction (AE) task which is the most crucial and difficult task in the ABSA domain. We propose a hybrid method that combines the strengths of the linguistic knowledge and those of deep learning methods to solve the problem of AE for the French language. We also enhance this method by means of a new pruning algorithm which is mainly based on an out-domain dataset. The developed hybrid method has significantly improved the current state of the art and has given encouraging results when applied on respectively the Amazon mobile phone reviews (86.39% of F-measure) and the SemEval-2016 restaurant dataset (76.62% of F-measure).

Keywords: Aspect extraction · Association rules · Sentiment analysis · Dependency relation · Pruning · Word embedding

1 Introduction

The internet's content has become an important mine of data that can be exploited in sociological, commercial, financial fields, etc. Therefore, because of the considerable increase in the amount of the produced information and the importance that society gives to the public opinion, sentiment analysis has been applied in several domains: political, medical, financial, etc. In this study, we were mainly interested by the task of sentiment analysis concerning the commercial domain. In fact, international companies aim at improving the quality of their products and satisfying the customer's demands. To attain these objectives, researchers developed many sentiment analysis methods. They classified this process into three levels: document level, sentence level and aspect level. The first and second levels consist in extracting the prevailing general sentiments in all documents or sentences. However, the third one offers more fine-grained sentiment analysis. Although the three levels are important and give a better view about the users' opinions, the third level remains the most important and the most difficult to realize.

© The Author(s), under exclusive license to Springer Nature Switzerland AG 2022
M. Ceci et al. (Eds.): ISMIS 2022, LNAI 13515, pp. 85–94, 2022.
https://doi.org/10.1007/978-3-031-16564-1_9

Unlike the other types of analysis, aspect-based sentiment analysis does not only inform us about the general customers' opinion concerning the product, but it also determines their opinions about the different characteristics of a certain product. We consider the following review:

This phone has a high storage capacity.

In this example, aspect-level sentiment analysis is performed to extract the positive sentiment given to the aspect "capacity of storage" of entity "phone". This task requires first the extraction of aspects and, then, the detection of the customers' sentiment about each aspect. AE analyzes explicit and implicit aspects. The former are those mentioned explicitly in reviews. They are generally nouns or noun phrases. However, the latter are mentioned implicitly in the reviews and expressed in an indirect way through indices. We take the following examples. In the first one, customer expresses his/her opinion explicitly about the aspect "color" which is an explicit aspect. On the other hand, in the second example, customer did not mention explicitly the aspects "appearance" and "price", but he/she gives his/her opinions about them through the opinion words "pretty" and "cheap". In this study, we focus only on the extraction of explicit aspects.

My phone color is very nice.
This cell phone is very pretty and cheap.

Although there are a lot of research works that has been conducted to solve this problem for the English language, it is not the case for the French language. This is essentially due to the complexity of the linguistic components in the French language and the lack of necessary resources and tools for processing. To concretize the aspect extraction task, four main approaches are used: linguistic knowledge-based, machine learning-based, deep learning-based and hybrid. In this paper, we present a hybrid method that addresses the core tasks necessary to detect explicit aspects from French review sentences. This method combines the high precision achieved by the linguistic knowledge-based method and the high recall released by the deep learning-based method. The main contributions of our research are outlined below:

- Collecting a new French dataset composed of 2400 mobile phone reviews. The reviews in this dataset were manually annotated where 2000 of them are used to validate and 400 to evaluate our proposed hybrid method.
- Developing a new hybrid method for aspect terms extraction. This method achieves high precision and high recall values.
- Introducing a new pruning algorithm to increase the accuracy rate provided by the proposed method.

The remainder of the paper is organized as follows. Section 2 discusses the related works on aspect extraction methods. Sect. 3 presents the different steps of the proposed method. Section 4 describes the used dataset and the experimental results. The last section gives some concluding remarks and perspectives.

2 Related Works

Given the importance of aspect extraction task, it has been intensively studied by the research community. Early studies divided the aspect extraction approaches into four main approaches: machine learning-based approach [2, 4, 11], deep learning-based approach [8, 14, 17], linguistic knowledge-based approach [1, 5–7] and hybrid approach [3, 9, 12, 16].

The machine learning has been widely used in the aspect extraction task as it allows classifying data. It is defined as a sub-task of artificial intelligence and consists of two phases: the learning phase and the test phase. The former is applied to train the model using a training dataset. However, the test phase utilizes a test dataset to validate this model. Although there are many machine learning algorithms, two of them have been largely employed in this field: the LDA (Latent Dirichlet Allocation) and CRF (Conditional Random Field) algorithms. Among the works that adopted these algorithms, we can cite the study of [11] where an Interdependent LDA (ILDA) model was proposed to extract the aspects. This model benefits from the dependency between the opinion words and the aspects. [2] improved the work of [11] and introduced another method called Sentence Segment LDA (SS-LDA) based on the LDA model, association rules and frequency. [4] applied the CRF algorithm and several features (e.g. POS (part of speech tag) and the distance between words, etc.) to identify the aspect terms.

The deep learning based-approach is an enhancement of the traditional machine learning algorithms. Its architecture was inspired from the human brain. Recently, deep learning algorithms have proven their efficiency in the aspect identification task. For this reason, many studies used these algorithms as it is the case [17] where authors employed the CNN (convolutional neural network) algorithm to identify the aspect terms. This algorithm aims at outputting a probability distribution over each aspect in the sentences. [14] enriched the CNN algorithm by the domain knowledge to extract the aspect terms. [8] combined many deep learning algorithms and used the Bi-LSTM, CRF and CNN algorithms to extract the aspect terms.

The linguistic knowledge-based approach is one of the most used approaches in the aspect extraction task. Researches, in this domain, use a set of linguistic knowledge (constraints, rules and grammar) to solve this task. Among them, we mention [7] who proposed a rule-based method to identify the frequent and infrequent aspect terms. However, [5] concentrated on the dependency relation between aspect terms and opinion words. [6] enhanced the work of [5] and exploited the dependency relations to detect, in addition to the single-word aspect terms, the multi-word aspect terms. They also introduced a pruning method to eliminate the non-aspect terms.

A hybrid approach takes advantage of the strengths of the existing approaches (linguistic knowledge-based approach, machine learning-based approach and deep learning-based approach) and enhances the aspects extraction task. It was employed by many authors such as [9] who applied deep learning-based and rules-based methods to identify the aspect terms. In the first step, the researchers used the word embedding (deep learning-based technique) technique to detect the aspects. After that, they utilized a set of rules to identify the rest of the aspects. In the same context, [3, 12] and [16] created a set of rules and coupled them with the deep learning algorithms to detect the aspect terms. Firstly, they applied the rules to annotate the dataset necessary in order to train the deep learning-based model. Then, they trained these models on the training dataset to extract aspects.

In this study, we adopt a hybrid method to solve the problem of aspect terms extraction in the French language.

3 Proposed Method

In this section, we present the proposed hybrid method for aspect terms extraction. It is based mainly on three steps. In the first one, a linguistic knowledge-based method was used to extract a reduced list of highly-precise aspect terms. Although this method achieved a high precision value, that of recall value remained low. To overcome this weakness, we applied a deep learning-based method to calculate the semantic similarity between words in the dataset as aspect words generally appear in similar contexts. Finally, to ameliorate the obtained result, we applied a pruning algorithm used to remove the non-aspect terms by applying an out-domain dataset.

3.1 Step 1: Linguistic Knowledge-Based Method for Aspect Terms Extraction

In this section, we describe in detail the proposed linguistic knowledge-based method for aspect terms extraction. We first use a syntactic dependency relation method to extract a list of highly-recalled aspect candidates. Then, the syntactic dependency relations are combined with the association rules to extract the highly-precise aspect terms.

Extraction of the Highly-Recalled Aspect Candidates List. This step aims essentially at finding and extracting the highly recalled aspect candidates. Some researchers [7, 13] assumed that product aspects appearing in datasets are nouns. Relying on this assumption, we suggest a method based on a set of rules to extract nouns with a high probability to be considered as aspects. Since the aspect terms and opinion words are strongly correlated in the same sentence, we utilized the syntactic dependency parser tool Stanford and a set of patterns to identify the different syntactic relations between these words. Although Stanford relations play an important role in identifying aspects terms' candidates, they are not all efficient in executing this task. Therefore, we conducted an empirical study (on 2000 mobile phone reviews) regarding all the Stanford relations and their specific impact on the aspect terms extraction. We concluded that only seven relations ("nsubj", "amod", "nmod", "obl", "obj", "conj" and "appos") were reasonably qualified to accomplish the aspects' extraction task. Then, they were combined with a set of patterns in order to create a set of rules applied to extract the aspect words shown in Table 1. The rules are well explained in Table 1. For example, if there is, in a sentence, the "nsubj" relation with a *NOUN*, as the first word, and an *Opinion Word* as the second one, the first word will be extracted as an aspect candidate term. We consider the following review:

La **qualité** est excellente. (The **quality** is excellent.)

det (La/The-DET, qualité/quality-NOUN), **nsubj** (qualité/quality-NOUN,

excellente/excellent-ADJ), **cop** (est/is-AUX, excellent/excellent-ADJ).

There is a *nsubj* relation between the noun "qualité" and the opinion word "excellente". Thus, according to the rule, the "qualité" word is considered as an aspect. In this work, we detect the opinion words using the opinion lexicon FEEL[1].

This set of rules was applied to extract more than 96% of aspects terms that are placed in **ACL1** (Aspects Candidate List) list. However, the precision value remained low (58.48%) due to the prominent extraction of a high percentage of the non-aspects terms. So, in the following step, another list of highly-precise aspects was extracted from the list of highly-recalled aspects.

Table 1. Aspect candidate's extraction rules.

Relations	Patterns	Rules
nsubj/obl/obj	(Noun, Opinion-Word)	If the relation is **nsubj**, **obj** or **obl** and the first word is *Noun* and the second word is *opinion-word*, then the first word will be aspect term
nmod/conj/appos	(Noun, Noun)	If the relation is **nmod**, **conj** or **appos** and the first word is *Noun* and the second word is *Noun*, then the first and the second words will be aspect terms
amod	(Opinion-Word, Noun)	If the relation is **amod** and the first word is *opinion-word* and the second word is *Noun*, then the second word will be aspect term

Extraction of the Highly Precise Aspect Candidates List. We extracted, in this step, highly-precise aspect terms by applying first the association rule mining method proposed by [10] to extract the nouns that frequently occur together in the same review. The use of this method originates from the fact that those aspects of products mostly appear together in the same review because customers usually share the review while mentioning several product's features. For this reason, a transaction file, where each line contains all the nouns existing in the same review, was created. Then, the Apriori algorithm was applied on this file with a high support value equal to 0.01[2] and a high confidence value equal 0.3 (this value is chosen according to an empirical study). In the association rules, the support value reflects the frequency of the appearance of all aspects together in the dataset. Therefore, a high support value between X and Y shows a strong link between them. However, the confidence value reflects the number of times a given rule turns out to be practically true. So, a list (**ACL2**) of the most frequent and correlated nouns in the dataset was formed. Taking this rule (qualité *(quality)* -> photo *(picture)*) generated by the Apriori algorithm with a support value equal to 0.02 and confidence

---

[1] The French lexical dictionary FEEL (French Expanded Emotion Lexicon) is composed of 14,128 opinion words: 8424 of them are positive and 5704 of them are negative.

[2] The value of support is chosen according to an empirical study that we effected on 2000 reviews. This study proves that the highly correlated aspect terms appear together for at least 20 times out of a total of 2000 transactions i.e. support 20/2000 = 0.01.

value equal to 0.5. According to the algorithm, the terms "qualité" *(quality)* and "photo" *(picture)* are highly correlated in the dataset, so we add these two terms to the **ACL2** list since they can be potential aspect terms. The size of this list (**ACL2**) would, therefore, increase by combining the dependency relations and the association rules. The aspect candidates extracted by Stanford dependency relations in the previous step were first examined. It was clear that the "nsubj" relation has the highest accuracy. This relation succeeded in extracting an important number of correct aspect terms. Thus, it was used as a prominent source to extract the right aspects. After that, we apply the association rules mining method, for the second time, with a support value (0.008 according to an empirical study) to produce more rules. Then, we verify for each term in the "nsubj" relation if it is correlated with any highly-precise term in **ACL2**. If it is the case, then the "nsubj" term will be added to a new list (**ACL3**). The resulting aspect terms list is called **ACL4** and it contains the aspect terms from both **ACL2** and **ACL3** (**ACL2 + ACL3**). Although this method helped to extract a highly precise aspect terms list (precision equal to 83.58%), the recall value remained low (33.53%).

The main objective of the next step is to ameliorate the recall and precision rates by applying a deep learning-based method.

3.2   Step 2: Deep Learning-Based Method for Aspect Terms Extraction

This section focuses on extracting rest of aspects by using the semantic similarity between words. This similarity can be captured by employing several methods (e.g. PMI) and different resources (e.g. WordNet, Probase, etc.). We essentially concentrate on the word embedding technique based on the neural networks architectures proving a high efficiency in the extraction of words appearing in similar contexts.

The word embedding is an NLP (Natural Language Processing) technique widely used to represent the words by vectors of real numbers. These vectors of words take into consideration the semantic similarity between words and represent the words appearing in similar contexts by relatively-close vectors. In this work, we employed the word embedding technique, already trained by neural networks, in order to compute the semantic similarities between aspects term's candidates. There are many publicly available word vectors trained on a huge deal of information extracted from many sources (e.g. Wikipedia, news, etc.). However, previous researches, such as [7], showed that the use of this type of word embedding can sometimes give bad findings and degrade the efficiency of the proposed aspects extraction methods. This result is due to the fact that these available embeddings are trained on different domains and not on a particular domain. Therefore, the use of a specific domain word embedding is more pertinent in the detection of semantic similarity between aspects than multi-domain embedding.

Despite the large number of available datasets in the English language for the sentiment analysis, the French language was not given much importance in this field. In order to overcome this limitation, we collected a large domain-specific dataset and used it to train our embedding by applying the Word2Vec[3] model. For the mobile phone's

---

[3] The genism library was utilized to train our Word2Vec model on 100 epochs with a vector size of 100.

domain reviews, the input represents the reviews we collected from Amazon website (around 19600 reviews). For the restaurant domain, 10000 Yelp restaurant reviews have been collected and used to training the model. After that, a similarity measure (Cosine measure) was employed to calculate the degree of similarity between the highly-precise aspect terms (**ACL4**) and the highly-recalled aspect candidates' terms (**ACL1**). The aspect terms resulting in this step are placed in **ACL5**.

We put the final list of aspect terms, resulting from the proposed hybrid method in **ACL6 (ACL5 + ACL4)**.

3.3 Step 3: Improvement of the Aspect Final List with an Out-Domain Pruning Algorithm

The developed method provided acceptable precision results and excellent recall results. So, we proposed a pruning algorithm based on the out-of-domain dataset to improve the precision of the extracted aspects by eliminating incorrect aspects. This pruning algorithm is suggested to detect whether a word is an aspect term or not based on its domain relevance degree. This idea comes from the assumption that aspect words are probably much more used in their specific domain than any other domain. To attain this purpose, an out-domain dataset was utilized to calculate the domain specificity of each term in our aspect list. The evaluation of this domain specificity depends on two frequency values and aims at determining the number of times in which the term appears in a domain-related dataset, relatively to how often the term appears in an out-domain dataset. As it is shown in the Eqs. (1) and (2), the domain-dependency DD (related/not related) of each term t was computed according to the total number of terms existing in the same dataset.

$$DD_R(t) = \frac{Freq(t, Dr)}{\sum_1^n T} \tag{1}$$

$$DD_{NR}(t) = \frac{Freq(t, Dnr)}{\sum_1^n T} \tag{2}$$

Dr: the related-domain dataset. **T:** the total number of terms in the dataset. **Dnr:** the non-related-domain dataset.

Then, the values DD_R and DD_{NR} were compared. If the DD_R was equal or superior to DD_{NR}, the term would be considered as an aspect since it is related to the specific domain. Otherwise, it would be deleted from the list of aspect terms.

This method enhanced the precision value obtained by applying the introduced method and allowed removing a big number of the non-aspect terms. However, it can face the problem of the appearance of words (e.g. "Game") that can be aspects in two different domains. This word is an infrequent aspect term (does not usually appear as an aspect term) in the domain of mobile phones, but it is an important and frequent aspect term in the domain of football. In this case, it would be eliminated from the aspect terms list though it is a correct aspect term.

4 Experiments and Results

In this section, we evaluate the performance of the proposed aspect extraction method and we compare it with a set of state-of-the-art benchmarks. We also present the used datasets.

4.1 Dataset

The proposed method has been experimented and evaluated on mobile phone (400 reviews) and SemEval-2016 restaurant (335 reviews) datasets. In fact, we collected 22000 Amazon mobile phone reviews on French language. We used 19600 reviews (out of a total of 22000) to train the word embedding technique and create words' vectors. The rest of the reviews (2400 reviews) were manually annotated in our laboratory by two annotators: the first one is a member of our laboratory and the second one is a linguist from the faculty of Letters and Human Sciences of Sfax. From these 2400 reviews, 2000 were used to validate our method and 400 were used to evaluate the performance of our method. In the other hand, we collected 10000 reviews from Yelp website to train the word embedding technique and then, we evaluated our method on SemEval-2016 restaurant (335 reviews) reviews. In addition to the mobile phone dataset, another football dataset (1302909 words) was employed to apply the pruning algorithm.

4.2 Evaluation of Our Proposed Method for Aspect Terms Extraction

In this section, we evaluate our proposed hybrid method for aspect terms' extraction. it is shown by Table 2, the obtained results demonstrate that the LKBM (Linguistic-Knowledge-based Method) and DLBM (Deep Learning-Based Method) methods provided good results. In fact, the LKBM (ACL4) achieved a very high precision rate (83.58% for mobile phone domain and 82.05% for restaurant domain) and low recall rate (33.53 for mobile phone domain and 21.05% for restaurant domain). However, the DLBM (ACL5) provided us with an acceptable precision (70.96% and 70.22%) and recall values (62.78%, 60.52%) for both mobile phone and restaurant domains.

In order to ameliorate the obtained results and take advantage of the strengths of the two methods, the latter were combined (ACL6). As it is exposed in Table 2, the proposed method HAEM (Hybrid Aspect Extraction Method) gave very encouraging results (F-measure equal to 84.51% for mobile phone and 77.02% for restaurant) for both studied domains. It also improved significantly the findings obtained by LKBM and DLBM and achieved a high recall value and precision rate. In addition, to enhance the precision value and remove the non-aspect terms, we suggested a pruning algorithm. This algorithm enhanced significantly the precision values by 3% and 2% respectively for mobile phone and restaurant domains, respectively and removes a significant number of non-aspect words.

In the other hand, we compared our proposed method with two other works [8, 12] (presented in Sect. 2) that had been dealing with the aspect extraction task in French language. The experiments in both [8] and [12] were performed on a French SemEval-2016 restaurant dataset. The obtained results summarized in Table 3 revealed that our method outperformed the other benchmarks in terms of aspects' extraction.

Table 2. Evaluation of the proposed aspect-extraction method according to Precision (P), Recall (R) and F-measure (F1).

	Mobile phone			Restaurant		
	P	R	F1	P	R	F1
Step 1–1 of LKBM	58.48%	97.00%	72.96%	39.17%	94.07%	55.31%
Step 1–2 of LKBM	83.58%	33.53%	47.86%	82.05%	21.05%	33.50%
Step 2: DLBM	70.94%	62.87%	66.66%	70.22%	60.52%	65.01%
HAEM	75.23%	96.40%	84.51%	72.94%	81.58%	77.02%
+Pruning	78.32%	95.20%	**86.39%**	74.15%	79.27%	**76.62%**

Table 3. Comparison of our method with benchmarks.

	P	R	F1
HAEM	75.00%	77.96%	**76.45%**
[8]	69.73%	71.76%	70.73%
[12]	-	-	61.20%

5 Conclusion

In this research paper, we proposed a new hybrid method to extract the aspect terms. In this method, we combined the linguistic knowledge based-methods with the deep learning-based methods to obtain better results. We first extracted a list of highly-recalled aspect terms using a set of rules. The latter detected the aspect terms based on the dependency relations between aspects terms and opinion words. After that, this list was exploited to create another list containing only the highly-precise aspect terms. These terms were extracted based on the association rules and the dependency relations methods. Finally, the deep learning technique (word embedding) was applied to form another list of aspect terms semantically similar to the highly-precise aspect terms. In addition, to ameliorate the performance of the proposed method, a new pruning algorithm was proposed. This algorithm improved mainly the ratability of our method and augmented the precision value by 3% and 2% for both mobile phone and restaurant domains.

The suggested method achieved high performance and surpassed the other benchmarks with a value of F-score equal to **86.39%** and **76.62%** for both mobile phone and restaurant domains.

In our future work, we plan to enhance the precision of our method by applying a new efficient pruning method using a set of rules and statistics. Then, we will be more interested in detecting the implicit aspect terms that have become an important sub-task in the aspect extraction task. Finally, we will suggest a method to analyze the sentiment of users concerning the aspect terms.

References

1. Ameen, B., Zohair, A., Ali, D., Rabeeh, A., Hussain, D.: Aspect based sentiment analysis for polarity estimation of customer reviews on Twitter. Comput. Mater. Continua **67**(2), 2203–2225 (2020)
2. Baris, O., Akcayol, M.: A new topic modeling based approach for aspect extraction in aspect based sentiment analysis: SS-LDA. Expert Syst. Appl. **168**, 114231 (2021)
3. Chauhan, G.S., Meena, Y.K., Gopalani, D., Nahta, R.: A two-step hybrid unsupervised model with attention mechanism for aspect extraction. Expert Syst. Appl., 161–113673 (2020)
4. Chen, Z., Mukherjee, A., Liu, B.: Aspect extraction with automated prior knowledge learning. In: Proceedings of the 52nd Annual Meeting of the Association for Computational Linguistics, pp. 347–358 (2014)
5. Guang Q., Bing, L., Jiajun, B., Chun, C.: Expanding domain sentiment lexicon through double propagation. In: 31 International Joint Conference on Artificial Intelligence, vol. 9, pp. 1199–1204 (2009)
6. Manju, V., Gupta, D.: An unsupervised hierarchical rule-based model for aspect term extraction augmented with pruning strategies. Procedia Comput. Sci. **171**, 22–31 (2020)
7. Minqing, H., Liu, B.: Mining and summarizing customer reviews. In: Proceedings of the Tenth ACM SIGKDD International Conference on Knowledge Discovery and Data Mining, pp. 168–177 (2004)
8. Nihel, P., Erwan, K.: Analyse de sentiments à base d'aspects par combinaison de réseaux profonds: application à des avis en Français . In: conférence sur le Traitement Automatique des Langues, pp. 303–310 (2018)
9. Qian, L., Bing, L., Yuanlin, Z., Doo, K., Zhiqiang, G.: Improving opinion aspect extraction using semantic similarity and aspect associations. In: Thirtieth AAAI Conference on Artificial Intelligence (2016)
10. Rakesh, A., Srikant, R.: Fast algorithms for mining association rules. In: Proceedings of 20th International Conference Very Large Data Bases, VLDB, pp. 478–499 (1994)
11. Samaneh, M., Ester, M.: Opinion digger: an unsupervised opinion miner from unstructured product reviews. In: Proceedings of the 19th ACM International Conference on Information and Knowledge Management, pp. 1825–1828. (2020)
12. Sebastian, R., Ghaffari, P., Breslin, J.: Insight-1 at semeval-2016 task 5: Deep learning for multilingual aspect-based sentiment analysis. arXiv preprint arXiv:1609.02748 (2016)
13. Zhang, C., Li, Q., Song, D., Wang, B.: A multi-task learning framework for opinion triplet extraction. arXiv preprint arXiv:2010.01512 (2020)
14. Liang, Y., Meng, F., Zhang, J., Chen, Y., Xu, J., Zhou, J.: A dependency syntactic knowledge augmented interactive architecture for end-to-end aspect-based sentiment analysis. Neurocomputing **454**, 291–302 (2021)
15. Wu, C., Wu, F., Wu, S., Yuan, Z., Huang, Y.: A hybrid unsupervised method for aspect term and opinion target extraction. Knowl.-Based Syst. **148**, 66–73 (2018)
16. Bhamare, B.R., Prabhu, J.: A supervised scheme for aspect extraction in sentiment analysis using the hybrid feature set of word dependency relations and lemmas. PeerJ Comput. Sci. **7**, e347 (2021)
17. Ruidan, H., Wee, S.L.: An interactive multi-task learning network for end-to-end aspect-based sentiment analysis. arXiv preprint arXiv:1906.06906 (2019)

Exploring the Impact of Gender Bias Mitigation Approaches on a Downstream Classification Task

Nasim Sobhani[✉] and Sarah Jane Delany

Technological University Dublin, Dublin, Ireland
nasim.x.sobhani@mytudublin.ie, sarahjane.delany@tudublin.ie

Abstract. Natural language models and systems have been shown to reflect gender bias existing in training data. This bias can impact on the downstream task that machine learning models, built on this training data, are to accomplish. A variety of techniques have been proposed to mitigate gender bias in training data. In this paper we compare different gender bias mitigation approaches on a classification task. We consider mitigation techniques that manipulate the training data itself, including data scrubbing, gender swapping and counterfactual data augmentation approaches. We also look at using de-biased word embeddings in the representation of the training data. We evaluate the effectiveness of the different approaches at reducing the gender bias in the training data and consider the impact on task performance. Our results show that the performance of the classification task is not affected adversely by many of the bias mitigation techniques but we show a significant variation in the effectiveness of the different gender bias mitigation techniques.

Keywords: Gender bias · Training data · Classification

1 Introduction

NLP systems are trained on natural language content and it has been shown that they can display bias learned from the training data. Prior work has shown gender bias in core NLP tasks such as co-reference resolution [5,13] and language modelling [13] and in word embeddings which are used to represent text data [3,4,12]. Gender bias has also been demonstrated in more practical applications of NLP, such as abusive language detection [7,16] and sentiment analysis [11].

A variety of bias mitigation techniques have been proposed. These techniques include approaches which manipulate the training data itself including removing all gendered words, known as data scrubbing [6], or swapping gendered words with their gender equivalent [25]. Techniques which attempt to de-bias word embeddings have also been proposed, those that remove bias from static word embeddings after they have been generated [3] and those that alter the training process to generate de-biased word embeddings [26].

© The Author(s), under exclusive license to Springer Nature Switzerland AG 2022
M. Ceci et al. (Eds.): ISMIS 2022, LNAI 13515, pp. 95–105, 2022.
https://doi.org/10.1007/978-3-031-16564-1_10

In this work we compare different gender bias mitigation techniques on training data in two ways. We look at the effect that the techniques have on reducing the gender bias in the data and we evaluate the impact of the techniques on a downstream task that a model which is trained on the data is built to achieve. The techniques we consider include those that aim to neutralise the gender through data manipulation and augmentation and the approach of using de-biased word embeddings as the representation for the data.

To measure the gender bias on training data requires identifying gender in some way in the training data. A challenge faced is identifying training datasets that include appropriate labelling for the downstream task and labelling to facilitate measuring gender bias. We use the benchmark BiasBios dataset published by [6] which has dual labelling; the target class is occupation but each instance in the dataset, which represents an individual biography, is also labelled by gender. In addition we use the technique which is named by [21] as Gender Bias Evaluation Testsets (GBETs) to generate a test dataset for a hate speech classification task. GBETs are designed to check that NLP systems avoid making mistakes due to gender bias. Our results show significant variation in the effectiveness of the different gender bias mitigation techniques on the gender bias of the training data although the impact on the performance of the classification task is less significant.

2   Related Work

Techniques used to remove the gender bias in training data primarily involve (i) manipulation and augmentation of the training data and (ii) using debiased word embeddings as the representation. There are two primary ways that the data is manipulated, firstly by removing gender indicators from the data, known as data scrubbing [6], and secondly by augmenting the data with additional examples that are gender-swapped. Scrubbing has been shown to have effect on reducing gender bias in classification while preserving the overall model accuracy [17]. Gender swapping has been shown to be successful in reducing gender bias in classification [16] and coreference resolution [25] and, although it is easy to implement, it requires paired lists of gender specific terms and it doubles the size of the training data which is computationally expensive.

Counterfactual Data Augmentation (CDA) [13] was proposed to improve basic gender swapping. In addition to swapping gendered words which co-refer to a proper noun, for example Queen Elizabeth, are not swapped. CDA also includes the appropriate swapping of "her", "he" and "him" to maintain the correct grammar of sentences. Counterfactual Data Substitution (CDS) [9] was proposed to avoid duplicating the full dataset using gender swapping, and involves substituting fifty percent of the data with gender-swapped versions. The substitution is done probabilistically on a per document basis rather than within document to avoid grammatical errors.

The different approaches for de-biasing word embeddings can be grouped into post-processing approaches that debias the embedding after it has been generated and those that attempt to train and generate embeddings with minimal bias. GN-Glove (Gender-Neutral Global Vectors) is an example of the latter [26], training debiased word embeddings from scratch with gender as the protected attribute. A common post-processing method for debiasing word embedding uses the gender subspace or direction that captures the bias [3]. Gender-neutral words (pre-defined) are altered to be zero in the gender subspace by projecting them orthogonally to the gender subspace. Then predefined equality sets of words which differ only in the gender component (e.g. grandfather, grandmother) are altered to be equidistant from the gender neutral words. However, it has been shown though that while this approach substantially reduces bias, it is not fully removed, only hidden, and can be recovered [8]. A disadvantage of these post processing approaches is that sets of gender neutral and equality words are required prior to the de-biasing process. Debiasing embeddings can have negative effects on gender bias in downstream tasks and has been shown to actually increase gender bias, although classification accuracy was also increased [17].

The predictions from a unbiased or fair NLP model should not be influenced by gender mentions in the input text content. Differences in system performance for inputs where the text content varies only by gender can indicate that the system is not fair. This can be achieved by gender swapping the test instances to see whether the NLP system will perform differently on test data that is gender specific. This approach has been used in coreference resolution [13].

Generating a synthetic test set with test instances that isolate gender, also called Gender Bias Evaluation Testsets (GBETs) [21], has been more commonly used to evaluate gender bias. The GBET dataset can be mined from existing natural language data [24] but, more commonly, the GBET dataset is generated from sentence templates that reflect the NLP task and include gender identification words. Pairs of sentences are generated from the sentence template each with a specific gender identity. Differences in the NLP system performance across the pairs demonstrate the existence of gender bias in the training dataset. The extent of the difference can reflect the extent of gender bias in the system. Although GBETs have a few limitations including non naturalistic text and lack of coverage [2] they have been used in a variety of different NLP tasks including sentiment analysis [11], abusive language detection [7,16], coreference resolution [5,18] and to evaluate bias in language models [15].

There are a variety of measures in the literature used to measure fairness or bias for algorithmic classification problems [22] and to detect gender bias in NLP methods [20]. Most of the recent work on evaluating gender bias in NLP systems use variations on Hardt et al.'s work on equalised odds and equal opportunity [10]. These measures are group measures and use the gender distributions in the training data rather than the democratic parity measure which insists on equal outcomes for both genders regardless of prevalence or ground truth. Based on the equalised odds definition of fairness where the predictions are independent of gender but conditional on the ground truth or actual outcomes in the training data, error rate equality differences [7,16] capture the extent of unintended bias. Other work focuses on equality of opportunity which considers only the desirable

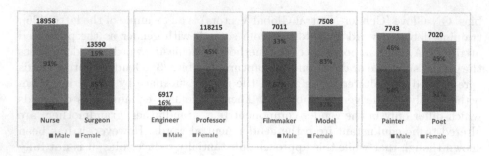

Fig. 1. Four datasets extracted from BiasBios each with biographies of two occupations, showing the class (occupation) distributions and the gender distributions across each occupation.

outcome [17]. This means that the true positive rate of the system should be independent of gender but conditional on the actual outcomes.

3 Approach

Our aim is to compare different data bias mitigation techniques for handling gender bias in training data and to evaluate the impact of these techniques on the performance of a downstream classification task. The bias mitigation techniques we consider are scrubbing, CDA and CDS and we also consider the impact of different word embeddings.

We use two datasets, one that is already labelled for gender, the BiasBios dataset [6] a dataset of biographies across different occupations with a gender label for each biography, and one where we utilise a Gender Bias Evaluation Testset (GBET) to measure the gender bias. Biographies for pairs of occupations were selected from BiasBios that are likely to demonstrate gender bias but that also have different class distributions. These pairs included surgeon-nurse, engineer-professor, model-filmmaker, poet-painter. Figure 1 shows the occupation pairs and gender distributions across these datasets. The surgeon-nurse dataset has high gender imbalance, nurses are 10 times more likely to be female than male, but surgeons are almost 6 times less likely to be female than male Fig. 1(a). The first step in pre-processing this data is to remove the first sentence of each biography due to the existence of the occupation word [6]. Noise removal, involving removing tags and replacing contractions (*don't* is replaced with *do not* etc.), and normalisation, converting all text to lower case, is performed. The text is tokenised into words and stop words and all punctuation are removed.

Our second dataset is a Hate Speech dataset of tweets [23] where the downstream classification task is to predict whether the tweet is abusive or not. This dataset, which has a size of almost 17K instances, was collected from Twitter and consists of 3,383 samples of sexist content, 1,972 samples of racist content and 11,559 neutral samples. We transform the dataset to a binary classification problem by concatenating the sexist and racist contents to "abusive" and assigning neutral samples to "non abusive" labels. This data was pre-processed in a similar way to the BiosBias datasets and split into 80%20% train and test splits for experimentation. As this dataset does not have any gender label, to measure gender bias a synthetic unbiased test set is generated using the identity term template method proposed by [7]. This approach generates an unbiased test set for each gender by filling in sentence templates with gender identity terms that provide the gender label for the test instance. The template is also filled in with offensive and non-offensive nouns and adjectives that provide the target label for the test instance. Following work by [16] the templates, the identity term pairs and the nouns and adjectives we used are given in Table 1. We generated 1332 test samples in total, 666 pairs giving equal sets of male and female instances and equal distribution across abusive and non-abusive also.

We used a simplified version of the deep neural network model used by [6] for classification. Our model included an embedding layer as an input layer, followed by a bidirectional long short term memory (BiLSTM) layer to encode the input sentence, a dropout layer and a linear output layer with cross entropy to compute the loss which comes after a dense layer with Relu activation. All hyper parameters were tuned and their optimal values determined.

We measured classification performance using average class accuracy (ACA) due to the imbalanced class distributions in both datasets. We evaluated gender bias using True Positive Rate Gap $(TPR_{gap}) =| TPR_{male} - TPR_{female} |$ [17] which measures the differences in the gender specific true positive rates.

4   Results

Figure 2 shows the average class accuracy across different embeddings using different bias mitigation techniques on the four occupation datasets. The baseline model is trained without applying any data manipulation techniques to the

Table 1. The templates, identity term pairs and the nouns & attributes used to generate test data for the HateSpeech classification task.

Templates	Identity Pairs
You are a < adjective > < identity term >	woman/man girl/boy
< verb > < identity term >	female/male sister/brother
Being a < identity term > is < adjective >	daughter/son wife/husband
I am a < identity term >	gal/guy mother/father
I hate < identity term (plural form) >	grandmother/grandfather

Nouns & Adjectives & Verbs	Target
disgusting, filthy, nasty, rotten, horrible, terrible awful,stupid, moronic, dumb, ugly,repulsive, vile idiotic,shitty, fucked, kill, murder, hate, destroy	Abusive
great, fun, nice, neat, happy, good, best, fantastic wonderful, lovely, excellent, incredible, friendly gracious, kind, caring, hug, like, love, respect	Non-Abusive

training data. All the explicit gender indicators are removed using data scrubbing. Results are also reported for the CDA and CDS data augmentation techniques. Three different word representations were used to analyze the impact of embeddings on gender bias. These include (i) pretrained word2vec embeddings [14], labelled *Word2Vec*, (ii) pretrained hard debiased word2vec embeddings [3], labelled *Debiased-Word2Vec* and (iii) pretrained conceptnet embeddings [19] labelled *ConceptNet*. The latter two embeddings were selected as they were the top embeddings found to have least bias by [1].

In general, the classification performance with bias mitigation does not vary significantly from the baseline. CDA has the best performance across all embeddings and datasets which is most likely due to the increase in training data as a result of duplicating the training set. CDS and scrubbing tend to have a negative impact on classification performance except for the ConceptNet embeddings.

Figure 3 shows the gender gap TPR_{gap} for each of the four occupation datasets. Results show that applying any data manipulation approach, scrubbing, CDA, or CDS significantly reduces the bias compared to the baseline. This pattern is evident across all three embeddings and all datasets. It is particularly apparent in the occupations which have a significant imbalance in gender distribution including nurse and surgeon Fig. (3a), engineer Fig. (3b), filmmaker and model Fig. (3c). Across all types of embeddings the CDA data augmentation technique performs the best. Professor, poet and painter are occupations that have more or less equal gender distributions and show a low gender gap indicating a low level of bias. As can be expected with low bias, the bias mitigation techniques do not have a significant impact on reducing the gender gap but do not have any negative impact either.

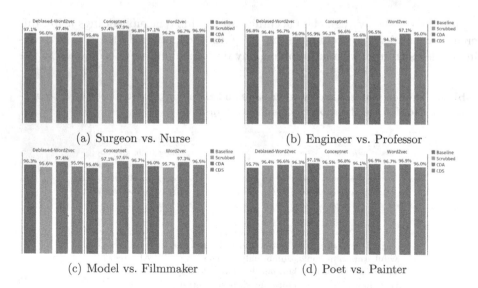

(a) Surgeon vs. Nurse

(b) Engineer vs. Professor

(c) Model vs. Filmmaker

(d) Poet vs. Painter

Fig. 2. Classification performance (ACA) for the four binary occupation datasets

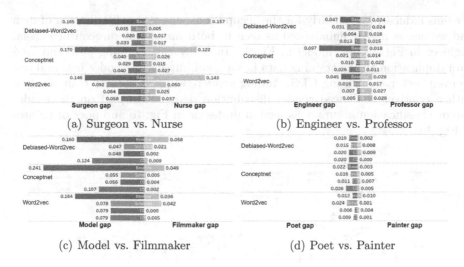

(a) Surgeon vs. Nurse

(b) Engineer vs. Professor

(c) Model vs. Filmmaker

(d) Poet vs. Painter

Fig. 3. True Positive Rate gap (TPR_{gap}) for each of the occupation datasets.

Surprisingly, using debiased-word2Vec embeddings on the original training data (without any data manipulation) does not reduce the bias significantly in any case, and actually increases the gender bias for both classes in the surgeon-nurse dataset as seen in Fig. 3a. However, combining debiased word2vec with any of the data manipulation techniques does significantly decrease the bias, across all datasets, more so that using the original word2vec. This suggests that using de-biased word embeddings alone is not adequate to mitigate gender bias.

While CDA has shown good performance both in terms of classification performance and reducing gender bias, the required duplication of the data means it is computationally expensive. CDS was proposed to alleviate this challenge. However, while CDS performs well compared with the baseline for all occupations, it is not as effective at reducing bias as CDA, particularly for occupations where a material gender gap exists. The only exception to this is for word2vec embeddings where the gender gap is comparable.

Inspired by CDA and CDS, we explored augmenting the training data by adding a proportion of the original dataset, gender-swapped, to the original dataset. We randomly selected 20%, 50%, and 70% percent of the dataset, applied CDA to this proportion of the dataset and added it to the training data. To counteract the random element in the data sampling, we repeated the process twice with two different random selections for each proportion and reported the average. Figure 4 shows the average class accuracy and TPR_{gap} results for different proportions of data duplication in addition to the TPR_{gap} for CDS (which is labelled as $GAP\text{-}CDS$). As the results show the data duplication amount does not have a significant impact on the classification performance. However, increasing the proportion of data duplication has a direct impact on the gender bias. As gender-swapped data is added to the training data, the gender bias reduces, particularly for the occupations where the distribution of men and women is highly imbalanced as seen in both nurse and surgeon in Fig. 4a, engineer in Fig. 4b and model in Fig. 4c. In many cases adding 50% and 70% data duplication to the training data has a good impact on gender bias and can be even better than using CDS. Adding gender-swapped data to training data with a relatively balanced gender distribution does not impact on the gender gap or classification accuracy as seen in professor in Fig. 4b and poet or painter in Fig. 4d.

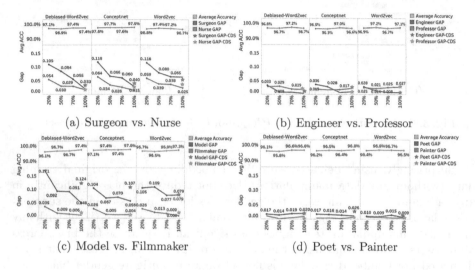

(a) Surgeon vs. Nurse   (b) Engineer vs. Professor

(c) Model vs. Filmmaker   (d) Poet vs. Painter

Fig. 4. ACA and TPR_{gap} results for different proportions of data duplication. The CDS TPR_{gap} is labelled as $GAP\text{-}CDS$

We evaluated the impact of the bias mitigation techniques on the Hate Speech data using the GBET template approach to generate synthetic test data to measure gender bias. Table 2a shows the average class accuracy on both the 20% test split in the original data and the synthetic test dataset. Similar to our previous results bias mitigation techniques do not adversely impact on classification performance and an interesting result here is that classification performance is significantly improved by using CDA. Classification performance is significantly lower on the test dataset, and although the purpose of this data is to measure the gender bias rather than the classification performance, such poor performance might suggest that this data does not match well with the classification. We also looked at whether using different word embeddings for the text representation had any impact. Results in Table 2b show that the word embedding used does not have as much impact on gender bias as the data manipulation approaches.

Table 2. (a) ACA and (b) TPR_{gap} for the hate speech

		Original ACA	GBET ACA
Debiased Word2vec	Baseline	84.1%	59.5%
	Scrubbed	83.4%	58.7%
	CDA	94.8%	60.8%
	CDS	84.4%	58.9%
Conceptnet	Baseline	84.0%	60.0%
	Scrubbed	84.3%	59.1%
	CDA	95.5%	61.1%
	CDS	84.0%	58.3%
Word2vec	Baseline	83.5%	59.8%
	Scrubbed	84.7%	58.3%
	CDA	94.8%	58.9%
	CDS	85.1%	60.8%

(a)

(b)

5 Conclusion

In this work, we examined the impact of various bias mitigation techniques on downstream classification tasks. We looked at different data manipulation techniques including data scrubbing which removes explicit gender indicators from the training data, and CDA and CDS, two data augmentation approaches which use gender-swapping. We also looked at whether using different word embeddings for the text representation had any impact. We evaluated the impact on gender bias on datasets that are naturally labelled for gender. We also looked at a dataset that does not have a gender label and generated synthetic non-biased test datasets to allow an evaluation of gender bias.

Our findings show that while all the data manipulation approaches do reduce gender bias, the CDA data augmentation approach has the best impact generally. It does not impact on the classification performance of the downstream task and in one situation actually improved it.

Where training data did not exhibit much gender bias the bias mitigation techniques did not impact negatively on classification performance or gender bias. This suggest that these techniques can be used on training data for classification tasks where the gender bias is unknown in advance.

CDA has a significant limitation in that as it adds a full gender-swapped version of the training data it doubles the size of the training data. CDS, which was proposed to offset this limitation, does not perform as well as CDA in our experiments. We explored adding gender-swapped proportions of the training data rather than the full dataset. These also do reduce the bias in the training data without impacting on classification performance. This suggests that a smaller proportion of the training data could be used for CDA rather than the full dataset.

Word embeddings are a popular text representation in NLP systems and we included a number of word embedding models in our experimentation. The embeddings used were selected as they had been shown to have the least gender bias on a study of bias in word embeddings [1]. Our results show that the embedding used does not have as much impact on gender bias as the data manipulation approaches.

Acknowledgements. This publication has emanated from research conducted with the financial support of Science Foundation Ireland under Grant number 18/CRT/6183. For the purpose of Open Access, the author has applied a CC BY public copyright licence to any Author Accepted Manuscript version arising from this submission.

References

1. Badilla, P., Bravo-Marquez, F., Pérez, J.: Wefe: the word embeddings fairness evaluation framework. In: Proceedings of IJCAI (2020)
2. Blodgett, S.L., et al.: Stereotyping Norwegian salmon: An inventory of pitfalls in fairness. In: Proceedings of ACL (2021)
3. Bolukbasi, T., et al.: Man is to computer programmer as woman is to homemaker? debiasing word embeddings. In: Advances in NeurIPS (2016)
4. Caliskan, A., Bryson, J.J., Narayanan, A.: Semantics derived automatically from language corpora contain human-like biases. Science (2017)
5. Cao, Y.T., et al.: Toward gender-inclusive coref. resolution. In: Proceedings of ACL (2020)
6. De-Arteaga, M., othersRomanov, A., Wallach, H., et al.: Bias in bios: a case study of semantic representation bias in a high-stakes setting. In: Proceedings of FAT* (2019)
7. Dixon, L., Li, J., Sorensen, J., Thain, N., Vasserman, L.: Measuring and mitigating unintended bias in text classification. In: Proceedings of AAAI/ACM Conference on AIES (2018)
8. Gonen, H., et al.: Lipstick on a pig: debiasing methods cover up systematic gender biases in word embeddings but do not remove them. In: Proceedings of NAACL (2019)
9. Hall Maudslay, R., et al.: It's all in the name: mitigating gender bias with name-based counterfactual data substitution. In: Proceedings of EMNLP-IJCNLP (2019)
10. Hardt, M., et al.: Equality of opportunity in supervised learning. NIPS (2016)
11. Kiritchenko, S., Mohammad, S.: Examining gender and race bias in two hundred sentiment analysis systems. In: Proceedings of Conference on SEM (2018)
12. Kurita, K., Vyas, N., Pareek, A., et al.: Measuring bias in contextualized word representations. In: Proceedings of 1st workshop on Gender Bias in NLP (2019)
13. Lu, K., et al.: Gender Bias in Neural Natural Language Processing. arXiv (2018)
14. Mikolov, T., Sutskever, I., Chen, K., et al.: Distributed representations of words and phrases and their compositionality. In: Proceedings of NIPS (2013)
15. Nadeem, M., Bethke, A., Reddy, S.: StereoSet: Measuring stereotypical bias in pretrained language models. In: Proceedings of ACL (2021)
16. Park, J.H., et al.: Reducing gender bias in abusive language. In: EMNLP (2018)
17. Prost, F., Thain, N., Bolukbasi, T.: Debiasing embeddings for reduced gender bias in text classification. In: Proceedings of the 1st Workshop on Gender Bias in NLP (2019)
18. Rudinger, R., et al.: Social bias in elicited nli. In: Proceedings of ACL on Ethics (2017)
19. Speer, R., et al.: An open multilingual graph of general knowledge. In: AAAI (2017)
20. Stanczak, K., et al.: A survey on gender bias in nlp. arXiv preprint (2021)
21. Sun, T., et al.: Mitigating gender bias in nlp: Lit. review. In: Proceedings of ACL (2019)
22. Verma, S., et al.: Fairness definitions explained. In: Proceedings of Software Fairness (2018)

23. Waseem, Z., et al.: Hateful symbols or hateful people? predictive features for hate speech detection on Twitter. In: Proceedings of NAACL (2016)
24. Webster, K., Recasens, M., Axelrod, V., Baldridge, J.: Mind the GAP: a balanced corpus of gendered ambiguous pronouns. Trans. ACL (2018)
25. Zhao, J., Wang, T., Yatskar, M., Ordonez, V., Chang, K.W.: Gender bias in coreference resolution: Evaluation and debiasing methods. In: Proceedings of NAACL (2018)
26. Zhao, J., et al.: Learning gender-neutral word embeddings. In: EMNLP (2018)

A Semi-automatic Data Generator for Query Answering

Fabrizio Angiulli, Alessandra Del Prete, Fabio Fassetti, and Simona Nisticò(✉)

DIMES Department, University of Calabria, Rende, Italy
{fabrizio.angiulli,fabio.fassetti,simona.nistico}@dimes.unical.it

Abstract. Question Answering (QA) is a critical NLP task mainly based on deep learning models that allow users to answer questions in natural language and get a response. Since available general-purpose datasets are often not effective enough to suitably train a QA model, one of the main problems in this context is related to the availability of datasets which fit the considered context. Moreover, such datasets are generally in English, making QA system design in different languages difficult. To alleviate the above-depicted issues, in this work, we propose a framework which automatically generates a dataset for a given language and a given topic. To train our system in any language, an alternative way to evaluate the quality of the answers is needed, so we propose a novel unsupervised method. To test the proposed technique, we generate a dataset for the topic "computer science" and the language "Italian" and compare the performance of a QA system trained on available datasets and the built one.

1 Introduction

Natural Language Processing (NLP) models represent a class of Machine and Deep Learning solutions devoted to automated tasks involving texts written in natural language. Since such models need large amounts of data to train to obtain good performances, one commonly faced problem is the lack of training datasets. In the above-depicted context, it is convenient to have methods to collect training data.

In literature, many datasets have been created to solve tasks related to Question Answering. Their categorization is due to the type of Question Answering problem faced. The main category is Open-domain one, in which datasets are designed to train models to provide one-shot answers to factoid questions starting from many available general-purpose texts. We can further categorize Open-domain QA datasets by exploiting the type of QA task considered.

The first class of datasets that we will consider is the one for Extractive QA, one well-known member of this category is SQuAD 1.1 [7,8] by Stanford University. This dataset has questions and answers generated from Wikipedia, more details will be provided later. TiviaQA [4] is another data collection for extractive

© The Author(s), under exclusive license to Springer Nature Switzerland AG 2022
M. Ceci et al. (Eds.): ISMIS 2022, LNAI 13515, pp. 106–114, 2022.
https://doi.org/10.1007/978-3-031-16564-1_11

QA containing 650K question-answer-context triples collected by trivia enthusiasts and independently gathered by evidence documents. In average, each question have six answers. NewsQA [11] is a dataset composed by 100K human-generated question-answers pairs based on a set over 10K CNN news. All the datasets cited until now can also be employed for Reading Comprehension (RC) tasks. Another subcategory is that of *conversational* datasets, where different rows are not only referred to the same context but also belong to the same dialogue. For this category, the most relevant datasets are CoQA (*Conversational Question Answering dataset*) [9] and QuAC (*Question Answering in Context*). [2] main goal is to train a predictor able to detect the question context and suitably answer even if questions are meaningless if isolated by the context, like "Where?" or "Why?". In the second case, the model input is not a pair (context, question), but a sequence (context, history). Looking at abstractive QA, where answers are generated rather than retrieved, we find NarrativeQA [10] a dataset with stories and related questions. The text sources are Project Gutenburg and movie scripts repositories, mainly imsdb. It contains 46K samples. TweetQA dataset [15], which belongs to the same family, is composed of information retrieved from twitter and contains 17K samples. The last subcategory we will consider for Open-domain question answering is one of the datasets for information retrieval QA (IR), in which we find the WikiQA [16] dataset. Besides open domain datasets, there are several datasets for specific domains like FiQA (*Financial Opinion Mining and Question Answering*) [5] or JEC-QA (*Legal Question Answering*) [17].

Focusing our attention again on the SQuAD dataset, we have that each dataset entry contains the *context, question* and *answer* fields. Different rows can be related to the same context and since, as we have already said, this dataset has been designed for extractive QA, the answer is a text segment coming from the context. Consequently, the *answer* field is a dictionary containing texts of correct answers and the initial position of the answer span. There are, typically, three or four correct answers for each question. The accuracy of models trained using this dataset is measured through EM (Exact Match) and macro-average F1 scores.

As for datasets in the Italian language, in [3] a translation of SQuAD is proposed. This dataset is available both in JSON format and on the *Hugging Face* library [14], which is very popular for QA, since it allows it to work easily with tokenizers, models and pipelines, to get datasets ready to use and contains many pre-trained models.

However, there are no domain-specific datasets for question-answering in the Italian language. Thus, we propose a modular pipeline for domain-specific question-answering dataset generation. This generation pipeline can also apply to other languages. The rest of the paper is organized as follows: Sect. 2 presents the proposed framework, Sect. 3 shows the experimental results and introduces an alternative evaluation metric to measure answers quality, and finally Sect. 4 concludes this work.

2   Proposed Method

In this Section, the method proposed in this work is described.

As previously stated, the proposed framework aims to automatically generate datasets for the Question Answering (QA) task, which goal is achieved through the construction of the complete pipeline sketched in Fig. 1.

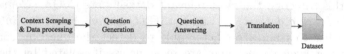

Fig. 1. Pipeline of the proposed approach.

The problem addressed by the proposed method is the lack of datasets in languages different from English and particularly domain-specific datasets for question generation. For this reason, the pipeline is designed to generate datasets in many languages and related to various user-defined topics. Framework modules can also be easily customized. The strategy adopted is to extract the information from plain text.

According to Definition 1, the main steps to be addressed to build a dataset are (*i*) context definition, (*ii*) question generation and (*iii*) golden answer generation. Except for the first step, where contexts are chosen and acquired, the other activities require specific models. The structure depicted by Definition 1 is one of the Extractive QA datasets.

Definition 1 (Dataset for Question Answering). *A dataset $\mathcal{D}$ suited for the extractive QA task is a set of entries $\langle c, q, a \rangle$, where c is a context, q is a question and a is a gold answer.*

Due to the absence of models able to generate answers and questions in the Italian language, we have chosen to acquire texts in the English language and add a translation phase to obtain the final dataset. This solution make this methodology applicable also to other languages.

In Fig. 1 are reported main steps of the proposed pipeline, which are described next. For fine-tuning, experiments, pipeline implementation and models testing, HuggingFace's Transformers library is employed.

Context Scraping and Raw Data Processing. Here the pipeline extracts the text from a source. Typically this operation is performed through API. At this point, the data retrieved is a raw text that must be pre-processed. Methods for tokenization and features pre and post-processing are applied to it to obtain a clean text. It is used then to exact the contests representing the starting point for questions and answers extraction. Since models like BERT have problems working with long texts, reference text dimension reduction is needed. If the maximum embedding dimension is overtaken, then the sentence is truncated The text is chunked into parts with a dimension of up to 512 tokens to extract the contexts. In the splitting operation, strong punctuation is considered to avoid sentence breaking.

Question Generation. This phase consists of generating questions starting from the text, which, in this case, is the context scraped during the first step of the pipeline.

Three requirements should be satisfied: (*i*) humans should be as less involved as possible, so questions should be generated automatically; (*ii*) questions should have good quality; (*iii*) the generated question has to be referred to the context from which the answer is extracted.

To deal with this problem, we use answer-agnostic question answering solutions, which do not require human selection of answer span, attaining (*i*). Because questions here must be extracted from the context, the task most related to our problem is closed-book question answering. The model chosen is T5 [6], which solves the previously mentioned problem allowing us to satisfy (*ii*), thanks to the good performances demonstrated. To attain also (*iii*) the base version of this model is employed. The mentioned model generates a pool of questions starting from the context, each separated by a token. So, there will be about five entries for the same context in the produced dataset.

Gold Answer Extraction. This step is devoted to the extraction of gold answers and. In particular, two transformers models trained on SQuAD 1.1 [8] are combined to perform this task. Since the model extracts a pool of questions, there is a need to select the best one to insert into the dataset. We will refer to this operation by the name of Gold Answer Extraction. To perform this operation, we use QA models. This task represents one of the main critical issues of the proposed method, so a careful choice of models and methods is needed to select the best answers for each row of the dataset. Among the two types of models used for QA, we use *Answer Extraction* models because the dataset used is in SQuAD format, so it is used for generative Question Answering. The pipeline can also be adapted for generative datasets creation. To accomplish this issue, we choose to use the MiniLM [13] model since, in our experiments, it shows the best dimension-accuracy trade-off. For the answer span research inside contexts, we need to consider three requirements: (*i*) we need answers as accurate as possible, and to satisfy this, we need models with high performances in terms of EM and F1 scores. (*ii*) We must have correct answers also for specific questions. Given that we consider the scientific field, we need to deal also with questions related to specific terms, formulas and complexity values. (*iii*) Since we need to prevent the possible bottleneck, we must use efficient and thin models. To satisfy these desiderata, we focus our attention on compressed models, which are less complex, maintaining at the same time good competitive performances.

To satisfy (*ii*), we place MiniLM near a model pre-trained on scientific data. The choice has fallen on SciBERT [1]. The main difference between the two models is inside vocabulary since there is only a 42% of overlap between the two dictionaries. This evidence suggests a substantial difference between the recurrent terms in the scientific domain and general texts. Another difference is that MiniLM uses WordPiece tokenization while SciBERT uses SentencePiece tokenization, which brings different token semantic representations.

Since both models are employed in the pipeline, we need a criterion to choose between the different answers. The decision is performed by exploiting the model prediction score. Generally speaking, the best one is selected using a standard heuristic score. This score is computed as the sum between the starting and the final span probability values. Rows with a score of less than 20% are dropped away to improve the mean accuracy of the answers.

Translation. A text-to-text model is employed to obtain the in-language version of the dataset. The role of this model is to perform the translation. It is an optional model and has to be changed according to the desired language. In the experiments, an English-to-Italian translator is employed. The choice to divide the translation into other pipeline steps is performed to ensure flexibility concerning different languages.

3   Experiments

In this section, we evaluate the quality of the dataset produced by the pipeline to assess the efficacy of the proposed methodology. Since, without good data, it is not possible to have good deep learning models, the strategy adopted is to measure the quality of the proposed methodology result in quality terms of the model trained using it as a dataset.

The first operation for the experimentation is to find a model to use for dataset comparison. There are a few Italian language models, and none of them has been applied before for the question answering task on the SQuAD-it dataset. For this reason, we have fine-tuned some selected models to select the one to use during experiments. As it is possible to see from the Table 1, Electra Italian XXL is the model with the best performance both in terms of EM and F1 score, so, from this point on, we will use it as a reference.

Table 1. Performance of models for Italian language trained on SQuAD-it.

Model	EM	F1 Score
DrQA-it	56.1	65.9
UmBERTo	60.5	72.41
Italian BERT	63.76	75.3
Italian ELECTRA xxl	**65.6**	**77.13**

Table 2. Models evaluation on MoQA dataset

Fine-Tuning	EM	F1
SQuAD-it	**42.58**	64.28
MoQA	11.81	44.61
SQuAD-it+MoQA	19.58	**72.55**

Table 3. EM and F1 scores before translation

Fine-Tuning	EM	F1
SQuAD	80.44	88.43
SQuAD+MoQA	**84.62**	**91.08**

Table 4. Caption

Fine-Tuning	α		
	1.0	2.0	3.0
SQuAD	0.35	0.55	0.85
MoQA	0.41	0.33	0.13
SQuAD+MoQA	**0.58**	**0.66**	**0.86**

3.1 Quantitative Analysis

Here results of training the Italian ELECTRA XXL model with different datasets are shown to assess the improvement introduced by the dataset obtained through this pipeline. During this analysis, we will consider three datasets: SQuAD-it, MoQA, and a dataset obtained by combining the two previously-cited datasets.

Table 2 shows the validation results of fine-tuned models on the MoQA dataset. The validation is always performed on the MoQA validation set. The first insight contained in these results is that there is an improvement if we combine the MoQa dataset with SQuAD-it instead of considering only the MoQA dataset. It is reasonable since SQuAD-it brings the model to extract semantically correct answers while MoQA makes the model able to detect domain-specific language and peculiarities. Without SQuAD-it, MoQA lacks general-purpose language knowledge. Another observation is that the models trained only on SQuAD-it perform better in Exact Match terms. One possible motivation is the presence of only 1 golden answer inside the dataset. EM and F1 values could be higher if we would insert 2 or 3 alternative answers for each question, as happens in SQuAD. It is possible to mitigate this issue by extending the dataset in such a way as to have more than one answer. Another possible motivation for this issue involves translation, which is needed since it is impossible to extract answers directly from the context in the considered language. Because of this process, the context not always contains the answer span, so EM and F1 are underestimated.

To investigate the impact of the translation, we computed the EM and F1 scores reached using the dataset resulting before the translation. In order to see how the value of these metrics changes, we perform the validation on the MoQA's English version validation set.

Since the metrics values reported in Table 3 are better before the translation process, our suspicion is confirmed. SQuAD+MoQA EM and F1 on validation are better than the SQuAD ones, so the problem is not the quality of the answers generated.

3.2 A Better Estimation for QA Model Performances

In agreement with [12], a way to understand model internal behaviour is to analyze its latent space. Generally, the word's embedding is influenced by its position, so it is reasonable that near sentence tokens will be near also in latent space.

Given an answer a, let $a_1, \ldots, a_n$ denote its tokens and let $\epsilon(\cdot)$ be the embedding of a token. An embedding is a latent vector computed in the hidden states of the model for each token of the input sentence. We do not consider punctuation, padding and BERT's special tokens in our calculation. It is important to remember that a token is not a word but the minimum unit produced by the model tokenizer for it.

The *aggregation index* t of a is defined as $t(a) = \sum_{i,j} \|\epsilon(a_i) - \epsilon(a_j)\|_2$.

Given an answer a and a context c, with a little abuse of notation, let $t(c)$ denotes the aggregation index of the context computed as the maximum aggregation index of possible answers having the same length of a in terms of number of tokens. Thus, letting c_i^n be the sentence in c starting from i and composed by n tokens, we have $t(c) = \max_i t(c_i^n)$.

Exploiting aggregation index and the above equation, the *normalized aggregation index* of an answer a in the context c can be defined as: $\tau(a) = 1 - \frac{t(a)}{t(c)}$.

Given two answers a and a', the *matching index* between a and a' is defined as: $s(a, a') = \frac{1}{2} \sum_{a_i} \left\| \epsilon(a_i) - \arg\min_{\epsilon(a'_j)} \|\epsilon(a_i) - \epsilon(a'_j)\|_2 \right\|_2 + \frac{1}{2} \sum_{a'_j} \left\| \epsilon(a'_j) - \arg\min_{\epsilon(a_i)} \|\epsilon(a_i) - \epsilon(a'_j)\|_2 \right\|_2$,

and the *normalized matching index* between answers a and the associated golden answer $g(a)$ in the context c is defined as $\sigma(a) = 1 - \frac{s(a, g(a))}{n \cdot \max_{c_i, c_j \in c} \|\epsilon(c_i) - \epsilon(c_j)\|_2}$.

Given an answer a and its related golden answer $g(a)$, the latent quality score ξ of a is computed as:

$$\xi(a) = \frac{\alpha - \tau(g(a))}{\alpha} \cdot \tau(a) + \frac{\tau(g(a))}{\alpha} \cdot \sigma(a), \tag{1}$$

where α is an hyperparameter which controls the contributions provided by aggregation and matching indices. In absence of further information, it can be set to 2 for balancing.

To quantify the distance between points in the latent space of the last layer of MiniML, we compute the Average Cosine Similarity between the answers token. After, the distribution of correctly and wrongly predicted replies is compared. The data considered here is the SQuAD validation set. The metric chosen is unsupervised since it does not depend on the data's nature or the model, and further can also be used when golden answers are unavailable. To remove the noise, the PCA is applied. We keep the 95% of variance. This metric is computed between all answer span tokens. We discard single-token answers since they have a similarity equal to 1.

What results from this analysis is that the similarity has different distributions on the correctly predicted samples and not correctly predicted ones. While in the first case, there is a mean similarity of about 0.7 and less variance, in the second case, the mean similarity is about 0.4.

A New Evaluation. Stated that translation brings to an underestimation of the quality of the dataset when EM and F1 are used as metrics, we now perform an evaluation using the metric defined in this section. The setting adopted uses a subset of the MoQA validation set composed of 1000 samples for evaluation. It is performed considering three different values for the α hyper-parameter in Eq. (1).

What is possible to observe from Table 4 is that for all the alpha values considered, the model trained on the SQuAD+MoQA dataset always reaches the best result.

4 Conclusion

In this paper, we have proposed a pipeline to create domain-specific datasets for question-answering models for languages different from English. We have also assessed the quality of the proposed methodology by exploiting a state-of-art QA model.

References

1. Beltagy, I., Cohan, A., Lo, K.: Scibert: Pretrained contextualized embeddings for scientific text. arXiv preprint arXiv:1903.10676, 1(1.3), 8 (2019)
2. Choi, E., et al.: QUAC: question answering in context. arXiv preprint arXiv:1808.07036 (2018)
3. Croce, D., Zelenanska, A., Basili, R.: Neural learning for question answering in Italian. In: Ghidini, C., Magnini, B., Passerini, A., Traverso, P. (eds.) AI*IA 2018. LNCS (LNAI), vol. 11298, pp. 389–402. Springer, Cham (2018). https://doi.org/10.1007/978-3-030-03840-3_29
4. Joshi, M., Choi, E., Weld, D.S., Zettlemoyer, L.: Triviaqa: a large scale distantly supervised challenge dataset for reading comprehension. arXiv preprint arXiv:1705.03551 (2017)
5. Maia, M., et al.: Www 2018 open challenge: financial opinion mining and question answering (2018)
6. Raffel, C., et al.: Exploring the limits of transfer learning with a unified text-to-text transformer. arXiv preprint arXiv:1910.10683 (2019)
7. Rajpurkar, P., Jia, R., Liang, P.: Know what you don't know: unanswerable questions for squad. arXiv preprint arXiv:1806.03822 (2018)
8. Rajpurkar, P., Zhang, J., Lopyrev, K., Liang, P.: Squad: 100,000+ questions for machine comprehension of text. arXiv preprint arXiv:1606.05250 (2016)
9. Reddy, S., Chen, D., Manning, C.D.: Coqa: a conversational question answering challenge. TACL **7**, 249–266 (2019)
10. Kǒ ciský, T., et al.: The NarrativeQA reading comprehension challenge. TACL, TBD:TBD (2018)
11. Adam Trischler, et al.: Newsqa: a machine comprehension dataset. arXiv preprint arXiv:1611.09830 (2016)
12. van Aken, B., Winter, B., Löser, A., Gers, F.A.: How does Bert answer questions? A layer-wise analysis of transformer representations. In: CIKM, pp. 1823–1832 (2019)
13. Wang, W., Wei, F., Dong, L., Bao, H., Yang, N., Zhou, M.: Minilm: deep self-attention distillation for task-agnostic compression of pre-trained transformers. arXiv preprint arXiv:2002.10957 (2020)

14. Wolf, T.: Transformers: State-of-the-Art Natural Language Processing
15. Xiong, W., et al.: Tweetqa: a social media focused question answering dataset. arXiv preprint arXiv:1907.06292 (2019)
16. Yang, Y., Yih, W., Meek, C.: Wikiqa: a challenge dataset for open-domain question answering. In: Proceedings of the 2015 Conference on Empirical Methods in Natural Language Processing, pp. 2013–2018 (2015)
17. Zhong, H., Xiao, C., Tu, C., Zhang, T., Liu, Z., Sun, M.: JEC-QA: a legal-domain question answering dataset. arXiv preprint arXiv:1911.12011 (2019)

Explainability

XAI to Explore Robustness of Features in Adversarial Training for Cybersecurity

Malik AL-Essa[1,2]([✉]) [iD], Giuseppina Andresini[1,2] [iD], Annalisa Appice[1,2] [iD], and Donato Malerba[1,2] [iD]

[1] Department of Computer Science, University of Bari "Aldo Moro", Bari, Italy
{malik.alessa,giuseppina.andresini,annalisa.appice,
donato.malerba}@uniba.it
[2] CINI - Consorzio Interuniversitario Nazionale per l'Informatica, Bari, Italy

Abstract. Adversarial training is an effective learning approach to harden deep neural models against adversarial examples. In this paper, we explore the accuracy of adversarial training in cybersecurity. In addition, we use an XAI technique to analyze how certain input features may have an effect on decisions yielded with adversarial training giving the security analyst much better insight into robustness of features. Finally, we start the investigation of how XAI can be used for robust features selection within adversarial training in cybersecurity problems.

Keywords: Cybersecurity · Deep learning · XAI · Adversarial training · Features selection

1 Introduction

During the last decade, cybersecurity literature has given a prominent role to deep learning as a powerful artificial intelligence paradigm to achieve advanced classification capabilities in several cybersecurity problems [2,4,9]. Adversarial training [22] is an increasingly popular adversarial learning approach to mitigate the vulnerability of deep neural models to adversarial attacks. It incorporates adversarial samples with correct class targets in the training stage to train a new classification model that is more robust than the attacked model based on given metrics. On the other hand, eXplainable Artificial Intelligence (XAI) [14] is a preeminent sub-field of artificial intelligence that aims to enable humans to understand decisions of black-box artificial systems (such as deep neural models).

In this paper, we explore how an adversarial training approach may contribute to learn robust deep neural models in two cybersecurity problems, i.e. malware detection and network intrusion detection. In addition, we use a post-hoc, global XAI technique to explain how adversarial samples may modify features that mainly attract attention of deep neural models. A few, recent, cybersecurity studies [13,17] have already investigated the XAI within the adversarial learning field. However, these studies have used XAI as offensive means to compromise the confidentiality and privacy of deep neural models trained for cyber-threat detection. Conversely to these previous studies, we explore the XAI for

M. Ceci et al. (Eds.): ISMIS 2022, LNAI 13515, pp. 117–126, 2022.
https://doi.org/10.1007/978-3-031-16564-1_12

cybersecurity achievements under the umbrella of defensive adversarial learning. In particular, we take advantage of XAI techniques to explain the effect of adversarial samples on features relevance in decisions yielded with a deep neural model learned with adversarial training. In addition, we start the investigation of how XAI-based features selection can help to harden deep neural models against adversarial samples possibly improving the accuracy of unseen data.

The paper is organized as follows. Related works are presented in Sect. 2, while the proposed methodology is described in Sect. 3. The results of the evaluation are discussed in Sect. 4. Finally, Sect. 5 refocuses on the purpose of the research, draws conclusions and illustrates future developments.

2   Related Work

Adversarial learning has recently attracted great attention in cybersecurity, where various studies have mainly focused on the offensive perspective. In [9,18] several techniques to generate adversarial samples are studied to evade a variety of models learned for both network intrusion detection and malware detection. Meanwhile, a few recent studies have started the investigation of the defense perspective in cybersecurity applications of adversarial learning. In particular, adversarial training techniques have been recently used by security practitioners due to their ability to achieve high empirical robustness, to scale to deep neural networks and to perform equally well for different attack models [6]. For example, adversarial training is explored with generative adversarial networks in problems of network intrusion detection [23] and malware detection [20].

Deep neural models are black-boxes, while easier-to-explain models are becoming increasingly desirable also in cybersecurity. The recent study in [22] has started to investigate how XAI techniques can be applied to produce explanations for the decisions of deep neural models, trained for both malware detection and vulnerability discovery applications. XAI techniques are also used in [5,21] to identify the most relevant input features for detecting network intrusions. An XAI analysis is conducted in [1] to monitor features that become relevant to detect each category of network intrusions. XAI is used in [3] with the attention mechanism, in order to improve the accuracy of deep neural models trained for network intrusion detection. On the other hand, a few recent cybersecurity studies have investigated XAI coupled with adversarial learning. In [17] an XAI technique is used to express the minimal modifications required to change the output of the black-box for any mis-classified sample. In [13] an XAI technique is used to formulate black-box attacks against network intrusion classifiers. Notably, our study continues the investigation of the XAI coupled with the adversarial training, but it stays under the umbrella of the defense perspective. In addition, it starts the exploration of how explanations of decisions yielded with adversarial training can be also leveraged for robust feature selection.

3 Proposed Methodology

Let us consider a dataset $T = \{(\mathbf{x}_i, y_i)\}_{i=1}^N$ of N training samples where $\mathbf{x} \in \mathbf{X} \subseteq \mathbb{R}^d$ is a d-dimensional vector of input features that describe cyber-data samples (e.g., malware apps or network traffic flow traces), whereas $y \in Y$ is the value of the target variable Y that may assume K distinct classes where: *normal* samples and various types of *threats*, depending on those historically detected and labeled. We define a six-stepped methodology for cyber-threat detection:

S1: We learn a deep neural model $M_\theta \colon \mathbb{R}^d \mapsto Y$ with parameter θ estimated from T.

S2: We produce an adversarial set $\mathcal{A}$ from T using M_θ.

S3: We learn a deep neural model $M_{\theta'} \colon \mathbb{R}^d \mapsto Y$ with parameter θ' estimated from $T \oplus \mathcal{A}$ through the adversarial training approach.

S4: We use an XAI technique to explain relevance of features in decisions of $M_{\theta'}$ yielded for samples of T and $\mathcal{A}$, respectively.

S5: We select features that appear at the intersection of the top-k positions of the two ranks identified according to the relevance feature values measured in step S4 on T and $\mathcal{A}$, respectively. Let k be a user-defined parameter, while h be the number of features selected in the intersection.

S6: We learn a deep neural model $M_{\theta''} \colon \mathbb{R}^h \mapsto Y$ with parameter θ'' estimated from the projection of $T \oplus \mathcal{A}$ on the h features selected in the step S5.

Let us denote: B – the configuration that learns M_θ (step S1), T+A – the configuration that learns $M_{\theta'}$ (step S3) and T+A+XAIFS – the configuration that learns $M_{\theta''}$ (step S6). The performance of T+A+XAIFS depends on the input parameter k that conditions the number h of features selected with the XAI module in the step S5.

As an adversarial sample generator, we consider three possible techniques: FGSM [12], BIM [19] and PGD [15]. These are white-box adversarial sample generators that are based on the gradient formula $g(\mathbf{x}) = \nabla_\mathbf{x} J(\theta, \mathbf{x}, y)$, where $\nabla_\mathbf{x}$ denotes the gradient computed with respect to $\mathbf{x}$, and $J(\theta, \mathbf{x}, y)$ denotes the loss function of the neural model M_θ. Specifically, FGSM determines the minimum perturbation ϵ to add to a training sample $\mathbf{x}$ to create an adversarial sample that maximizes the loss function. According to this theory, for each $(\mathbf{x}, y) \in \mathcal{D}$, we generate a new sample $(\mathbf{x}^{\mathbf{adv}}, y) \in \mathcal{A}$ such that $\mathbf{x}^{\mathbf{adv}} = \mathbf{x} + \epsilon sign(g(\mathbf{x}))$. As in [6], we select the perturbation ϵ as a small value in the range between 0 and 0.1, in order to scale the noise and ensure that perturbations are small enough to remain undetected to the human eye, but large enough to fool the attacked neural model. Note that BIM extends FGSM in a straightforward manner by applying FGSM multiple times with small step size, and clipping values of intermediate results after each step to ensure that they are in an ϵ-neighborhood of the original sample. PGD is also a multi-step variant of FGSM on the negative loss function with l_∞ computed in place of l_2. Both BIM and PGD spend more training time than FGSM since they both perform multiple iterations to generate perturbations. On the other hand, adversarial training with FGSM takes much less training time since the FGSM takes one step to generate perturbations, but fails to increase adversarial robustness [15].

As XAI module we use DALEX framework [7] that integrates post-hoc XAI techniques for both global and local explanations. We use the global explanation technique of DALEX, which measures the global relevance of input features on decisions of a deep neural model. Any global XAI technique may be used in this step. DALEX uses a permutation-based variable-relevance black-box algorithm. For each feature, its effect is removed by permuting the values of the feature and a loss function compares the performance before and after. Intuitively, if a feature is important, randomly permuting its values will cause the loss to increase. Based upon this theory, we are able to sort the input features according the feature relevance computed by DALEX on a specific dataset (the training set or the adversarial set) so that the top-ranked feature is the one with the highest loss increase computed by DALEX.

4   Empirical Evaluation

We analyzed the facets of defined methodology in two cybersecurity datasets described in Sect. 4.1. We illustrate the implementation used in the evaluation in Sect. 4.2, while we discuss the experimental results in Sect. 4.3.

4.1   Dataset Description

Two cybersecurity datasets are used in this evaluation: MalDroid20 – an android malware dataset – and CICIDS17 – a network traffic security dataset.

MalDroid20 [16][1] includes samples of recent Android apps collected from several sources. The dataset spans over five distinct categories: Adware, Banking malware, SMS malware, Riskware and Benign. According to the study in [16], each app is described by 40 features that represent the top-40 static and dynamic features extracted with CopperDroid and ranked according to the Mutual Information. The dataset is balanced. In this work, we used a stratified division of the dataset in the training set (70%) and the testing set (30%). Notably, [16] used the same strategy to divide the dataset. However, they performed experiments in the semi-supervised setting using both the (labeled) training set and the (unlabeled) testing set to train the model used to predict the testing data. Differently, we worked in the supervised setting and processed the (labeled) training set only to learn a model used to predict the testing set.

CICIDS17 is a 5-day network traffic data created on 2017 by the Canadian Institute for Cybersecurity. This dataset has been recently revised by [10] who has removed meaningless artifacts and dataset errors by reducing the number of input features from 78 to 72.[2] This revised dataset was used for binary classification in [5], while we used it for multi-class classification. In this work, we

---

[1] https://www.unb.ca/cic/datasets/maldroid-2020.html.
[2] downloads.distrinet-research.be/WTMC2021.

performed stratified sampling without replacement to extract two independent sets of 100000 samples, as training and testing sets. The processed dataset comprised 8 types of attacks with DoS GoldenEye, FTP-Patator, SSH-Patator, DoS slowloris, and DoS Slowhttptest as minority classes. In accordance with the original distribution of the data, we extracted 80% of benign samples and 20% of attacks for both the training and testing set.

4.2 Implementation Details

The methods considered in this evaluation were implemented in Python 3.9 using Keras 2.7 library. For each dataset we optimized the hyper-parameter of the neural networks (i.e., mini-batch size in $\{2^5, 2^6, 2^7, 2^8, 2^9\}$, learning rate between 0.0001 and 0.001, number of neurons in $\{2^5, 2^6, 2^7, 2^8, 2^9, 2^{10}\}$ and dropout between 0 and 1) using the tree-structured Parzen estimator algorithm, as implemented in the Hyperopt library, by using 20% of the training set as a validation set. We selected the configuration of the parameter that achieved the best validation loss. All the neural networks considered in this study were defined with 3 fully-connected layers (the number of neurons was chosen with the hyper-parameter optimization), one dropout layer and one batch-normalization layer, to prevent the overfitting. The output probabilities were obtained using the softmax activation function in the last layer, and the ReLU activation function was used in all the other hidden layers. We performed the gradient-based optimization using the Adam update rule and initialized the weights following the Xavier scheme. Furthermore, a maximum number of epochs equal to 150 was set, retaining the best models, using an early stopping approach that achieves the lowest loss on a validation set. Finally, we used the Adversarial Robustness Toolbox library[3] to produce the adversarial samples and the DALEX Python package[4] for measuring the global feature relevance[5]

4.3 Results

The empirical validation was done to answer the following questions: (Q1) To what extent adversarial training influences the accuracy of the classification model? (Q2) How do explanations disclose information on robustness of features to adversarial samples? (Q3) Are explanations helpful in selecting robust features for adversarial training? Experiments were run with adversarial samples generated with $\epsilon = 0.0001$ for MalDroid20 and $\epsilon = 0.01$ for CICIDS17.

Adversarial Training Analysis (Q1). We measured the accuracy performance of B and T+A by varying the adversarial sample generator of T+A among FGSM, BIM and PGD. In all the experiments, we measured standard multi-class accuracy metrics such as WeightedF1, MacroF1 and overall accuracy - OA - on

---

[3] https://adversarial-robustness-toolbox.readthedocs.io/.
[4] https://github.com/ModelOriented/DALEX.
[5] The source code is available at https://github.com/malikalessa/ISMIS.

Table 1. WeightedF1, MacroF1 and OA of B and T+A with FGSM, BIM and PGD

Method	MalDroid20			CICIDS17		
	WeightedF1	MacroF1	OA	WeightedF1	MacroF1	OA
B	0.830	0.800	0.831	0.920	0.640	0.911
T+A (FGSM)	0.880	0.860	0.871	0.960	0.640	0.946
T+A (BIM)	0.870	0.850	0.871	0.960	0.730	0.952
T+A (PGD)	0.890	0.870	0.887	0.950	0.650	0.940

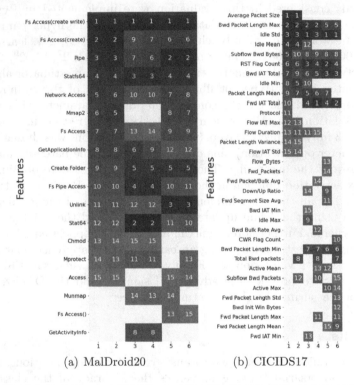

(a) MalDroid20   (b) CICIDS17

(1)T+A (FGSM) explains T (2)T+A (FGSM) explains A (3)T+A (BIM) explains T (4) T+A (BIM) explains A (5)T+A (PGD) explains T (6)T+A (PGD) explains A

Fig. 1. Top-15 features ranked by DALEX on the training set and the adversarial set with the deep neural model learned by T+A with FGSM, BIM and PDG

the testing set of each dataset. Table 1 collects the accuracy metrics measured in both the MalDroid20 and CICIDS17 datasets. These results show that the use of the adversarial training approach contributes to gain accuracy in both problems. In addition, this improvement is commonly achieved independently of the adversarial sample generator used.

Feature Explanation Analysis (Q2). We analyzed the information contained in the visual explanation maps produced for both MalDroid20 and CICIDS17. The heatmaps in Fig. 1 depict the rank of the average relevance of the features that drew the attention of the deep neural models learned with T+A to classify samples of the training set and the adversarial set, respectively. For each dataset, the heatmap reports the union of the top-15 features ranked by DALEX by varying the adversarial sample generator among FGSM, BIM and PGD. Notably, the set of the top-ranked features that each deep neural model mostly attended in the training set largely overlaps the set of the top-ranked attended in the adversarial set. In fact, there are several features that are equally relevant in classifying both training and adversarial samples independently of the algorithm used to perturb the original training samples (for example, *fs_access(create write)* and *fs_access(create)* in Maldroid20, as well as *Bwd Packet Length Max* and *Idle Std* in CICIDS17). In both the datasets, the number of features that are simultaneously ranked as the top-15 relevant features to explain the decisions in both the training set and the adversarial set is greater in FGSM than in BIM and PGD. This result suggests that the iterative perturbation mechanism implemented by both BIM and PGD can lead to identify more vulnerabilities in features than the one-step mechanism of FGSM.

XAI-based Feature Selection Analysis (Q3). We completed this evaluation by exploring the effect of the XAI-based feature selection on the accuracy of T+A+XAIFS. Figure 2 shows the OA measured on the testing set of both Mal-Droid20 and CICIDS17, as well as the number of features (h) actually selected for learning the deep neural model in both B, T+A and T+A+XAIFS. The performance of T+A+XAIFS is evaluated with k ranging among 15 and 35. These results show that the use of explanations for feature selection allows us to gain significant accuracy in CICIDS17, where the OA grows about 4% from T+A to T+A+XAIFS independently of the adversarial sample generator tested. This improvement is achieved with a few features ($k = 15$ and 13 out of 72 features selected in T+A (FGSM), $k = 15$ and 9 out of 72 features selected in both T+A (BIM) and T+A (PGD)). On the other hand, a limited gain in OA is observed in MalDroid20 with $k = 35$ and 35 out of 40 features selected with explanations. We recall that CICIDS17 is an imbalanced dataset, while MalDroid20 is a balanced. Figure 3 collects the F1 per class of both T+A and T+A+XAIFS with BIM in CICIDS17 (with $k = 15$) and in MalDroid20 (with $k = 35$). These results show that the use of explanations for feature selection allows the adversarial training approach to gain accuracy in detecting almost all the classes. The only exception is observed for the class *benign* of MalDroid20. In addition, the gain in accuracy is greater on the minority classes of CICIDS17. Similar conclusions are drawn analyzing the F1 per class of T+A and T+A+XAIFS with FGSM or PGD. In short, this analysis per class highlights that the adversarial training contributes to gain accuracy in both balanced and imbalanced cybersecurity problems, while the XAI-based feature selection contribute in further improving the performance of adversarial training in the presence of imbalanced data.

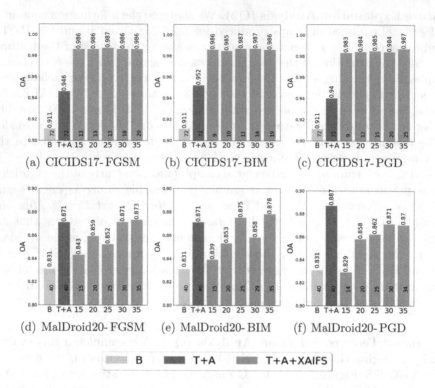

Fig. 2. OA of B, T+A and T+A+XAIFS by varying FGSM, BIM and PGD. The number on the top of each bar denotes the OA measured on the testing set by the configuration. The number inside each bar denotes the number h of features actually selected in the configuration.

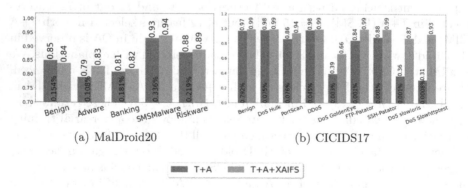

Fig. 3. F1 per class of T+A and T+A+XAIFS with BIM. The number on the top of each bar denotes the F1 measured per class on the testing set of the configuration. The number inside each bar denotes the support percentage of the class in the dataset.

5   Conclusion

In this work, we explore the performance of a methodology formulated to combine deep learning, adversarial training, XAI and feature selection for accurate and explainable multiclass classification in cybersecurity applications. We show the effectiveness of the various components of the proposed methodology in two cybersecurity datasets. The future direction for this work is to extend the evaluation by considering additional cybersecurity datasets and to investigate the use of multi-input neural networks to process simultaneously original samples, adversarial samples and explanations, without requiring any separate feature selection step. As cyber-data are commonly collected in a streaming scenario [5] we plan to explore the proposed methodology in combination with data windowing mechanisms to be able of making the learned model more robust to novelties [8]. Finally, we plan to start the investigation of how XAI-based feature selection can help to harden deep neural models against adversarial samples in new domains (e.g., securing smart home environments [11] from adversarial attacks).

Acknowledgment. The research of Malik AL-Essa is funded by PON RI 2014-2020 - Machine Learning per l'Investigazione di Cyber-minacce e la Cyber-difesa - CUP H98B20000970007. We acknowledge the support of the project "Modelli e tecniche di data science per la analisi di dati strutturati" funded by the University of Bari "Aldo Moro".

References

1. AL-Essa, M., Appice, A.: Dealing with imbalanced data in multi-class network intrusion detection systems using xgboost. In: PKDD/ECML Workshops (2), vol. 1525, pp. 5–21. Springer, Cham (2021). https://doi.org/10.1007/978-3-030-93733-1_1

2. Andresini, G., Appice, A., Paolo Caforio, F., Malerba, D.: Improving cyber-threat detection by moving the boundary around the normal samples. In: Maleh, Y., Shojafar, M., Alazab, M., Baddi, Y. (eds.) Machine Intelligence and Big Data Analytics for Cybersecurity Applications. SCI, vol. 919, pp. 105–127. Springer, Cham (2021). https://doi.org/10.1007/978-3-030-57024-8_5

3. Andresini, G., Appice, A., Caforio, F.P., Malerba, D., Vessio, G.: Roulette: A neural attention multi-output model for explainable network intrusion detection. Expert Syst. Appl., 117144 (2022)

4. Andresini, G., Appice, A., Mauro, N.D., Loglisci, C., Malerba, D.: Exploiting the auto-encoder residual error for intrusion detection. In: Proceedings of EuroS&P Workshops 2019, pp. 281–290. IEEE (2019)

5. Andresini, G., Pendlebury, F., Pierazzi, F., Loglisci, C., Appice, A., Cavallaro, L.: INSOMNIA: towards concept-drift robustness in network intrusion detection. In: Proceedings of AISec@CCS 2021, pp. 111–122. ACM (2021)

6. Bai, T., Luo, J., Zhao, J., Wen, B., Wang, Q.: Recent advances in adversarial training for adversarial robustness. In: Proceedings of IJCAI 2021, pp. 4312–4321. ijcai.org (2021)

7. Biecek, P.: DALEX: explainers for complex predictive models in R. J. Mach. Learn. Res. **19**(84), 1–5 (2018)

8. Ceci, M., Appice, A., Loglisci, C., Caruso, C., Fumarola, F., Malerba, D.: Novelty detection from evolving complex data streams with time windows. In: Rauch, J., Raś, Z.W., Berka, P., Elomaa, T. (eds.) ISMIS 2009. LNCS (LNAI), vol. 5722, pp. 563–572. Springer, Heidelberg (2009). https://doi.org/10.1007/978-3-642-04125-9_59

9. Demetrio, L., Coull, S.E., Biggio, B., Lagorio, G., Armando, A., Roli, F.: Adversarial exemples: A survey and experimental evaluation of practical attacks on machine learning for windows malware detection. ACM Trans. Priv. Secur. **24**(4), 27:1–27:31 (2021)

10. Engelen, G., Rimmer, V., Joosen, W.: Troubleshooting an intrusion detection dataset: the CICIDS2017 case study. In: IEEE EuroS&P Workshops (2021)

11. Ferilli, S., De Carolis, B., Pazienza, A., Esposito, F., Redavid, D.: An agent architecture for adaptive supervision and control of smart environments. In: Proceedings of PECCS 2015, pp. 160–167. SciTePress (2015)

12. Goodfellow, I.J., Shlens, J., Szegedy, C.: Explaining and harnessing adversarial examples. In: 3rd International Conference on Learning Representations, ICLR 2015, Conference Track Proceedings (2015)

13. Kuppa, A., Le-Khac, N.A.: Adversarial XAI methods in cybersecurity. IEEE Trans. Inf. Forensics Secur. **16**, 4924–4938 (2021)

14. Lakkaraju, H., Kamar, E., Caruana, R., Leskovec, J.: Faithful and customizable explanations of black box models. In: Proceedings of the 2019 AAAI/ACM Conference on AI, Ethics, and Society, pp. 131–138 (2019)

15. Madry, A., Makelov, A., Schmidt, L., Tsipras, D., Vladu, A.: Towards deep learning models resistant to adversarial attacks. In: 6th International Conference on Learning Representations, ICLR 2018, Conference Track Proceedings (2018)

16. Mahdavifar, S., Alhadidi, D., Ghorbani, A.A.: Effective and efficient hybrid android malware classification using pseudo-label stacked auto-encoder. J. Netw. Syst. Manag. **30**(1), 22 (2022)

17. Marino, D.L., Wickramasinghe, C.S., Manic, M.: An adversarial approach for explainable AI in intrusion detection systems. In: IECON 2018–44th Annual Conference of the IEEE Industrial Electronics Society, pp. 3237–3243 (2018)

18. Pierazzi, F., Pendlebury, F., Cortellazzi, J., Cavallaro, L.: Intriguing properties of adversarial ml attacks in the problem space. In: 2020 IEEE Symposium on Security and Privacy (SP), pp. 1332–1349 (2020)

19. Wang, J.: Adversarial examples in physical world. In: Proceedings of the Thirtieth International Joint Conference on Artificial Intelligence, IJCAI-21, pp. 4925–4926 (2021)

20. Wang, J., Chang, X., Wang, Y., Rodríguez, R.J., Zhang, J.: LSGAN-AT: enhancing malware detector robustness against adversarial examples. Cybersecurity **4**(1), 1–15 (2021). https://doi.org/10.1186/s42400-021-00102-9

21. Wang, M., Zheng, K., Yang, Y., Wang, X.: An explainable machine learning framework for intrusion detection systems. IEEE Access **8**, 73127–73141 (2020)

22. Warnecke, A., Arp, D., Wressnegger, C., Rieck, K.: Evaluating explanation methods for deep learning in security. In: 2020 IEEE EuroS&P, pp. 158–174. IEEE (2020)

23. Yin, C., Zhu, Y., Liu, S., Fei, J., Zhang, H.: Enhancing network intrusion detection classifiers using supervised adversarial training. J. Supercomput. **76**(9), 6690–6719 (2019). https://doi.org/10.1007/s11227-019-03092-1

Impact of Feedback Type on Explanatory Interactive Learning

Misgina Tsighe Hagos[1,3]([✉]) [iD], Kathleen M. Curran[1,2] [iD],
and Brian Mac Namee[1,3] [iD]

[1] Science Foundation Ireland Centre for Research Training in Machine Learning,
Dublin, Ireland
misgina.hagos@ucdconnect.ie
[2] School of Medicine, University College Dublin, Dublin, Ireland
kathleen.curran@ucd.ie
[3] School of Computer Science, University College Dublin, Dublin, Ireland
brian.macnamee@ucd.ie

Abstract. Explanatory Interactive Learning (XIL) collects user feedback on visual model explanations to implement a Human-in-the-Loop (HITL) based interactive learning scenario. Different user feedback types will have different impacts on user experience and the cost associated with collecting feedback since different feedback types involve different levels of image annotation. Although XIL has been used to improve classification performance in multiple domains, the impact of different user feedback types on model performance and explanation accuracy is not well studied. To guide future XIL work we compare the effectiveness of two different user feedback types in image classification tasks: (1) instructing an algorithm to ignore certain spurious image features, and (2) instructing an algorithm to focus on certain valid image features. We use explanations from a Gradient-weighted Class Activation Mapping (GradCAM) based XIL model to support both feedback types. We show that identifying and annotating spurious image features that a model finds salient results in superior classification and explanation accuracy than user feedback that tells a model to focus on valid image features.

Keywords: Explanatory interactive learning · Deep learning ·
Interactive machine learning · User feedback

1 Introduction

The need to involve humans (or experts) in the model training process, referred to as Human-in-the-Loop Learning (HITL), has inspired research on interactive and active learning [5]. Interactive Machine Learning (IML) [3] is a type of machine learning that adds human interaction to the model training process, as opposed to typical machine learning which aims to use training algorithms alone. One example of IML is active learning [1,14] in which a human participates in the model training process by providing labels for unlabelled instances.

© The Author(s), under exclusive license to Springer Nature Switzerland AG 2022
M. Ceci et al. (Eds.): ISMIS 2022, LNAI 13515, pp. 127–137, 2022.
https://doi.org/10.1007/978-3-031-16564-1_13

Even though the training process of IML and active learning is interactive, it falls short in involving humans at a detailed level of interaction. Humans are usually only involved in instance class labelling—a relatively low level task. A recent trend in interactive learning, referred to as Explanatory Interactive Learning (XIL) [16], proposes richer interaction with humans by accepting user feedback on visual explanations (motivated by recent advances in Explainable Artificial Intelligence (AI) [4]). Explainable AI is a research area that focuses on providing understandable interpretations of AI models, which are usually considered as a black-box, to end users. User feedback in the form of annotations can be collected on explanations and used for model and explanation refinement [11,13].

In classification tasks where the class labels are mutually exclusive, mutual relations between image regions can be ignored and the expected user feedback can be narrowed down to the object in an image and the confounding image region. Assuming correct classification of instances, the two most common types of user feedback solicited in XIL are: (1) *Missing Region* feedback: in which users identify regions that the model is currently ignoring, but should be focused upon; and (2) *Spurious Region* feedback: in which users annotate regions that the model is currently focusing on but should have been ignored since they represent spurious signals. This gives rise to an obvious, but as yet unaddressed question: *which type of feedback is more effective in XIL scenarios?* While the XIL literature explores various rich feedback collection mechanisms for model training; analysis and comparison of different user feedback types is largely ignored. Although it might seem obvious to collect both types of feedback, user interaction is time consuming and expensive. Moreover, different to active learning in which the cost of labelling instances is largely uniform across unlabelled instances [14], different feedback types in XIL have different impacts on user experience and the cost associated with soliciting feedback, since the expected feedback involves annotating image features which are not always uniform across input images. Given that hundreds or thousands of images are usually required for model training, the process of feedback collection in the form of annotation can take hours or days. Another aspect of XIL that is usually ignored in the literature is reporting its impact on the explanation localization accuracy of models. Since the basic idea behind XIL is using model explanations as a medium of interaction with annotators and a way to identify if a model is focusing on spurious features or if it is ignoring important regions of an image, the model training that follows should have a positive impact on the accuracy of the explanations too. In short, a model trained using XIL should be better at focusing on important regions and ignoring wrong image regions than models trained without feedback.

In this paper, we compare the effectiveness of the two most common types of user feedback used in XIL on model performance and explanation accuracy. To do this we use Gradient-weighted Class Activation Mapping (Grad-CAM) [12] as a feature attribution based model explanation to visualize salient regions of images. We train models on two decoy versions of the Fashion MNIST (FMNIST) dataset [17] using cross entropy classification loss and explanation loss that is computed between missing and spurious region feedback and GradCAM explanations. Furthermore, we compare the impact of using classification loss only and adding feedback to model training on model performance and explanation localization accuracy. The main contributions of this paper are:

1. The first comparison in the XIL literature of the impact of different user feedback types on the performance of XIL algorithms;
2. A class-wise decoy version of the FMNIST dataset is created and will be provided as a resource for future XIL research;
3. Our experiments demonstrate that collecting spurious region feedback is more valuable than collecting missing region feedback in XIL.

2 Related Work

As shown in Fig. 1, XIL methods can be categorized based on the approaches they use during model training, model explanation, and user feedback collection. We explore existing work within XIL through these three lenses.

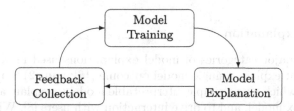

Fig. 1. The Explanatory Interactive Learning (XIL) loop.

2.1 Model Training

The two most common approaches to model training in XIL are model retraining and model fine-tuning.

- Model-retraining. This approach utilises user feedback to produce new training examples, or counter examples, to mitigate against the impact of spurious features [11]. In this approach, an expert is presented with explanations of a model's output showing features that a model finds useful for its prediction. Based on the provided explanations the expert provides feedback on whether the features are valid or not. Counter-examples are then fabricated by removing spurious features and adding valid features into a sample dataset for model refinement.
- Model fine-tuning (explanation losses.) In this category, an explanation loss penalty is added to the cost function (See Eq. 1) that is optimised during model training in addition to the loss terms that penalize a model for wrong predictions. This approach is summarised in Eqs. 2 and 3 using GradCAM

explanations, where $M_n \in \{0, 1\}$ is the ground truth annotation and *norm* normalizes the Grad-CAM output, θ holds a model's parameters, with input X, labels y, predictions $\hat{y}$, and a parameter regularization term λ. Techniques such as Right for Right Reasons using Integrated Gradients (RRR-IG) [10], Right for the Right Reasons using GradCAM (RRR-GC) [11], and Right for Better Reasons (RBR) [15] modify a model through explanation and training losses. Explanation losses can be computed between a feature annotations ground truth dataset and model generated explanations as can be seen in Eq. 2 [11].

$$Classification\ Loss\ = \sum_{n=1}^{N} \sum_{k=1}^{K} -y_{nk} \log \hat{y}_{nk} \tag{1}$$

$$Explanation\ Loss = \sum_{n=1}^{N} (M_n * norm(GradCAM_\theta(X_n)))^2 \tag{2}$$

$$Loss\ =\ Classification\ Loss\ + Explanation\ Loss\ +\ \lambda \sum_i \theta_i^2 \tag{3}$$

2.2   Model Explanation

There are two major categories of model explanations used in XIL: (1) Local explanations that explain a single model outcome [16]; and (2) Surrogate model based learning, which uses a simple interpretable model to explain a more sophisticated black-box model, and to drive interaction with users [8]. While surrogate models are effective for understanding the overall behaviour of a model, they may miss unique features that can be observed if local explanations are used.

2.3   Feedback Collection

Better feedback collection mechanisms will increase user involvement in XIL. Feedback in non-image domains, for example Recommender Systems (RS) and Natural Language Processing (NLP), can be more transparent since their explanations can be presented in a conversational natural language format and users can provide feedback using template questionnaires [7,9]—for example, Dalvi et al. (2022) used users' textual feedback on explanations to refine a model trained on a multiple choice questions dataset [2]. Models that learn and predict concepts [6] can simplify the feedback collection process because feedback is expected to be one of the learned concepts and a user only needs to detect if a wrong concept is being used for classification.

Due to the complex nature of image data, it is often more effective to use visual feedback mechanisms. Missing region and spurious region feedback are the two most commonly used types of user feedback in image-based XIL under the assumption of correct classification of instances. While techniques such as RRR-IG [10], RRR-GC [11] and RBR [15] use spurious region feedback to fine-tune

Fig. 2. Sample images of class *bag* from FMNIST (left) and random decoy FMNIST (right). Intensity values of pixels at the top right corner of the image from decoy FMNIST are exaggerated for presentation purposes.

a model to ignore spurious features, Human Importance-aware Network Tuning (HINT) trains a model to focus on valid image objects [13].

Although model explanations have been used for rich user interaction, the most effective types of user feedback that lead to high performance in XIL remain unknown. We compare the effectiveness of missing region and spurious region feedback, in terms of model performance and explanation accuracy, using two decoy versions of the FMNIST dataset. This has a potential to set the standard for future designs of explanation-based interactive machine learning.

3 Methods

This section describes the datasets used, the experimental setup, and the model training process employed in the experiments described in this paper.

3.1 Dataset for XIL

To demonstrate the effectiveness of XIL to refine models, Teso and Kersting [16] used a decoy version of the FMNIST dataset [17]. The decoy FMNIST dataset is made up of FMNIST images with 4×4 squares of high-intensity pixels added to randomly selected image corners (example images are shown in Fig. 2). Locations of these confounders are class independent. In our work, in addition to experimenting with the decoy FMNIST dataset [16], we also created a class dependent decoy version. Our decoy version of FMNIST contains confounders in the same image region across the same class. For easier reference, we refer to the Teso and Kersting [16] version as Random Decoy FMNIST, and our version as Class-wise Decoy FMNIST.

3.2 Experimental Setup

Each of the two datasets contain 60,000 28×28 pixel images for model training. A test dataset of 10,000 unseen images each containing spurious region and object ground truth annotations is used to measure model performance. The spurious and missing region feedback are image masks of the added confounding regions and objects in images, respectively. We used accuracy to measure classification performance, and dice scores to assess explanation accuracy between model explanations and ground truth annotations.

Table 1. Summary of feedback type influence on model performance.

Feedback type	Random decoy FMNIST			Class-wise decoy FMNIST		
	None	Missing Region	Spurious Region	None	Missing Region	Spurious Region
Explanation accuracy against object annotations	0.27	0.44	0.65	0.21	0.44	0.70
Explanation accuracy against spurious region annotations	0.05	0.03	0.02	0.04	0.03	0.04
Classification accuracy	87.65	85.20	85.53	88.00	84.50	86.21

3.3 Model Training

We trained six different models using the two datasets with the Adam optimizer and a decaying learning rate starting with $1e^{-3}$. Two models are trained using only classification loss (cross entropy loss) using the Random and Class-wise Decoy FMNIST datasets. Two other models are trained using a combination of classification loss and explanation loss based on missing region feedback using each of the two datasets. The last two models are trained using a combination of classification loss and explanation loss based on spurious region feedback on the two datasets. After experimenting with different architectures and comparing their performances, a convolutional neural network with 3 convolutional layers followed by two fully connected layers was selected. We used a Right for the Right Reasons using GradCAM (Eq. 2) explanation loss that is computed between GradCAM model explanations and missing region or spurious region feedback and L2 regularization to address overfitting.

4 Results

This section describes the results of the experiments performed, first exploring those based on the Random Decoy FMNIST dataset and then those based on the Class-wise Decoy FMNIST dataset.

4.1 Random Decoy Fashion MNIST

Explanation Localization Accuracy. Figures 3, 4 and 5 compare the performance of models trained using classification loss alone, and models trained with explanation losses using spurious region and missing region feedback on models' explanation accuracy using dice scores. Our target is to maximize dice scores that are computed against object annotations towards one, and minimize dice scores computed against spurious region towards zero. Average dice scores of explanations of the models trained on random decoy FMNIST compared against the test dataset of spurious region and object ground truth annotations is shown in Table 1. In both cases, the model which used spurious region feedback to compute explanation losses for training achieved superior performance.

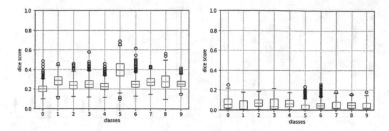

Fig. 3. Dice score evaluation of a model trained on the random decoy dataset with classification loss only using object (left) and spurious region (right) ground truth annotations.

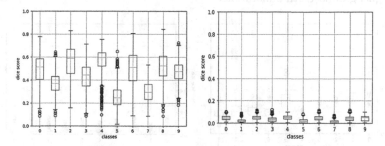

Fig. 4. Dice score evaluation of a model trained on the random decoy dataset with classification loss and missing region feedback explanation loss using object (left) and spurious region (right) ground truth annotations.

Classification Accuracy. A summary of classification performance of the models trained on the random decoy dataset is displayed in Table 1. There is a slight performance loss in models that added feedback to their training compared to the model that only used classification loss. Compared against state of the art XIL methods, RRR-IG achieved the highest accuracy of 89.40, while RBR, HINT and RRR-GC scored 87.60, 58.20 and 78.60, respectively. All our models that used GradCAM explanations to compute explanation losses achieved superior performance scores compared to RRR-GC which also used GradCAM for model training.

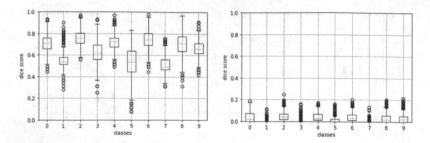

Fig. 5. Dice score evaluation of a model trained on random decoy with classification loss and spurious region feedback explanation loss using object (left) and spurious region (right) ground truth annotations.

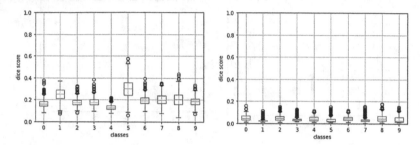

Fig. 6. Dice score evaluation of a model trained on the class-wise decoy dataset with only classification loss using object (left) and spurious region (right) ground truth annotations.

4.2   Class-wise Decoy Fashion MNIST

Explanation Localization Accuracy. A dice score comparison between training with classification loss only, and training with added explanation losses using spurious region and missing region feedback on a model's explanation accuracy on the Class-wise Decoy FMNIST data is displayed in Figs. 6, 7 and 8. Average dice scores for explanations of models trained on the Class-wise Decoy FMNIST dataset compared against the test dataset of spurious region and object ground truth annotations is shown in Table 1. While the model trained with spurious region feedback achieved the highest average dice score when compared against object ground truth annotations, the model trained with missing region feedback achieved the best against spurious region annotations.

Classification Accuracy. Average accuracy scores for all experiments are shown in Table 1. Except for the model that was trained using missing region feedback, the remaining two models achieved superior performance compared to their counterpart models that were trained on the random decoy FMNIST.

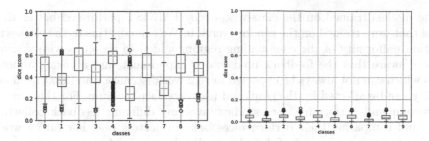

Fig. 7. Dice score evaluation of a model trained on the class-wise decoy dataset with classification loss and missing region feedback explanation loss using object (left) and spurious region (right) ground truth annotations.

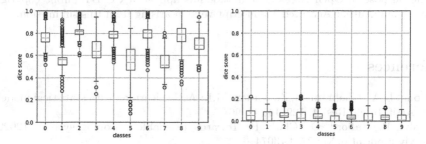

Fig. 8. Dice score evaluation of a model trained on class-wise decoy with classification loss and spurious region feedback explanation loss using object (left) and spurious region (right) ground truth annotations.

5 Conclusion

In this work, in addition to using a publicly available random decoy version of the FMNIST dataset, we generated a class-wise decoy version to compare effectiveness of different feedback types on a model's classification performance and explanation localization accuracy. We studied two feedback types: missing region and spurious region feedback. Apart from a slight classification performance loss when compared to a model trained using classification loss only, we achieved improved classification performance when compared against RRR-GC method that used GradCAM as explanations. In general, the models that utilized explanation losses achieved better explanation localization accuracy than models that only used classification losses. We believe the significant gains in explanation localization accuracy performance outweighs the slight loss of classification performance that these models suffered. More importantly, we were able to observe that using spurious region feedback is more valuable method to increase a model's classification and explanation accuracy than using missing region feedback. We believe our approach can be extended to other cases and investing on collecting spurious region feedback instead of missing region feedback has the potential to improve models and reduce associated cost. Most of the models trained on the class-wise decoy FMNIST performed better than those that were trained on the random version. We accredit this to the class-wise location uniformity of the confounding regions added to the class-wise version. We are aware that the feedback used in our experiments are accurate annotations which can not always be expected in real world scenarios and that feedback quality can be affected by the employed user interface medium. For this reason, we recommend performing user studies involving different feedback collection tools and comparing their performances before selecting one. For future work, we plan to compare impact of different feedback types with a user study.

Acknowledgements. This publication has emanated from research conducted with the financial support of Science Foundation Ireland under Grant number 18/CRT/6183. For the purpose of Open Access, the author has applied a CC BY public copyright licence to any Author Accepted Manuscript version arising from this submission.

References

1. Cohn, D.A., Ghahramani, Z., Jordan, M.I.: Active learning with statistical models. J. Artif. Intell. Res. **4**, 129–145 (1996)
2. Dalvi, B., Tafjord, O., Clark, P.: Towards teachable reasoning systems (2022). arXiv preprint arXiv:2204.13074
3. Fails, J.A., Olsen Jr, D.R.: Interactive machine learning. In: Proceedings of the 8th International Conference on Intelligent user Interfaces, pp. 39–45 (2003)
4. Kenny, E.M., Delaney, E.D., Greene, D., Keane, M.T.: Post-hoc explanation options for XAI in deep learning: the *Insight Centre for Data Analytics* perspective. In: Del Bimbo, A., Cucchiara, R., Sclaroff, S., Farinella, G.M., Mei, T., Bertini, M., Escalante, H.J., Vezzani, R. (eds.) ICPR 2021. LNCS, vol. 12663, pp. 20–34. Springer, Cham (2021). https://doi.org/10.1007/978-3-030-68796-0_2
5. Kim, B.: Interactive and interpretable machine learning models for human machine collaboration. Ph.D. thesis, Massachusetts Institute of Technology (2015)
6. Koh, P.W., Nguyen, T., Tang, Y.S., Mussmann, S., Pierson, E., Kim, B., Liang, P.: Concept bottleneck models. In: International Conference on Machine Learning, pp. 5338–5348. PMLR (2020)
7. Madaan, A., Tandon, N., Rajagopal, D., Yang, Y., Clark, P., Sakaguchi, K., Hovy, E.: Improving neural model performance through natural language feedback on their explanations. arXiv preprint arXiv:2104.08765 (2021)
8. Popordanoska, T., Kumar, M., Teso, S.: Machine guides, human supervises: Interactive learning with global explanations. arXiv preprint arXiv:2009.09723 (2020)
9. Rago, A., Cocarascu, O., Bechlivanidis, C., Lagnado, D., Toni, F.: Argumentative explanations for interactive recommendations. Artif. Intell. **296**, 103506 (2021)
10. Ross, A.S., Hughes, M.C., Doshi-Velez, F.: Right for the right reasons: training differentiable models by constraining their explanations. arXiv preprint arXiv:1703.03717 (2017)
11. Schramowski, P., et al.: Making deep neural networks right for the right scientific reasons by interacting with their explanations. Nature Mach. Intell. **2**(8), 476–486 (2020)

12. Selvaraju, R.R., Cogswell, M., Das, A., Vedantam, R., Parikh, D., Batra, D.: Grad-cam: visual explanations from deep networks via gradient-based localization. In: Proceedings of the IEEE International Conference on Computer Vision, pp. 618–626 (2017)
13. Selvaraju, R.R., Lee, S., Shen, Y., Jin, H., Ghosh, S., Heck, L., Batra, D., Parikh, D.: Taking a hint: leveraging explanations to make vision and language models more grounded. In: Proceedings of the IEEE/CVF International Conference on Computer Vision, pp. 2591–2600 (2019)
14. Settles, B.: Synthesis lectures on artificial intelligence and machine learning. Active learning **6**(1), 1–114 (2012)
15. Shao, X., Skryagin, A., Schramowski, P., Stammer, W., Kersting, K.: Right for better reasons: training differentiable models by constraining their influence function. In: Proceedings of Thirty-Fifth AAAI Conference on Artificial Intelligence (AAAI) (2021)
16. Teso, S., Kersting, K.: Explanatory interactive machine learning. In: Proceedings of the 2019 AAAI/ACM Conference on AI, Ethics, and Society, pp. 239–245 (2019)
17. Xiao, H., Rasul, K., Vollgraf, R.: Fashion-mnist: a novel image dataset for benchmarking machine learning algorithms. arXiv preprint arXiv:1708.07747 (2017)

Learning and Explanation of Extreme Multi-label Deep Classification Models for Media Content

Marco Minici[1,2], Francesco Sergio Pisani[1], Massimo Guarascio[1(✉)],
Erika De Francesco[3], and Pasquale Lambardi[3]

[1] ICAR-CNR, Via P. Bucci, 8/9c, Rende, Italy
{marco.minici,francescosergio.pisani,massimo.guarascio}@icar.cnr.it
[2] Università degli Studi di Pisa, Pisa, Italy
marco.minici@phd.unipi.it
[3] Relatech S.p.a., Via Anguissola, Milano, Italy
{erika.defrancesco,pasquale.lambardi}@relatech.com

Abstract. Providing rich and accurate metadata for indexing media content is a crucial problem for all the companies offering streaming entertainment services. These metadata are typically used to improve the result of search engines and to feed recommendation algorithms in order to yield recommendation lists matching user interests. In particular, the problem of labeling multimedia content with informative tags (able to accurately describe the topics associated with such content) is a relevant issue. Indeed, the labeling procedure is time-consuming and susceptible to errors process as it is usually performed by domain experts in a fully manual fashion. Recently, the adoption of Machine Learning based techniques to tackle this problem has been investigated but the lack of clean and labeled training data leads to the yield of weak predictive models. To address all these issues, in this work we define a Deep Learning based framework for semi-automatic multi-label classification integrating model prediction explanation tools. In particular, Model Explanation techniques allow for supporting the operator to perform labeling of the contents. A preliminary experimentation conducted on a real dataset demonstrates the quality of the proposed solution.

Keywords: Extreme multi-label classification · Image analysis · Interpretable machine learning · Prediction explanation

1 Introduction

Nowadays, the entertainment industry represents one of the most profitable and widespread business sectors, with constant growth in terms of the number of users. With estimated revenues amounting to about 2 trillion dollars worldwide, providing effective research services is a crucial task for the companies operating in multimedia content delivery. In particular, the rise of streaming services and on-demand content fostered the interest for AI-based solutions capable to facilitate the research and identification of content matching the user interests. Just as an example, Recommender Systems (RS) are technologies widely adopted by big players (e.g., Netflix, Disney+, Amazon, etc.) to suggest items in their catalogs able to arouse users' interest.

Besides them, the technologies that allow for enriching content metadata with informative labels (or tags) act a key role as they can be leveraged to improve the RS performances and simultaneously enable more effective research by means of the traditional research engines. Basically, these labels are used to group content exhibiting common features and provide aggregated views for the users. However, the labeling task is time-consuming and prone to the error process since it is manually performed by domain experts. Indeed, the lack of a commonly shared taxonomy can lead to yield repeated labels describing the same concept. Moreover, the assignment of a label to content is subjective and depends on the skill and perception of the expert.

In this context, Artificial Intelligence (AI) techniques can be leveraged to automate such a process by limiting the human factor and, as a consequence, reducing the classification error. However, effectively addressing this problem requires the definition of specific solutions able to cope with different hard issues e.g., unbalancing of the classes, lack of labeled data, the capability of the models to process different types of data (text and images), and providing multi-class predictions on a high number of labels.

The usage of the Deep Learning (DL) paradigm [10] represents a reliable solution to tackle all these issues. Indeed, DL-based models allow for extracting predictive models by combining raw low-level data, gathered from a wide variety of sources (e.g., wikidata, IMDB, etc.). These models are learned with a hierarchical scheme: several layers of non-linear processing units are stacked in a single network and each subsequent layer of the architecture can extract features with a higher level of abstraction compared to the previous one. Therefore, DL-based approaches allow for extracting data abstractions and representations at different levels, they also represent a good choice for analyzing raw data provided in different formats and by different types of sources.

Although the employment of Deep Learning models is considered a promising approach to deal with the above described hard issues, the role of the domain expert remains crucial since he/she is responsible for the final labeling. Therefore, explaining the predictions yielded by the classification model is a further important mechanism to support the operator. Moreover, the model explanations can be used by the analyst to "debug" the model behavior and discover the causes for misclassifications.

In this paper, we define a solution able to combine different types of data (gathered from different publicly available data sources) for classifying media data and enrich them with informative labels, named EMCEE (**E**xtreme **M**ulti-label **C**lassifier and **E**xplain**E**r). An explanation module (integrated into the framework) allows for generating visual artifacts that enable the interpretation of the recommendations generated by the model. In Fig. 1, we depicted the overall learning and explanation scheme. After an *Information Retrieval* stage in which data are gathered and wrapped in a single view, these raw data are provided as input to *Machine Learning* block. Our solution relies on a hierarchical

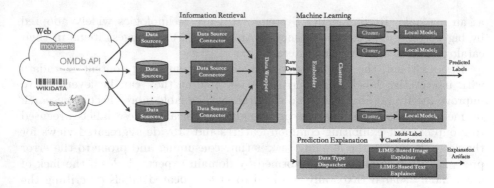

Fig. 1. Overview of the learning and explanation process.

DL framework: on top, an ensemble of pre-trained models (*Embedder*) are fine-tuned and used to map the input (text and/or images) in a low-dimensional latent space. Here, the underlying intuition is that media content with similar labels generates similar vector representations (*embeddings*). Then, a clustering algorithm (*Clusterer*) is used to group similar contents and yield sub-samples of the original dataset. Finally, each sub-sample is used to train a local model focused on a limited set of labels. This approach allows for generating more accurate predictions for specific tags. Finally, data and models can be provided as input to the explanation module in order to inspect the behavior of the model and understand the root causes of the prediction for uncertain cases.

Although our approach is general and capable to cope with different types of data by adding specific models to the Embedder, in this work we focus on processing (classification and explanation) only image data. To the best of our knowledge, our solution is the first tentative to add an explanation for the media content labeling process. An experimental evaluation conducted on a real dataset containing movie posters demonstrates the quality of our approach in labeling the movie with related tags.

The rest of this paper is organized as follows: in Sect. 2 we provide an overview of the main approaches proposed in the literature to tackle the automatic content tagging problem. In Sect. 3, we describe the framework used to address the problem and the deep learning architecture used to learn the multi-label classification model, and the procedure to generate the explanations; while in Sect. 4 we discuss the experimental results. Section 5 concludes the work and introduces some new research lines.

2   Related Work

The problem of classifying movies is not new in the literature and can be considered a general classification task on heterogeneous (video, images, audio, text) data. [15] process user reviews to extract relevant tags for movies. Afterward,

Table 1. Analysis of current literature on Genre/Tag classification.

Approach	Dataset	Number of tags	DL architecture	Data Type	XMLC	Multi-Modal	Metric	Result
Kar et al. [8]	MPST	71	LSTM	Text	y	n	Micro F1	0.37
Arevalo et al. [1]	MM-IMDb	26	Multimodal Fusion with Pre-Trained nets	Text, Image	n	y	Micro F1	0.63
Arevalo et al. [1]	MM-IMDb	26	Multimodal Fusion with Pre-Trained nets	Text, Image	n	y	Macro F1	0.54
Wehrmann et al. [14]	LMTD	22	Multimodal Convolutional NN	Audio, Image	n	y	Micro AUC-PR	0.65
Wehrmann et al. [14]	LMTD	22	Multimodal Convolutional NN	Audio, Image	n	y	Macro AUC-PR	0.74
Fish et al. [4]	MMX-Trailer-20	20	Multimodal classifiers	Audio, Image	n	y	F1-weighted	0.60

they propagate these tags to less popular products according to the movie similarity based on multiple attributes (e.g.: title, summary). Hence, this work draws from the collaborative recommendation paradigm, while our proposal exploits deep metric learning and content-based techniques to solve the tag sparsity problem. Arevalo et al. [1] employs a neural architecture - inspired by recurrent units such as LSTM - named Gated Multimodal Unit (GMU) to effectively combine features coming from the poster image and the plot synopsis. They focus on solving the multi-modal fusion problem rather than the movie tagging itself. Indeed, their dataset contains fewer tags than ours.

The work that most resembles our approach is [8], which makes use of plot synopses to predict tags in the realm of movies. They focus on modeling the plot text as an emotion flow - i.e.: a series of consecutive states of emotion. Their main conclusion is that incorporating the emotion flow increases the tag prediction quality with respect to naive approaches.

[14] analyze movie trailers for performing multi-label genre classification. They explore the extraction of the audio and image features to establish spatiotemporal relationships between genres and the entire trailer. Similar to our approach, different learners are combined. Standalone models are trained separately for the image and the audio input, then they are fused using a weighted average. Anyhow, as stated by the authors, the main limitation of the work relies on the use of only nine common movie genres.

[4] highlight how a single movie genre holds back a large semantic that can be exploited to have a fine-grained description of the movie. The proposed model merges the embeddings yielded by four pre-trained multi-modal 'experts' processing the audio and video of the movie. The goal of the learning process is to improve the quality of the embeddings, i.e. the similarity between each movie clip and one of the 20 tag genres.

Table 1 summarizes the most significant approaches among those described above. Compared to these approaches, there are some major differences with regard to the problem we aim to tackle: first, the tagging task is relative to a high number of labels. Second, this large number of labels exhibits a long-tail distribution as discussed later in the paper. To the best of our knowledge, our solution is the first approach that can handle large amounts of labels (XMLC - eXtreme Multi-Label Classification), process different types of data and provide interpretable explanation of the predictions.

3   Framework

In this section we illustrate our solution and the main components of the proposed DL based architecture. As highlighted in Sect. 1, we adopted a hierarchical approach composed of three main components shown in Fig. 2: *(i)* an *Embedder*, devoted to summarizing the original input into a vector representation (with size N); *(ii)* a cluster module (*Clusterer*) that allows for identifying media with similar contents and extracting focused sub-samples of the original dataset; and *(iii)* the local models that perform the final predictions.

3.1   Hierarchical Deep Multi-label Classification

Embedder. As introduced in Sect. 1, the current implementation of our classification and explanation framework allows for analyzing image data. Although in literature several approaches have been proposed to effectively process image data, an emerging research topic focused on the usage of *Transformer* [13] architectures (typically used to process text data) for the image classification as their accuracy and scalability. In a nutshell, the Transformer models integrate an *Attention Mechanism* aiming at increasing the importance of a part of the input sentence while decreasing other sections. Basically, this mechanism permits the removal of uninformative parts of the raw input and reduces the effect of noisy data. In our *EMCEE* , the backbone used for our embedding phase is the architecture proposed in [3], named *Vision Transformer* (ViT). Similar to a traditional Transformer, ViT attempts to discover relations among the single information units composing the input (words for traditional Transformer models, pixels for ViT). However, since the size of the input for an image is typically higher than the text, the model is trained against sub-portions of the images (*patches*). In more detail, we used a pre-trained instance against ILSVRC-2012 ImageNet dataset with $1k$ classes and $1.3M$ images of ViT are adopted in *EMCEE* to yield the poster embeddings.

Notably, our ViT instance is further fine-tuned by adopting a *Deep Metric Learning* [9] based approach: three instances of the same architecture sharing the same weights are trained against triplets $\langle anchor, positive, negative \rangle$. Basically, the term anchor refers to the reference input whereas positive and negative are other image instances respectively similar and dissimilar to the anchor. The objective is to minimize the distance between the anchor and the positive example while, simultaneously, the distance between the anchor and the negative one is maximized. A customized version of the *triplet loss* for multi-label tasks is employed in the learning phase. Specifically, we adopted a semi-hard negative mining approach that filters out negative instances which share more tags with the anchor w.r.t. the positive ones. At prediction time, only a model is used to

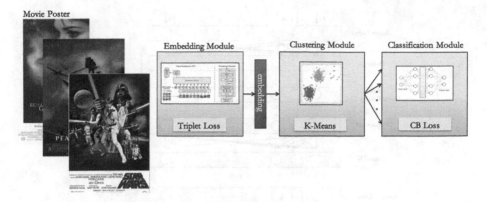

Fig. 2. Our multi-label classification approach includes three main stages: embedding, clustering, and classification. The respective implementations can be modified to address different goals - e.g.: different data modalities, training strategies, or models.

compute the vector representation of the input data. The main benefit of this approach relies on the possibility of combinatorically increasing the input size and handling the lack of labeled examples.

Clusterer. A clustering algorithm is involved in the framework to group similar movies, thus allowing the deployment of local classification models. Specifically, each movie poster is mapped in a latent space (i.e., a numeric vector) via the embedder described above, then the clustering is performed only on the basis of this compact representation. Each cluster includes movies that share a minimal number of tags with respect to the whole label space. Hence, this phase further alleviates the extreme-classification problem. New instances, to be classified, are assigned to the closest cluster on the basis of a suitable distance metric (in our case, *euclidean distance* is adopted). As shown in Fig. 2, we adopted the K-Means algorithm as our clusterer.

Local Models. In our framework, local models take the form of neural networks too. Specifically, the DNN-based architecture, shown in Fig. 3, is used to provide more accurate predictions also for minority classes. Basically, it can be thought of as a Feed-Forward Neural Network composed of more instances of a specific pattern that includes: (i) a fully-connected dense layer equipped with Rectified Linear Unit (ReLU) activation function [11], for each node composing the layer, (ii) a batch-normalization layer for improving stability and performances of the current dense layer [7], (iii) and a dropout layer for reducing the overfitting problem [6]. Several instances of this base component can be stacked in a single model, in particular, in our experimentation, we tested a solution with three instances. The output layer of the architectures is equipped with a *sigmoid* activation function [5] and a variable number of neurons (generically denoted with C in the figure) depending on the number of labels falling on the cluster associated with the local model. Basically, the output layer provides a class probability

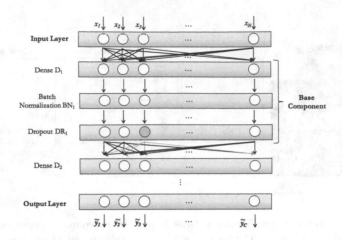

Fig. 3. Neural architecture of a generic local model. It is fed with a compressed representation (with size N) of the Image and yield the probability score to belong to each of the C classes.

for each label. To address the Class Imbalance Problem, each local model is trained by using the Class-Balanced (CB) loss proposed in [2]. Here, the main idea consists of weighting loss inversely with the effective number of samples per class.

3.2   From Prediction to Explanation

Our framework integrates a public Python implementation of the post-hoc explanation method *LIME* (Locally Interpretable Model-agnostic Explanations) proposed in [12]. Although different frameworks (e.g., *SHAP* and *Grad-CAM*) have been proposed in literature to make interpretable black-box models, LIME is adopted in our solution in order to limit the computational costs required to yield the explanations. The main idea of the technique consists in approximating the behavior of a given Black Box Model M around the instance x (for which we want an explanation of the classification yielded by M) by perturbing x and training a surrogate local model. In more detail, a (sparse) linear classification model is learned through Ridge regression on several artificial instances (result of the original input perturbation). Every generated example is labeled with the class predicted by M for it, and it is weighted according to its proximity to x (measured on the basis of a suitable similarity measure), in order to increase the attention of the model to the examples closer to x. Notably, for image explanations LIME works in a slightly different way since the perturbation on single pixels could be unmeaningful. Therefore, image variants are generated by segmenting the image into macro-regions (superpixels) and turning them off or on. These regions are typically identified on the basis of the color homogeneity in such regions, although other strategies could be implemented. Let g a surrogate model able to explain the model behavior, f a classification function that maps the input with the probability to belong to a class, and π_x the neighborhood of x, then the loss function $\mathcal{L}(g, f, \pi_x)$ estimates the error of g in reproducing f behavior in the neighborhood π_x. Then, the explanation e is computed as follow $e = argmin_g \mathcal{L}(g, f, \pi_x) + \Omega(g)$ where $\Omega(g)$ measures the complexity of the explanation ($\Omega(g)$ depends on the type of model used e.g., it could be the depth for a decision tree).

4 Experimental Results

In this section, first, we provide some numerical results in terms of predictive accuracy of the approach, then we show how the explanation allows for confirming/rejecting the recommendations generated by the model.

4.1 Numerical Results

In order to assess the quality of our approach in labeling movies with related tags, we conducted a preliminary experimentation on a real dataset extracted by fusing data from different sources. In particular, first we illustrate the dataset and the challenges to address for providing accurate predictions in this scenario; then we describe the adopted evaluation protocol and metrics; finally, we show an ablation study aiming to highlight the benefits of the proposed solution.

$EMCEE$ has been evaluated on a novel media content dataset gathering different types of information on a movie catalog (e.g., movie plot, trailer, poster, synopsis, tags, etc.). Specifically, we focused our analysis on visual data (i.e., posters) and tags extracted from multiple open sources such as Wikipedia, Wikimedia, and TMDB in order to provide also a visual interpretation of the predictions. When available, a poster image is associated with the movie. The final dataset contains ~14k movies and 134 different tags, with an average of ~3 tags per movie. We can observe that the data distribution exhibits a long tail shape, indeed a restricted number of tags (mainly the genres) occurs more frequently than other specific ones (which can be considered as keywords summarizing some aspects of the movie), resulting in a Gini Index equal to 0.67. Basically, 113 out of 134 tags have a number of occurrences less than the average value.

The experiments are performed on a DGX machine equipped with V100 GPUs. The dataset has been partitioned (in a stratified fashion) into training and test sets respectively with 70%/30% percentages. $Adam$ is used as optimizer while, as mentioned above, we exploit two loss functions to handle the imbalance problem i.e., the triplet loss and the CB-loss. The first one is used during the Embedder learning phase while the last is used for training the local models. As regards the Clusterer, the number of groups k has been empirically determined to 20. In more details, we tested different values of k and selected the best trade-off between variance within the clusters and number of clusters. As a result, each local model can focus the learning on a limited number of tags, in particular, the average number of tags per cluster is ~23.

An ablation study has been conducted to demonstrate the quality of the *EMCEE* approach: the idea to evaluate the contribution to the predictive performances of the different components building up the framework i.e., ViT, Clustering, and loss. Specifically, in Fig. 4 we reported the results in terms of *F1-Score* by excluding the last two components. First, we can observe that a single component is not able to exhibit good predictive performances whereas the overall approach allows for boosting the initial result. Moreover, *EMCEE* approach allows us to predict some examples as belonging to minority tags that are totally ignored in the other cases.

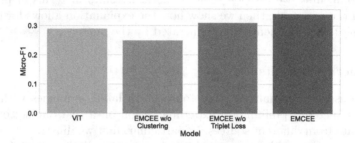

Fig. 4. Experimental Results. *EMCEE* shows an higher Micro-F1 with respect to a ViT trained on the same dataset. We also report an ablation study highlighting the contribution of both clustering and Triplet Loss to the *EMCEE* performances.

4.2   Explanation Prototypes

We next show three examples illustrating the exploitation of the explanation capabilities of our framework in a simulation scenario where the hierarchical model, named hereinafter M, is used to predict the tags of the posters. Figure 5 shows some tag recommendations yielded by the model for the tag *sailing* (under-represented tag in the training set). The first two columns depict two positive examples for which the interpretation allows for confirming the model classification. Indeed, we can observe that the tool seems to label these movies as *sailing* since two ships are present in both the posters, and in particular the model has been able to reveal the presence of the hull among the objects included within the pictures without any previous pre-training on this concept. The third movie presents a false positive i.e., a case for which *EMCEE* miss-classifies the tag *sailing*. In more details, the explanation emphasizes a rigid airship - which is deceiving the model - but helps the operator to reject the classification.

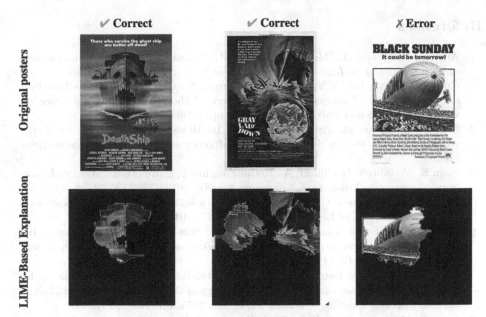

Fig. 5. Explanation of the prediction yielded for three movies sharing the same tag i.e., *sailing*. In the first row, the original posters provided as input to the framework are shown, the second one contains the LIME-based Explanations.

5 Conclusions and Future Work

Enriching metadata with informative labels is a crucial problem for the companies operating in the media content delivery field. However, automating this task requires addressing different challenging issues (i.e., noisy data, class imbalance, and lack of labeled examples for the training phase). Moreover, the final labeling is delegated to the operator, therefore the interpretation of the classification model output is likewise important to make aware the expert in its decisions. In this work, we defined a hierarchical DL-based solution for extreme multi-label classification that integrates model explanation tools. An experimentation conducted on a real dataset demonstrates the quality of the approach.

As a pointer for further research, we aim at boosting the overall performance of the proposed approach by integrating information coming from unlabeled data in a semi-supervised or self-supervised way. Also, active learning schemes can be fruitfully exploited by implementing ad-hoc oracle labeling strategies. Finally, we are interested to extend the experimentation for a fully multi-modal scenario by including heterogeneous data e.g., movie plots and subtitles.

Acknowledgements. This work was partially supported by PON I&C 2014-2020 FESR MISE, Catch 4.0.

References

1. Arevalo, J., Solorio, T., Montes-y Gómez, M., González, F.A.: Gated multimodal units for information fusion. arXiv preprint arXiv:1702.01992 (2017)
2. Cui, Y., Jia, M., Lin, T.Y., Song, Y., Belongie, S.: Class-balanced loss based on effective number of samples. In: Proceedings of the IEEE/CVF Conference on Computer Vision and Pattern Recognition, pp. 9268–9277 (2019)
3. Dosovitskiy, A., et al.: An image is worth 16x16 words: transformers for image recognition at scale. In: International Conference on Learning Representations (2020)
4. Fish, E., Weinbren, J., Gilbert, A.: Rethinking movie genre classification with fine-grained semantic clustering. arXiv preprint arXiv:2012.02639 (2020)
5. Guarascio, M., Manco, G., Ritacco, E.: Deep learning. Encyclopedia Bioinform. Comput. Biol. ABC of Bioinform. 1–3, 634–647 (2018)
6. Hinton, G.E., Srivastava, N., Krizhevsky, A., Sutskever, I., Salakhutdinov, R.: Dropout: a simple way to prevent neural networks from overfitting. J. Mach. Learn. Res. 15, 1929–1958 (2014)
7. Ioffe, S., Szegedy, C.: Batch normalization: Accelerating deep network training by reducing internal covariate shift. In: Proceedings of the 32nd International Conference on Machine Learning - Volume 37. ICML 2015, pp. 448–456 (2015)
8. Kar, S., Maharjan, S., Solorio, T.: Folksonomication: predicting tags for movies from plot synopses using emotion flow encoded neural network. In: Proceedings of the 27th International Conference on Computational Linguistics, pp. 2879–2891 (2018)
9. Kaya, M., Bilge, H.S.: Deep metric learning: a survey. Symmetry 11(9) (2019). https://doi.org/10.3390/sym11091066
10. Le Cun, Y., Bengio, Y., Hinton, G.: Deep learning. Nature 521(7553), 436–444 (2015)
11. Nair, V., Hinton, G.E.: Rectified linear units improve restricted Boltzmann machines. In: Proceedings of the 27th International Conference on Machine Learning. ICML 2010, pp. 807–814 (2010)
12. Ribeiro, M.T., Singh, S., Guestrin, C.: "why should i trust you?" explaining the predictions of any classifier. In: Proceedings of the 22nd ACM SIGKDD International Conference on Knowledge Discovery and Data Mining, pp. 1135–1144 (2016)
13. Vaswani, A., et al.: Attention is all you need. In: Proceedings of the 31st International Conference on Neural Information Processing Systems, pp. 6000–6010 (2017)
14. Wehrmann, J., Barros, R.C.: Movie genre classification: a multi-label approach based on convolutions through time. Appl. Soft Comput. 61, 973–982 (2017)
15. Wu, C., et al.: Exploiting user reviews for automatic movie tagging. Multimedia Tools Appl. 79(17), 11399–11419 (2020)

An Interpretable Machine Learning Approach to Prioritizing Factors Contributing to Clinician Burnout

Malvika Pillai[1]([:envelope:]) [iD], Karthik Adapa[1,2] [iD], Meagan Foster[1,2] [iD], Ian Kratzke[2], Nadia Charguia[3], and Lukasz Mazur[1,2]

[1] Carolina Health Informatics Program, University of North Carolina (UNC) Chapel Hill, Chapel Hill, NC, USA
mpillai@live.unc.edu
[2] Division of Healthcare Engineering, UNC Chapel Hill, Chapel Hill, NC, USA
[3] Department of Psychiatry, UNC Chapel Hill, Chapel Hill, NC, USA

Abstract. Clinician burnout is a multi-factorial problem, and there are limited studies utilizing a theoretical model to assess factors contributing to clinician burnout. A survey of demographic characteristics and work system factors was administered to 278 clinicians (participation rate: 55%). We compare four classifiers with four feature selection methods to predict clinician burnout. We used SHapley Additive exPlanations (SHAP) and permutation importance to prioritize key factors contributing to clinician burnout and interpret the predictions. Random forest had the highest AUC of 0.82 with work system factors only. Six work system factors (*administrative burden, excessive workload, inadequate staffing, professional relationship, intrinsic motivation, and values and expectations*) and one demographic factor (*race*) had the highest impact on predicting clinician burnout. Identifying and prioritizing key factors to mitigate clinician burnout is essential for healthcare systems to allocate resources and improve patient safety and quality of care.

Keywords: Clinician burnout · Medicine · Machine learning · Explainable AI

1 Introduction

Recognition of burnout among healthcare clinicians has increased over the past ten years. Before the COVID-19 pandemic, more than 50% of US clinicians reported experiencing symptoms of burnout, "a syndrome characterized by high emotional exhaustion, high depersonalization (feelings of detachment and lack of empathy), and low personal achievement" [1]. The COVID-19 pandemic has placed tremendous mental and physical stress on clinicians and has markedly worsened the outstanding concerns about clinician well-being and burnout [2]. As of April 2021, 3607 US healthcare workers died during the COVID-19 pandemic, with clinicians constituting ~80% of all deaths [3]. Clinician burnout has wide-ranging consequences, including but not limited to increased medical

M. Ceci et al. (Eds.): ISMIS 2022, LNAI 13515, pp. 149–161, 2022.
https://doi.org/10.1007/978-3-031-16564-1_15

errors, [4] lack of motivation, absenteeism, poor morale, turnover, and reduced productivity leading to a deterioration in the quality of healthcare services [5–7]. A recent US-based study estimated that \$4.6 billion in annual costs is attributable to burnout, primarily related to staff turnover and reductions in clinical hours [8]. Thus, there is an urgent need to examine the key factors contributing to clinicians' burnout during the COVID-19 pandemic.

Although demographic factors such as gender and age are often independent predictors of burnout, studies suggest that clinician burnout is a complex multifactorial problem [9]. Recently, the US National Academy of Medicine (NAM) proposed a systems-based framework and identified multiple evidence-based work system factors contributing to clinician burnout [10]. This theoretical framework demonstrates how these numerous interacting work system factors are often mediated by individual characteristics such as gender, age, and race resulting in clinician burnout. Although the association between job-related factors and clinician burnout is well established, [11, 12] there is limited research utilizing a theoretical model of clinician burnout to explore the relationship between distinct work system factors and burnout. Further, to the best of our knowledge, no previous study has utilized the NAM framework to examine the relative importance of key work system factors contributing to clinician burnout during the COVID-19 pandemic. Identifying and prioritizing specific and potentially modifiable work system factors in a healthcare system is crucial to identifying interventions likely to yield the highest return on investment and guide the optimal allocation of scarce resources to mitigate and prevent clinician burnout.

Although machine learning has been advancing for several years, it has only recently been used for behavioral sciences and, more specifically, in predicting burnout in healthcare workers. Lee et al. used k-means to group about one thousand nurses working in a medical center in Taiwan into two classes (burnout and non-burnout states). Then they applied the convolutional neural network deep learning method to build a predictive model to estimate 38 parameters for the burnout sample [13]. Kurbatov et al. used k-means unsupervised clustering (k-means analysis) and supervised clustering (k-means cluster group) on survey data (53 respondents) to identify and predict burnout in surgical residents [14]. Nishi et al. used an ensemble model comprising elastic net, average, median, and generalized linear model (GLM) blender on survey data (422 responses) to predict Japanese physician's well-being and identified high-impact predictive factors for physicians' well-being [15]. Thus, to the best of our knowledge, no previous study has utilized machine learning approaches to examine the role of work system and demographic factors in predicting US clinicians' burnout during the COVID-19 pandemic. Further, we use feature selection and feature importance methods to identify key factors that best predict US clinicians' burnout to provide evidence for targeting interventions to reduce US clinicians' burnout and improve the quality of care.

2   Methods

2.1   Data Collection and Study Measures

A composite survey was created to assess the following: demographic factors (clinical position, gender, race, and marital status), burnout using the 2-item Maslach Burnout

Inventory (MBI) [16], and severity ratings of 21 evidence-based work system factors based on the NAM's system-based framework. The survey was designed using Qualtrics Online Survey Software and administered to clinicians in a large academic medical center's oncology, primary care, and surgery departments. For this study, we used the Centers for Medicare & Medicaid Services (CMS)' definition of a clinician - those who provide "principal care for a patient where there is no planned endpoint of the relationship; expertise needed for the ongoing management of a chronic disease or condition; care during a defined period and circumstance, such as hospitalization; or care as ordered by another clinician [17]." Per CMS' definition, clinicians may be physicians, nurses, pharmacists, or other allied health professionals. The survey was administered between November 2020 and May 2021 with a participation rate of 55% (Table 1). The study was approved by the UNC-Chapel Hill Institutional Review Board.

Table 1. Number and type of survey responses

Feature	Feature Categories	No. (%)
Burnout	With burnout	110 (61.45%)
	Without burnout	69 (38.55%)
Clinical position	Attending physicians	70 (39.11%)
	Nurses	89 (49.72%)
	Training physicians (residents)	17 (9.50%)
	Pharmacists	3 (1.68%)
Gender	Male	32 (17.88%)
	Female	132 (73.74%)
	Non-binary	3 (1.68%)
	Transgender male	3 (1.68%)
	Transgender female	1 (0.56%)
	Prefer to self-describe	2 (1.12%)
	Prefer not to disclose	6 (3.35%)
Race (R)	Caucasian	133 (74.30%)
	African American	10 (5.59%)
	Latino or Hispanic	1 (0.56%)
	Asian	8 (4.47%)
	Native American	2 (1.12%)
	Native Hawaiian or Pacific Islander	2 (1.12%)
	Other	10 (5.59%)
	Prefer not to disclose	13 (7.26%)

(*continued*)

Table 1. (*continued*)

Feature	Feature Categories	No. (%)
Marital status	Single	26 (14.43%)
	Married	119 (66.48%)
	Divorced	10 (5.59%)
	Separated	6 (3.35%)
	Widowed	4 (2.23%)
	Other	3 (1.68%)
	Prefer not to disclose	11 (6.15%)
Work system factors (WS)		
Job demands (JD)	Excessive workload	177 (98.88%)
	Unmanageable work schedules	175 (97.77%)
	Inadequate staffing	174 (97.21%)
	Time pressure	174 (97.21%)
	Inefficient workflows	173 (96.65%)
	Interruptions and disruptions	173 (96.65%)
	Inadequate technology	172 (96.10%)
	Moral distress	175 (97.77%)
	Patient factors	172 (96.10%)
	Administrative burden	173 (96.65%)
Job resources (JR)	Lack of recognition for QI activities	173 (96.65%)
	Lack of dedicated time	173 (96.65%)
	Lack of support for research	172 (96.10%)
	Professional relationships	172 (96.10%)
	Organizational culture	173 (96.65%)
	Physical work environment	171 (95.53%)
	Values and expectations	171 (95.53%)
	Job control	174 (97.21%)
	Intrinsic motivation	169 (94.41%)
	Extrinsic motivation	175 (97.77%)
	Work-life integration	175 (97.77%)

Categorical: Burnout, clinical position, gender, race, marital status; Ordinal: Work system factors.

2.2 Dataset Preparation

The input variables were the 21 work system factors and four demographic characteristics. Further, the 21 work system factors were divided into ten job demand factors and 11 job resource factors. The outcome variable for the analysis was burnout as measured by emotional exhaustion (EE) (1 to 6) and depersonalization (DP) (1 to 6). An EE and DP summative score > 3 correlates best with a more inclusive definition of burnout [18]. However, for this analysis, we considered a score > 3 on EE and DP individually as a more restrictive definition of burnout to categorize the clinicians into two classes: with ($\geq$3 EE & $\geq$3 DP) and without burnout (<3 EE & <3 DP) to enable a binary classification.

The dataset (179 cases) was separated with a stratified split, where 143 cases were used as a training set, and 36 cases were used as the test set. The test set was set aside to evaluate performance after all model training and parameter tuning. The training set was further divided with a split stratified by burnout category into a train set (114 cases) and a development set (29 cases). The development set was used to rank the models in terms of their accuracy as defined by the area under the receiver operating characteristic curve (ROC-AUC) and decide which model to use for testing. 5-fold cross-validation (CV) was used for training and hyperparameter tuning to avoid overfitting.

2.3 Feature Selection

Feature selection is a preprocessing technique that detects relevant features and discards redundant attributes that do not contribute to predicting clinician burnout and removing them. We used the following four feature selection methods: mutual information, chi-square, recursive feature elimination, and manual selection. Manual selection consisted of selecting variables based on NAM groupings and past literature. All feature selection methods were compared across the three classifiers. Mutual information is the measurement of mutual dependence between two random variables, where the larger the measurement, the more dependent one variable is on another. The top ten features with a mutual information score > 0.05 were included in the analysis. The mutual information score threshold was selected by training classifiers with 5-fold cross-validation and setting the threshold related to the highest accuracy. Chi-square feature selection consists of testing the independence between variables, and we aim to select the features that are significantly dependent on the outcome variable. The top ten features ranked by p-value < 0.01 were included in the analysis. In recursive feature elimination (RFE), features were iteratively selected while optimizing the area under the receiver operating characteristic curve (AUC) performance. After each iteration, the less relevant features were removed, and the key factors that best predicted clinicians' burnout were identified.

2.4 Classification

We used random forest (RF), linear support vector machine (linear SVM), decision tree (DT), and logistic regression (LR), four widely used classifiers for binary classification to predict clinicians' burnout. Initially, the classifiers were trained with factors based on the NAM framework: work system factors (WS) & demographic characteristics (all factors), work system factors & race (R), work system factors only, job demand (JD) factors only, job resource (JR) factors only, job demand factors + race, job resources factors + race (manual feature selection). AUC was used to evaluate the models. Random forest included 100 estimators, and the linear support vector machine had a scaled gamma value. Logistic regression had an L2 penalty, the maximum number of iterations was set to 10,000, and a Limited-memory Broyden-Fletcher-Goldfarb-Shanno (LBFGS) solver.

2.5 Feature Importance

Feature importance (variable importance), in this work, refers to estimating how much each work system factor and demographic characteristic contributed to predicting clinicians' burnout. Thus, feature importance helps determine the key predictors of clinicians' burnout on the test set (i.e., after training) to *explain the learned model* as opposed to feature selection methods mainly applied before or during model training to select the principal features of the final input data. Studies suggest that feature importance estimates might be biased if only a single method is employed, as it might not consider the synergy between orthogonal features [19]. Utilizing multiple feature importance methods makes the final feature importance estimates robust and helps explain the results of machine learning models more accurately. We computed permutation importance (PI) and used Shapley Additive Explanations (SHAP) to evaluate and understand key predictors of clinicians' burnout on the test set for the final selected classifier. We used PI and SHAP because they are model-agnostic, reasonably efficient, and reliable techniques. PI measures variable importance by observing the effect on model accuracy while randomly shuffling each predictor variable [20]. We choose PI over Mean Decrease Impunity (MDI) using Gini importance, another model-agnostic approach, as MDI disproportionately increases the importance of continuous or high-cardinality categorical variables [21]. We did not choose local interpretable model-agnostic explanations (LIME) as it does not have a good global approximation of feature importance and only provides feature importance for individual instances. Furthermore, LIME is sensitive to small perturbations in the input, leading to different feature importance for similar input data [19]. Although widely used, we also recognize that PI can result in misleading results as highly correlated features affect PI [22]. We use SHAP explanations based on the cooperative game theory approach to explain the machine learning model predictions with SHAP values, calculated as a weighted average of features' marginal contribution [23]. SHAP is a consistent and accurate feature attribution method that shows the overall importance of features via a mean SHAP values metric and how the models perform with and without a feature for every combination of features [23]. Thus, we used PI and SHAP as feature importance methods but will primarily use mean SHAP values to interpret the test predictions with the final selected algorithm.

3   Results

3.1   Feature Selection

Table 2 shows the top ten features with a mutual information (MI) score >0.05 and the top ten features from chi-squared (CS) with a p-value <0.01. Features that are similar in both MI and CS are marked in **bold**.

The number of features with which RFE showed the highest CV accuracy for each classifier is shown in Fig. 1. The three classifiers had varying numbers of optimal features-

Table 2. Mutual information and chi-squared features

MI features	CS features
Time pressure	**Time pressure**
Intrinsic motivation	**Intrinsic motivation**
Extrinsic motivation	**Extrinsic motivation**
Lack of dedicated time	**Lack of dedicated time**
Job control	**Job control**
Values and expectations	**Values and expectations**
Professional relationships	Lack of support for research
Work life integration	Administrative burden
Interruptions and disruptions	Moral distress
Excessive workload	Patient factors

random forest (7), decision tree (8), logistic regression (21), and linear SVM (21). Only two features, organizational culture and intrinsic motivation, were chosen by all four classifiers.

3.2 Classification

The three classifiers were trained using the training set and development set. Table 3 shows the accuracies (i.e., AUC) of classifiers using and not using feature selection methods (i.e., using all features) with the highest accuracy for each classifier highlighted in bold, and the classifier with the highest average cross-validation AUC is highlighted in **bold** and underlined. Random forest achieved the highest AUC with chi-squared feature selection (0.80). Also, random forest consistently outperformed all other classifiers with and without feature selection (highlighted in **bold** and *italicized*). The highest AUC of decision tree was with work system factors and work system factors + race (0.68). The highest AUC for logistic regression was with job demands (JD), job demands + race (JD + R), and work system factors + race (WS + R) (0.66). Linear SVM achieved 0.62 highest AUC with job demands.

Table 3. Model cross-validation (5-fold) average AUC

Models	All	JD	JR	JD+ R	JR+ R	WS	WS+ R	MI	CS	RFE
RF	*.79*	*.69*	*.71*	*.70*	*.75*	*.76*	*.77*	*.76*	**.80**	*.78*
DT	.63	.64	.58	.66	.65	**.68**	**.68**	.61	.64	.64
LR	.64	**.66**	.53	**.66**	.54	.63	**.66**	.56	.58	.62
Linear SVM	.58	**.62**	.51	.41	.48	.59	.60	.46	.43	.61

Table 4 shows the average AUC of classifiers using and not using feature selection methods on the development set out of 100 runs. Random forest achieved the highest AUC with chi-squared feature selection (0.71) (highlighted in **bold** and underlined). Also, random forest consistently outperformed all other classifiers with and without feature selection, like in CV (highlighted in **bold** and *italicized*).

Table 4. Average AUC for all classifiers on the development set

Models	All	JD	JR	JD+ R	JR+ R	WS	WS+ R	MI	CS	RFE
RF	*0.66*	*0.69*	*0.66*	*0.66*	*0.64*	*0.67*	*0.67*	*0.68*	<u>**0.71**</u>	*0.69*
DT	0.52	0.57	**0.63**	0.54	0.58	0.55	0.54	0.61	0.66	0.54
LR	**0.65**	0.60	0.64	0.62	0.62	0.62	0.61	0.56	0.55	0.61
Linear SVM	0.54	0.52	0.42	0.51	0.43	**0.55**	0.52	0.45	0.49	**0.55**

Since random forest consistently outperformed other classifiers on the development set, only random forest was evaluated on the test set. Table 5 shows the test set AUC for random forest with and without feature selection methods. Random forest with work system factors had the highest test AUC. Among feature selection methods, mutual information and RFE achieved identical test AUC. Chi-squared feature selection, which consistently had the highest CV and development AUC, had the lowest test AUC compared to other feature selection methods and without feature selection methods.

Table 5. Test set AUC for random forest

Models	All	JD	JR	JD+ R	JR+ R	WS	WS+ R	MI	CS	RFE
RF	0.76	0.72	0.78	0.71	0.75	**0.82**	0.80	0.77	0.69	0.77

3.3 Feature Importance

SHAP provides both global and individual explanations. The SHAP summary plot brings all data to a single plot and helps identify how impactful each feature is on the model output for cases in the test set. On the y-axis, the features are sorted by decreasing order of importance. The top features are the key contributors to clinician burnout prediction, and the bottom ones are the least impactful predictors. The x-axis shows the mean SHAP values (the average impact on model output). Thus, the features are sorted by the sum of the SHAP value magnitudes across all samples. Figure 1 (top) shows that professional relationships, administrative burden, intrinsic motivation, excessive workload, race, and inadequate staffing with high mean SHAP values are the key predictors for the without (class 0, blue color) and with burnout groups (class 1, red color).

PI generates an ordered list of features along with their importance values. Figure 1 (bottom) is a PI plot with the y-axis highlighting the feature importance value, estimated by the mean accuracy decrease across 20 model runs, and the x-axis showing different features. The longer blue bars indicate higher importance than the shorter blue bars. Each bar also shows the average feature importance value and the standard deviation of importance values across all random forest iterations. Professional relationships, inadequate staffing, intrinsic motivation, values and expectations, patient factors, and

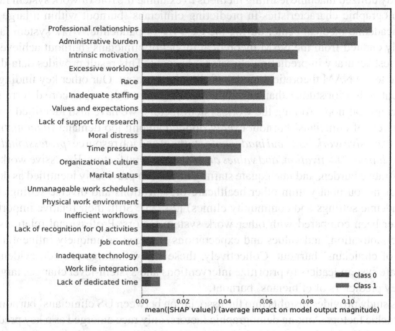

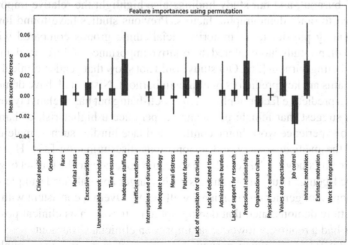

Fig. 1. SHAP summary plot (top) and permutation importance plot (bottom)

excessive workload have the highest impact on model prediction. The negative values for PI, as shown under the dashed blue line (e.g., job control, race, organizational culture, moral distress), occur when features do not contribute to predictions. This is more common with small datasets such as in this study because there is more room for chance.

4 Discussion

This study utilized machine learning methods to examine the role of work system factors and demographic characteristics in predicting clinicians' burnout within a large integrated academic medical center during the COVID-19 pandemic. Work system factors manually curated from the NAM theoretical model of clinicians' burnout achieved the highest test accuracy in predicting clinicians' burnout. This result provides data-driven evidence to the NAM theoretical model of clinician burnout. Our other key findings are consistent with prior studies that have identified job-related factors as core determinants of clinicians' burnout. Among the work system factors, we have also identified six primary drivers of clinicians' burnout evenly divided among job demand (*administrative burden, excessive workload, and inadequate staffing*) and job resource (*professional relationship, intrinsic motivation, and values and expectations*) factors. Excessive workload, administrative burden, and inadequate staffing have been previously identified as drivers of clinicians' burnout within other healthcare organizations and diverse settings (e.g., non-academic settings and community clinics) [24, 25]. Still, their relative importance has never been compared with other work system factors. Professional relationships, intrinsic motivation, and values and expectations emerged as uniquely influential predictors of clinicians' burnout. Collectively, these study findings provide evidence to healthcare system leaders to prioritize interventions and system-level changes targeting these key predictors of clinicians' burnout.

This study provides insights into the association between US clinicians' burnout and demographic factors. Among demographic factors, only race emerged as a key predictor of clinicians' burnout, and our study findings also highlight the relative importance of race compared to other demographic factors. Previous studies have found lower rates of burnout among physicians in minority racial/ethnic groups compared with white physicians. Still, no study has explored the relative importance of different demographic factors in predicting burnout [26]. Our study could not show that gender, clinical position, and marital status are key predictors of clinician burnout. Nishi et al. have demonstrated that gender is a predictive factor of physicians' well-being but has a relatively low impact [15]. Studies suggest that female physicians experience a higher risk of burnout, are more likely to experience work-home conflict, and face hurdles such as male-dominant structures of the medical society and unconscious discrimination [27]. However, the sample used in this study is 73% female, so we could not evaluate the impact of gender on clinicians' well-being. Thus, future studies with a larger and balanced sample size must explore the impact of gender on clinicians' well-being. Overall, consistent with previous studies, our study demonstrated that demographic factors such as clinical position and marital status had a relatively lower or no impact on clinicians' burnout.

During COVID-19, clinician burnout has emerged as a significant public health issue. As healthcare systems devise strategies to recruit, retain and support clinicians, our research provides an important contribution by providing an evidence-based prioritization of work system factors and demographic characteristics. By identifying six work system factors and one demographic factor as the most influential predictors of clinician burnout among 25 work system factors and demographic characteristics, our findings are likely to help guide the allocation of resources to where they may have the greatest impact in mitigating and preventing clinician burnout. Most of our findings are consistent with Nishi et al.'s recent study that used Japanese physician survey data to develop an ensemble of machine learning models with the highest mean AUC of 0.72. Some of the key differences between our study and their study are that they did not use a theoretical model to determine the work system factors contributing to clinicians' burnout, did not evaluate performance on a test set, and used only one feature importance method (i.e., PI) to assess key factors predicting burnout.

This study has several limitations that need acknowledgment. First, our results are based on a small sample of clinicians at a single US academic medical center with unique institutional characteristics, which may limit generalization to other practice settings. Second, the clinicians who responded to this survey do not represent all the major medical specialties, groups, and subsets of clinicians, which may add potential response bias. Third, the study's objective was to identify key predictors of US clinicians' burnout; thus, the findings should not be interpreted as causal. Therefore, our study findings, although promising, cannot be generalized without further investigation.

5 Conclusion

The increasing prevalence of burnout adversely affects clinicians, patients, and healthcare systems. Although demographic characteristics such as race play a role, the primary drivers of clinician burnout are related to work system factors. The COVID-19 pandemic has placed an acute strain on the US clinician workforce, and healthcare systems are under greater pressure to implement effective burnout mitigation and prevention strategies. Our findings provide evidence-based prioritization of key factors contributing to clinicians' burnout, and healthcare systems may consider targeting these factors to optimize scarce resources. Future studies should extend this analysis with a larger dataset, different demographic sub-populations, urban-rural locations, academic and non-academic settings, medical specialties, and team structures to understand the potential differences in feature importance across diverse clinician populations and practice environments.

References

1. Rotenstein, L.S., Torre, M., Ramos, M.A., et al.: Prevalence of burnout among physicians: a systematic review. JAMA **320**, 1131–1150 (2018). https://doi.org/10.1001/jama.2018.12777
2. Leo, C.G., Sabina, S., Tumolo, M.R., et al.: Burnout among healthcare workers in the COVID 19 era: a review of the existing literature. Front. Publ. Health **9**, 750529 (2021). https://doi.org/10.3389/fpubh.2021.750529
3. Lost on the frontline: US healthcare workers who died fighting Covid-19—US news—The Guardian. https://www.theguardian.com/us-news/ng-interactive/2020/aug/11/lost-on-the-frontline-covid-19-coronavirus-us-healthcare-workers-deaths-database. Accessed 16 June 2022

4. Shanafelt, T.D., Balch, C.M., Bechamps, G., et al.: Burnout and medical errors among American surgeons. Ann. Surg. **251**, 995–1000 (2010). https://doi.org/10.1097/SLA.0b013e3181bfdab3
5. Shanafelt, T.D.: Finding meaning, balance, and personal satisfaction in the practice of oncology. J. Support Oncol. **3**(157–62), 164 (2005)
6. Shanafelt, T.D., Hasan, O., Dyrbye, L.N., et al.: Changes in burnout and satisfaction with work-life balance in physicians and the general US working population between 2011 and 2014. Mayo Clin. Proc. **90**, 1600–1613 (2015). https://doi.org/10.1016/j.mayocp.2015.08.023
7. Dewa, C.S., Loong, D., Bonato, S., et al.: How does burnout affect physician productivity? A systematic literature review. BMC Health Serv. Res. **14**, 325 (2014). https://doi.org/10.1186/1472-6963-14-325
8. Han, S., Shanafelt, T.D., Sinsky, C.A., et al.: Estimating the attributable cost of physician burnout in the United States. Ann. Int. Med. **170**, 784–790 (2019). https://doi.org/10.7326/M18-1422
9. De Hert, S.: Burnout in healthcare workers: prevalence, impact and preventative strategies. Local Reg. Anesth. **13**, 171–183 (2020). https://doi.org/10.2147/LRA.S240564
10. National Academies of Sciences, Engineering, and Medicine; National Academy of Medicine; Committee on Systems Approaches to Improve Patient Care by Supporting Clinician Well-Being: Taking Action Against Clinician Burnout: A Systems Approach to Professional Well-Being. National Academies Press (US), Washington (DC) (2019)
11. West, C.P., Shanafelt, T.D., Kolars, J.C.: Quality of life, burnout, educational debt, and medical knowledge among internal medicine residents. JAMA **306**, 952–960 (2011). https://doi.org/10.1001/jama.2011.1247
12. Shanafelt, T.D., Noseworthy, J.H.: Executive leadership and physician well-being: nine organizational strategies to promote engagement and reduce burnout. Mayo Clin. Proc. **92**, 129–146 (2017). https://doi.org/10.1016/j.mayocp.2016.10.004
13. Lee, Y.-L., Chou, W., Chien, T.-W., et al.: An app developed for detecting nurse burnouts using the convolutional neural networks in Microsoft excel: population-based questionnaire study. JMIR Med. Inform. **8**, e16528 (2020). https://doi.org/10.2196/16528
14. Kurbatov, V., Shaughnessy, M., Baratta, V., et al.: Application of advanced bioinformatics to understand and predict burnout among surgical trainees. J. Surg. Educ. **77**, 499–507 (2020). https://doi.org/10.1016/j.jsurg.2019.11.008
15. Nishi, M., Yamano, M., Matoba, S.: Prediction of well-being and insight into work-life integration among physicians using machine learning approach. PLoS ONE **16**, e0254795 (2021). https://doi.org/10.1371/journal.pone.0254795
16. Maslach, C., Jackson, S.E., Leiter, M.P.: Maslach burnout inventory. psycnet.apa.org (1997)
17. Crum, E.: Clinicians and payers expect to wait and see before embracing CMS MIPS value pathways. Am J. Manag. Care **27**, SP245–SP246 (2021). https://doi.org/10.37765/ajmc.2021.88735
18. Li-Sauerwine, S., Rebillot, K., Melamed, M., et al.: A 2-question summative score correlates with the Maslach burnout inventory. West J. Emerg. Med. **21**, 610–617 (2020). https://doi.org/10.5811/westjem.2020.2.45139
19. Rengasamy, D., Rothwell, B.C., Figueredo, G.P.: Towards a more reliable interpretation of machine learning outputs for safety-critical systems using feature importance fusion. Appl. Sci. **11**, 11854 (2021). https://doi.org/10.3390/app112411854
20. Alvarez-Melis, D., Jaakkola, T.S.: On the robustness of interpretability methods. arXiv. https://doi.org/10.48550/arxiv.1806.08049 (2018)
21. Strobl, C., Boulesteix, A.-L., Zeileis, A., Hothorn, T.: Bias in random forest variable importance measures: illustrations, sources and a solution. BMC Bioinform. **8**, 25 (2007). https://doi.org/10.1186/1471-2105-8-25

22. Hooker, G., Mentch, L., Zhou, S.: Unrestricted permutation forces extrapolation: variable importance requires at least one more model, or there is no free variable importance. arXiv https://doi.org/10.48550/arxiv.1905.03151 (2019)
23. Orlenko, A., Moore, J.H.: A comparison of methods for interpreting random forest models of genetic association in the presence of non-additive interactions. BioData Min. **14**, 9 (2021). https://doi.org/10.1186/s13040-021-00243-0
24. Edwards, S.T., Helfrich, C.D., Grembowski, D., et al.: Task delegation and burnout trade-offs among primary care providers and nurses in Veterans affairs patient aligned care teams (VA PACTs). J. Am. Board Fam. Med. **31**, 83–93 (2018). https://doi.org/10.3122/jabfm.2018.01.170083
25. Helfrich, C.D., et al.: The association of team-specific workload and staffing with odds of burnout among VA primary care team members. J. Gen. Intern. Med. **32**(7), 760–766 (2017). https://doi.org/10.1007/s11606-017-4011-4
26. Garcia, L.C., Shanafelt, T.D., West, C.P., et al.: Burnout, depression, career satisfaction, and work-life integration by physician race/ethnicity. JAMA Netw. Open **3**, e2012762 (2020). https://doi.org/10.1001/jamanetworkopen.2020.12762
27. Villanueva, A.M., Kaye, D., Abdelhak, S.S., Morahan, P.S.: Comparing selection criteria of residency directors and physicians' employers. Acad. Med. **70**, 261–271 (1995). https://doi.org/10.1097/00001888-199504000-00008

A General-Purpose Method for Applying Explainable AI for Anomaly Detection

John Sipple[1,2(✉)] and Abdou Youssef[2]

[1] Google, Mountain View, California, USA
sipple@google.com
[2] The George Washington University, Washington, DC, USA
{jsipple,ayoussef}@gwu.edu

Abstract. The need for explainable AI (XAI) is well established but relatively little has been published outside of the supervised learning paradigm. This paper focuses on a principled approach to applying explainability and interpretability to the task of unsupervised anomaly detection. We argue that explainability is principally an *algorithmic* task and interpretability is principally a *cognitive* task, and draw on insights from the cognitive sciences to propose a general-purpose method for practical diagnosis using explained anomalies. We define Attribution Error, and demonstrate, using real-world labeled datasets, that our method based on Integrated Gradients (IG) yields significantly lower attribution errors than alternative methods.

Keywords: Anomaly detection · Interpretability · Explainable AI

1 Anomaly Detection and Interpretability

Imagine you are a technician that maintains a system with thousands of networked devices. Chances are high that at any given time, somewhere in the fleet there are faulty devices that require your attention. As a technician, you would like an anomaly detector to provide a rich explanation about the symptoms, allowing you to detect, diagnose, prioritize and fix faulty devices. This paper is motivated by the unmet need to provide the technician *explained anomalies*. We combine insights from the cognitive sciences with explainable AI (XAI), and propose a general-purpose approach to aid the technician in detecting the anomaly and understanding the fault behind the anomaly. We show comparative results with sensor failure on a Variable Air Volume (VAV) device, and a fuel pressure failure on a general aviation aircraft engine.

Anomaly Detection (AD) is the machine learning task of detecting observations that do not conform to expected or normal behavior [8,25,27,36]. AD is susceptible to both false positive and false negative errors [2]. Adjudicating false positive errors is often time consuming and may even make an AD solution ineffective. A nondescript anomaly score provides little insight to understanding what caused the anomaly and choosing the best treatment. We hypothesize

M. Ceci et al. (Eds.): ISMIS 2022, LNAI 13515, pp. 162–174, 2022.
https://doi.org/10.1007/978-3-031-16564-1_16

that integrating XAI techniques with AD can reduce the human workload in processing a stream of anomalies.

Relatively little about explainable AD has been published. An explainable AD method was developed by combining an autoencoder with the SHAP explainability method [3]. DIFFI [7] is an explainability technique for Isolation Forest (IF) [14] that generates attributions based on dimensions that contribute significantly to isolating anomalous points and causing data imbalances by random data splits. MADI [32] combines a neural network classifier that distinguishes between observed and randomized samples, and uses Integrated Gradients (IG) [35]. The One-Class Deep Taylor Decomposition (OC-DTD), uses a One-Class Support Vector Machine (OC-SVM) to train ("neuralize") a feed-forward neural network [25]. Then, it applies Deep Taylor Decomposition [4] that yields variable attributions for anomalies [11]. Explaining anomalies in image data, FCDD [15], applies a convolutional upsampling technique to create an heatmap of anomalous regions detected by a neural network variant of OC-SVM.

This work makes the following contributions to the field of explainable AD: **a.** creates a clear distinction between explainability and interpretability; **b.** proposes a methodology that leads to meaningful anomaly explanations combining the cognitive sciences with XAI methods; and **c.** defines Attribution Error and provides two real-world datasets for evaluating anomaly explanations.

2   Conventions and Basic Definitions

A *system* is an observable, complex, and stateful process or object, which may be biological, physical, social, economic, etc. We define *expert* broadly as a stakeholder of the system who applies a knowledge base to interpret the explanation to achieve a specific goal. Device s in system S generates periodic observations $x_{s,t}$ in $\mathbb{R}^D$ at time t, where we omit subscripts s and t for simplicity. The baseline set, or normal points, are indicated with an apostrophe, $X' = x'_1, x'_2, \dots$. An anomaly detector is a classifier model F that scores each observation between *anomalous* (0) and *normal* (1): $F : \mathbb{R}^D \to [0,1]$. An *explanation function*, $B : \mathbb{R}^D \to [0,1]^D$ maps an observation x to an explanation, *attribution*, or *blame*, $B(x) = b$, where $b_d \in [0,1]$ and $\sum_{d=1}^{D} b_d \le 1$. With respect to a specific baseline point x', we shall include the baseline point as an argument, $B(x, x')$. An *interpretation function* is a mapping from explanations to the powerset of discrete *diagnostic conditions*: $I : [0,1]^D \to \mathcal{P}(C)$, where C is a set of all diagnostic conditions known to the expert, $C = c_1, c_2, \dots$.

3   Explainability vs. Interpretability

Recent works on XAI have proposed various definitions of explainability and interpretability. [10] describes explainability as models that are able *to describe the internals of a system that is understandable to humans.* [13] argues that interpretability reflects on *trust, causality, transferability, informativeness, and fair*

Table 1. Differences between explainability and interpretability.

Explainability: *Generate a prediction and explanation*	**Interpretability**: *Align the explanation with a knowledge base, and complete a task*
Role of Transmitter	Role of Receiver
Algorithmic Task	Cognitive Task
Expressed by Model	Generated and used by Expert
Spatial, Temporal Correlation	Root cause and Causation
Observation to Attribution	Attribution to Diagnostic Condition
Expresses symptoms	Diagnoses and treats

and ethical decision making. [9] define interpretability as *the ability to explain or to present in understandable terms to humans.* [19] equates interpretability with explainability as a mode in which an observer can understand a decision. According to [6] explainability is the model's ability *to provide a description of how a model's outcome came to be, and interpretability refers to a human's ability to make sense, or derive meaning, from a given stimulus so that the human can make a decision.* Similar to [6], we propose that explainability and interpretability are two distinct ideas. The communication model with an information source, a message, a transmitter, a noisy channel, a receiver, and a destination [28] provides a framework for describing the difference between explainability and interpretability, Table 1. The information source are the devices that stream multidimensional observations. The transmitter is the anomaly detector that generates an anomaly score and an explanation. Receivers are the *experts*, who apply their knowledge of the world to make judgements of the anomalies. Hence, *explainability* is the *algorithmic task* of extracting an explanation from the model, and *interpretability* is the *cognitive task* of combining the explanation with the expert's knowledge base, and forming a hypothesis.

3.1   Explainability

There are several ways to describe methods of XAI. *Global methods* provide general explanations generally true about all model predictions. *Local methods* [1,26] provide explanations for individual observations. Since they are specific to individual observations, *local explainability* generally is more suitable for interpreting anomalies. Some local model explainability techniques, such as LIME [23] and SHAP [16], treat models as black boxes and perturb the inputs and observe the prediction and can be applied to any classifier model, and are called *model-agnostic*. In contrast, *model-specific* explainability techniques often place specific requirements on the model, such as the gradients or network architecture, that make them compatible. Examples include IG, Layer-wise Relevance Propagation (LRP), and DeepLIFT.

Variable attribution, or *blame*, $b_d = B(x_d)$, quantifies the importance of the value on dimension d, x_d, in directionality (sign) and weight (magnitude) on a prediction. In non-linear models, these variables are local approximations of a hyperplane fit along the steepest gradient in feature space. Some local methods, such as IG [35] and Deep LIFT [31], require a neutral baseline point.

3.2 Interpretablity

Interpreting the model's explanation requires the expert to align the observation, prediction, and explanation with the system's context and a relevant knowledge base [24]. To be successful, the expert must associate the model's explanation with a particular cause or fault, and then decide to take a corrective action.

Anomaly Detection Expert. We propose there are three kinds of experts that use eplainable AI: *physicus*, *technicus*, and *secularus*, described in extended manuscript [33] Appendix A. Technicus has a detailed knowledge base of the system, and aims to alter or influence the system in response to known conditions [26]. The explanations should describe the anomaly in its original state and enable the technicus to easily choose the most suitable treatment for the anomaly. The natural objective of XAI is to generate explanations that enable effective interpretation for the expert. Therefore, it is helpful to consider how the human mind recognizes, generalizes and categorizes, and reasons about causality in order to better design explanations.

Generalization. Empirical studies in cognitive sciences revealed that humans assign objects to categories based on some notion of distance or dissimilarity between multidimensional stimuli [22]. One prominent generalization theory developed by [29,30] suggested that distance can be well approximated with Euclidean L2 or City Block L1 distance measures, and stimuli are internalized in some internal metric space called *psychological spaces*. Experiments with human subjects also indicated that similarity generally follows an exponential decay with distance.

Prototypes and Exemplars. It is well established that when categorizing stimulus of a previously unseen object, humans will tend to use previously observed objects as a reference for making their decision. This gave rise to Prototype and Exemplar Theories [12,21], where the former suggests that humans remember an anchoring object (i.e., a prototype) when categorizing, and the latter suggests that humans tend to retain multiple examples in memory for a category or concept.

Contrastive Explanations. One of the most important conclusions from a review of explanations for AI drawn from the social sciences, was that good explanations are *contrastive* [18]. Humans perform better in understanding explanations when presented with a contrastive, counterfactual event, and prefer explanations with *why did event P happen, instead of counterfactual event Q*.

4   An Approach for Interpreting Anomalies

In this section, we combine AD, explainability and interpretability concepts. The proposed approach provides an explanation with both a contrastive baseline point x' and a variable attribution or blame, $B(x, x')$. We assume that we have already trained a differentiable anomaly detector $F(x) = [0, 1]$, from which gradient ∇F can be computed. The historic data for the unlabeled training dataset X is assumed to be available, and is representative of new observations from system S. Each of the five steps is presented as a separate subsection next.

4.1   Choose an Exemplar Baseline Set

As discussed in Sect. 3, research in psychology suggests that humans retain and comprehend concepts with the support of exemplars, or reference points. We select baseline reference points from the training data that represent normal. The normal baseline sample from the training set have highest scores: $F(x) \approx 1$. Normal observations are suitable candidates for the baseline set.

Proposition 1. If the training data is a statistically representative sample of the inference set, then points with the highest normal scores are suitable exemplars.

In many complex systems, devices operate in different configurations (heating/cooling, idle/active, takeoff/landing, etc.), and distribute observations across multiple modes. Multiple modes occur when these device configurations generate probability mass functions with multiple local maxima. In cases where there are multiple distinct baseline modes, and when the number of normal points is unbalanced across modes, as illustrated with modes A and B in Fig. 1 (left), there is no guarantee that a baseline will represent all modes. Furthermore, it is

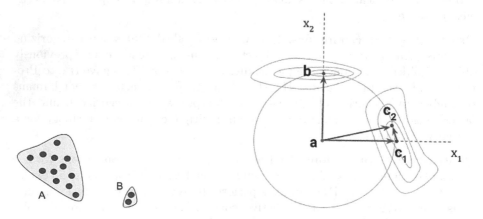

Fig. 1. (Left) Multimodal distribution of the data, where mode A has more observations than mode B. (Right) Selection of baseline points for anomaly a with three equidistant point from two modes.

usually not efficient to oversample baseline points from dense regions. For these reasons, we propose a clustering-based approach for selecting N points randomly from each cluster, presented in extended manuscript [33] Appendix B.

4.2 Choose a Measure of Dissimilarity

We must apply distance function for 1. selecting the baseline set, and 2. for selecting the baseline point for each anomaly. As discussed in Sect. 3.2, based on empirical evidence on human categorization of variables are correlated in complex systems, adopt a Euclidean L2 or City Block L1 distance function.

Proposition 2. Euclidean L2 or City Block L1 distances are suitable for choosing a contrastive baseline point based on empirical results from the cognitive sciences.

4.3 Select a Baseline Point

Guided by Occam's Razor, we seek the simplest, most efficient explanation. Transforming an anomalous point into any baseline point yields an explanation, regardless of its distance from the anomaly. Since each dimension has been normalized, an equal-length displacement in any direction incurs the same cost. Therefore, a baseline point from the nearest mode leads to the most efficient explanation. We apply explainability methods that use baselines as exemplars and choose the nearest exemplar, such as IG. The distance between each anomalous point and the baseline can be computed with the distance function, and the point with the minimum distance is chosen as the representative baseline point.

Definition 1. *Two baseline points x'_1, x'_2 result in equivalent attributions if for any anomalous point x, the blame attributions $B(x, x'_1)$ and $B(x, x'_2)$ map to the same diagnostic condition, i.e., $I(B(x, x'_1)) = I(B(x, x'_2))$.*

As will be shown in Theorem 1 below, two baseline points drawn from the same mode and separated by a small perturbation result in the same interpretation, i.e., result in equivalent attributions. This justifies our next proposition:

Proposition 3. *A small perturbation in the baseline point yields an equivalent attribution.*

Before the formalization and proof of this proposition, we provide a simple illustration. Consider the two-dimensional case shown in Fig. 1 (right) with anomalous observation a. Suppose that three possible baseline points were selected: point b from one mode, and points c_1, c_2 are from a different mode, and $F(a) = 0$, $F(b) = F(c_1) = F(c_2) = 1$.

The attribution of a with respect to b will be the vector $[0, 1]$, but the attribution of a with respect to c_1 will be $[1, 0]$. The attribution of a with respect to c_2 will be $[1, 0]$, i.e., nearly equal to the attribution with respect to c_1. While the attributions on b and c_1, c_2, are different, as long as all the points are equidistant from a and have the same classification score, they are valid attributions.

Theorem 1. *The reference points c_1 and c_2 will generate equivalent attributions if the following conditions are met: the points (a) are close $c_1 \approx c_2$, (b) are equidistant from anomaly a, and (c) have the same classification score $F(c_1) = F(c_2)$. That is, under those conditions, the difference of attributions on each dimension will tend to zero as c_1 and c_2 become infinitesimally close,*

$$\lim_{\|c_1 - c_2\| \to 0} (IG_d(a, c_1) - IG_d(a, c_2)) = 0.$$

Proof is provided in the extended manuscript [33] Appendix C.1.

Theorem 1 implies that a sparse sampling of reference points in the neighborhood $x \in X : F(x) \approx 1$ as baseline points can be used to represent normal modes.

4.4   Choose a Path from the Anomaly to the Baseline

According to [35], any path connecting the anomaly x and its baseline point x' is valid. However, to be consistent with findings from cognitive tests performed on human cohorts, described in Sect. 3.2, it is most appropriate to prefer the L1 path (City Block) or L2 path (Euclidean). According to the cognitive tests, human subjects preferred the L2 path when significant correlation is manifested in two or more dimensions.

Proposition 4. Both L1 and L2 paths connecting the anomaly and baseline are acceptable for interpretability based on empirical results from the cognitive sciences.

4.5   Apply an Explanation Function with the Baseline

As defined in Sect. 2, an explanation function accepts an AD model F, an observation x, and the model's score of the observation $F(x)$, and returns a variable attribution on the input data, $B(x)$. In this section, we propose multiple desirable properties of an explanation function, and then evaluate various methods for suitability. First, an explanation function must be *contrastive*, as defined by [18]. In this work, we extend the definition with an approach that provides constructive explanations. In AD, observation x is anomalous, so the suitable explanation would be based on a contrastive normal point x', selected as discussed in Sect. 4.3 from the exemplar set. Second, an explanation function must be *conservative* (as in Definition 2, [20]) or, equivalently, *complete* (as in the Completeness Axiom of [35] or Efficiency Axiom in [34], where the sum of attributions equals the difference in classification scores between the baseline and observed points: $\sum_{d \in D} B_d(x, x') = F(x') - F(x)$. Third, an explanation function for local explanations must satisfy *sensitivity*, first in the sense of [35], which states that any dimension that changes the prediction when its value is altered should also have a non-zero attribution. Further, any dimension that induces no change to the prediction when perturbed (i.e., dummy variable) should have a zero attribution. Fourth, the explanation should satisfy *proportionality*[1]; this is elaborated next.

Definition 2. *Given a path P from x to x', dimension u has a stronger influence on F than dimension v over P if the average (or overall) rate of change of F along dimension u is greater than the average (or overall) rate of change of F along dimension v.*

If P is the path defined by $z = x + \alpha(x' - x)$ for $\alpha \in [0, 1]$, between x and x', then the overall rate of change of F along dimension u over path P is $IG_u(x, x') \equiv \int_0^1 (x' - x)_u \nabla_u F(x + \alpha(x' - x)) \, d\alpha$. This is justified by taking $G(\alpha) = F(z) = F(x + \alpha(x' - x))$, and observing that

$$G'(\alpha) = \sum_u \nabla_u F \times \frac{\partial z_u}{\partial \alpha} = \sum_u \nabla_u F \times (x' - x)_u$$

that is, the sum of dimensional rates of change of F over path P. Therefore, the instantaneous rate of change of F at a point $z = x + \alpha(x' - x)$ along dimension u is $\nabla_u F \times (x' - x)_u$, and thus the overall rate of change along dimension u is the *integral* of the instantaneous rate of change of F over the path.

Therefore, Definition 2 becomes: Given a path $P = (x + \alpha(x' - x))$ from x to x', dimension u has a *stronger influence* on F than dimension v over P if

$$\int_0^1 (x' - x)_u \nabla_u F(x + \alpha(x' - x)) \, d\alpha$$
$$> \int_0^1 (x' - x)_v \nabla_v F(x + \alpha(x' - x)) \, d\alpha \tag{1}$$

that is, dimension u has a *stronger influence* on F than dimension v over P if $IG_u(x, x') > IG_v(x, x')$.

Definition 3. *An explanation function $B(x, x')$ is proportional on observation $x \in \mathbb{R}^D$ if for any dimension u that has a stronger influence on F than another dimension v over a path P from a point x to a point x', then $B_u(x, x') > B_v(x, x')$.*

In other words, dimensions that have a stronger influence on the anomaly prediction F should have greater attributions than dimensions that have a weaker influence. Proportionality enables ranking and thresholding of dimensions.

We evaluated LIME [23], SHAP [16], Layerwise Relevance Propagation (LRP) [4,5], Deep Taylor Decomposition (DTD) [11], One Class Deep Taylor Decomposition (OC-DTD) [11] and Integrated Gradients (IG) against contrastive, conservative, sensitivity and proportionality desiderata. Details and jstification are provided in extended manuscript [33] Appendix D. LIME, LRP, and

---

[1] Proportionality in this work is different from Proportionality defined in [35], since Proportionality in the latter refers to a condition under which the dimensional components of the distance between a baseline point and an observation is proportional to the attribution.

most implementations of SHAP are not contrastive [17]. DTD, OC-DTD make approximate the gradient as a linaerly, and are not guaranteed to satisfy proportionality. IG does not specify a fully defined method for choosing a baseline point. Instead, its authors provided general guidelines for selecting a good baseline, where a good baseline point should have nearly opposite scores as the point of interest (i.e., if $F(x) \approx 1$, then $F(x') \approx 0$, or vice-versa). Therefore, IG provides a contrastive explanation against the set of all baseline points that have significantly contrastive scores. Because it applies a path integral between x and x', IG accurately approximates the gradient, is conservative, and meets sensitivity criteria [35]. In addition, by computing the path gradient for each dimension, IG also satisfies proportionality (as shown in Theorem 2). Therefore, because it satisfies each desideratum, we propose IG is a suitable explanation function for AD.

Theorem 2. *Integrated Gradients satisfies proportionality, as defined in Definition 3.*

Proof is provided in extended manuscript [33] Appendix C.2.

5   Experiments

In this section, we fix the anomaly detector F and compare the proposes approach, based on IG against two alternative explainability methods, SHAP and LIME.

Attribution Labels: To our knowledge, there are no standard datasets curated for evaluating blame attributions for AD; therefore, we adopt the following labeling method. In addition to the familiar true binary anomaly class label, $A(x) = [anomalous, not\,anomalous]$, we append a vector $\beta(x)$ of dimension D, where $\beta_d(x) = 1/n_A$ if x_d is one of n_A relevant, explanatory dimensions of the defect selected by expert technicians for fault diagnosis, and 0 otherwise; with $\sum_{d \in D} \beta_d(x) = 1$.[2] To score the attribution for method i, $B^{(i)}(x)$, against the label $\beta(x)$ for anomalous observation x, we compute the **Blame Attribution Error** as the mean absolute difference: $\epsilon^{(i)}(x) = \sum_{d \in D} \left| B_d^{(i)}(x) - \beta_d(x) \right| / |D|$ for every point x where $A(x) = anomalous$.

VAV Dataset (406 anomalous observations): A Variable Air Volume (VAV) device provides ventilation and heating using airflow dampers and a hot water heat exchanger, and are commonly installed in commercial office spaces. There are 12 dimensions reported by the VAV, including temperature and airflow setpoints and measurements. Due to an intermittent failure of its the temperature sensor, during normal business hours the VAV's airflow sensor reports values significantly higher than the airflow setpoint. The hot water valve is consistently

---

[2] We apply equal weighting to each relevant dimension in $\beta(x)$, because our technician labelers have found it impractical to assign relative importance weights/preference to the relevant dimensions.

open, reporting around 100%, and the zone air temperature is below its setpoint. We trained one anomaly detector (AUC = 0.99) based on [32] using unlabeled historic observations from a cohort of 203 similar VAVs.

Aircraft Dataset (448 anomalous observations): This dataset contains 16 real-valued dimensions extracted from a small general aviation aircraft flight data recorder, on which the fuel pump partially failed in flight. During normal operations, the fuel pressure ranges between 50 and 80 psi, with only short bursts above 80. During the failure, the fuel pressure achieved pressures between 80 and 120 psi. We trained an anomaly detector based on [32] (AUC = 0.94) with unlabeled earlier flight recordings of the same aircraft.

Table 2. Mean and Standard Deviations of attribution error values % for labeled anomaly explanation datasets. Highlighted values are the top-scoring detectors based on a 5% significance threshold (Mann-Whitney U test).

	IG	SHAP	LIME
Aircraft	**6.0 ± 1.9**	7.4 ± 2.8	12.8 ± 0.2
VAV	**7.5 ± 2.8**	13.1 ± 1.0	13.8 ± 0.1

We compare the proposed method using IG, with two model-agnostic explanation functions, SHAP and LIME and compute the average and standard deviation of the explanation errors, shown in Table 2. In both datasets, IG yielded significantly lower attribution errors than SHAP and LIME. LIME appears to be vulnerable to larger because it of the local approximation around x, and in many cases the gradient in the neighborhood of the anomaly may be very small and noisy resulting in unstable attributions. We hypothesize that IG achieves the lowest anomaly attribution error because it traverses the entire gradient from the anomaly to the nearest baseline point avoiding local approximations.

6   Discussion and Future Work

This paper considers how XAI can be applied to AD to provide insightful, contrastive explanations that enable the expert to understand an anomalous observation. We proposed that explainability is an *algorithmic* transformation from prediction to the explanation, and interpretability is a *cognitive* transformation from explanation to the diagnostic condition. We have considered numerous explainability methods, and reviewed what aspects are important for a good explanation, inspired by the social sciences. We propose a novel method of choosing baseline points that enable contrastive explanations. We propose a method for evaluating blame attributions and show that IG has the lowest attribution error in two real-world datasets.

While this paper proposes an approach to scoring blame attributions, additional work is needed in evaluating the accuracy and importance of a contrastive normal baseline. It will be beneficial to demonstrate the general utility of this relationship across different types of systems and AD models, such as [33] Appendix E. With advances in wearable technology, is it possible for an explainable anomaly detector to predict illnesses by simply wearing a few commodity sensors?

Acknowledgements. The authors would like to thank Klaus-Robert Müller and Ankur Taly for instructive and practical advice and for their detailed technical reviews, and the anonymous reviewers for identifying gaps and suggesting improvements.

References

1. Adadi, A., Berrada, M.: Peeking inside the black-box: a survey on explainable artificial intelligence (xai). IEEE Access **6**, 52138–52160 (2018)
2. Aggarwal, C.C.: Outlier Analysis, 2nd edn. Springer, Cham (2016). https://doi.org/10.1007/978-3-319-47578-3
3. Antwarg, L., Miller, R.M., Shapira, B., Rokach, L.: Explaining anomalies detected by autoencoders using shap (2020)
4. Bach, S., Binder, A., Montavon, G., Klauschen, F., Müller, K.R., Samek, W.: On pixel-wise explanations for non-linear classifier decisions by layer-wise relevance propagation. PLoS ONE **10**(7), 1–46 (2015)
5. Binder, A., Montavon, G., Lapuschkin, S., Müller, K.-R., Samek, W.: Layer-wise relevance propagation for neural networks with local renormalization layers. In: Villa, A.E.P., Masulli, P., Pons Rivero, A.J. (eds.) ICANN 2016. LNCS, vol. 9887, pp. 63–71. Springer, Cham (2016). https://doi.org/10.1007/978-3-319-44781-0_8
6. Broniatowski, D.: Psychological foundations of explainability and interpretability in artificial intelligence (2021–04-12 04:04:00 2021)
7. Carletti, M., Terzi, M., Susto, G.A.: Interpretable anomaly detection with diffi: Depth-based isolation forest feature importance (2020)
8. Chandola, V., Banerjee, A., Kumar, V.: Anomaly detection: a survey. ACM Comput. Surv. **41**(3) (2009). https://doi.org/10.1145/1541880.1541882
9. Doshi-Velez, F., Kim, B.: Towards a rigorous science of interpretable machine learning (2017)
10. Gilpin, L.H., Bau, D., Yuan, B.Z., Bajwa, A., Specter, M.A., Kagal, L.: Explaining explanations: an approach to evaluating interpretability of machine learning. CoRR abs/1806.00069 (2018)
11. Kauffmann, J., Müller, K.R., Montavon, G.: Towards explaining anomalies: a deep Taylor decomposition of one-class models. Pattern Recogn. **101**, 107198 (2020). https://doi.org/10.1016/j.patcog.2020.107198
12. Lieto, A., Radicioni, D.P., Rho, V.: Dual PECCS: a cognitive system for conceptual representation and categorization. J. Exp. Theor. Artif. Intell. **29**(2), 433–452 (2017)
13. Lipton, Z.C.: The mythos of model interpretability. CoRR abs/1606.03490 (2016). http://arxiv.org/abs/1606.03490
14. Liu, F.T., Ting, K.M., Zhou, Z.: Isolation forest. In: 2008 Eighth IEEE International Conference on Data Mining, pp. 413–422, December 2008. https://doi.org/10.1109/ICDM.2008.17
15. Liznerski, P., Ruff, L., Vandermeulen, R.A., Franks, B.J., Kloft, M., Müller, K.R.: Explainable deep one-class classification (2021)

16. Lundberg, S.M., Lee, S.I.: A unified approach to interpreting model predictions. In: Guyon, I., et al. (eds.) Advances in Neural Information Processing Systems, vol. 30. Curran Associates, Inc. (2017)

17. Merrick, L., Taly, A.: The explanation game: explaining machine learning models with cooperative game theory. CoRR abs/1909.08128 (2019)

18. Miller, T.: Explanation in artificial intelligence: insights from the social sciences. Artificial Intelligence 267 (2017). https://doi.org/10.1016/j.artint.2018.07.007

19. Miller, T.: Contrastive explanation: a structural-model approach. Knowl. Eng. Rev. **36**, e14 (2021). https://doi.org/10.1017/S0269888921000102

20. Montavon, G., Bach, S., Binder, A., Samek, W., Müller, K.: Explaining nonlinear classification decisions with deep Taylor decomposition. CoRR abs/1512.02479 (2015)

21. Navarro, D.J.: On the interaction between exemplar-based concepts and a response scaling process. Math. Psychol. **51**(2), 85–98 (2007)

22. Nosofsky, R.M.: Attention, similarity, and the identification-categorization relationship. J. Exp. Psychol. **115**(1), 39–61 (1986)

23. Ribeiro, M.T., Singh, S., Guestrin, C.: "why should I trust you?": explaining the predictions of any classifier. CoRR abs/1602.04938 (2016)

24. Ricoeur, P.: Interpretation Theory: Discourse and the Surplus of Meaning. Texas Christian University Press, Fort Worth (1972)

25. Ruff, L., et al.: A unifying review of deep and shallow anomaly detection. CoRR abs/2009.11732 (2020)

26. Samek, W., Wiegand, T., Müller, K.: Explainable artificial intelligence: understanding, visualizing and interpreting deep learning models. CoRR abs/1708.08296 (2017)

27. Schölkopf, B., Platt, J.C., Shawe-Taylor, J.C., Smola, A.J., Williamson, R.C.: Estimating the support of a high-dimensional distribution. Neural Comput. **13**(7), 1443–1471 (2001). https://doi.org/10.1162/089976601750264965

28. Shannon, C.E., Weaver, W.: The Mathematical Theory of Communication. University of Illinois Press, Urbana (1949)

29. Shepard, R.N.: Stimulus and response generalization: a stochastic model relating generalization to distance in psychological space. PsychometrikaR **22**, 325–345 (1957)

30. Shepard, R.N.: Toward a universal law of generalization for psychological science. Science **237**(4820), 1317–1323 (1987). https://doi.org/10.1126/science.3629243

31. Shrikumar, A., Greenside, P., Kundaje, A.: Learning important features through propagating activation differences (2017)

32. Sipple, J.: Interpretable, multidimensional, multimodal anomaly detection with negative sampling for detection of device failure. In: III, H.D., Singh, A. (eds.) Proceedings of the 37th International Conference on Machine Learning. Proceedings of Machine Learning Research, vol. 119, pp. 9016–9025. PMLR, 13–18 July 2020

33. Sipple, J., Youssef, A.: A general-purpose method for applying Explainable AI for anomaly detection (extended manuscript) (2022). https://arxiv.org/abs/2207.11564

34. Sundararajan, M., Najmi, A.: The many Shapley values for model explanation. In: III, H.D., Singh, A. (eds.) Proceedings of the 37th International Conference on Machine Learning. Proceedings of Machine Learning Research, vol. 119, pp. 9269–9278. PMLR, 13–18 July 2020

35. Sundararajan, M., Taly, A., Yan, Q.: Axiomatic attribution for deep networks. CoRR abs/1703.01365 (2017)
36. Tax, D.M.J., Duin, R.P.W.: Uniform object generation for optimizing one-class classifiers. J. Mach. Learn. Res. **2**, 155–173 (2002)

More Sanity Checks for Saliency Maps

Lars Holmberg[(✉)] [ID], Carl Johan Helgstrand, and Niklas Hultin

Malmö University, Malmö, Sweden
lars.holmberg@mau.se

Abstract. Concepts are powerful human mental representations used to explain, reason and understand. In this work, we use theories on concepts as an analytical lens to compare internal knowledge representations in neural networks to human concepts. In two image classification studies we find an unclear alignment between these, but more pronounced, we find the need to further develop explanation methods that incorporate concept ontologies.

Keywords: Explainable AI · Understandable AI · Human-centric AI

1 Introduction

The focus of this work is on Explainable Artificial Intelligence (XAI) methods producing explanations valid for end-users. In our approach, we do not primarily regard XAI as a tool for Machine Learning (ML) experts, instead, we concentrate on how XAI methods can complement average case metrics, such as class probability, and increase end-users understanding of reasons for a promoted decision. This is important since ML-based systems are increasingly used in settings impacting individuals and society as a whole [4,5]. Our central research goal is to explore and evaluate if neural networks in combination with XAI methods can produce explanations aligned with human understanding.

We answer this via image classification experiments in which our human participants compare the usefulness of humanly created explanations to explanations created by XAI methods. The explanations compared are in the form of saliency maps wherein areas important for a proposed classification are highlighted. In a second experiment, we select pretrained ML-models and contrast their consistency to humans[1] By using concepts as an analytic lens we show that there are important challenges ahead for XAI if we expect the methods to produce evidence that can underpin explanations useful for end-users.

The rest of the article is organised as follows: First, we present a more detailed definition and description of concepts and their role in human reasoning. Then,

---

This work was partially financed by the Knowledge Foundation through the Internet of Things and People research profile.

[1] Additional material and code to reproduce our experiments can be found at https://github.com/k3larra/XAI-F.

M. Ceci et al. (Eds.): ISMIS 2022, LNAI 13515, pp. 175–184, 2022.
https://doi.org/10.1007/978-3-031-16564-1_17

we present our study and our results. In the discussion sections, we use our theoretical approach to deepen and add perspective to the results. Finally, we offer concluding remarks and propose further research avenues.

2   Related Work

Neural networks are black boxes from an explanatory perspective, not only as a consequence of their architecture, but also, as a result of the incalculable amount of data used for training. The situation in some sense bears similarities to humans, an obstruction we partly overcome by our unique ability to understand, reason and explain. Central to this ability are *concepts*, mental representations that serve as fundamental building blocks for thoughts and beliefs that we additionally use to generalise across domains [11,12].

Human beliefs, related to concepts and their properties, can be both false and incomplete and contain both causal and descriptive factors [7]. Our mental representation for a concept build on central tendencies of a concept category [14]. A label presented by an ML-system only points to a concept by presenting a string of characters. When we as humans explain differences and similarities between objects we use both subordinate and superordinate concepts related to the concept we aim to explain. For example, the difference and similarities between a horse and donkey can be explained by using concrete subordinate concepts like ears and abstract superordinate concepts like a definition of species. Concepts are part of an ontology and they come with knowledge effects that make it possible for humans to reason [12] and, for example, explain why polar bears do not pose a threat to penguins.

Neural networks create Internal Knowledge Representations (IKR) during training as a consequence of the tension between labels and training data. These multidimensional IKRs tend to be more complex towards the final layers of the network and more generic in the earlier layers [9]. IKRs bear similarities with concepts and revealing IKRs locally, in relation to a singular classification, or globally, in relation to the neural network's overall knowledge representations, is one focal point in XAI (Explainable Artificial Intelligence). XAI is a growing research field with no agreed taxonomy and several XAI methods exist that aim to present reasons for a promoted decision [1,3]. In this work, we focus on two of the most well-cited methods, Grad-CAM and Occlusion [10]. We will in the work presented here focus on two aspects of these methods: if they lift out IKRs aligned with concepts, and if learned IKRs are consistent between comparable pretrained neural network models.

Examples presented to show the usefulness of a specific XAI method or approach are often selected for best-case performance and confirmation bias follows as an obvious risk [8]. In this paper, we hypothesise that: if theories around concepts are used actively we lower the risk of confirmation bias originating from reflexively confounding a label with a concept. Our approach differs from, to us, known work in that we focus on experiments with end-users and existing well-tested pretrained neural networks [2].

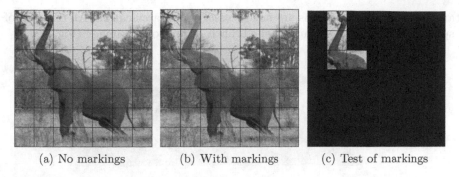

| (a) No markings | (b) With markings | (c) Test of markings |

Fig. 1. User interface for human saliency map creation. The elephant was used as an instructional image before the map creation started with other images. (Color figure online)

3   Comparing Human's Saliency Maps to XAI's

In this study, we first assigned, to a small group of designers (n = 5), the task of via a web interface create saliency maps. The mean outcome of the design groups' saliency maps were then calculated, visualised and used as part of the study (row C in Fig. 2). Following this, we let a larger group of users (n = 25) compare three types of saliency maps: averaged designer maps and maps created by two XAI methods, Grad-CAM [15] and Occlusion [16]. We selected these XAI methods since saliency maps indistinguishable from them with relative ease can be created by humans through an interface that allows them to give importance scores to square-shaped areas in an image (See Fig. 1). The highest (greenest) level indicates that the square is, on its own, sufficient to make a classification whereas a lower level indicates that the square needs to be combined with other squares to be sufficient. An unmarked square indicates that the square is not necessary for the classification.

For the study, we selected images classified by a ResNet101 model trained on ImageNet1k [6] with over 99% class probability. The images selected picture clearly identifiable animals, two mammals (Arabian camel and gorilla) and one bird (toucan) (See top row in Table 4). We here hypothesise that humans, even if they don't know the species by name, can, due to knowledge effects, identify subordinate concepts that are sufficient for classification. Images with high top-class probability were selected to increase the likelihood that IKRs created during training of the network, and exposed by the XAI-methods, potentially are aligned with subordinate concepts.

Many saliency maps produced by XAI methods mark areas as contributing positively or negatively to the classification by using, for example, red and green colours. To make our results clearer we omitted negative attribution and solely focused on positive and, consequently, explanations based on association and not on contrast. Spurious correlations, sometimes picked up by ML systems, are in this study implicitly covered by the non-marked areas since humans understand that these correlations are not necessary for classification. An example of a spurious correlation is marking the branch in the toucan image (Fig. 2).

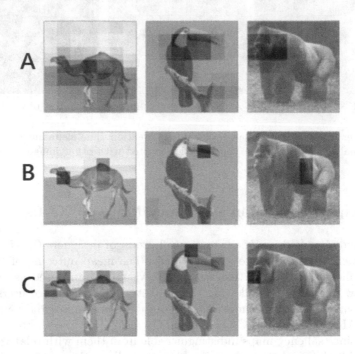

Fig. 2. Saliency maps. Row A is created by the XAI method Grad-CAM, row B by Occlusion and row C is based on averaged human annotations. (Color figure online)

The design team experimented to find a balance between the grid size, colouring, marking levels and user interface design. With a 7×7 grid, green opaque markings and a possibility to blackout non-marked squares we believe that we stroke a balance between usability and utility for the forthcoming study (Fig. 1).

Study Setup. For the study we selected 25 participants of various age, gender and educational background (11 female and predominantly younger persons). To make the results more generally valid we selected a relatively mundane domain (animals) in which we didn't expect significant differences between demographic groups (this was verified for gender and education). The central phases in the study were as follows: first the participants tried out the marking tool on an image picturing an elephant (See Fig. 1), then, they created saliency maps for the Arabian camel, the gorilla and the toucan. The participants were then presented the saliency maps in Fig. 2 with the task of ordering them based on how well they perceived that the markings matched the identification of the animal. To measure how similar saliency maps are Eq. 1a is used and to measure how disparate maps are Eq. 1b and 1c is used.

Table 1. Study participants (n = 25) preferred selection of XAI method.

Method	First selection	Second selection	Third selection
Human average	**80%**	12%	8%
Grad-CAM	16%	**64%**	20%
Occlusion	4%	24%	**72%**

Similarity Score (Eq. 1a). A normalised and averaged value over the participant's (n = 25) markings gives $\bar{X}_i$. The positive $\bar{X}_i$ is subtracted from the corresponding positive normalised saliency map marking (b_i). The absolute average is then calculated over all squares (j = 49) and used as a similarity score (S_{score}). Results are presented in Table 2 where a similarity score is calculated for the set {Grad-CAM (Fig. 2 row A), Occlusion (row B), study designers averaged saliency map (row C)}.

$$a) \; S_{score} = \sum_{i=1}^{j} \frac{|\bar{X}_i - b_i|}{j} \quad b) \; D_{score} = \sum_{i=1}^{j} \frac{\sigma_i}{j} \quad c) \; D_{scorex} = \sum_{i=1}^{x} \frac{\operatorname*{argmax}_{x}(\sigma_i)}{x} \quad (1)$$

Dispersion Score (Eq. 1b). To get dispersion (D_{score}) for a set of saliency maps we compute mean of the standard deviation (σ) for all squares (j = 49) at the same position in the set {study designers (n = 5), study participants (n = 25)}. We thereby get a measurement of the saliency map agreement between the participants in a selected set. We complement this by calculating the average of standard deviation for the 10% ($D_{score10}$) of the squares with the highest σ (Eq. 1c). This gives a metric that to some degree is independent of the number of unmarked squares in the image.

Result. In Table 1 the participants ranking of saliency maps is presented and the outcome was that 80% preferred the design teams map. We then created, by using the participant's 25 × 3 maps, averaged saliency maps for the camel, gorilla and toucan. By using Eq. 1a we calculated how similar the participants maps are to the Grad-CAM maps, the Occlusion maps the design team's maps. Through the result presented in Table 2 we can see that the participant's maps are most similar to the designer's average, then to Occlusion and that they are least similar to Grad-CAM. This indicates that the subordinate concepts humans find important for classification are more misaligned with IKRs learned by the pretrained ML-model used than with their human counterparts. Additionally, we calculated the dispersion for the groups to get a measurement of how well humans agree on their markings. Dispersion is calculated both for the whole map (Eq. 1b) and for the 10% of the squares the participants disagree most on (Eq. 1c). The result is presented in the first four data columns in Table 3, the result shows a non-negligible disagreement on which squares are most important.

Table 2. Similarity score (see Eq. 1 a), the study participants averaged saliency map compared to the three saliency maps in Fig. 2. (Lower value indicates more similarity).

Animal	Grad-CAM (A)	Occlusion (B)	Designers average (C)
African Camel	0.26	0.16	0.07
Gorilla	0.20	0.16	0.05
Toucan	0.22	0.18	0.11
Av. S_{score}	**0.23**	**0.17**	**0.08**

Discussion. The study was conducted by 25 study participants that were asked to rank saliency maps created by humans and by XAI methods based on how well they perceived the markings matched the identification of the animal. We found that humans prefer the humanly created saliency maps compared to maps created by Grad-CAM and Occlusion. The saliency map created by Grad-CAM is in our study least similar to the humanly created but it still ranked higher than Occlusion. One possible explanation is that Grad-CAM in comparison marks more squares than Occlusion and that, to humans, an omitted marking of a relevant subordinate concept is perceived as more negative than an extra marking.

This study has many subjective ingredients: interface design, selection of images, the number of study participants, selection of ML-model and selection of XAI methods. The instructions to the participants on how to value the squares are also not directly comparable to how the XAI methods measure and how these markings are presented by the visualisation. Despite this, we find the results useful since they point towards that IKRs exposed by Occlusion and Grad-CAM are not aligned to human subordinate concepts. The expectations for a study like this, performed in a mundane domain, using a model trained under optimal conditions on a large amount of training data and tested under ideal conditions suggests that there are still many open challenges for neural networks before they can be used in more, from an explanatory perspective, challenging domains.

4   Dispersion for Saliency Maps

In this part of the study, we turn to models pretrained on ImageNet1k. We compare dispersion for models and humans and investigate in what way ML-model's IKRs relate to human subordinate concepts. For this part of the study, we will only use the XAI method Occlusion [16] since it is a model agnostic method that produces a saliency map indifferent to the neural network's architecture.

Results. We calculate the dispersion between the five ML-model's saliency maps using Eq. 1b and 1c (two ResNet models (101 and 152), GoogLeNet, Inception v3 and EfficientNet). We then compare the models dispersion scores with scores for the human saliency maps (Table 3). The dispersion is considerable within all the sets and there is no real agreement on which squares are most important. A dispersion of 0.5 is maximum, and a $D_{score10}$ of 0.41, as for 10% of the 49 squares for the ML-models, indicates that there are at least 5 squares with a strong disagreement. For the ML-models D_{score}s are higher throughout the study, except D_{score} for the camel. To get a deeper understanding of what constitutes the differences we need to inspect and compare the visual content of the squares for IKRs and subordinate concepts.

Table 3. Dispersion score (see Eq. 1b and c).

Animal	Designer n = 5		Participant n = 25		ML n = 5	
	D_{score}	$D_{score10}$	D_{score}	$D_{score10}$	D_{score}	$D_{score10}$
African Camel	0.08	0.31	0.14	0.34	0.05	0.39
Gorilla	0.06	0.30	0.21	0.33	0.16	0.42
Toucan	0.08	0.33	0.12	0.35	0.20	0.43
Av. D_{score}	**0.07**	**0.31**	**0.16**	**0.34**	**0.14**	**0.41**

In Table 4, on the first row of saliency maps, the Occlusion averaged saliency map (Eq. 1a) for the ML-models (n = 5) is visualised. The last row visualise the corresponding human created saliency map (n = 5). Between these rows, marked with red colour intensity, the average absolute difference between the human's and the ML-model's saliency maps is visualised.

To exemplify, the camel's hump (3,5) is important both for humans and ML-models and is therefor not marked with a red square. For the camel ((3,1)(3,2)(4,1)), the gorilla (3,1) the toucan (1,4) the mentioned squares stands out as more important for the classification to humans than they do to the ML-models. Additionally, in some of the images such as the toucan image, there are many markings that are spurious and hard to understand in what way they contribute to the classification.

Discussion. In this study, we compared averaged saliency maps from five humans to averaged maps from five pretrained ML-models (Trends presented are the same if we instead use the larger group in our study). The main takeaway here is that humans focus on subordinate concepts that are known to generalise between domains, like head-related subordinate concepts. For some distinct subordinate concepts, like the camel's hump and the toucan's bill, the IKRs are well aligned to human concepts. We can also from the images identify spurious correlations. Additionally, we can expect that a human can explain why a specific square is selected, for example, that the bird in the experiment has an extraordinarily large bill compared to other birds. Since IKRs are not organised using any known ontology, we cannot expect any comparable explanation to be produced from the model. Additionally, we can only speculate around a saliency map attention to, for example, the hump of the camel, if it indicates that the model has learned anything comparable to the concept hump. Since it is, in a

Table 4. XAI explanations. The class probability in per cent is the average class probability for the five used ML-models. The row Occlusion average visualises positive squares for the classification from five pretrained models. The row Human average visualises squares marked as positive by the human design team. The row |Human-ML| visualises markings in which the human averaged saliency map differs from the ML-models averaged saliency maps.

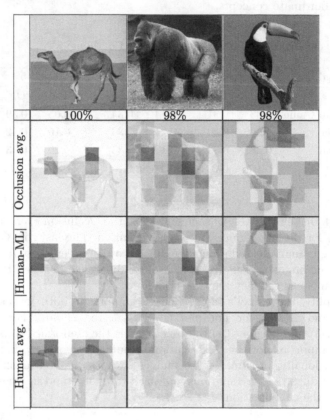

human context, questioned if concepts can be learned inductively [13] it is reasonable to believe that humanly agreed concept ontologies has to be part of the system as knowledge priors.

5   Conclusion

In this work, we analyse pretrained neural networks through the lens of human understandable concepts. We judge this approach as a valid perspective based on the fundamental importance concepts have for human reasoning, understanding, and our ability to explain. The work presented here concludes that even if neural networks can, in an impressive fashion, connect labels with raw input data, connecting internal knowledge representations in the neural network to human understandable subordinate concepts comprise many open challenges. Especially salient is the manner many concepts carry a wealth of knowledge effects. This, to us, indicates that a more basic approach towards the goal of XAI via concepts can be a path forward. We will, therefore, in the next phase of our research, focus on concepts with lesser knowledge effects, for example, basic shapes, lines, triangles, squares, etc., since these concepts are useful across many domains. This in turn can gradually direct the training of the network towards, internally, learning more complex subordinate concepts useful for explanations.

References

1. Adadi, A., Berrada, M.: Peeking inside the black-box: a survey on explainable artificial intelligence (XAI). IEEE Access **6**, 52138–52160 (2018). https://doi.org/10.1109/ACCESS.2018.2870052
2. Adebayo, J., Gilmer, J., Muelly, M., Goodfellow, I., Hardt, M., Kim, B.: Sanity checks for saliency maps. In: Advances in Neural Information Processing Systems, vol. 2018-Decem, pp. 9505–9515 (2018). https://goo.gl/hBmhDt
3. Barredo Arrieta, A., et al.: Explainable Artificial Intelligence (XAI): concepts, taxonomies, opportunities and challenges toward responsible AI. Inf. Fusion **58**, 82–115 (2020). https://doi.org/10.1016/j.inffus.2019.12.012
4. Bender, E.M., Gebru, T., McMillan-Major, A., Shmitchell, S.: On the dangers of stochastic parrots: can language models be too big? In: FAccT 2021 - Proceedings of the 2021 ACM Conference on Fairness, Accountability, and Transparency, pp. 610–623 (2021). https://doi.org/10.1145/3442188.3445922
5. Couldry, N., Mejias, U.A.: The Costs of Connection: How Data Are Colonizing Human Life and Appropriating It for Capitalism, 1st edn. Stanford University Press (2019). https://doi.org/10.1515/9781503609754
6. Deng, J., Dong, W., Socher, R., Li, L.J., Kai Li, Li Fei-Fei: ImageNet: a large-scale hierarchical image database. In: IEEE Conference on Computer Vision and Pattern Recognition, pp. 248–255. IEEE, Miami (2009). https://doi.org/10.1109/cvpr.2009.5206848
7. Genone, J., Lombrozo, T.: Concept possession, experimental semantics, and hybrid theories of reference. Philos. Psychol. **25**(5), 717–742 (2012). https://doi.org/10.1080/09515089.2011.627538
8. Ghorbani, A., Wexler, J., Zou, J., Kim, B.: Towards automatic concept-based explanations. In: Advances in Neural Information Processing Systems (2019)
9. Lecun, Y., Bengio, Y., Hinton, G.: Deep learning. Nature **521**(7553), 436–444 (2015). https://doi.org/10.1038/nature14539
10. Linardatos, P., Papastefanopoulos, V., Kotsiantis, S.: Explainable AI: a review of machine learning interpretability methods (2021). https://doi.org/10.3390/e23010018
11. Margolis, E., Laurence, S.: Concepts. In: The Stanford Encyclopedia of Philosophy. Metaphysics Research Lab, Stanford University (2021)
12. Murphy, G.: The Big Book of Concepts. MIT press (2004). https://doi.org/10.7551/mitpress/1602.001.0001
13. Murphy, G.L.: Is there an exemplar theory of concepts? Psychon. Bull. Rev. **23**(4), 1035–1042 (2015). https://doi.org/10.3758/s13423-015-0834-3
14. Rosch, E.H.: Natural categories. Cogn. Psychol. 4(3), 328–350 (1973). https://doi.org/10.1016/0010-0285(73)90017-0

15. Selvaraju, R.R., Cogswell, M., Das, A., Vedantam, R., Parikh, D., Batra, D.: Grad-CAM: visual explanations from deep networks via gradient-based localization. Int. J. Comput. Vision **128**(2), 336–359 (2019). https://doi.org/10.1007/s11263-019-01228-7
16. Zeiler, M.D., Fergus, R.: Visualizing and understanding convolutional networks. In: Fleet, D., Pajdla, T., Schiele, B., Tuytelaars, T. (eds.) ECCV 2014. LNCS, vol. 8689, pp. 818–833. Springer, Cham (2014). https://doi.org/10.1007/978-3-319-10590-1_53

Intelligent Systems

Deep Reinforcement Learning
for Automated Stock Trading: Inclusion
of Short Selling

Eeshaan Asodekar[✉], Arpan Nookala, Sayali Ayre, and Anant V. Nimkar

Sardar Patel Institute of Technology, Mumbai, India
{eeshaan.asodekar,ganesharpan.nookala,sayali.ayre,
anant_nimkar}@spit.ac.in

Abstract. Multiple facets of the financial industry, such as algorithmic trading, have greatly benefited from their unison with cutting-edge machine learning research in recent years. However, despite significant research efforts directed towards leveraging supervised learning methods alone for designing superior algorithmic trading strategies, existing studies continue to confront significant hurdles like striking the optimum balance of risk and return, incorporating real-world complexities, and minimizing max drawdown periods. This research work proposes a modified deep reinforcement learning (DRL) approach to automated stock trading with the inclusion of short selling, a new thresholding framework, and employs turbulence as a safety switch. The DRL agents' performance is evaluated on the U.S. stock market's DJIA index constituents. The modified DRL agents are shown to outperform previous DRL approaches and the DJIA index, in terms of absolute returns, risk-adjusted returns, and lower max drawdowns, while giving insights into the effects of short selling inclusion and proposed thresholding.

Keywords: Machine learning · Deep reinforcement learning ·
Actor-critic framework · Markov Decision Process · Automated stock
trading

1 Introduction

The optimal stock trading strategy problem, and the search for a robust and perennially profitable automated stock trading strategy have intrigued researchers and industry practitioners alike, even in decades prior to modern advances in machine learning and the exponential growth of affordable computing power. The primary aim of the optimal stock trading strategy is the maximization of investment returns, on the basis of projected returns and the underlying risk of a given set of stocks, by optimizing the weighted allocation of capital. A reliable and sustainably profitable equity trading strategy is instrumental to asset management companies and Quant funds.

© The Author(s), under exclusive license to Springer Nature Switzerland AG 2022
M. Ceci et al. (Eds.): ISMIS 2022, LNAI 13515, pp. 187–197, 2022.
https://doi.org/10.1007/978-3-031-16564-1_18

The traditional approach proposed by Markowitz [10] consists of two phases. The expected returns and covariance matrix for a stock are computed. Then, optimal weights are computed by either maximizing the return for a target risk level for the portfolio, or by minimizing the risk level for a target level of return. However, if the investor desires to update the decisions taken at each step and consider factors such as transaction cost, this technique might be challenging to execute. Another disadvantage of this approach is its inability to incorporate technical indicators, which are used to identify stock trading opportunities based on statistical trends. A second approach to stock trading is to construe it as a Markov Decision Process (MDP) [13] and then narrow it down on the optimal strategy using Dynamic Programming. However, the scalability of this paradigm is limited due to the mammoth state spaces needed to encompass the stock market and its complexities using dynamic programming. High-frequency trading (HFT) is another approach that gained a foothold in the last decade. HFT uses sophisticated computer algorithms to trade humongous quantities of stock in fractions of a second, based on analysis of multiple markets and financial indicators. However, because of increased competition, reduced volatility and trading volumes, stricter regulations, and high maintenance costs, there has been a significant decline in the profitability of HFTs. These factors aggravated conditions for HFTs, which were already operating within a zero-sum game.

A recent approach is the application of machine learning (ML) [7] and deep learning (DL) algorithms [16] to predict stock prices and infer market conditions. Although the results are nearly satisfactory and have improved over the years, there are significant deficiencies in this approach of utilizing supervised learning methods alone to forecast stock prices with the intention of making a profit. Firstly, accurately forecasting stock prices does not automatically translate into the system making a profit. Secondly, it is excruciatingly complex to train a supervised learning model to consider all the complications, uncertainties and latency of the financial markets. Thirdly, profitable trading strategies vary significantly according to market situations, and thus the supervised learning system must master not a single, but multiple strategies, therefore there is not a single strategy that the model must optimize for.

This paper proposes a modified deep reinforcement learning (DRL) approach to the optimal stock trading strategy problem. The main contributions of this work are as follows. Firstly, we propose a modified DRL approach with the addition of short selling to the action space of the agents, thereby increasing the accrued alpha. Secondly, in order to maximize the net profit, a thresholding control is implemented, which significantly decreases the possibility of the model acting on weak or stray sell signals. Lastly, we also discuss the use of the turbulence as a safety switch to avert excruciating losses during extreme events like market crashes and recessions.

The rest of the paper is organized as follows. Section 2 presents a literature survey the various approaches to the optimal stock trading strategy problem. Section 3 gives background on the three DRL algorithms used. Section 4 presents the proposed modified DRL approach to automated stock trading. Section 5 delineates the performance metrics and presents the results along with discussion. Section 6 gives concluding remarks and future scope of work.

2   Related Works

It is only in the latter half of this decade that prodigious computing power has become available to the world. As a result, most algorithmic trading models built by mathematicians, physicists, and financial experts to tackle the optimal stock trading strategy problem do not use machine learning. Such traditional trading models use methods based on complex mathematical analysis, and utilize mean reversion and trend following methods.

Most academic research on applying machine learning (ML) [7] methods to algorithmic trading focus on forecasting or prediction of financial securities using deep learning (DL) methodologies [16]. The overarching idea of this approach is that if the price or value of a specific financial security or asset is known in advance, to a reasonable degree of accuracy, then a party may make strategically beneficial and optimum trading decisions to profit from market movements. Various supervised learning approaches [16] with significant modifications since their inception have already been investigated.

Alternatively, fewer research works [1] have explored Reinforcement Learning (RL) approaches, with some researchers proposing significant modifications to the vanilla approaches [4] to tackle the challenge of algorithmic trading and portfolio management. Moody et al. [12] asserted the superiority of reinforcement learning techniques over traditional supervised learning techniques, and have employed direct reinforcement learning and recurrent reinforcement learning models to train intraday trading agents. Researchers have also employed an adaptive RL model [5] as a basis for automated trading systems.

Deep reinforcement learning (DRL) methodologies [2,3], which are a nascent branch of machine learning, are now being explored to tackle the automated stock trading problem in recent times by researchers. Deng et al. [6] introduced contemporary DL into the traditional DRL framework and have made the system technical indicator free. Another research work explored the DRL approach by training a Deep Deterministic Policy Gradient (DDPG) agent to learn a stock trading strategy [18] with promising results. More recently, an Ensemble Strategy [17] combining three actor-critic DRL algorithms was proposed by researchers.

3   Background

This section gives a distilled description of the three DRL algorithms employed to achieve the objective of inter-day profit maximization, and the four technical indicator inputs given to the DRL agents.

3.1   Deep Reinforcement Learning Algorithms

1. Advantage Actor Critic (A2C)

   A2C [11] is a synchronous, deterministic, actor-critic algorithm that employs an advantage function to reduce the variance of the policy gradient, thereby making the model more robust. It waits for each actor to complete their segment of experience before updating, averaging over all actors.
2. Deep Deterministic Policy Gradient (DDPG)

   DDPG [9] learns a Q-function and a policy at the same time. It is specifically adapted for environments with continuous action spaces, thereby making it a felicitous algorithm for stock trading, and also maximizes investment returns. It learns the Q-function using off-policy data and the Bellman equation and then utilizes the Q-function to learn the policy.
3. Proximal Policy Optimization (PPO)

   PPO [14] is an on-policy algorithm, which guarantees that the new updated policy does not differ significantly from the preceding policy. This results in less variance in training at the expense of some bias, but it assures smoother training and ensures that the agent does not incur cataclysmic drops in performance. Thus, PPO is robust, stable, and executes with celerity.

3.2   Technical Indicators

Technical indicators are pattern based signals derived by analyzing historical stock data. Each of these indicators gives unique insights into stock price movements based on statistical trends.

1. Moving Average Convergence Divergence (MACD): A strong indicator of trend-following momentum, it juxtaposes multiple moving averages of the price of a stock, each of a different time frame.
2. Commodity Channel Index (CCI): An indicator of irrational behaviour, representing potential opportunities wherein a stock is being overbought due to unreasonable optimism or being oversold due to undue pessimism.
3. Relative Strength Index (RSI): A momentum indicator which evaluates recent stock price changes to signal a bullish or bearish momentum.
4. Average Directional Index (ADX): A tool used to gauge the strength of stock prices' bullish or bearish trends.

4   Deep Reinforcement Learning Approach to Automated Stock Trading

The following section details the formulation of the deep reinforcement learning paradigm applied to automated stock trading and expounds the proposed modifications to the DRL agents.

4.1 Stock Trading Problem Formulation

To account for the stochasticity and the uncertainties of the real-world financial markets, a Markov Decision Process is used as the framework as follows:

- State $s = [b, c, p]$: an information vector, which encapsulates the current cash balance b, the current holdings c, and the stock price p, for all N stocks under consideration, where $b \in Q$, $c \in Z$, $p \in Z^+$
- Action a: a vector of actions allowed for the agent, for all the N stocks under consideration. The allowed actions on each stock include buying, holding, selling, shorting, and covering a short.
- Reward r : The reward signal is the difference between the value of the portfolio holding on the nth and the $(n+1)^{th}$ day i.e. inter-day profit
- Policy $\pi(s)$: The current trading strategy, at state s, which is the probability distribution of actions available to the agent at state s.
- Q-value $Q_\pi(s, a)$: The expected reward for taking a specific action a, at state s, by following policy π.

At any given time step n, the agent has the following actions at its disposal for all the N stocks under consideration:

- Buying $p[d]$ shares results in $c_{t+1}[d] = c_t[d] + p[d]$, and $d = 1, ..., N$
- Holding, $c_{t+1}[d] = c_t[d]$ shares
- Selling $q[d] \in [1, c[d]]$ shares results in $c_{t+1}[d] = c_t[d] - q[d]$, where $q[d] \in Z^+$
- Shorting $r\prime[d]$ shares results in $c_{t+1}[d] = -(r_t[d] + r\prime[d])$
- Covering $u[d]$ shares, thereby reducing or eliminating the shorting position to $c_{t+1}[d] = -(r_t[d] - u[d])$.

Thus, depending upon the action taken by the agent in the time step t, the portfolio value at $t+1$ would be a spectrum of values, all depending upon the combination of the actions taken by the agent in the preceding time step, t.

4.2 Dataset

The agents are trained on the U.S. equity market's DJIA index's constituents, from January 2000 to December 2016 (in-sample) and the performance is evaluated from January 2017 to June 2021 (out-of-sample). The input data to the DRL agents was the daily end of day data (viz. open, high, low, adjusted close, and volume data) and the four technical indicators (viz. MACD, CCI, RSI, and ADX) as given in Sect. 3.2. The given time-span of the data is specifically chosen such that the agents get the experience of all the stages of the equity market cycle; from the early phase booms, to consolidation and steady growth in the mid-phase, to a crash or a meltdown towards the ending of the cycle.

4.3 Trading Objectives

The primary underlying objective of the agent is inter-day net profit maximization. The reward function is defined such that the reward at time step $t+1$ is the difference between the value of the agent's long and short positions from time step t and $t+1$, thereby maximizing daily positive change in portfolio value.

4.4 Financial Markets' Constraints

The trading costs and commissions are incorporated in the proposed scheme, and it is assumed to be 0.1% for each transaction, to reflect the trading costs incurred in the financial markets. Also, the constituents of the *Dow Jones Industrial Average* (DJIA) are deliberately chosen since they have sufficient market liquidity.

4.5 Shorting Thresholding

To decrease the likelihood of the agents acting on stray or weak selling(and thus, shorting) signals, a thresholding mechanism is put into place. k is a value used by the agents as a decision parameter and ranges from 1 to -1. A positive value of k results in a long position being taken or a short being covered, wherein the size of the position is determined by the magnitude of k. Whereas a negative value of k results in a shorting position being taken or a long position being sold. The thresholding mechanism allows the agent to take a shorting position only when the k value goes beyond a certain cutoff value.

4.6 Turbulence as a Safety Switch

Turbulence [8] is frequently used as an indicator for market volatility and extreme price fluctuations, which occurs during sudden events like wars, bubbles bursting, and financial crises. The agents are hard-coded to liquidate all positions when the turbulence reaches a certain threshold and the effect of which are delineated in the results. Mathematically, turbulence is represented as:

$$d_t = (y_t - \mu) \sum\nolimits^{-1} (y_t - \mu)' \tag{1}$$

where d_t denotes the turbulence for a particular time period t (scalar), y_t denotes a vector of asset returns for period t (1 × n vector), μ denotes the average vector of historical returns (1 × n vector), $\sum$ denotes historical returns' covariance matrix (n × n matrix).

5 Results and Discussion

This section describes the performance metrics of evaluation, and showcases the results obtained and the significant alpha delivered by the three modified DRL agents in comparison to vanilla DRL agents and benchmarks. All models are given access to one million USD of capital at the beginning of their evaluation period. We also present a rationale on the underlying factors contributing to the superior performance of the modified DRL agents proposed.

5.1 Performance Metrics

Performance Metrics used in the evaluation are: firstly, the Cumulative returns, which is the difference between the final value of the portfolio and the initial value during evaluation. Secondly, the annual returns are the geometric average of the returns earned by the agent every year, during evaluation. Thirdly, the Sharpe ratio [15], which gives key insights into risk-adjusted returns, and is used widely in the industry. Lastly, Max Drawdown, which is the biggest loss experienced by a portfolio from its maximum to its minimum value before a new maximum value is reached again.

5.2 Performance Evaluation of Modified DRL Agents

The agents' training is continued in the performance evaluation stage to help acclimatize to ongoing market conditions. It can be observed from Table 1 that the three DRL agents with the proposed modifications outperform the vanilla agents, the ensemble approach [17], and the DJIA index in terms of the cumulative returns, annual returns, Sharpe ratio, and max drawdowns (Fig. 1). The benchmark approaches give cumulative returns of 70%–91%, while vanilla DRL methods

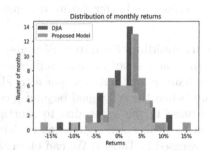

Fig. 1. Monthly returns of Proposed Model (A2C and k=-0.4) vs DJIA

give 72%–81%, and the proposed DRL models give 96%–146%. A similar outperformance can be seen from Table 1 for the remaining three performance metrics. All the three modified DRL agents beat the DJIA and the ensemble method, during the bull market periods as indicated by higher returns, as well as during bear market periods, as indicated by lower max drawdowns.

Table 1. Performance of the 3 modified DRL agents over the evaluation period

Models	Proposed models			Vanilla DRL models			Benchmarks	
	PPO	A2C	DDPG	PPO	A2C	DDPG	DJIA index	Ensemble
Cumulative return (%)	121.97	96.71	146.61	72.84	75.33	81.26	70.32	91.46
Annual return (%)	20.21	16.82	23.04	13.39	13.77	14.64	13.19	16.513
Sharpe ratio	1.03	1.12	0.93	0.66	0.76	0.64	0.77	0.83
Max drawdown (%)	−21.82	−18.32	−33.39	−37.07	−28.27	−40.51	−34.78	−32.684

Short Selling Inclusion. The inclusion of short selling adds two more action dimensions to the actions space of the DRL agent, and the results imply its effective adaptation by automated trading DRL agents. It can be observed from the results showcased in Table 1, all the agents with proposed modifications beat their vanilla counterparts and the ensemble approach, both of which did not have the shorting action, thereby asserting the value of the addition of shorting and covering in the action space of the agent.

Comparison of the Three DRL Agents. The DDPG agent was the best performing in terms of returns, but the A2C agent was the best performing in terms of the Sharpe ratio and the max drawdowns. It was noticed that as soon as the markets started to rebound after the March 2020 Covid-19 recession, A2C and DDPG agent's performance paled in comparison to that of the PPO agent, which was a departure form the previous trend. This leads to the inference that the A2C and DDPG agents give better returns during the middle and the matured phases of an equity market cycle. However, the PPO agent is seen to be performing better during the early to mid-phase of the equity market cycle.

Thresholding Shorting Signals. To decrease the likelihood of the agents acting on a stray or weak selling(and thus, shorting) signal, the thresholding mechanism was put into place to allow the agent to take a shorting position only when the sell signal beyond a certain cutoff value k. There is a significant decrease in the losses due to shorting positions which the model acted upon due to weaker sell signals, as evidenced by the results of this modification as showcased in Table 2. We can observe increasing annual returns and cumulative returns, and decreasing max drawdowns as we decrease the k value from -0.2

Table 2. Variation of returns and max drawdown as shorting signals are thresholded

k	A2C			PPO			DDPG		
	Annual return	Cumulative return	Max drawdown	Annual return	Cumulative return	Max drawdown	Annual return	Cumulative Return	Max drawdown
−0.9	13.46%	76.68%	−31.61%	13.97%	76.69%	−31.61%	20%	123.44%	−36.42%
−0.8	14%	109.16%	−28.61%	18.47%	109.16%	−28.61%	15.19%	85.06%	−37.17%
−0.7	25.03%	55.87%	−45.18%	10.74%	55.87%	−45.18%	15.33%	86.08%	−24.30%
−0.6	13.29%	17.21%	−58.37%	3.72%	17.21%	−58.37%	20.31%	123.64%	−36.94%
−0.5	11.42%	72.84%	−37.07%	13.39%	72.84%	−37.07%	13.06%	70.64%	−36.94%
−0.4	20.89%	54.07%	−41.95%	10.44%	54.07%	−41.95%	10.94%	57.11%	−37.72%
−0.3	17.64%	80.40%	−38.89%	14.52%	80.40%	−38.89%	23.04%	146.61%	−33.39%
−0.2	13.77%	39.79%	−44.30%	8%	39.79%	−44.30%	14.64%	81.26%	−40.11%

Table 3. A2C turbulence thresholding with shorting cutoff $k = 0.4$

Volatility threshold	Annual Return	Cumulative return	Max drawdown	Annual volatility	Sharpe ratio
50	0.06%	0.28%	−0.65%	0.49%	0.13
100	10.10%	52.05%	−11.30%	9.38%	1.07
150	16.82%	96.71%	−18.32%	14.91%	1.12
200	19.89%	120.28%	−31.47%	20.57%	0.99
250	16.42%	93.80%	−23.65%	19.39%	0.88
300	17.50%	101.75%	−28.19%	20.66%	0.88

to -0.6; beyond which the returns start to decrease because of an excessively high threshold for shorting, making the agents long-only in most situations.

Experimentation with Turbulence as a Safety Switch. The agents are hard-coded to liquidate all positions if the turbulence in the equity markets exceeds a certain set limit. If the limit is set to be too large, the agents are allowed to trade even when the markets are in a downturn incurring excruciating losses. At the same time, if the level is too low the agents liquidate all positions when the markets get even a little turbulent in a long bull run. Few selected results are presented in Table 3, which lead to the conclusion that a threshold of 100 to 150 serves the best in acting as a safety switch whilst still allowing the agents to take advantage of market movements, but a threshold in the range of 50 proves to be very restrictive and sensitive. A threshold of 200 and beyond is equivalent to not putting any safety switch for liquidation, as those levels of turbulence are seldom met, and significant losses would have already been incurred in the downturn when these levels are breached. It can be observed from Table 3 that the best results in terms of the returns, max drawdown, volatility, and Sharpe ratio are obtained when the turbulence threshold is set at 150, beyond which although there is an increase in the returns but volatility increases significantly, and a decreased Sharpe ratio is obtained. Thus, it can be inferred that the use of the market volatility as a safety switch proves to be critical especially during panic periods and the early phases of a recession, forcing the agents to cut their losses. This modification proves to be significant in reducing the max downturn during equity market crashes, thereby working as an effective safety switch.

6   Conclusion

In this paper, we have proposed a modified deep reinforcement learning (DRL) approach for the agents to learn a robust and consistently profitable automated stock trading strategy. The modifications being the addition of shorting and covering to the action space of the DRL agents, along with thresholding of the shorting action, and the use of the turbulence as a safety switch for three DRL algorithms, namely Proximal Policy Optimization (PPO), Advantage Actor Critic (A2C), and Deep Deterministic Policy Gradient (DDPG). The inclusion of shorting in the action space of the agents allows them to gain a significant edge over the previously researched DRL approaches, and their non-shorting counterparts as well as the benchmark indices. The thresholding of the sell signals for shorting significantly decreases the losses caused when the agents act on stray or weak selling signals. Also, the use of turbulence as a safety switch greatly cuts the losses that would have accumulated when a excruciating market correction occurs. All the modifications lead to higher returns, Sharpe ratio, and lower max drawdowns compared to past approaches.

In the future, we aim to explore the incorporation of fundamental analysis and macroeconomic parameters, use more esoteric ML techniques such as Autoencoders, and use Natural language Processing to give equity analysts' research reports as input. We would also like to train and evaluate the agents in an intra-day trading setting, and increase the stock-space to larger indexes' constituents, and expose the agents to the data of the financial markets of the major economies of the world. All with the goal of moving towards a closed loop, perpetually profitable automated stock trading system.

References

1. Bertoluzzo, F., Corazza, M.: Testing different reinforcement learning configurations for financial trading: introduction and applications. Procedia Econ. Fin. **3**, 68–77 (2012)
2. Boukas, I., et al.: A deep reinforcement learning framework for continuous intraday market bidding. Mach. Learn. **110**(9), 2335–2387 (2021). https://doi.org/10.1007/s10994-021-06020-8
3. Chen, L., Gao, Q.: Application of deep reinforcement learning on automated stock trading. In: 2019 IEEE 10th International Conference on Software Engineering and Service Science (ICSESS), pp. 29–33 (2019)
4. Dang, Q.-V.: Reinforcement learning in stock trading. In: Le Thi, H.A., Le, H.M., Pham Dinh, T., Nguyen, N.T. (eds.) ICCSAMA 2019. AISC, vol. 1121, pp. 311–322. Springer, Cham (2020). https://doi.org/10.1007/978-3-030-38364-0_28
5. Dempster, M., Leemans, V.: An automated fx trading system using adaptive reinforcement learning. Expert Syst. Appl. **30**(3), 543–552 (2006)
6. Deng, Y., Bao, F., Kong, Y., Ren, Z., Dai, Q.: Deep direct reinforcement learning for financial signal representation and trading. IEEE Trans. Neural Networks Learn. Syst. **28**(3), 653–664 (2017)
7. Jiao, Y., Jakubowicz, J.: Predicting stock movement direction with machine learning: An extensive study on s&p 500 stocks. In: 2017 IEEE International Conference on Big Data (Big Data), pp. 4705–4713 (2017)
8. Kritzman, M., Li, Y.: Skulls, financial turbulence, and risk management. Financ. Anal. J. **66**(5), 30–41 (2010)
9. Lillicrap, T.P., et al.: Continuous control with deep reinforcement learning. In: ICLR, Conference Track Proceedings (2016)
10. Markowitz, H.: Portfolio selection. J. Finan. **7**(1), 77–91 (1952)
11. Mnih, V., et al.: Asynchronous methods for deep reinforcement learning. In: Proceedings of the 33rd International Conference on Machine Learning. Proceedings of Machine Learning Research, vol. 48, pp. 1928–1937. PMLR (2016)
12. Moody, J.E., Saffell, M.: Learning to trade via direct reinforcement. IEEE Trans. Neural Networks **12**(4), 875–89 (2001)
13. Neuneier, R.: Optimal asset allocation using adaptive dynamic programming. In: Touretzky, D., Mozer, M., Hasselmo, M. (eds.) Advances in Neural Information Processing Systems, vol. 8. MIT Press (1995)
14. Schulman, J., Wolski, F., Dhariwal, P., Radford, A., Klimov, O.: Proximal policy optimization algorithms. arXiv:abs/1707.06347 (2017)
15. Sharpe, W.F.: The sharpe ratio. J. Portfolio Manag. **21**(1), 49–58 (1994)
16. Shen, J., Shafiq, M.O.: Short-term stock market price trend prediction using a comprehensive deep learning system. J. Big Data **7**(1), 1–33 (2020). https://doi.org/10.1186/s40537-020-00333-6

17. Yang, H., Liu, X.Y., Zhong, S., Walid, A.: Deep reinforcement learning for auto-mated stock trading: an ensemble strategy. In: Proceedings of the First ACM International Conference on AI in Finance, ICAIF 2020 (2020)
18. Xiong, Z., Liu, X.Y., Zhong, S., Yang, H., Walid, A.: Practical deep reinforcement learning approach for stock trading. In: NeurIPS Workshop on Challenges and Opportunitiesfor AI in Financial Services: the Impact of Fairness, Explainability, Accuracy, and Privacy (2018)

Scaling Posterior Distributions over Differently-Curated Datasets: A Bayesian-Neural-Networks Methodology

Alfredo Cuzzocrea[1]($\boxtimes$), Selim Soufargi[1], Alessandro Baldo[2], and Edoardo Fadda[2,3]

[1] iDEA Lab, University of Calabria, Rende, Italy
{alfredo.cuzzocrea,selim.soufargi}@unical.it
[2] ISIRES, Turin, Italy
alessandro.baldo@isires.org
[3] DISMA, Politecnico di Torino, Turin, Italy
edoardo.fadda@polito.it

Abstract. This paper provides an introduction to an innovative methodology for scaling posterior distributions over differently-curated datasets. The proposed methodology is based on Bayesian Neural Networks, improved by effective sampling algorithms. These algorithms finally realize a suitable model setup for improving the scaling effect. Theoretical results are presented and discussed in details, as well as a modern case study focused on stock quotation prediction that confirms the successful application of our proposed methodology to emerging *big data analytics* settings.

Keywords: Scaling posterior distributions · Bayesian Neural Networks · Intelligent big data systems

1 Introduction

Nowadays, *big data analytics* (e.g., [31,36]) is gaining the momentum in emerging research. Indeed, this also because of the tremendous speed-up in a wide family of real-life applications, ranging from smart cities (e.g., [2]) to intelligent transportation systems (e.g., [39]), from social networks (e.g., [5]) to epidemic intelligence (e.g., [18]), and so forth.

Under this so-delineated scientific setting, data mining and, more recently, machine learning and deep learning are playing the role of fundamental mechanisms on top of which a plethora of big data analytics procedures are developed (e.g., [22]).

Bayesian statistics is one of the most relevant branch of the field, and nowadays is gaining more attention due to its crossovers with the deep learning field. It is founded on the Bayes' theorem which, given a prior distribution on models' parameters θ (namely $p(\theta)$), a set of data $\mathcal{D}$, the likelihood $p(\mathcal{D}|\theta)$, and the

M. Ceci et al. (Eds.): ISMIS 2022, LNAI 13515, pp. 198–208, 2022.
https://doi.org/10.1007/978-3-031-16564-1_19

marginal likelihood (or model evidence) $p(\mathcal{D})$ it computes the posterior distribution $p(\theta|\mathcal{D})$ as follows:

$$p(\theta|\mathcal{D}) = \frac{\overbrace{p(\mathcal{D}|\theta)}^{\text{likelihood}}\,\overbrace{p(\theta)}^{\text{prior}}}{\underbrace{p(\mathcal{D})}_{\text{marginal likelihood}}} \tag{1}$$

Analytically computing Eq. (1) is often unfeasible, since a closed-form solution is only possible when prior and likelihood are conjugate, thus leading to known form for both the posterior and the marginal likelihood term. Hence, a typical approximation of the posterior considers $p(\mathcal{D})$ as a normalization constant. As any distribution, it is possible identify the best set of parameters maximizing the overall probability. In the case of the posterior, this point-wise estimator is called *Maximum A Posteriori* (MAP), as defined by:

$$\hat{\theta}^{\text{MAP}} = \underset{\theta}{\text{argmax}}\; p(\theta|\mathcal{D}) \tag{2}$$

In order to simplify the mathematical computation, the logarithm is applied to both terms of Equation (1) and, as stated above, the marginal likelihood is not considered. Thus, with a slight abuse of notation, we can write as follows:

$$\log p(\theta|\mathcal{D}) = \log p(\mathcal{D}|\theta) + \log p(\theta) \tag{3}$$

Recently, a debate on Bayesian treatment of deep learning has been at the center of the attention of some studies [1,38], presenting extensive experimental campaigns about the effects which, adjusting posterior distributions, has on the ability of a model to cope with data uncertainty.

In [38], the concept of "cold" posterior is introduced, representing a scaling of the left hand side of posterior by means of a temperature parameter T, as follows:

$$\log p_{\text{cold}}(\theta|\mathcal{D}) = \frac{1}{T}\log p(\mathcal{D}|\theta) + \frac{1}{T}\log p(\theta) \tag{4}$$

In the paper, the authors show that for deep neural networks the Bayes posterior (at temperature $T = 1$) works poorly but by cooling the posterior using a temperature $T < 1$ it is possible to significantly improve the prediction performance. Their experimental setup employs "cold" posteriors to optimize large deep learning models for image [13] and text classification through *Monte Carlo Markov Chain* (MCMC) sampling frameworks, providing noticeable benefits, and proving to reach performances similar to the non-probabilistic state-of-the-art architectures.

However, [1] argues whether these theories are generalizable to every Bayesian learning task, moving the main critic to the effectiveness and relevance of the results, since based on so-called curated datasets: data collections where uncertainty is negligible. Thereby, presenting an altered version of scaled posteriors, namely the "tempered" posterior as follows:

$$\log p_{\text{tempered}}(\theta|\mathcal{D}) = \frac{1}{\lambda}\log p(\mathcal{D}|\theta) + \log p(\theta), \quad \text{with } \lambda \propto T \qquad (5)$$

It is shown how solely impacting the likelihood term leads to the same expected results.

As a consequence, such position contrasts the thesis that, with a good initialized prior [26,35], any handcrafted modification of the posterior should not be needed. Rather, tempered posteriors allow to have large-magnitude improvements on performances, denoting a weak/marginal role of the prior distribution.

Starting from this recent debate, our study aims at assessing whether these theses are theoretically valid, extending the focus over the dark corner of non-curated datasets, where a degree of noise is present inside data [25].

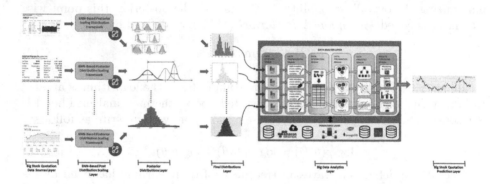

Fig. 1. Case study: big data ecosystem for stock prediction via BNN-based posterior distribution scaling.

2   Case Study: Stock Quotation Prediction

Figure 1 shows a case study where our proposed framework can be successfully used. Here, the Bayesian-Neural-Networks framework for posterior distribution scaling is exploited within the context of a *big data ecosystem oriented to support stock quotation prediction*, by taking advantages from well-understood big data management and analytics technologies. As shown in Fig. 1, the proposed framework can be deployed on top of a Cloud-based computing infrastructure ("*at node premises*"), and the big data prediction phase can fully take advantages from a typical collaborative environment where multiple ensembles can smoothly cooperate to improve the final prediction accuracy.

The reference big data ecosystem is a multi-layer solution where the following layers can be identified (see Fig. 1):

1. *Big Stock Quotation Data Sources Layer*: it is the layer where the input big stock quotation data are located;

2. *BNN-based Post Distribution Scaling Layer*: it is the Cloud-based infrastructure where our proposed framework is implemented;
3. *Posterior Distributions Layer*: it is the layer where the output scaled posterior distributions are produced;
4. *Final Distributions Layer*: it is the layer where the target scaled posterior distributions are represented in form of normalized (data) distributions, ready for further processing;
5. *Big Data Analytics Layer*: it is the layer where the big data analytics architecture is implemented, which is in charge of supporting the final prediction phase;
6. *Big Stock Quotation Prediction Layer*: it is the layer where the final distributions that model the stock quotation prediction are located.

The proposed big data ecosystem exposes several research challenges to be investigated. First, the main research problem, i.e. posterior distribution scaling, is addressed and solved by our proposed Bayesian-Neural-Networks framework. Nevertheless, being an intrinsic big data architecture, additional challenges arise, due to the particular nature of such data sources. Among others, *scalability* of solutions play a major role. In fact, it is well understood that classical machine learning algorithms cannot be used "*as it is*", but specialized solutions are necessary in order to tame the well-known 3V characteristics of big data (i.e., volume, velocity and variety), like highlighted in recent studies (e.g., [15,32]).

3   Related Work

By analyzing active literature, there are several proposals that are relevant to our work. In this Section, we report on some of the most interesting ones.

To reproduce the kernel Hilbert space (RKHS), [21] proposes *a new nonparametric method for robustly embedding conditional and posterior distributions*. Robust embedding is achieved by RKHS eigenvalue decomposition. By keeping only the main eigenvectors, the noise in the data is systematically ignored. The non-parametric conditional and posterior distribution embeddings obtained by our method are applicable to a wide range of Bayesian inference problems. In this paper, authors apply this method to the problem of heterogeneous face detection and zero-shot object detection. Experimental validation shows that the proposed method yields better results than the comparison algorithms.

[30] extends the asymptotic approximation of higher-order posterior distributions to *pseudo-posterior distributions*, such as pseudo-likelihood functions and posterior distributions based on appropriate priors, which have proven particularly useful when full likelihood is not analytically or computationally feasible. In particular, from a theoretical point of view, a Laplace approximation is derived for the scalar parameter of interest even if there is an interference parameter for the pseudo-posterior distribution and its tail region. Starting from these higher-order approximations from a computational point of view, the Higher Order Tail Region (HOTA) algorithm useful for approximating marginal posterior distributions and related quantities is discussed. The main advantage of the HOTA

algorithm compared to the standard Monte Carlo method using Markov chains is that it provides independent samples with little computational overhead. The computing methodology is illustrated by two examples.

[27] moves the focus on *calculating the risk-based misclassification error density distribution for an ensemble*, an important but difficult task. The Bayesian method provides one way to estimate this density distribution. In this paper, a Bayesian modeling approach is used to calculate the posterior distribution of misclassification error densities for both binary and non-binary classifiers. Actual data sets and control samples are used to describe the calculation of the posterior distribution of misclassification errors. These posterior error distributions are very useful for comparing ensembles and providing estimates of risk-based misclassification costs.

[23] explores the problem of *finding efficient methods for calculating or estimating the posterior distribution* as one of the most important model parameters in Bayesian statistics. In fact, Bayesian inference has become very popular in applications after the advent of efficient algorithms such as Monte Carlo Markov chains. The practicality of the posterior distribution depends heavily on the combination of the likelihood function and the prior distribution. In some cases it is possible to obtain a closed formula for the posterior distribution. This paper presents a calibration-like method for explicitly calculating the posterior distributions for three important models: normal, Poisson, and Bernoulli, based on the distortion function theory. The article ends with some applications of the stock market.

[34] studies how to use a Gaussian process emulator to *approximate an observed parameter map or negative log-likelihood in a Bayesian inverse problem*. The authors prove the Helinger distance error range between the true posterior distribution and various approximations based on a Gaussian process emulator. The proposed analysis includes approximations based on the mean of the prediction process as well as approximations based on the full Gaussian process emulator. However, the results show that the Hellinger distance between the actual posterior value and its approximation can be limited by the error moment of the emulator. The numerical results confirm the theoretical conclusions.

Finally, [20] explores how the idea of covariance localization in numerical weather prediction can be used in Markov Chain Monte Carlo (MCMC) sampling of multivariate posterior distributions arising from Bayesian inverse problems. Localizing the inverse problem requires (i) applying the expected "local" structure by ignoring the small off-diagonal elements of the prior precision and covariance matrix; (ii) limit the impact of observations on the surrounding environment. For linear problems, we can specify the condition that the posterior moments of the local problem are close to those of the original problem. Authors describe the physical interpretation of local structural assumptions and discusses the general higher-order concepts of the MCMC functional space and the higher-order concepts of other local problems. Gibbs sampling is a natural choice of the MCMC algorithm for local inverse problems and shows that the convergence rate is dimension-independent for local linear problems. Nonlinear problems can also be effectively solved by localization, and authors present the Metropolis-within-Gibbs localization sampler as a simple example of this idea. Several linear and nonlinear numerical examples demonstrate localization in the context of the MCMC sampler for inverse problems.

4   Innovative Sampling Methods

Posterior distributions can often be summarized by a point-wise estimator: the MAP, defined as the mode of the distribution in Eq. (2).

Scaling a posterior means perturbing the shape of this distribution, either concentrating the mass under the MAP [29], or flattening the distribution, making it uninformative. Choosing a temperature parameter T equal to 1 conceptually resembles the standard Bayesian approach. In this context, such parameter is instead chosen to be between 0 and 1, therefore encompassing the first scenario. As T tends to zero, the distribution becomes more peaked and narrower, thus determining how the model copes with uncertainty in data.

Under a more practical point of view, there is a strong equivalence in performing a more tailored data augmentation, where each data point is over-counted proportionally to $1/T$ [28].

In particular, "cold" posteriors are characterized by maintaining unaltered the position of the MAP, thus cancelling out any secondary modes characterizing the distribution. Due to this, multi-modal distributions, which are very likely to occur in high-dimensional spaces, like those ones characterizing neural networks, are completely smoothed out, and reduced to have only one peak.

The effect of the "tempered" posteriors is instead less straightforward to visualize. Only scaling the likelihood distribution, the over-counting effect is still preserved. However, beyond resulting narrower, the overall posterior is no more centered on the original MAP estimate. While seemingly disruptive, this new distribution proves effective in maximizing the certainty inside data.

Bayesian Neural Networks are a powerful tool, allowing to approximate high-dimensional, analytically intractable posterior distributions. Their robustness, combined with a more faithful quantification of uncertainty, make them widely adopted, especially whether domains are sensitive (i.e. all those fields where a measure of confidence has to be considered). However, they generally lead to minor performances than the non-probabilistic counterpart, due to the several fine-tuning techniques which could not be applied in the Bayesian framework.

Despite this context, adjusting posterior distributions has proved to narrow the gap in performances with state-of-the-art models, leveraging on approximation methods used to converge to the true posterior distribution.

Sampling algorithms play a pivotal role in directing the training procedure. In this paper, the focus is mainly on MCMC algorithms: a family of sophisticated methods [7,9,16,17,33,37] relying on a concept of energy-based models, minimizing the randomicity of the walk toward the target distribution.

In particular, we resort to the SGHMC technique: a scalable version of the *Hamiltonian Monte Carlo* (HMC) [14] which is an iterative method based on an acceptance/rejection procedure inherited from the simpler Metropolis-Hastings (MH) framework (Algorithm 1) that provides high acceptance ratios from distant proposal distributions.

Algorithm 1. Metropolis-Hastings

Posterior to approximate p_θ
Approximating model $f_\theta \propto p_\theta$
Proposal distribution $J(\hat{\theta}|\theta)$
loop
    Sample *the proposal*
$$\hat{\theta}_i \sim J$$
    Sample
$$r \sim \text{Uniform}(0,1)$$
    Accept $f_{\theta_i} = f_{\hat{\theta}_i}$ if:
$$r < \frac{J(\theta_{i-1}|\hat{\theta}_i)}{J(\hat{\theta}_i|\theta_{i-1})} \frac{f_{\hat{\theta}_i}}{f_{\theta_{i-1}}}$$
    else $f_{\theta_i} = f_{\theta_{i-1}}$
until *convergence or max iterations reached*

HMC defines a physical system and associates the log-posterior distribution to the *potential energy* in the following manner:

$$p(\theta|\mathcal{D}) \propto \exp(-U(\theta)) \tag{6}$$

Moreover, the *kinetic energy* of the system is defined by a set of *momentum* variables r and a mass matrix M:

$$K(r) = \frac{1}{2}r^T M^{-1} r \tag{7}$$

The objective driving the algorithm is the so-called *Hamiltonian function* (Eq. (8)): a joint distribution defining the acceptance rule, and which is used to update the model's parameters θ and the momentum variables r by solving the system in Eq. (9).

$$H(\theta, r) = U(\theta) + K(r) \tag{8}$$

$$\begin{cases} \dot{\theta} = M^{-1}r \\ \dot{r} = -\nabla_\theta U(\theta) \end{cases} \tag{9}$$

Importantly, the Hamiltonian dynamics preserve the total energy H from one iteration to the next one, thus leading to a continuous acceptance of new samples, and a smooth convergence to the true posterior. However, in practice, the impossibility to simulate a completely continuous system does not make it happen deterministically.

SGHMC inherits the same structure, allowing a more efficient computation of the gradients adopted in the update/sampling rule, which would be massively costly in large model instances. As Stochastic Gradient optimization, it aims to generalize them only considering a small fraction of data points, thus introducing the concept of batches $\tilde{\mathcal{D}} \subset \mathcal{D}$ (Eq. (10)).

$$\nabla \tilde{U}(\theta) = \frac{|\mathcal{D}|}{|\tilde{\mathcal{D}}|} \sum_{x \in \tilde{\mathcal{D}}} \nabla \log p(x|\theta) - \nabla \log p(\theta) \tag{10}$$

It then introduces a matrix B, taking into account the contribution of the gradient noise introduced by the batches, better directing the update of momentum variables, and, as a further improvement, a *friction* term, with the scope to avoid the costly Metropolis-Hastings step (11). As a consequence, consecutive samples obtained through this procedure present a quasi-completed de-correlation, and the algorithm leads to a faster convergence.

$$\begin{cases} d\theta = M^{-1}r \\ dr = -\nabla U(\theta) - \underbrace{BM^{-1}r}_{\text{friction}} + \mathcal{N}(0, 2B) \end{cases} \tag{11}$$

5   Conclusions and Future Work

Our study aims at complementing the knowledge on a new topic arising in the Bayesian community, by focusing on the problem of scaling posterior distributions on differently-curated datasets. However, the framework needs further consolidation by evaluating it on a larger variety of datasets, and models. Moreover, we consider the theories about the marginal role of prior distributions widely arguable, since the cold posterior setup has the effect of producing (also) an over-regularized model, by shrinking the prior distribution too. Further, the abstract concept of non-curated datasets needs a re-definition, or at least it should not be limited only to the nature of the labels. Indeed, being the uncertainty a product of a flaw in a human supervision process, an imprecise labeling of data would primarily be caused by a difficult interpretation of the images, or whether some external biases influence the recognition process itself.

Future studies should thus focus on the (very) deep models counterpart, in order to definitely solve the conflicting natures of tempered and cold posteriors and, thus, to address the effective role of the prior distribution in them (e.g., [24]). On the other hand, big data issues (e.g., [3,4,6,8,10–12,19]) must be taken into account with a deeper detail.

References

1. Aitchison, L.: A statistical theory of cold posteriors in deep neural networks (2021)
2. Al Nuaimi, E., Al Neyadi, H., Mohamed, N., Al-Jaroodi, J.: Applications of big data to smart cities. J. Internet Serv. Appl. **6**(1), 1–15 (2015). https://doi.org/10.1186/s13174-015-0041-5

3. Audu, A.-R.A., Cuzzocrea, A., Leung, C.K., MacLeod, K.A., Ohin, N.I., Pulgar-Vidal, N.C.: An intelligent predictive analytics system for transportation analytics on open data towards the development of a smart city. In: Barolli, L., Hussain, F.K., Ikeda, M. (eds.) CISIS 2019. AISC, vol. 993, pp. 224–236. Springer, Cham (2020). https://doi.org/10.1007/978-3-030-22354-0_21

4. Bellatreche, L., Cuzzocrea, A., Benkrid, S.: F&A: a methodology for effectively and efficiently designing parallel relational data warehouses on heterogenous database clusters. In: Bach Pedersen, T., Mohania, M.K., Tjoa, A.M. (eds.) DaWaK 2010. LNCS, vol. 6263, pp. 89–104. Springer, Heidelberg (2010). https://doi.org/10.1007/978-3-642-15105-7_8

5. Bello-Orgaz, G., Jung, J.J., Camacho, D.: Social big data: recent achievements and new challenges. Inf. Fusion **28**, 45–59 (2016)

6. Bonifati, A., Cuzzocrea, A.: Efficient fragmentation of large XML documents. In: Wagner, R., Revell, N., Pernul, G. (eds.) DEXA 2007. LNCS, vol. 4653, pp. 539–550. Springer, Heidelberg (2007). https://doi.org/10.1007/978-3-540-74469-6_53

7. Brooks, S., Gelman, A., Jones, G.L., Meng, X.L.: Handbook of Markov Chain Monte Carlo. Chapman and Hall/CRC, Boca Raton (2011)

8. Ceci, M., Cuzzocrea, A., Malerba, D.: Effectively and efficiently supporting roll-up and drill-down OLAP operations over continuous dimensions via hierarchical clustering. J. Intell. Inf. Syst. **44**(3), 309–333 (2013). https://doi.org/10.1007/s10844-013-0268-1

9. Chen, T., Fox, E.B., Guestrin, C.: Stochastic gradient Hamiltonian monte Carlo (2014)

10. Cuzzocrea, A., Darmont, J., Mahboubi, H.: Fragmenting very large XML data warehouses via k-means clustering algorithm. Int. J. Bus. Intell. Data Min. **4**(3/4), 301–328 (2009)

11. Cuzzocrea, A., Furfaro, F., Greco, S., Masciari, E., Mazzeo, G.M., Saccà, D.: A distributed system for answering range queries on sensor network data. In: 3rd IEEE Conference PerCom 2005. Workshops, 2005. pp. 369–373. IEEE Computer Society (2005)

12. Cuzzocrea, A., Furfaro, F., Saccà, D.: Enabling OLAP in mobile environments via intelligent data cube compression techniques. J. Intell. Inf. Syst. **33**(2), 95–143 (2009)

13. Heek, J., Kalchbrenner, N.: Bayesian inference for large scale image classification. CoRR abs/1908.03491 (2019)

14. Hoffman, M.D., Gelman, A.: The no-u-turn sampler: adaptively setting path lengths in Hamiltonian monte Carlo. J. Mach. Learn. Res. **15**, 1593–1623 (2011)

15. Koulali, R., Zaidani, H., Zaim, M.: Image classification approach using machine learning and an industrial Hadoop based data pipeline. Big Data Res. **24**, 100184 (2021)

16. Li, C., Chen, C., Carlson, D., Carin, L.: Preconditioned stochastic gradient Langevin dynamics for deep neural networks (2015)

17. Ma, Y.A., Chen, T., Fox, E.B.: A complete recipe for stochastic gradient MCMC (2015)

18. Milinovich, G.J., Magalhães, R.J.S., Hu, W.: Role of big data in the early detection of Ebola and other emerging infectious diseases. Lancet Glob. Health **3**(1), 20–21 (2015)

19. Morris, K.J., Egan, S.D., Linsangan, J.L., Leung, C.K., Cuzzocrea, A., Hoi, C.S.H.: Token-based adaptive time-series prediction by ensembling linear and non-linear estimators: a machine learning approach for predictive analytics on big stock data. In: 17th IEEE International Conference on ICMLA 2018, pp. 1486–1491. IEEE (2018)

20. Morzfeld, M., Tong, X.T., Marzouk, Y.M.: Localization for MCMC: sampling high-dimensional posterior distributions with local structure. J. Comput. Phys. **380**, 1–28 (2019)

21. Nawaz, M.Z., Arif, O.: Robust kernel embedding of conditional and posterior distributions with applications. In: 15th IEEE ICMLA 2016, pp. 39–44. IEEE Computer Society (2016)

22. Ngiam, K.Y., Khor, W.: Big data and machine learning algorithms for health-care delivery. Lancet Oncol. **20**(5), 262–273 (2019)

23. Nguyen, D.T., Nguyen, S.P., Pham, U.H., Nguyen, T.D.: A calibration-based method in computing Bayesian posterior distributions with applications in stock market. In: Kreinovich, V., Sriboonchitta, S., Chakpitak, N. (eds.) TES 2018. SCI, vol. 753, pp. 182–191. Springer, Cham (2018). https://doi.org/10.1007/978-3-319-70942-0_10

24. Ollier, V., Korso, M.N.E., Ferrari, A., Boyer, R., Larzabal, P.: Bayesian calibration using different prior distributions: an iterative maximum A posteriori approach for radio interferometers. In: 26th European Conference, EUSIPCO 2018, pp. 2673–2677. IEEE (2018)

25. Ovadia, Y., et al.: Can you trust your model's uncertainty? evaluating predictive uncertainty under dataset shift (2019)

26. Pearce, T., Tsuchida, R., Zaki, M., Brintrup, A., Neely, A.: Expressive priors in Bayesian neural networks: kernel combinations and periodic functions (2019)

27. Pendharkar, P.C.: Bayesian posterior misclassification error risk distributions for ensemble classifiers. Eng. Appl. Artif. Intell. **65**, 484–492 (2017)

28. Perez, L., Wang, J.: The effectiveness of data augmentation in image classification using deep learning (2017)

29. Ramamoorthi, R.V., Sriram, K., Martin, R.: On posterior concentration in mis-specified models. Bayesian Anal. **10**(4), 759–789 (2015)

30. Ruli, E., Ventura, L.: Higher-order Bayesian approximations for pseudo-posterior distributions. Commun. Stat. Simul. Comput. **45**(8), 2863–2873 (2016)

31. Rajaraman, V.: Big data analytics. Resonance **21**(8), 695–716 (2016). https://doi.org/10.1007/s12045-016-0376-7

32. Shokrzade, A., Ramezani, M., Tab, F.A., Mohammad, M.A.: A novel extreme learning machine based KNN classification method for dealing with big data. Expert Syst. Appl. **183**, 115293 (2021)

33. Springenberg, J.T., Klein, A., Falkner, S., Hutter, F.: Bayesian optimization with robust Bayesian neural networks. In: Advances in Neural Information Processing Systems, vol. 29, pp. 4134–4142 (2016)

34. Stuart, A.M., Teckentrup, A.L.: Posterior consistency for gaussian process approximations of Bayesian posterior distributions. Math. Comput. **87**(310), 721–753 (2018)

35. Tran, B.H., Rossi, S., Milios, D., Filippone, M.: All you need is a good functional prior for Bayesian deep learning (2020)

36. Tsai, C.-W., Lai, C.-F., Chao, H.-C., Vasilakos, A.V.: Big data analytics: a survey. J. Data **2**(1), 1–32 (2015). https://doi.org/10.1186/s40537-015-0030-3

37. Welling, M., Teh, Y.W.: Bayesian learning via stochastic gradient Langevin dynamics. In: Proceedings of the 28th International Conference on ICML 2011, pp. 681–688. Omnipress (2011)
38. Wenzel, F., et al.: How good is the Bayes posterior in deep neural networks really? (2020)
39. Zhu, L., Yu, F.R., Wang, Y., Ning, B., Tang, T.: Big data analytics in intelligent transportation systems: a survey. IEEE Trans. Intell. Transp. Syst. **20**(1), 383–398 (2018)

Ensembling Sparse Autoencoders for Network Covert Channel Detection in IoT Ecosystems

Nunziato Cassavia[1] , Luca Caviglione[2] , Massimo Guarascio[1(✉)] ,
Angelica Liguori[3] , and Marco Zuppelli[2]

[1] Institute for High Performance Computing and Networking, Via Pietro Bucci, 8/9C, 87036 Rende, Italy
{nunziato.cassavia,massimo.guarascio}@icar.cnr.it
[2] Institute for Applied Mathematics and Information Technologies, Via de Marini, 6, 16149 Genova, Italy
{luca.caviglione,marco.zuppelli}@ge.imati.cnr.it
[3] University of Calabria, Via Pietro Bucci, Arcavacata, Italy
angelica.liguori@dimes.unical.it

Abstract. Network covert channels are becoming exploited by a wide-range of threats to avoid detection. Such offensive schemes are expected to be also used against IoT deployments, for instance to exfiltrate data or to covertly orchestrate botnets composed of simple devices. Therefore, we illustrate a solution based on Deep Learning for the detection of covert channels targeting the TTL field of IPv4 datagrams. To this aim, we take advantage of an Autoencoder ensemble to reveal anomalous traffic behaviors. An experimentation on realistic traffic traces demonstrates the effectiveness of our approach.

Keywords: Deep autoencoder · Ensemble method · Covert channel · Intelligent cyber attack detection system

1 Introduction

Owing to their flexibility, Internet of Things (IoT) technologies are commonly used to remotely operate large-scale infrastructures or to control physical systems. Alas, the resource-constrained nature of many IoT nodes jointly with inaccurate development and configuration practices lead to countless security and privacy flaws [11]. For instance, the Mirai malware allowed to launch attacks against many international organizations [3], and IoT devices can be inspected to derive insights of a specific deployment [14]. Indeed, a major trend exploits information hiding and steganography to make malware difficult to detect and able to bypass classical network security tools [9].

Even if information hiding can be used to implement various advanced techniques (e.g., anti-forensics and multi-stage loading schemes), its prime adoption concerns the creation of network covert channels, i.e., parasitic communication paths cloaked within legitimate traffic flows. Since network covert channels proven their effectiveness in the creation of Command & Control (C&C) communications as well as to bypass firewalls, guaranteeing the security of modern IoT ecosystems also requires to consider such a

© The Author(s), under exclusive license to Springer Nature Switzerland AG 2022
M. Ceci et al. (Eds.): ISMIS 2022, LNAI 13515, pp. 209–218, 2022.
https://doi.org/10.1007/978-3-031-16564-1_20

class of offensive mechanisms [9]. Unfortunately, the tight coupling between the hiding mechanism and the abused protocol makes their mitigation difficult to generalize [16]. Recently, Machine Learning revealed to be effective for working towards more abstract mitigation frameworks, also for the case of covert communications [6]. Yet, steganographic malware often remains unnoticed and its understanding requires a major reverse engineering [9].

To face all these issues, this work addresses the problem of detecting network covert channels targeting IoT ecosystems. In more detail, an incremental learning scheme based on an ensemble of unsupervised neural network architectures (autoencoders) is adopted to spot the presence of a cloaked communication attempt. The proposed solution allows for learning effective detection models in absence of a labeled training set and can be incrementally updated in order to be deployed also on devices with limited computational and storage resources. Moreover, prior works dealing with the detection of covert channels targeting IoT scenarios mainly focus on timing channels and SCADA applications, see, e.g., [2]. Thus the contribution of this work is the design of an AI-based framework to spot channels targeting fields within the header of a protocol in a more general setting. Moreover, another contribution is the creation of a dataset containing covert channels prepared by modifying real traces collected in the wild. Since security of IoT ecosystems is often implemented in resource-constrained devices (e.g., home gateways) emphasis has been put on the footprint required by the proposed approach.

The remainder of the paper is structured as follows. Section 2 provides details on the considered attack scenario and its modeling, Sect. 3 introduces our detection approach, and Sect. 4 showcases numerical results. Finally, Sect. 5 concludes the paper and outlines possible future research directions.

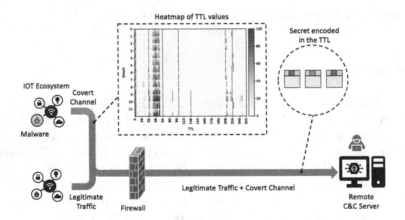

Fig. 1. Attack scenario: a malware sends data towards a remote command and control facility via a network covert channel in the TTL field of IPv4 traffic.

2   Attack Scenario and Threat Modelling

In this section, first we introduce the attack scenario and then we discuss the dataset prepared to model a covert channel targeting an IoT ecosystem.

2.1   Attack Scenario

Figure 1 depicts the general attack scenario. Specifically, we consider a malware abusing an IoT node to create a network covert channel. The latter can be used for exfiltrating information towards a C&C server or exchanging commands without being detected by a firewall. Despite almost any protocol and traffic feature can be manipulated for hiding data, the resource-limited nature of IoT nodes poses constraints on the complexity of the covert channel [16]. Hence, the hiding mechanism should be simple to not disclose the presence of the malware due to additional delays or anomalous energy consumption. We then consider a malware cloaking data within the TTL field of the IPv4 header.

Specifically, the TTL of the traffic of the compromised IoT node is manipulated by the malware to transport arbitrary information. To not appear suspicious, the malware should not directly write the secret data in the field [15]. Rather, it should encode the bits 1 and 0 by increasing or decreasing the observed TTL of a suitable threshold or by exploiting the most popular values as "high" and "low" signals. The attacker should design a proper encoding by evaluating the "clean" traffic conditions and adapt accordingly the hiding mechanism. Figure 1 depicts also an heatmap reporting values obtained in a real set up[1] during an observation window of 12 h. As shown, the values for the TTL aggregate in two main ranges 32–64 and 208–224. Other values are present in the traffic but have an intermittent behavior, e.g., datagrams with a TTL equal to 128 are present only for 3 h thus limiting the duration of the covert communication. In general, the data hidden in the TTL should not represent an anomaly, i.e., disrupt the heatmap characterizing clean traffic conditions.

2.2   Dataset for Modelling the Covert Channel

To model the attack leveraging a network covert channel and quantify the performance of our approach, we built a benchmark dataset starting from real traffic traces. To avoid burdening the data, we removed IPv6, ICMP, DNS and NTP conversations as well as multicast/broadcast traffic. To prevent unwanted signatures, we also removed traffic generated by non-IoT devices, e.g., smartphones and laptops. The final dataset had a duration of an entire week and contained flows exchanged by 28 different IoT nodes, such as smart speakers and home hubs.

A realistic attack template has been obtained by modeling the presence of a malware compromising a specific IoT device. Typically, this requires to exploit a CVE for granting access to the device or by leveraging a configuration error (e.g., weak/default credentials) [11]. In this work, we considered a threat abusing a smart camera to smuggle

---

[1] In this work, we used the collection of IoT traffic made available in [14]. Heatmaps have been computed by using the 24-h slice of data captured from September 22, 2016 at 16:00 to September 23, 2016 at 16:00, whereas for the performance evaluation we used traces containing traffic collected from September 22, 2016 at 16:00 to September 29, 2016 at 16:00.

sensitive data towards a remote host controlled by the attacker. To this aim, we assumed that the Dropcam IoT device has been under control of the malware for 3 days. To create the covert communications, we rewrote the flows generated by the Dropcam directly in the original traffic captures [17]. To not make the detection trivial, we encoded bits 1 and 0 in TTL values equal to 64 and 100, respectively. To prevent that bursts of manipulated datagrams would reveal the presence of the channel, we randomly interleaved packets containing hidden data with legitimate/unaltered ones [15]. The secret information transmitted via the covert channel has been modeled with randomly-generated strings: this is representative of an attacker using some obfuscation technique, such as encryption or scrambling [10]. To bear with the exfiltration of several contents (e.g., username+password pairs or configuration files), each day of attack contained a different volume of hidden data, i.e., we considered the exfiltration of 69, 80, and 64 kbit of data. As a result, the compromised IoT node manipulates the 18%, 1%, and 12% of the overall daily traffic, respectively.

3   Deep Ensemble Learning Scheme

In this section, we illustrate a solution based on Deep Learning (DL) to spot the presence of covert channels within traffic flows. Our detection model takes the form of an ensemble of unsupervised neural network models. The main benefit in using the unsupervised approach relies on the capability of the models to raise alarms also on never seen attacks: this represents a frequent scenario when dealing with covert channels, since they are often undocumented and unknown *a priori*. We first illustrate the detection mechanism for a single model and then we describe how this technique can be extended to learn effective and scalable ensembles.

3.1   Detection Through a Single Autoencoder

The idea behind our solution is to adopt a neural encoder-decoder architecture trained against traffic data. Basically, an autoencoder is an unsupervised (i.e., trained without any information concerning the nature of the attack/normal behavior) neural network model performing two main operations: first, it compresses the input data (i.e., a number of statistics computed over the traffic generated by the IoT network and described in Sect. 4.1) within a latent space, then it reconstructs the original information provided as input. In our setting, the model is only trained against the normal behavior. The underlying intuition is that the legitimate input data should be (almost) correctly reconstructed by the autoencoder, in other words, the encoding/decoding phases should not introduce a heavy distortion in the output. By contrast, outliers and anomalous values in the input will yield a deviant output.

The usage of the reconstruction error as a measure of outlierness to discover abnormal behaviors has already been proposed in the literature, but the adoption of unsupervised techniques (and in particular of encoder-decoder architectures) for revealing covert channels is quite unexplored [1,5,6]. As discussed in [4,8], autoencoders are considered as a valid solution to the problem of effectively summarizing the main information of a given input into a low-dimensionality representation. In essence, these neural network models aim at yielding as output a duplicate similar to the input data.

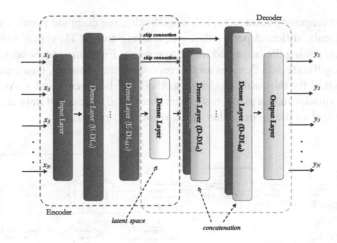

Fig. 2. Neural architecture (Sparse Autoencoder) used to perform the detection of covert channels targeting the TTL of IPv4 traffic.

In this work, we employ the neural model shown in Fig. 2. Basically, it includes two main components, named *Encoder* and *Decoder*, respectively. Let $\mathbf{x} = \{x_1, \ldots, x_N\}$ be a set of numeric features (in our case, a number of traffic flow statistics computed for a time slot). The former sub-network is devoted to map the input data with a latent space (*encoding*), i.e., learning a function $\mathbf{z} = enc(\mathbf{x})$, whereas the second one yields the overall network output by reconstructing the input from the features extracted by the encoder $\mathbf{y} = dec(\mathbf{z})$ (*decoding*). Gradient descent is employed to learn the model weights by minimizing a suitable loss function. In our approach, the *Mean Square Error*, i.e., $Loss_{MSE}(\mathbf{x}) = \frac{1}{N} \sum_i \|x_i - y_i\|_2$, is used as loss.

Notably, the architecture of Fig. 2 exhibits two main differences w.r.t. a standard encoder-decoder model: *(i) Skip Connections* are used to boost the predictive performances of the model and to reduce the number of iterations required for the learning algorithm convergence, and *(ii)* a hybrid approach including the usage of *Sparse Dense Layers* is adopted to make the autoencoder more robust to noise, especially since attacks often exhibit slight differences compared with normal behaviors. Both, encoder and decoder are composed of M hidden layers, therefore we adopted a symmetric architecture. In more detail, the adoption of the skip connections simplifies the learning process of the network by providing as input to each layer of the decoder ($D\text{-}DL_i$), except for the shared latent space, both the previous ($D\text{-}DL_{i-1}$) and the correspondent encoder layer ($E\text{-}DL_{M-i+1}$). As regards the Sparse Layers, they are used to generate a wider number of discriminative features, which allow for extracting a more representative latent space.

In Fig. 3 we illustrate the detection process of covert channels targeting the TTL field of IPv4 datagrams. Without loss of generality, we assume to monitor an "infinite datastream", i.e., the traffic of the various IoT nodes feeds our detection mechanism in a continuous manner. At pre-fixed time intervals (corresponding to a time slot in Fig. 3), we compute a number of statistics to describe the behavior of the TTL fields composing the aggregate traffic flow. This can be done without impairing the overall traffic and by using limited computing resources [13]. Specifically, we compute metrics such as the min, average, max, different percentiles, etc., starting from TTL values gathered from the packets composing the inspected traffic aggregate. First, an autoencoder, pretrained only against legitimate data flows, is used to reproduce the statistics, then reconstruction error is calculated for the current example as the MSE between x and y. As a last step, if the error is smaller than a given *outlierness threshold*, the current data are labeled as "normal" and update the model, otherwise a warning is raised.

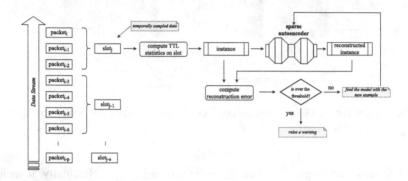

Fig. 3. Detection mechanism for revealing the presence of network covert channels.

3.2   Learning and Combining Different Detectors

A main limitation of the above described approach relies on the necessity to learn the neural network model against the whole training set (that could be unfeasible in IoT networks with tight computational resources). Moreover, in real scenarios the limited resources of the device where the detector is deployed and the presence of concept drifts in the observed behaviors [7] can affect the predictive performances of the autoencoder. To mitigate such issues, we devised an incremental learning scheme based on an ensemble of encoder-decoder architectures shown in Fig. 4. Basically, we consider the case where only a limited number of training examples $\mathscr{D}$ can be gathered and stored in a data chunk (named D_i in the figure).

Our ensemble solution relies on building up a series of k base DNN detectors (denoted as $M_i, M_{i-1}, \ldots, M_{i-k}$) sharing the same neural architecture described above. These autoencoders are trained from disjoint *data chunks* (denoted as $D_i, D_{i-1}, \ldots, D_{i-k}$, respectively), which are fed with data instances gathered in different temporal intervals. Specifically, the learning process is loosely inspired to *Transfer Learning* (TL). The main idea of TL consists in re-using DL models, learned on different domains/datasets or for tackling different tasks, by fine-tuning them for addressing the own learning problem. Differently from a standard TL approach, in our solution the model M_i is trained (i.e., fine tuned) from the weights of the model M_{i-1} and the sample D_i, and so way for all the models composing the ensemble. In this way the detection model will be able to gradually adapt to normal concept drifts that can occur, for instance, due to the deployment of new devices in the network that can modify the network statistics. This approach can also reduce the risk of *catastrophic forgetting* [12] that affects DL models (and also different types of shallow architectures) when learned incrementally. The final anomaly score for each instance is hence computed as the median value of the k reconstruction errors (denoted as Deviance Score DS in the Figure) $\{DS_i, \ldots, DS_{i-k}\}$ yielded by the base models.

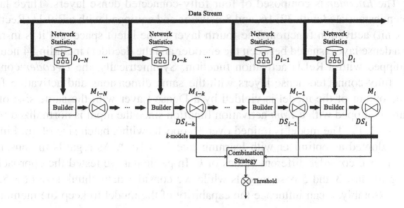

Fig. 4. Incremental Deep Ensemble model approach.

4 Performance Evaluation

In this section, we illustrate experiments for evaluating the performances of our deep ensemble-based approach to spot covert channels in IoT scenarios. First, we describe the data pre-processing procedures, model parameters and evaluation metrics, then we present the obtained results.

4.1 Pre-processing, Parameters and Evaluation Metrics

To assess the quality of the proposed approach in detecting the presence of network covert channels within traffic aggregates, we developed a prototype in Python based on the TensorFlow[2] library. From the traffic dataset described in Sect. 2.2 we extracted the following fields: a progressive timestamp, the number of incoming packets within a given time slot, the average and median values of observed TTLs, the values of the 10th, 25th, 75th and 90th percentile, minimum and maximum TTLs, as well as a label indicating the presence of the attack (i.e., for testing purposes). Recalling that our approach exploits a "slotted" architecture (see Fig. 3), in this work we consider a time slot with a duration of 5 s.

Data have been partitioned in training and test sets by using a temporal split: *(i)* the data gathered in the first 96 h only contains legitimate traffic and has been used for

---

[2] TensorFlow machine learning library. Available online at: https://www.tensorflow.org/ [Last Accessed: June 2022].

the learning phase of the ensemble, whereas *(ii)* the remaining instances compose the test set. As a result, the training and the test set contains 69,116 and 51,837 instances, respectively. Moreover, input data are further pre-processed through a normalization procedure: a *MinMax* normalization has been adopted to map each feature in the range $\{-1,1\}$ to improve the stability of the learning process.

As described in Sect. 3, the base model composing the ensemble is an autoencoder model. The *Encoder* is composed of four fully-connected dense layers. Three layers have been instantiated with 32, 16, and 8 neurons and equipped with a ReLU (Rectified Linear Unit) activation function. The fourth layer is the latent space and it is instantiated as a dense layer (shared between the encoder and the decoder) including 4 neurons and equipped with a ReLU activation function. Symmetrically, the *Decoder* consists of three fully-connected dense layers with the same dimensions and activation function. The output of the model is yielded by a Dense Layer with the same size of the input, and equipped with a *Tanh* activation function since the input is normalized in the range in $\{-1,1\}$. The model is trained over 16 epochs with a batch size of 16. Finally, *Adam* is adopted as optimizer with learning rate $lr = 1e^{-4}$. As regards the ensemble parameters, we consider different values of k. In particular we tested the approach by including the last 3 and 5 base models while we consider data chunk size of $\sim$5,000 instances. Notably, k can influence the capability of the model to keep the memory of past behaviors.

To evaluate the quality of our solution a number of metrics have been computed. Let us define TP as the number of positive cases correctly classified, FP as the number of negative cases incorrectly classified as positive, FN as the number of positive cases incorrectly classified as negative, and TN as the number of negative cases correctly classified. Then, the following metrics have been considered:

- *Accuracy*: defined as the fraction of cases correctly classified, i.e., $\frac{TP+TN}{TP+FP+FN+TN}$.
- *Precision* and *Recall*: metrics used to estimate the detection capability of a system since they provide a measurement of accuracy in identifying attacks and avoiding false alarms. Specifically, *Precision* is defined as $\frac{TP}{TP+FP}$, while *Recall* as $\frac{TP}{TP+FN}$.
- *F-Measure*: condenses the overall system performances and is calculated as the harmonic mean of *Precision* and *Recall*.

Lastly, to perform experiments, we used a machine with 32 Gb RAM, an Intel i7-4790K CPU @4.00 GHz and a 1Tb SSD drive.

4.2 Numerical Results

In Table 1 we report the results of our experimentation by comparing the performances of a single autoencoder with respect to the ensemble model. The performance metrics are computed by ranging different ensemble sizes and sensitivity thresholds. As regards the anomaly score, it is estimated by computing the reconstruction error for each instance contained in the training set and extracting the values corresponding to 90th, 95th and 99th percentiles. As expected, for all the model types, the usage of a looser threshold (e.g., the 90th percentile) allows for improving the probability of detection (i.e., the recall) but at the price of a higher number of false alarms. By contrast, a higher

Table 1. Experimental results for different outlier thresholds and ensemble size.

Model type	Ensemble size	Detection threshold	Accuracy	Precision	Recall	F-Measure
Single Model	–	90th perc.	0.882	0.743	**0.993**	0.850
Single Model	–	95th perc.	0.921	0.822	0.976	0.893
Single Model	–	99th perc.	0.936	0.942	0.865	0.902
Ensemble	3	90th perc.	0.894	0.771	0.979	0.863
Ensemble	3	95th perc.	0.947	0.902	0.948	0.924
Ensemble	3	99th perc.	**0.955**	**0.950**	0.915	**0.932**
Ensemble	5	90th perc.	0.890	0.764	0.977	0.858
Ensemble	5	95th perc.	0.933	0.863	0.954	0.906
Ensemble	5	99th perc.	0.952	0.944	0.911	0.927

threshold (e.g., the 99th percentile) allows to limit the number of FP but a lesser recall value is obtained.

Moreover, the adoption of the ensemble strategy improves the overall predictive capabilities of the detection systems: the best result (reported in bold in the table) with the higher value of F-Measure (i.e., the best trade-off between recall and precision) is obtained with an ensemble size equal to 3 and by considering the threshold value corresponding to the 99th percentile. Finally, the slight reduction of the predictive performances when increasing the ensemble size appears to be mainly due to the evolving nature of traffic characterizing IoT ecosystems. In fact, asynchronous activations of nodes, external triggers, or periodical synchronizations account for broad changes in traffic conditions. Therefore, recent data in some cases could be more informative for revealing an attack as the "past history" could not be representative of the actual traffic exchanged over the network.

5   Conclusions and Future Work

In this work, we showcased the use of ensemble of autoencoders for detecting network covert channels targeting IoT scenarios. Our approach has been designed to be lightweight and required a limited number of training examples at time to be effective. Numerical results demonstrated its effectiveness: the method can achieve a probability of detection (i.e., recall) of ∼91% while exhibiting a good precision ∼95%.

Future works aim at considering other types of network covert channels. To this aim, part of our ongoing research is devoted to develop some form of "intermediate" representations that can be used to develop a more general mechanisms for facing different threats, e.g., protocol-agnostic representations.

Acknowledgment. This work has been partially supported by the Horizon 2020 Program within the framework of CyberSec4Europe (Grant Agreement No. 830929).

References

1. Ahmad, Z., Khan, A.S., Shiang, C.W., Abdullah, J., Ahmad, F.: Network intrusion detection system: a systematic study of machine learning and deep learning approaches. Trans. Emerg. Telecommun. Technol. **32**(1), e4150 (2021)
2. Alcaraz, C., Bernieri, G., Pascucci, F., Lopez, J., Setola, R.: Covert channels-based stealth attacks in industry 4.0. IEEE Syst. J. **13**(4), 3980–3988 (2019)
3. Antonakakis, M., et al.: Understanding the Mirai botnet. In: 26th USENIX Security Symposium, pp. 1093–1110 (2017)
4. Bengio, Y., Pascal, L., Dan, P., Larochelle, H.: Greedy layer-wise training of deep networks. In: Advances in Neural Information Processing Systems (NeurIPS), vol. 19, pp. 153–160. MIT Press (2007)
5. Darwish, O., Al-Fuqaha, A., Brahim, G.B., Jenhani, I., Vasilakos, A.: Using hierarchical statistical analysis and deep neural networks to detect covert timing channels. Appl. Soft Comput. **82**, 105546 (2019)
6. Elsadig, M.A., Gafar, A.: Covert channel detection: machine learning approaches. IEEE Access **10**, 38391–38405 (2022)
7. Folino, G., Guarascio, M., Papuzzo, G.: Exploiting fractal dimension and a distributed evolutionary approach to classify data streams with concept drifts. Appl. Soft Comput. **75**, 284–297 (2019)
8. Hinton, G., Salakhutdinov, R.: Reducing the dimensionality of data with neural networks. Science **313**(5786), 504–507 (2006)
9. Mazurczyk, W., Caviglione, L.: Information hiding as a challenge for malware detection. IEEE Secur. Priv. **13**(2), 89–93 (2015)
10. McLaren, P., Russell, G., Buchanan, B.: Mining malware command and control traces. In: 2017 Computing Conference, pp. 788–794 (2017)
11. Neshenko, N., Bou-Harb, E., Crichigno, J., Kaddoum, G., Ghani, N.: Demystifying IoT security: an exhaustive survey on IoT vulnerabilities and a first empirical look on Internet-scale IoT exploitations. IEEE Commun. Surv. Tutorials **21**(3), 2702–2733 (2019)
12. Parisi, G.I., Kemker, R., Part, J.L., Kanan, C., Wermter, S.: Continual lifelong learning with neural networks: a review. Neural Networks **113**, 54–71 (2019)
13. Repetto, M., Caviglione, L., Zuppelli, M.: bccstego: a framework for investigating network covert channels. In: The 16th International Conference on Availability, Reliability and Security, pp. 1–7 (2021)
14. Sivanathan, A., et al.: Classifying IoT devices in smart environments using network traffic characteristics. IEEE Trans. Mob. Comput. **18**(8), 1745–1759 (2018)
15. Zander, S., Armitage, G., Branch, P.: Covert channels in the IP time to live field. Swinburne University of Technology Report (2006). https://researchrepository.murdoch.edu.au/id/eprint/35012/1/covert. Accessed Jun 2022
16. Zander, S., Armitage, G., Branch, P.: A survey of covert channels and countermeasures in computer network protocols. IEEE Commun. Surv. Tutorials **9**(3), 44–57 (2007)
17. Zuppelli, M., Caviglione, L.: pcapstego: a tool for generating traffic traces for experimenting with network covert channels. In: The 16th International Conference on Availability, Reliability and Security, pp. 1–8 (2021)

Towards Automation of Pollen Monitoring: Image-Based Tree Pollen Recognition

Elżbieta Kubera[1]([✉])(iD), Agnieszka Kubik-Komar[1]([✉])(iD),
Alicja Wieczorkowska[2]([✉])(iD), Krystyna Piotrowska-Weryszko[3]([✉])(iD),
Paweł Kurasiński[1]([✉])(iD), and Agata Konarska[3]([✉])(iD)

[1] Department of Applied Mathematics and Computer Science, University of Life Sciences in Lublin, Lublin, Poland
{elzbieta.kubera,agnieszka.kubik,pawel.kurasinski}@up.lublin.pl
[2] Polish-Japanese Academy of Information Technology, Warsaw, Poland
alicja@poljap.edu.pl
[3] Department of Botany and Plant Physiology, University of Life Sciences in Lublin, Lublin, Poland
{krystyna.piotrowska,agata.konarska}@up.lublin.pl

Abstract. Pollen monitoring helps predict the risk of pollen-induced allergies. Traditionally, this monitoring is performed based on the biological material obtained from volumetric Hirst's traps. A palynological specialist analyzes the obtained microbiological specimen under the microscope, and recognizes and counts pollen grains of various taxa. This is a tedious task, and automatic detection and counting of pollen grains in digital microscopic images can support specialists in their work. YOLOv5 and Faster R-CNN are the state-of-art deep neural networks used for object detection in many fields of computer vision. In the presented research, these detectors were applied to analyze specimen with pollen grains of four taxa, typical of early spring in Central and Eastern Europe. The obtained results enabled the selection of the detector that should be the first choice in pollen grains recognition tasks. Statistical analysis of differences in the distribution of the recognition quality measures also supports the conclusions.

Keywords: Pollen detection · Image recognition · Deep learning

1 Introduction

Pollen grains are one of the main causes of allergies. In order to detect pollen of various plants in the air, monitoring stations are carrying out the detection of pollen all over the world, especially in Europe [1]. Such monitoring can help improve the prevention of allergic diseases. Annual asthma treatment cost in the United States alone is nearly $ 18 billion, whereas the treatment cost for other

© The Author(s), under exclusive license to Springer Nature Switzerland AG 2022
M. Ceci et al. (Eds.): ISMIS 2022, LNAI 13515, pp. 219–229, 2022.
https://doi.org/10.1007/978-3-031-16564-1_21

allergies is about $ 7 billion [2]. Therefore, numerous pollen monitoring stations are monitoring ambient air quality, using various systems [3].

The most commonly used method of pollen monitoring is based on a volumetric spore trap of the Hirst design [4], constructed to continuously sample the air. The data from the samplers are usually collected diurnally and daily, and the total grain counts per cubic meter of air are calculated [5,6]. In the Hirst's trap, a clock mechanism moves a transparent plastic band coated with adhesive. The analysis of the samples is time-consuming, as the particles collected are viewed under a microscope, and in the pollen season trees can produce even over 12,000 grains per cubic meter of air [7]. Therefore, hours of a specialist's work are often needed to elaborate the results, and automation of this process could help.

1.1 Background

Automation of pollen monitoring is a difficult task. Automated real-time systems, which are based on different principles than the Hirst trap, include such systems as automated multispectral imaging flow cytometry combined with deep learning [8], metabarcoding method of species differentiation based on DNA data [9], and labeling procedures based on the combination of fluorescein and propidium iodide [11]. However, the results obtained from the automated systems often differ from the manually elaborated data [12]. A Global Pollen Project even provides tools for crowdsourcing to aid the identification of pollen grains and provides access to a free database of pollen images and their metadata [10].

In Poland, all pollen monitoring centers use Hirst's traps, and the automation of the process of data elaboration would facilitate palynological experts' work, and allow comparative analysis of data collected in the last decades.

Automated identification of pollen images from a microscope, providing occurrence data for each taxon, has been a research topic since the nineties [13–15]. In the last years, deep neural networks have been extensively used for this purpose, including convolutional networks [16–20], but it is difficult to compare the results, as various architectures and data sets are used in the experiments.

Pollen grains from microscope images often contain grains of various species, so single-label identification of species cannot be applied to the whole image. In this paper, we apply detectors for object localization, thus no manual segmentation of the image into the regions corresponding to particular grains is necessary. We decided to use neural networks, including YOLO, which is a state-of-the-art, real-time object detection system. YOLO detectors have been applied in many papers, see [21–23] for reference. Examples of usage include detection of cells in 2D and 3D fluorescence microscopy images, stomata in leaves of plants, and white blood cells in leukemia. YOLOv3 has also been used in automated detection of pollen grains, see [24], with very good results obtained for 16 taxa, using a prototype device (similar to Hirst's trap), installed in Austria.

The remainder of the paper is structured as follows. Section 2 describes the database used in the experiment. In Sect. 3 we present selected methods of object detection: YOLO and Faster R-CNN, together with measures of their evaluation, as well as statistical methods used for their comparison. The results are summarized in Sect. 4, and discussed in Sect. 5, which contains also conclusions of the presented work.

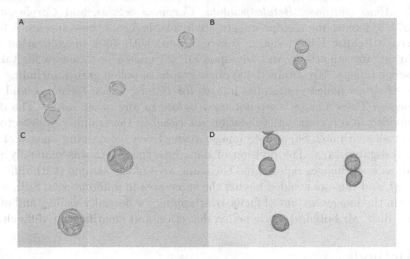

Fig. 1. Microscope images of the analyzed pollen grains: A-*Alnus*, B-*Betula*, C-*Carpinus*, D-*Corylus*

2   Biological Data

For the investigations presented in this paper, we selected pollen grains representing species which are especially important in allergology, namely birch (*Betula*), hazel (*Corylus*), alder (*Alnus*), and additionally hornbeam (*Carpinus*) grains, typically present in pollen traps at the same time. Images of the pollen grains for these species are shown in Fig. 1. The degree of pollen coloring is independent of taxa.

In the spring, the *Betula* pollen is the main cause of allergy in Central and Northern Europe. The allergy to *Betula* is usally accompanied by the allergy to the *Corylus* and *Alnus* pollen grains, because of a well-documented cross-reactions between the allergens of these trees. As a result, allergy symptoms in many patients last longer [25, 26]. Pollen grains of *Corylus* and *Alnus* are the first to appear in the air, and next *Betula* and *Carpinus*. However, all these pollen grains can be found in the traps in the same time, as the pollen seasons of these trees overlap. Unfortunately, the pollen grains look similar, especially in the case of *Betula* and *Corylus*, so a correct identification may be difficult. Still, general descriptions of pollen grains for the analysed trees are available. Namely, the pollen grains of *Carpinus* are the biggest of the analysed grains, whereas *Betula* pollen grains are the smallest, so they can be more easily identified. *Betula* and *Corylus* most often have three pores, whereas *Alnus* pollen grains have the highest number of pores.

For the presented analyses, we selected species representing *Alnus*, *Betula*, *Carpinus*, and *Corylus* genera, which are most commonly found in Poland, namely *Alnus glutinosa*, *Betula pendula*, *Carpinus betulus*, and *Corylus avellana* [27]. Reference microscope images are available for all these species. A light microscope Eclipse E400 (Nikon, Tokyo, Japan) with 400x magnification was used in our investigations, and Olympus DP 23 microscope camera for taking microscope images. We obtained 496 photographs of pollen grains, including 188 images of *Alnus* pollen grains, 103 images for *Betula*, 99 for *Carpinus*, and 106 for *Corylus*. These images were not used before in any other works. The data were divided into a train set, validation set (used in the training of detectors), and test set, with 362, 86, and 48 images respectively, containing images of the four investigated taxa. The division of data into these sets was manually performed, as some images represented the same area of the sample (with different focusing), and thus we avoided having the same area in different sets. Still, some objects in the images are out of focus, overlapping, or partially visible, and other objects (dust, air bubbles) made pollen detection and identification difficult.

3   Methods

3.1   Object Detection

Object detection in images consists in finding the location of an object of interest, and its classification into predefined classes. The position and size of the object is defined by a bounding box around this object. In a two-stage detection, regions of interest (RoI) are determined in the first step, and next classification is performed in each RoI. One-stage detectors determine the bounding box and perform classification for each object in one step.

Automatic pollen grain detection and classification from microscope images is difficult for several reasons. One of them is the similarity of grains representing different classes (e.g. *Betula* and *Corylus*), and another problem is a 2-dimensional representation of 3-dimensional objects, which introduces interclass differences between pollen grains. In our work, we applied two types of detectors, commonly used in the classification of objects in images.

YOLO (You Only Look Once) is a one-stage detector [28]. The input image is divided into a grid of cells, and within each cell the bounding box is determined, confidence level of each box, and predicted class probability. We decided to use YOLOv5 [29], i.e. the implementation of YOLO v3 in PyTorch. In this version, the models pre-trained using COCO database are available [30], so we could perform fine-tuning of this detector for our data. Therefore, we could obtain a high-quality detector faster. We selected yolov5s network architecture, as in our previous work we found no improvement when using bigger models [33].

Region-based Convolutional Neural Network (R-CNN) is a two-stage detector. The Faster R-CNN (FRCNN) model is the first detector from the R-CNN family, in which the search for RoIs is performed using a deep, fully convolutional neural network [31]. Our Faster R-CNN models were built using the Detectron2 platform [32] in Python.

3.2 Quality Measures for Model Evaluation

Standard measures applied in object detection describe how well the bounding box is matched to the object. The measures used in the object identification from COCO data set are based on mean Average Precision (mAP), namely mAP@.5 and mAP@[.5:.95]. The measure mAP@.5 is calculated for predicted bounding box and the ground-truth bounding box, based on IoU (intersection over union, which measures the overlap between two boundaries); IoU $\geq$ 0.5 means that it was a hit, otherwise it was a fail. The mAP@[.5:.95] corresponds to the average calculated for IoU thresholds, from 0.5 to 0.95, with step 0.05.

In the case of our research, we were not interested how exact the bounding box is, but what precision was obtained, which equals 100% if no object was incorrectly classified. Recall (how many grains were missed in the image when classifying a target class) is less important here, as a palynological specialist also omits grains when unsure about classification. F-score (F-measure), which is the harmonic mean of precision and recall, was also applied in our work.

3.3 Training Process

We started with three copies of YOLOv5 models and three copies of Faster R-CNN models pre-trained on COCO data set, and then each model was fine-tuned on our data, in 500 epochs. To avoid overfitting, which may happen when detection results for the validation test deteriorate in consecutive epochs, we used the model yielding the best results within all epochs for YOLOv5, and the model from the last epoch for FRCNN, as default in Detectron2. Therefore, overfitting was avoided in the case of YOLO, but still possible for FRCNN.

The input parameters used in the training phase were the same for all repetitions. The training procedure was a bit different for YOLOv5 and Faster R-CNN, due to different toolkits and evaluation measures used. As mentioned before, our main target measure is the precision of taxa recognition. Therefore, we have changed the default weights for calculating the fitness measure in YOLOv5 to 70% for precision and 30% for recall.

3.4 Statistical Analysis

The quality of both detectors we used, described by mean values of precision, recall and F-score, was analysed using paired samples Wilcoxon test [34] in Statistica [35]. This test, also known as Wilcoxon signed-rank test, is a non-parametric alternative to a paired t-test used to compare paired data. The quality measures were analyzed for each of the 48 test images, with a significance level of $\alpha = 0.05$.

In order to check if the differences among the detector's repetitions are significant the Quade test together with pairwise multiple comparisons of mean rank sums in the case of rejecting the null hypothesis was performed using PMCMR and PMCMRplus packages in R [36,37]. This test is an extension of the Wilcoxon test, to be used for more than two samples, as a nonparametric alternative to ANOVA with repeated measurements. The Quade test is recommended as more powerful than Frieman's in the case of less than five repeated measures [38].

4   Results

We built three YOLOv5 models and 3 Faster R-CNN models for pollen grains classification. Each detector was trained for 500 epochs. The results on the test set for all these models are presented in Fig. 2.

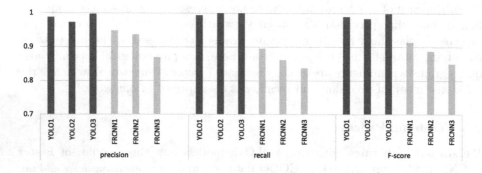

Fig. 2. Overall results from YOLO (YOLOv5) and FRCNN (Faster R-CNN).

Table 1. Paired samples Wilcoxon test results. N represents the number of non-zero differences between the compared detectors, T is the test statistics in the Wilcoxon test, Z - statistic with tied ranks correction, p is the statistical p-value, to be compared with the significance level α; * - $p<0.05$ ** - $p<0.01$

	Taxon	N	T	Z	p
Precision	*Alnus*	10	0.000	2.803	0.005**
	Betula	5	0.000	2.023	0.043*
	Carpinus	2	0.000	1.342	0.180
	Corylus	1			
Recall	*Alnus*	13	0.000	3.180	0.001**
	Betula	5	0.000	2.023	0.043*
	Carpinus	2	0.000	1.342	0.180
	Corylus	6	0.000	2.201	0.028*
F-score	*Alnus*	13	0.000	3.180	0.001**
	Betula	5	0.000	2.023	0.043*
	Carpinus	3	3.000	0.000	1.000
	Corylus	6	0.000	2.201	0.028*

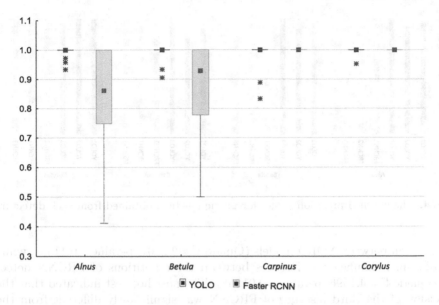

Fig. 3. Box-plot presenting precision of detectors YOLO (shown in red) and Faster R-CNN (in green) for the investigated 4 taxa classes. (Color figure online)

In order to compare the results obtained using YOLO and FRCNN models, mean values (calculated from the three instances for both types of detectors) of precision, recall, and F-score were computed for every test image. These values were analyzed using paired samples Wilcoxon test, as shown in Table 1.

The overall precision, recall and F-score distributions are significantly different for YOLO and FRCNN. The p-values from Wilcoxon test were equal to 0.0017 for precision, 0.000008 for recall, and 0.000017 for F-score. Figure 2 also shows that for YOLO we obtained higher average results than for FRCNN, and the differences within the results for YOLO models are smaller than for FRCNN.

Figure 3 shows precision for all 4 investigated taxa. For *Carpinus* and *Corylus* no difference in precision was found. For the *Alnus* and *Betula* classes, we found a significant difference between the precision of YOLO and FRCNN. This is consistent with the results of the Wilcoxon test, shown in Table 1, which indicate statistically significant differences between the distribution of the precision of YOLO and Faster R-CNN detection of *Alnus* (p = 0.005) and *Betula* (p = 0.043) pollen grains. For other taxa, no significant differences were found in the distribution of the precision of both detectors. For *Corylus* pollen grains, only one of the 6 investigated detector models yielded precision below 100%, and for one image only, so the Wilcoxon test could not have been calculated for this taxon.

The detailed precision values for each of the detection models and each taxon are shown in Fig. 4. Better reproducibility of the results can be observed for YOLO for each detector instance, but the detailed analysis of the recognition of particular taxa shows high precision of *Carpinus* and *Corylus* pollen grain identification for each Faster R-CNN model built. We found no significant differences

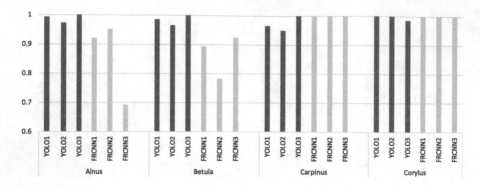

Fig. 4. The detailed precision values for particular taxa obtained from each detector.

in precision between YOLO models (Quade $F = 2.2201$, p-value $= 0.1143$), while we obtained significant differences between the repetitions of FRCNN detector (Quade $F = 4.1596$, p-value $= 0.01858$). The post hoc test indicated that the precision of the third instance of FRCNN was significantly different from the others (p-value $= 0.0230$ and p-value $= 0.0094$ for comparison between the 3rd model and the 1st model, and the 3rd and the 2nd, respectively). There were no significant differences in precision between the 1st and 2nd model of FRCNN (p-value $= 0.7359$). There were no significant differences in recall and F-score neither between repetitions of YOLO (Quade $F = 2.0432$, p-value $= 0.1353$ for recall, and Quade $F = 1.9454$, p-value $= 0.1486$ for F-score), nor FRCNN detector (Quade $F = 0.84796$, p-value $= 0.4315$ for recall, Quade $F = 1.041$, p-value $= 0.3572$ for F-score).

5   Summary and Conclusions

Pollen grain detection is the main step of the automation of pollen monitoring, based on the Hirst's trap. In the presented work, YOLO and Faster R-CNN were applied in the pollen grain detection task.

The results reported in this paper show that both YOLO and Faster R-CNN detectors can identify the target taxa with high precision, but the detailed statistical analysis reveals significant differences in the distribution of precision, recall, and F-score. The analysis of average values of these indicators suggests that YOLO is a better detector than Faster R-CNN to perform the pollen grain identification task. We also found that *Carpinus* was the best recognized class, independently of the detector used. This is probably caused by the difference in size between *Carpinus* pollen grains and other analyzed grains, as *Carpinus* pollen grains are bigger that the other grains. In general, YOLO yields more stable and repeatable results regarding the precision of the predictions, which was proven by statistical analysis results. The biggest differences between the investigated detectors were found for *Alnus* pollen grain identification.

The presented work corroborates our previous findings [33], indicating that YOLO should be a preferred detector in pollen grain identification, as it yields more stable precision, and higher values of all investigated measures.

References

1. European Academy of Allergy & Clinical Immunology: Worldwide Map of Pollen Monitoring Stations. https://www.eaaci.org/19-activities/task-forces/4342-pollen-monitoring-stations-of-the-world.html Last accessed 14 Jun 2022
2. Pawankar, R., Canonica, G.W., Holgate, S.T., Lockey, R.F.: World Allergy Organization (WAO) White Book on Allergy, vol. 3, pp. 156–157. WAO, Milwaukee (2011)
3. Buters, J., Antunes, C., Galveias, A., Bergmann, K.C., Thibaudon, M., Galán, C., Schmidt-Weber, C., Oteros, J.: Pollen and spore monitoring in the world. Clin. Transl. Allergy 8, 9 (2018). https://doi.org/10.1186/s13601-018-0197-8
4. Hirst, J.M.: An automatic volumetric spore trap. Ann. Appl. Biol. 39, 257–265 (1952)
5. Mandrioli, P., Comtois, P., Levizzani, V.: Methods in Aerobiology. Pitagora Editrice, Bologna, Italy (1998)
6. Galán, C., Cariñanos, P., Alcázar, P., Dominguez-Vilches, E.: Spanish Aerobiology Network (REA) Management and Quality Manual. Servicio de Publicaciones, Universidad de Córdoba, Córdoba, Spain (2007)
7. Piotrowska, K., Kubik-Komar, A.: The effect of meteorological factors on airborne Betula pollen concentrations in Lublin (Poland). Aerobiologia 28, 467–479 (2012)
8. Dunker, S., et al.: Pollen analysis using multispectral imaging flow cytometry and deep learning. New Phytologist 229(1), 593–606 (2021)
9. Bell, K.L., Burgess, K.S., Botsch, J.C., Dobbs, E.K., Read, T.D., Brosi, B.J.: Quantitative and qualitative assessment of pollen DNA metabarcoding using constructed species mixtures. Mol. Ecol. 28, 431–455 (2019)
10. Martin, A.C., Harvey, W.J.: The global pollen project: a new tool for pollen identification and the dissemination of physical reference collections. Methods Ecol. Evol. 8(7), 892–897 (2017). https://doi.org/10.1111/2041-210X.12752
11. Ascari, L., Novara, C., Dusio, V., Oddi, L., Siniscalco, C.: Quantitative methods in microscopy to assess pollen viability in different plant taxa. Plant Reprod. 33, 205–219 (2020)
12. Plaza, M., Kolek, F., Leier-Wirtz, V., Brunner, J., Traidl-Hoffmann, C., Damialis, A.: Detecting airborne pollen using an automatic, real-time monitoring system: evidence from two sites. Int. J. Environ. Res. Public. Health 19(4), 2471 (2022). https://doi.org/10.3390/ijerph19042471
13. Li, P., Flenley, J.R.: Pollen texture identification using neural networks. Grana 38, 59–64 (1999)
14. France, I., Duller, A.W.G., Duller, G.A.T., Lamb, H.F.: A new approach to automated pollen analysis. Quat. Sci. Rev. 19, 537–546 (2000)
15. Tello-Mijares, S., Flores, F.: A novel method for the separation of overlapping pollen species for automated detection and classification. Comput. Math. Methods Med. 2016, 5689346 (2016)
16. Battiato, S., Ortis, A., Trenta, F., Ascari, L., Politi, M., Siniscalco, C.: Pollen13K: a large scale microscope pollen grain image dataset. In: 2020 IEEE International Conference on Image Processing (ICIP), pp. 2456–2460. IEEE (2020)

17. Sevillano, V., Aznarte, J.L.: Improving classification of pollen grain images of the POLEN23E dataset through three different applications of deep learning convolutional neural networks. PLoS ONE **13**(9), e0201807 (2018)
18. Sevillano, V., Holt, K., Aznarte, J.L.: Precise automatic classification of 46 different pollen types with convolutional neural networks. PLoS ONE **15**, e0229751 (2020)
19. Astolfi, G., et al.: POLLEN73S: an image dataset for pollen grains classification. Ecol. Inf. **60**, 101165 (2020)
20. Razavian, A.S., Azizpour, H., Sullivan, J., Carlsson, S.: CNN features off-the-shelf: an astounding baseline for recognition. In: 2014 IEEE Conference on Computer Vision and Pattern Recognition (CVPR), DeepVision Workshop (2014)
21. Waithe, D., Brown, J.M., Reglinski, K., Diez-Sevilla, I., Roberts, D., Eggeling, C.: Object detection networks and augmented reality for cellular detection in fluorescence microscopy. J. Cell Biol. **219**(10), e201903166 (2020)
22. Casado-García, A., et al.: LabelStoma: a tool for stomata detection based on the YOLO algorithm. Comput. Electron. Agric. **178**, 105751 (2020)
23. Abas, S.M., Abdulazeez, A.M., Zeebaree, D.Q.: A YOLO and convolutional neural network for the detection and classification of leukocytes in leukemia. Indones. J. Electr. Eng. Comput. Sci. **25**, 200–213 (2022)
24. Cao, N., Meyer, M., Thiele, L., Saukh, O.: Automated pollen detection with an affordable technology. In: Proceedings of the 2020 International Conference on Embedded Wireless Systems and Networks EWSN 2020, pp. 108–119 (2020)
25. Puc, M.: Characterisation of pollen allergens. Ann. Agric. Environ. Med. **10**, 143–149 (2003)
26. Vik, H., Florvaag, E., Elsayed, S.: Allergenic significance of Betula (birch) pollen. In: D'Amato, G., Spieksma, F.T.M., Bonini, S. (eds.) Allergenic Pollen and Pollinosis in Europe, pp. 94–98. Blackwell Scientific Publications, London (1991)
27. Zając, A., Zając, M.: Atlas rozmieszczenia roślin naczyniowych w Polsce (Distribution atlas of vascular plants in Poland). Jagiellonian University, Poland (2001)
28. Redmon, J., Divvala, S., Girshick, R., Farhadi, A.: You only look once: unified, real-time object detection. In: Proceedings of the IEEE Conference on Computer Vision and Pattern Recognition, pp. 779–788 (2016)
29. YOLOv5 Documentation. https://docs.ultralytics.com/ Accessed 14 Jun 2022
30. COCO - Common Objects in Context. https://cocodataset.org/ Accessed 14 Jun 2022
31. Ren, S., He, K., Girshick, R., Sun, J.: Faster R-CNN: towards real-time object detection with region proposal networks. In: Cortes, C., et al. (eds.) Proceedings of the Advances in Neural Information Processing Systems, vol. 28. Curran Associates, Inc. (2015)
32. Wu, Y., Kirillov, A., Massa, F., Lo, W.Y., Girshick, R.: Detectron2 (2019). www.github.com/facebookresearch/detectron2 Accessed 14 Jun 2022
33. Kubera, E., Kubik-Komar, A., Kurasiński, P., Piotrowska-Weryszko, K., Skrzypiec, M.: Detection and recognition of pollen grains in multilabel microscopic images. Sensors **22**(7), 2690 (2022). https://doi.org/10.3390/s22072690
34. Conover, W.J.: Practical Nonparametric Statistics. Wiley, New York (1999)
35. StatSoft Inc., "Statistica" Data Analysis Software System, version 10 (2011)
36. Pohlert, T.: The pairwise multiple comparison of mean ranks package (PMCMR). R Package (2014)

37. Pohlert, T.: PMCMRplus: calculate pairwise multiple comparisons of mean rank sums extended. R Package (2018)
38. García, S., Fernández, A., Luengo, J., Herrera, F.: Advanced nonparametric tests for multiple comparisons in the design of experiments in computational intelligence and data mining: experimental analysis of power. Inf. Sci. **180**(10), 2044–2064 (2010)

Rough Sets for Intelligence on Embedded Systems

Katrina Nesterenko$^{(\boxtimes)}$ and Rory Lewis$^{(\boxtimes)}$

University of Colorado at Colorado Springs, Colorado Springs, CO 80918, USA
{kbrandau,rlewis5}@uccs.edu

Abstract. In the continuing effort to run artificial intelligence (AI) on resource-constrained embedded devices, we present experiments that test the viability of enabling an embedded system to run Rough Set Theory. We first show that the Fuzzy-Rough Nearest Neighbor (FRNN) algorithm, a classic Rough Sets methodology, can indeed be effectively run on an embedded device. Next we compare 10 iterations of four algorithms in terms of power consumption, speed, accuracy, and model size. Specifically, we analyze on an embedded system, KNN, Fuzzy-Rough Nearest Neighbour (FRNN) without weights, FRNN with weights and Fuzzy Rough One-Versus-One (FROVOCO). Herein, we present a step towards the goal of designing an AI microprocessor, without software, based entirely on Rough Sets.

Keywords: Embedded systems · Rough Set Theory

1 Introduction

For the last few decades we have witnessed a distinct separation between the hardware, software and machine learning communities [2,14]. The issue is that artificial intelligence and machine learning algorithms have always been designed to be implemented on software. However, in recent years there has been a paradigm shift and this was said no better than in December 2021s Association for Computing Machinery Communications magazine [1,20] that, in part states: "*We are experiencing a second pendulum swing back to specialized hardware. Catalysts include changing hardware economics, prompted by both the end of Moore's law . . . and the dizzying requirements of deploying machine learning to edge devices* [42]."

DISCERNING AI FROM ML AND DNNS. It is critical that one distinguishes AI, where training is controlled by humans, to its subsets of machine learning (ML) and deep learning such as Deep Neural Networks (DNNs), that learn from hidden layers void of human input or comprehension. DNNs have been implemented on non-critical embedded systems such as edge devices for Internet of Things (IoT) units [17,27] since 2017 [3,40]. However, embedded DNN system are not tolerated for systems involving i) human lives such as in medical institutions [5] and the military [8], or ii) critical resources such as satellites and missiles

M. Ceci et al. (Eds.): ISMIS 2022, LNAI 13515, pp. 230–239, 2022.
https://doi.org/10.1007/978-3-031-16564-1_22

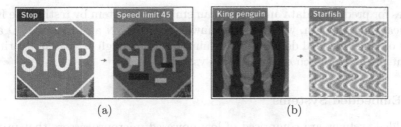

(a)   (b)

Fig. 1. Confused DNNs: (a) Stickers cause DNN to incorrectly identify a *Stop* sign as *Speed Limit 45* [38]. **(b)** DNN identifies mere patterns as familiar objects [29].

[7] because they are unreliable. Consider how Nature, a prestigious journal that rarely deep dives into the mathematics of computer science, recently reported [19] how easily neural networks can be fooled because their training is void of human sentience as illustrated in Figs. 1 & 2. Figure 1a, shows that it is possible to completely fool a DNN system designed by a heavily funded joint research project from Google and Facebook. The DNN correctly identifies the image of a *STOP* sign on the left-hand side. Sadly, by simply adding a few stickers onto the same *STOP* sign on the right-hand side, the researchers were able to convince the machine that

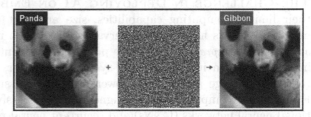

Fig. 2. Perception fooled. Noise added to right-hand image makes DNN identify it as a gibbon [19].

it was looking at a *Speed Limit 45* sign [38]. This of course is disastrous for self driving cars' AI systems! Similarly, Fig. 1b, shows that it was possible to make the same, incredibly powerful DNN system, see things that were not there, such as a penguin or a starfish in patterns of wavy lines [29]. Additionally, consider Google's top DNN system in Fig. 2 showing two pictures of a panda. Here, Google researchers added some Gaussian noise to the right-hand image, which to humans still looks like the original panda image on the left-hand side. However, this mere titillation of Gaussian noise completely confused one of the most powerful DNN systems to identify it as a gibbon!

THE ISSUE is that critical industrial sectors such as, but not limited to the medical and military sectors, are demanding that AI be deployed on edge devices [42]. Unfortunately, the computer science community is well aware, as illustrated above, that DNNs are not going to fulfill this need. As a resultant we need to '*go back*' to considering a concept barely studied or funded, integrating traditional AI onto embedded systems. Accordingly, this paper elucidates our current experiments implementing Rough Set Theory, which embraces incomplete information theories for predictive data analysis, on an embedded system by testing the KNN classification algorithm, weighted and unweighted FRNN and FROVOCO algorithms on an embedded device to determine whether rough set-based algorithms warrant further research on embedded systems.

1.1 Embedded Systems

Embedded systems are comprised of low-powered microprocessors that provide real-time programming instructions to a computer hardware system. We use the term "*embedded system*" to include digital signal processors (DSPs), application specific integrated circuits (ASICs), field programmable gate arrays (FPGAs), and graphic processing units (GPUs) [43]. Embedded software engineers write programming instructions, called "*firmware*", that the embedded system stores in read-only or flash memory chips. The architecture's sensors convert physical states into electrical signals, analog-to-digital converters and actuators.

THE CHALLENGE IN DEPLOYING AI ON EMBEDDED SYSTEMS lies in their limited computing capabilities, size and cost constraints [28] Although multi-core GPUs have become pervasive in the ML research communities [36], as illustrated *supra* §1, their large power requirements are a deal breaker when integrating them into low power systems. However, in these non critical systems where large amounts of power are not an issue, ML is having success. For example, computer vision designers have successfully implemented convolutional neural networks (CNNs) and recurrent neural networks (RNNs) on 32-bit RISC ARM processors for vision apps on mobile phones [30]. Other stand-out, ML algorithms on non-critical high powered embedded systems include: Random Forests (RFs) embedded systems use Logarithmic Number Systems (LNS), comprised of covariance matrix computation modules. [31]. ii) Support-Vector Machines (SVMs) embedded systems on an ARMv7 rev 2 to control two communications networks [16]. iii) Q-learning DNN (QS-DNN) heterogeneous embedded devices use an autonomous QS-DNN to optimize CNNs. [33]. iv) Convolutional Neural Networks (CNNs) eEmbedded systems use an approximated CNN with a STM32-F7 even on a Raspberry Pi 3B [4].

2   Algorithm Preliminaries

2.1   KNN

In 1951 Evelyn Fix and Joseph Hodges introduced the K-nearest neighbor [37] that comprises a non-parametric methodology [24] for pattern classification. This simple [35] and robust [10] classifier measures the distance between tested examples and the training examples [34]. KNN only has one hyperparameter k [15] and therefore yields a normalization for us to test variations of KNN in embedded systems. To accomplish classification, KNN invokes two steps. First, in the *Learning Phase*, a training set $\mathcal{D} = \{(x_t, y_t)\}_{t=1}^{N}$ is used where the input vectors x_t and y_t correspond to the $t-$th instance of y and $\mathcal{D}$ simply holds the data set

[23]. Once a value of k is chosen, it calculates the Euclidean of the training data points using the formula $\sqrt{\sum_{i-1}^{n}(x_i - y_i)^2}$ that finds the k-nearest neighbors and assigns a class containing the maximum number of nearest neighbors. Next, during the *Inference Phase*, when choosing the value of k, the optimal value of k is $\sqrt{N}$ where N is the total number of samples. So after storing the training set we need to normalize the parameters. For each query instance of x, a search is performed to retrieve KNN instances $\mathcal{N}(x_t) = \{(x_t^{(i)}, y_t^{(i)})\}_{t=1}^{k}$ that are closest to x it yields $\hat{y}$ as the weighted combination of the labels $y^{(1)}, ..., y^{(k)}$ based on a weighting function w along with the distance function d [23] as follows:

$$\hat{y} = f(x; \mathcal{D}) = \frac{\sum_{t=1}^{k} w(d(x, x^{(i)})) \cdot y^{(i)}}{\sum_{t=1}^{k} w(d(x, x^{(i)}))} \tag{1}$$

where k is the number of neighbors, d is the distance function where in this example we used Euclidean, and w is the weighting function. Some rules to consider for K is that when K is small, noise will have a higher influence on the result [13]. When K is large value make it computationally expensive [11]. If the number of classes is 2. we usually choose an odd value, and finally, as mentioned above, a true and tried approach to optimizing one's choice of K is to set $K = \sqrt{n}$.

2.2 FRNN

In 1990 Dubois and Prade [12] introduced Fuzzy Rough Sets (FRS) that provided a framework for feature subset selection called reducts [25]. The algorithm builds fuzzy approximations using the nearest neighbors algorithm [21] based on the pair of lower and upper approximations of a fuzzy set A in a universe $\mathbb{U}$ on which a fuzzy relation R is defined [18]. We define ℓ as a fuzzy implication and use $\ell(R(x, y), A(y))$ to express the extent that an element similar to x belongs to $\mathcal{A}$ [18]. We define the lower approximation as:

$$(R \downarrow A)(x) = \inf_{y \in \mathbb{U}} \ell(R(x, y), A(y)) \tag{2}$$

For the upper approximation, we define $\mathcal{T}$ a t-norm, $\mathcal{T}(R(x, y), A(y))$ to express the extent that an element similar to x belongs to $\mathcal{A}$, and define it as [9] [18]:

$$(R \uparrow A)(x) = \inf_{y \in \mathbb{U}} \mathcal{T}(R(x, y), A(y)) \tag{3}$$

Segueing from FRS to the Fuzzy-Rough Nearest Neighbor (FRNN) algorithm, after finding the k nearest neighbors in the test object j. We define a class C. If the sum of $R : (R \downarrow C)(y) + (R \uparrow C)y$ can reach to the maximum, then the test object belongs to class C [22]. In essence, If a value of fuzzy lower approximation is high, it shows that neighbors of newdata belong to a particular class C. Conversely, a high value of fuzzy upper approximation means that at least one neighbor belongs to that class.

2.3   FROVOCO

In 2018 Sarah Vluymans introduced the Fuzzy Rough One-Versus-One (OVO) combination (FROVOCO) classification algorithm for multi-class imbalanced datasets, where the imbalance ratio (IR) is severely skewed [41]. OVO's decomposition procedure transforms a multi-class problem into several binary subtasks, one for each pair of classes. In essence, it is designed for imbalanced data by balancing OVO decomposition with two global class afinity measures [26]. To classify a test instance, each binary classifier trained on a sub-task is fired and computes class confidence degrees for the element to both classes under its consideration. All values are grouped in a score-matrix $R(x)$ in the form of Eq. 4 and afterwards aggregated to one class prediction for the target.

$$R(x) = \begin{pmatrix} - & r_{12} & \cdots & r_{1m} \\ r_{21} & - & \cdots & r_{2m} \\ \vdots & & & \vdots \\ r_{m1} & r_{m2} & \cdots & - \end{pmatrix}, \tag{4}$$

3   Experiments

We begin these experiments with the expectation that the vanilla KNN algorithm will outperform the more complex fuzzy rough models on embedded systems. We start our embedded systems analysis comparing KNN models and FRNN models where the weights are set to zero. We then compare the same KNN models to FRNN

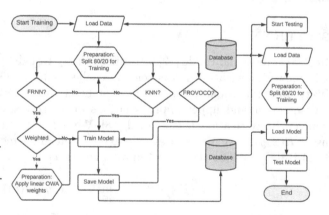

Fig. 3. Flowchart depicting the firmware training and testing loop.

models where the similarity weight is calculated using linear ordered weighted average (OWA) weights. Lastly, we make comparisons between KNN and a FROVOCO model on an embedded system. Our goal is to create a baseline for Rough Set algorithms on an embedded system, since to the best of our knowledge, this has yes to be accomplished. Our hypothesis is to verify whether Rough Set algorithms can be implemented on resource-constrained embedded systems, and if so, it will justify further Rough Set research on embedded microprocessors. For this reason, we are focused on comparing a few popular rough set algorithms using one dataset. In the future, we will use larger datasets with variations in the number of features and number of instances, etc. To test our hypothesis,

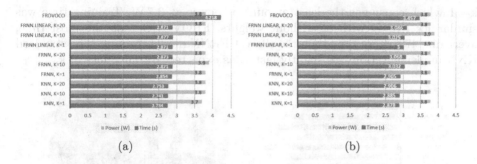

Fig. 4. Results: Comparison of model training and testing in terms of power consumption and average execution time. (a) Execution time for model training compared to the maximum recorded power consumption and (b) Execution time for model testing compared to the maximum recorded power consumption.

we implement algorithms from the fuzzy-rough-learn [26] python library on an NVIDIA Jetson Nano, as seen in Fig. 5b. We choose the Jetson Nano as its novel low-power embedded GPU lends itself to our future work implementing parallel processing of other rough set-based algorithms. Lenz, *et al.* [26] have implemented a small handful of fuzzy rough learning algorithms in an open-source python library, making fuzzy-rough-learn a good starting point for our investigation before moving on to more complex parallelized implementations of rough set-based algorithms.

We use the abalone dataset that has over 4,000 tuples and 9 attributes, with eight condition attributes and one decision attribute. The dataset focuses on approximating the age of abalones, a type of marine snail, traditionally done by drilling holes through their shells and counting the rings. The condition attributes include sex, length, diameter, height, whole weight, shucked weight, viscera weight, and shell weight, where the decision attribute is the number of rings. The rings range from 1 to 29, which we group into three bins: 1 to 10, 11 to 20, and 21 to 30. Specifically, we compare FRNN and FROVOCO to the Scikit Learn [32] implementation of KNN to assess the plausibility of using fuzzy rough classification algorithms on the NVIDIA Jetson Nano, a CUDA-enabled embedded device. Power consumption is measured by the programmable power supply as illustrated in Fig. 5b. The firmware tests depicted in Fig. 3 are executed 10 times for each model. The execution time is averaged over 10 runs, and the maximum recorded power consumption over all 10 runs is stored.

3.1 Model Training

Now that we have successfully run fuzzy rough algorithms on an embedded system, we assess the time and power consumption for training each model, averaged over 10 executions each, illustrated in Fig. 4a. One observes that the maximum power consumption is the same for all test cases with the exceptions of FRNN $k = 10$ which has the highest power consumption at 3.9 W, and KNN

$k = 1$ which consumes the least amount of power at 3.7 W. Training time was similar for FRNN with or without weights. FROVOCO took the most time, averaging 47% more time than the slowest FRNN model training. The standard KNN implementation was slightly faster, as expected, for all values of k.

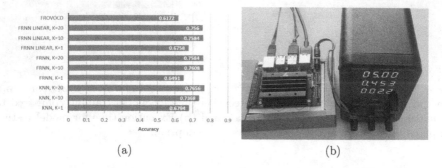

(a)   (b)

Fig. 5. (a) KNN, FRNN, and FROVOCO model accuracy are compared for all tested values of k after training on an 80/20 split of the abalone dataset, and (b) Physical experiment setup with the NVIDIA Jetson Nano connected to the programmable power supply.

3.2   Model Testing

Resource-constrained environments often trade faster execution times for lower power and memory usage [6]. We additionally compare the time it takes to load each model from a serialized file and test it with the remaining 20% of the abalone dataset. The results are illustrated in Fig. 4b. The standard KNN model once again outperforms the other models in regards to testing time. The power consumption across all 10 models was nearly identical in these tests as well, with the exception of FRNN with linear weights at $k = 1$ and $k = 10$ being 100 mW higher than the maximum recorded power consumption of all other models at 3.8W. We note that both FRNN models, with and without weights, with $k = 10$ slightly outperformed standard KNN $k = 10$ on an embedded system with approximately 76% accuracy each, while standard KNN was 74% accurate as seen in Fig. 5a. This warrants further investigation in certain research areas within resource-constrained environments where execution time is less important than accuracy [39]. Notably, we were indeed able to successfully implement the FROVOCO model on our embedded system where, as expected, it was the least accurate of all tested models at 62%. As mentioned above, FROVOCO was designed for balancing imbalanced datasets, and while the abalone dataset is chaotic and full of outliers, it is not imbalanced, so it stands to reason that FROVOCO did not perform well in terms of accuracy. The exported serialized models were the same size for all k values, with each of the KNN models being 503 kB and each of the weighted FRNN models being 2.07 MB. The non-weighted

FRNN models were also 2.07 MB, except for the k = 1 model, which was 1.08 MB. The FROVOCO model was the largest at 2.75 MB, which is expected as it is an ensemble classifier comprised of many smaller classifiers.

4   Future Research

Now that we have successfully implemented the aforementioned rough set-based algorithms on an embedded system, we will continue on to test the viability of running C4.5 on an embedded system with a database on a separate chip. The point of this will be to code the firmware to integrate i) indiscernibility matrices with ii) cached portions of an external embedded database and iii) a more advanced rough set algorithm. With these new experiments, if they are to succeed, we will have the basis of an embedded system comprising a central discernibility matrix and an array of rough set algorithms that is able to perform knowledge discovery in an interchangable database on a third-party embedded system. From there, we will begin to prototype our rough sets-based microprocessor using a hardware description language such as VHDL or Verilog on an FPGA. Our ultimate goal is to create an autonomous self-learning microprocessor based on Rough Set Theory that is capable of human-readable decision processing.

5   Conclusion

The success of our experiments proves that Rough Set Theory will be well-suited for this task due to its ability to embrace incomplete information systems for predictive data analysis. Additionally, we were able to compare these algorithms to the well-studied artificial intelligence algorithm KNN. After comparing the execution time and power consumption of these algorithms, we discerned that they are comparable to classical Knowledge Discovery in Databases (KDD) algorithms in certain cases and therefore conclude that our hypothesis of executing Rough Set Theory on embedded systems does indeed warrant further research. In fact, in some cases, the fuzzy rough algorithms consumed the same amount of power as KNN, and had better accuracy.

References

1. Association for Computing Machinery. https://www.acm.org/
2. Innovation through Collaboration: Building Research Communities. https://intel. ly/3Mz7pS9
3. Ajani, T.S., Imoize, A.L., Atayero, A.A.: An overview of machine learning within embedded and mobile devices-optimizations and applications. Sensors (Basel) 21(13), 4412 (2021). https://doi.org/10.3390/s21134412
4. Alippi, C., Disabato, S., Roveri, M.: Moving convolutional neural networks to embedded systems: the AlexNet and VGG-16 case. In: 2018 17th ACM/IEEE International Conference on Information Processing in Sensor Networks (IPSN), pp. 212–223. IEEE (2018)

5. Asan, O., Bayrak, A.E., Choudhury, A., et al.: AI and human trust in healthcare: focus on clinicians. J. Med. Internet Res. **22**(6), e15154 (2020)
6. Assayad, I., Girault, A., Kalla, H.: Tradeoff exploration between reliability, power consumption, and execution time for embedded systems: the TSH tricriteria scheduling heuristic. Int. J. Softw. Tools Technol. Transfer **15**(3), 229–245 (2013). https://doi.org/10.1007/s10009-012-0263-9
7. Boulanin, V.: The impact of artificial intelligence on strategic stability and nuclear risk: euro-atlantic perspectives (2019)
8. Clark, C.: Air combat commander doesn't trust project maven's AI yet. AIR WARFARE, NETWORKS / CYBER (2019). https://bit.ly/3MD5A6O
9. Cornelis, C., Jensen, R.: A noise-tolerant approach to fuzzy-rough feature selection. In: 2008 IEEE International Conference on Fuzzy Systems (IEEE World Congress on Computational Intelligence), pp. 1598–1605. IEEE (2008)
10. Cover, T., Hart, P.: Nearest neighbor pattern classification. IEEE Trans. Inf. Theory **13**(1), 21–27 (1967)
11. Dong, W., Moses, C., Li, K.: Efficient k-nearest neighbor graph construction for generic similarity measures. In: Proceedings of the 20th International Conference on World Wide Web, pp. 577–586 (2011)
12. Dubois, D., Prade, H.: Rough fuzzy sets and fuzzy rough sets. Int. J. Gen. Syst. **17**(2–3), 191–209 (1990)
13. García-Pedrajas, N., Ortiz-Boyer, D.: Boosting knn classifier by means of input space projection. Expert Syst. Appl. **36**(7), 10570–10582 (2009)
14. Gil, Y., Selman, B.: A 20-year community roadmap for artificial intelligence research in the us. arXiv preprint arXiv:1908.02624 (2019)
15. Gupta, C., et al.: Protonn: compressed and accurate knn for resource-scarce devices. In: International Conference on ML, pp. 1331–1340. PMLR (2017)
16. Haigh, K.Z., Mackay, A.M., Cook, M.R., Lin, L.G.: Machine Learning for Embedded Systems: A Case Study. BBN Technologies, Cambridge, MA, USA (2015)
17. Hayyolalam, V., Aloqaily, M., Özkasap, Ö., Guizani, M.: Edge intelligence for empowering IoT-based healthcare systems. IEEE Wirel. Comm. **28**(3), 6–14 (2021)
18. He, R., Xu, C., Li, D., Hou, W., Yu, X., Zhang, H.: A fuzzy-rough-based approach for uncertainty classification on hybrid info sys. In: 2018 IEEE 3rd International Conference on Image, Vision and Computing (ICIVC), pp. 791–796. IEEE (2018)
19. Heaven, D., et al.: Why deep-learning AIs are so easy to fool. Nature **574**(7777), 163–166 (2019)
20. Hooker, S.: The hardware lottery. Comm. ACM **64**(12), 58–65 (2021)
21. Hussein, A.S., Khairy, R.S., Najeeb, S.M.M., Alrikabi, H.T., et al.: Credit card fraud detection using FRNN and sequential minimal optimization with logistic regression. Int. J Interact. Mob. Technol. **15**(5), 24–42 (2021)
22. Jensen, R., Cornelis, C.: Fuzzy-rough nearest neighbour classification and prediction. Theo. Comput. Sci. **412**(42), 5871–5884 (2011)
23. Kang, S.: Knn learning with graph neural networks. Mathematics **9**(8), 830 (2021)
24. Kataria, A., Singh, M.: A review of data classification using KNN algorithm. Int. J. Emerg. Technol. Adv. Eng. **3**(6), 354–360 (2013)
25. Kumar, A., Prasad, P.S.: Scalable fuzzy rough set reduct computation using fuzzy min-max neural network preprocessing. IEEE Trans. Fuzzy Syst. **28**(5), 953–964 (2020)
26. Lenz, O.U., Peralta, D., Cornelis, C.: *fuzzy-rough-learn* 0.1: a Python library for machine learning with fuzzy rough sets. In: Bello, R., Miao, D., Falcon, R., Nakata, M., Rosete, A., Ciucci, D. (eds.) IJCRS 2020. LNCS (LNAI), vol. 12179, pp. 491–499. Springer, Cham (2020). https://doi.org/10.1007/978-3-030-52705-1_36

27. Merenda, M., Porcaro, C., Iero, D.: Edge machine learning for AI-enabled IoT devices: a review. Sensors **20**(9), 2533 (2020)
28. Nadeski, M.: Bringing ML to embedded systems. TX Instruments (2019)
29. Nguyen, A., Yosinski, J., Clune, J.: Deep neural networks are easily fooled: high confidence predictions for unrecognizable images. In: Proceedings of the IEEE Conference on Computer Vision and Pattern Recognition, pp. 427–436 (2015)
30. Nordby, J.: Machine learning on embedded devices (2020). http://bit.ly/2NmYgDA Accessed 15 Feb 2021
31. Osman, H.E.: On-chip object recognition system using random forests. In: 2008 IEEE/ASME International Conference on Mechtronic and Embedded Systems and Applications, pp. 77–82 (2008)
32. Pedregosa, F., et al.: Scikit-learn: machine learning in Python. J. Mach. Learn. Res. **12**, 2825–2830 (2011)
33. de Prado, M., Pazos, N., Benini, L.: Learning to infer: Rl-based search for dnn primitive selection on heterogeneous embedded systems. In: 2019 Design, Automation & Test in Europe Conference & Exhibition, pp. 1409–1414. IEEE (2019)
34. Prasatha, V., et al.: Effects of distance measure choice on KNN classifier performance-a review, p. 56. arXiv preprint arXiv:1708.04321 (2017)
35. Rendle, S., Freudenthaler, C., Gantner, Z., Schmidt-Thieme, L.: Bpr: Bayesian personalized ranking from implicit feedback. arXiv preprint arXiv:1205.2618 (2012)
36. Shadrin, D., Menshchikov, A., Ermilov, D., Somov, A.: Designing future precision agriculture: detection of seeds germination using artificial intelligence on a low-power embedded system. IEEE Sens. J. **19**(23), 11573–11582 (2019)
37. Silverman, B.W., Jones, M.C.: E. fix and J.L Hodges (1951): an important contribution to nonparametric discriminant analysis and density estimation: commentary on fix and Hodges. In: International Statistical Review/Revue Internationale de Statistique, pp. 233–238 (1989)
38. Szegedy, C., Zaremba, W., Sutskever, I., Bruna, J., Erhan, D., Goodfellow, I., Fergus, R.: Intriguing properties of neural nets. arXiv arXiv:1312.6199 (2013)
39. Taylor, B., Marco, V.S., Wolff, W., Elkhatib, Y., Wang, Z.: Adaptive deep learning model selection on embedded systems. ACM SIGPLAN Not. **53**(6), 31–43 (2018). https://doi.org/10.1145/3299710.3211336
40. Velasco-Montero, D., Fernández-Berni, J., Carmona-Galán, R., Rodríguez-Vázquez, Á.: Optimum selection of DNN model and framework for edge inference. IEEE Access **6**, 51680–51692 (2018)
41. Vluymans, S., Fernández, A., Saeys, Y., Cornelis, C., Herrera, F.: Dynamic affinity-based classification of multi-class imbalanced data with ovo decomposition: a fuzzy rough set approach. Knowl. Inf. Sys. **56**(1), 55–84 (2018)
42. Warden, P., Situnayake, D.: Tinyml: Machine Learning with Tensorflow Lite on Arduino and Ultra-Low-Power Microcontrollers. O'Reilly Media (2019)
43. Yonekawa, H., Nakahara, H.: On-chip memory based binarized convolutional deep neural network applying batch normalization free technique on an FPGA. In: 2017 IEEE International Parallel and Distributed Processing Symposium Workshops (IPDPSW), pp. 98–105 (2017). https://doi.org/10.1109/IPDPSW.2017.95, 00068

Context as a Distance Function in ConSQL

Hasan M. Jamil$^{(\boxtimes)}$ (iD)

Department of Computer Science, University of Idaho, Moscow, Idaho, USA
jamil@uidaho.edu

Abstract. In conversational query answering systems, context plays a significant role to accurately and meaningfully carry it forward. In many chatbots, such as in Expedia, the discussion quickly degenerates into circling back to restarting the conversation or to inviting a live agent to intervene because the bot could not grasp the context. Contexts shorten interactions by way of implied query constraints to narrow search and to not repeat them in subsequent queries. In this paper, we introduce a novel way of viewing contexts as a distance function via the concept of query relaxation. We demonstrate that a typed domain distance function is sufficient to model context in a conversation. Our approach is based on the idea of non-monotonic constraint inheritance in a context hierarchy.

Keywords: Contextual querying · Query relaxation · Non-monotonic constraint inheritance · Preferred tuples · Top-k querying

1 Introduction

The concept of context is often application specific and a generalization of it is difficult. Early research in databases (DB) and artificial intelligence (AI) context has been modeled sometimes indirectly without an explicit treatment. For example, cooperative query answering [2], indefinite and maybe queries [7], incomplete databases [6], null values [4], disjunctive databases [3], contextual query answering [8] and knowledge representation and reasoning [1]. Except for ConteLog [1], most of these systems treat context indirectly and often as a by product.

For example, consider the query against the table *Restaurants* below.

Q_1: Is there a five star rated Vietnamese restaurant in Moscow, Idaho?

TiD	Name	Location	Cuisine	Rating	Price	Dining
t_1	Pho Heaven	Moscow	Vietnamese	*	Inexpensive	Carryout
t_2	Hot Pot	Moscow	Korean	****	Very Expensive	Premise
t_3	Ao Dai	Pullman	Vietnamese	*****	Moderate	Both
t_4	Ravi Kabob	Arlington	Pakistani	*****	Inexpensive	Both
t_5	Little Tibet	Spokane	Tibetan	*****	Expensive	Both

In a commonsense world, we can imagine multiple equally acceptable responses depending on the how we interpret the intent of the query Q_1. In CoBase [2], the constants "Moscow", "Vietnamese" and "*****" serve as contexts in three axes. Using a type abstraction hierarchy (TAH), the domain values, or objects, are organized in a proximity hierarchy from fine to coarse. In such a hierarchy, "Vietnamese", "Korean" and "Chinese" will be similar (to "Oriental") types of cuisines. Therefore, they are also substitutable. "Tibetan" cuisine, on the other hand, is almost Oriental being close to Chinese cultural influence. "Mughlai" cusisine, however, is part of both "Indian" and "Paksitani" cuisine. Therefore, it can be included in a query response for Pakistani food.

Using similar TAH and similarity computing schemes, we can compute the similarities of the domain values in columns *Rating* and the *Location*, and finally compute a composite similarity score for each tuple based on these three attributes. We can even accommodate a weighted similarity score using user assigned relative importance of each of these query conditions. While the idea is simple, and processing such queries are not too difficult, the construction of the TAHs and their maintenance, however, are extremely complex. Slight change in membership often requires expensive reorganization. We refer the readers to [5] for a more detailed discussion on cooperative query answering in general and query relaxation in particular.

2   Context as a First-Class Citizen for Query Answering

While context plays a significant role in many systems, only the recently proposed ConteLog [1] modeled it as a first-class citizen by giving it a syntax, and a direct semantics. However, the model and the semantics assigned can only capture limited capabilities with a significant degree of user involvement. For example, in ConteLog, contextual queries of the form

Q_2: "Which side of the street will the Hall Building be if I came from the Dorval Airport along the De Maisonneuve Boulevard in Montreal?"

can be asked, and answered only if the location of the Hall Building is available relative to the streets, and the direction and position contexts are explicitly supplied in the query. However, in ways similar to CoBase's TAH, users must supply context information for ConteLog to be able to tease out interesting responses, and thus making the platform arguably similarly restrictive and cumbersome. We argue that a more hands off, and intuitive context modeling is preferred.

2.1   The Concept of Context in a Conversation

Technically, we view a conversation between a human agent, say Abebi, and a robot or a computer agent, say the Expedia chatbot, as a series of queries by Abebi and an at each step, efforts by Expedia to answer those. Usually, a conversation is initiated by the human agent, and answered by the chatbot until the human agent is satisfied or the robot exhausts all possibilities. Responses are often approximate, and can only be computed is some form of adjustments to the query conditions are made, i.e., relaxation [5].

In our view, the initial query serves as the context, the query conditions to be exact. Each subsequent query is considered a relaxation of the initial context or conditions, and viewed as a stack of queries, the most recent being at the top. Given a series of queries the conditions that are not relaxed flow upward (stay unchanged or inherited), and the conditions that change, override the previous conditions. We call it an inverted inheritance stack-tree. We now discuss the idea using the example in Fig. 1. As shown, while queries can be asked in a single stack and contexts modified, it is also possible to *switch* contexts by opening branch stacks to explore multiple possibilities, and hop branches.

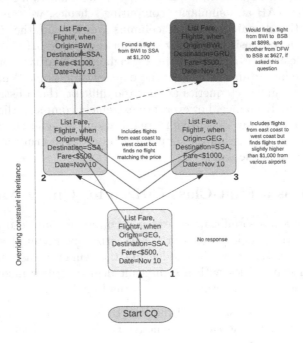

Fig. 1. Inverted context tree as a stack.

In the example in Fig. 1, the queries are asked in succession in the direction of the green arrow, and the description of responses that can be generated are shown on the side. In response to the initial query in the grey box number 1, no response could be found, and the *implied* context is set to the query conditions. The query response was empty largely because there were no flights from GEG (Spokane). However, as soon as the query in teal box 2 is asked by Abebi, in which the constraint *Origin* is relaxed from GEG to BWI (Baltimore), a significant number of flights became possible. However, the result is still empty. This is because by relaxing the *Origin* constraint, Abebi signaled that she is flexible on it and she is willing to accept a deviation up to a geographic distance from Spokane to Baltimore, which is about the entire continental USA in practical terms. However, the other constraints did not allow for a likely response.

The moment Abebi asked the query in the pink box 3 and relaxed the *Fare*, all the airports in USA with a fare less than $1,200 can now be searched. This relaxation allowed the discovery of the flight from BWI to SSA at $1,200 in the green box 4. It should be noted that there were cheaper flights on November 12, but were never considered because Abebi did not intend to relax the travel dates. On the other hand, Abebi never wanted to move the *Destination* from SSA (Salvador) to any other location to explore possibilities. Had she asked the question in the dark green box 5, she could have discovered the flights from BWI to BSB (Brasilia) at $898 and DFW (Dallas) to BSB at $627.

The guiding principle in our model is that every relaxation by the user estab-lishes a wider set of possibilities with a target to select the closest ones stated in initial query in the grey box 1 establishing a firm context that the user started with. The goal is to the find the tightest match. The mechanism used is the *probing* in the form of query relaxation to indicate what and by how much devi-ation a user is willing to accept from the original stated intent. The constraints at any stage of the interrogation or conversation is carried forward in an over-riding fashion (*Origin* in box 2 overrides the *Origin* in box 1). In box 3, the *Fare* overrides the *Fare* in box 1, but establishes a distinct context – a *context switch*.

2.2 Cumulative and Disjunctive Context Switch

The overall spirit captured in the conversation modeled in Fig. 1 in Sect. 2.1 is that Abebi is looking for a flight from Spokane to Salvador on November 10 at a price less than $500, but she could fly from anywhere in the United States to anywhere near Salvador but not too far from Sao Paulo. She is also flexible on the fare up to about $1,200. What is not negotiable is the date of travel on November 10. The query context can also include a larger picture. For example, the profile of Abebi could serve as a more defining context, and sit at the bottom of the query stack to influence the exploration. Let's assume that she is a Delta frequent flyer, has her home in Baltimore, and this a business trip for her paid for by her work. It is also the case that Abebi will have to be in Baltimore after she returns from Salvador to spend her winter vacation. The question is, how do these set of information change the search? Apparently, there are two distinct ways of applying the context based query refinement – *cumulative* and *disjunctive*. We now discuss these two semantics of contextual querying.

Cumulative Context Switching. In cumulative context querying, users interactively explore a very large information space to discover a closest match to a target goal considered as the initial context. Every successive query asked, called the *probe* queries, is intended to steer the system to explore an altered information space with a set of relaxed constraints by treating the probes as multi-dimensional constraint relaxations. For example, the exploration in Fig. 1 can be seen as the following conversation in natural English.

1. Abebi: Could I have the fare for a flight from Spokane to Salvador on November 10 at a fare no higher than $500? – (*initial context*)
2. Agent: No, Madam, there are no such flights at this price on this day.
3. Abebi: Could you try it from Baltimore? – (*context switch from Spokane to Baltimore*)
4. Agent: There are numerous flights from many other airports not too far from Spokane, but none has a fare within your price range.
5. Abebi: How about at about $1,200 from Spokane? – (*context switch from Baltimore to Spokane and Price from $500 to $1,200*)
6. Agent: There is a Delta flight from Baltimore to Salvador at $1,200 on November 10.
   optionally, if asked
7. Abebi: Do you think I will get a cheaper fare close to $500 if I went to Sao Paulo instead? – (*context switch from Salvador to Sao Paulo and Price from $1,200 to $500 back*)
8. Agent: Yes, Ma'am. You can fly from Baltimore to Brasilia at $898 on American, or fly from Dallas-Forth Worth to Brasilia at $627 on Copa Airlines.

During the entire conversation above, the agent was squarely focused on finding the closest match to the constraints expressed in the initial context in step 1. Every time Abebi changed (relaxed) a constraint, the agent reconsidered the initial context and all the other relaxations to find the closest match so far using the maximum relaxation distance. This is the essence of cumulative context switching – the entire tree is focused on the initial context.

Disjunctive Context Switching. In disjunctive context switching, on the other hand, the global view of the initial context is not preserved – only the top of the context stack in a single query stack remains the focus. For example, in the Fig. 1, box 2 and also in the conversation in the previous section in step 3, the context *Origin* changed from Spokane to Baltimore, and it stayed in the box 3 and in step 5 when the context *Price* changed keeping the *Origin* at Spokane. In disjunctive context switching, the relaxation in the other branch of the tree will not be active – only one branch at a time as a stack will be the active context. In other words, the agent's response at step 6 will be

6a. Agent: There is no flight from Spokane to Salvador on November 10 at this price.

Note that the agent is not considering the relaxation of the *Origin* expressed in the box 2 or step 3. Evidently, a disjunctive switch allows more selective and focused search suitable for weaving through an information space.

3 Contextual Query Language ConSQL

While a contextual conversational system can be implemented as a chatbot in natural English in text or voice, or as an interactive graphical interface, it will most likely interact with an SQL or NoSQL database in some fashion to carry out the search. In this section, we introduce an extension of SQL, called the ConSQL (pronounced consequal and stands for Contextual SQL) to model context in SQL.

3.1 Syntax of ConSQL

ConSQL supports conversations in one of the two contexts by a simple extension of basic SQL statements in a conversation tree similar to the one in Fig. 1. In fact, in a ConSQL database, conversations form a forest with multiple roots. The general syntax for a conversation has the following form.

> begin [cumulative|disjunctive]|end|backtrack|fresh conversation;

A conversation begins with begin conversation instruction either in cumulative or in disjunctive mode. It can end with end conversation instruction. An entirely new conversation can be started from inside an active conversation with fresh conversation instruction. Conversations are unnamed sessions and are not persistent. Therefore, once a conversation is abandoned (using an end conversation instruction), it cannot be re-entered – has to start afresh. However, a new an additional initial context can be established by issuing fresh conversation instruction without ending a conversation giving rise to conversation forests.

Once inside a conversation, a contextual query can being. While a query generally has a traditional SQL syntax, it is annotated with a context modifier of the following form with a well-defined grammar.

> context C_1 [parent C_2] as
> select $A_1, A_2 \ldots, A_n$
> from $r_1, r_2 \ldots, r_k$
> where θ;

In the above form, C_1 and C_2 are conversation wide distinct identifiers. The construction and issuance of these contextual queries must follow specific protocols and deserves a substantial discussion. However, in this article we want to be brief for the sake of brevity and only discuss essential components of ConSQL.

In a new or fresh conversation, the first contextual query must start with context C statement without the parent option since it must be the root, and must be the only context C statement until fresh conversation is issued. All subsequent statements must of the form context C parent C' to place the query under a node in the conversation or context tree. In both instances, a stack pointer points to the current context and all computations take place within its environment. To relocate the context pointer to alter the computational environment, backtrack conversation command can be issued to move the context toward the parent of the node. In the current edition of ConSQL, we do not see any need for a forward context pointer relocation. In particular when there are multiple candidates, e.g., box 1 in Fig 1.

While several syntactic shortcuts are possible, we do not overburden the syntax of ConSQL to make querying easier since such shortcuts can be supported in the user interfaces in some fashion. For example, currently we require that the query in step 3 be phrased in ConSQL as follows

> Q_4: context C parent D as
> select Flight#, Fare
> from *Flights*
> where *Origin*= "BWI" and *Destination*= "SSA" and *Fare*<500
>   and *Date*=11/10;

where the initial context is set up as

> Q_3: context D as
> select Flight#, Fare
> from *Flights*
> where *Origin*= "GEG" and *Destination*= "SSA" and *Fare*<500
>   and *Date*=11/10;

Again, note that it is possible to imagine several fancy syntactic alternatives. Basically what we want in a modified context is to alter a condition in the parent context to relax it. Therefore, even if we allowed the syntax

> Q_5: context C parent D as
> select Flight#, Fare
> from *Flights*
> where *Origin*= "BWI";

it would essentially mean the same query Q_4 since it inherits all the unaltered constraints of Q_3. In principle, we also allow adding new relations in the from clause in a subordinate context, i.e., monotonic inflation of information space so that all previously listed constraints can be enforced. For this reason also, we require the full syntax (not the Q_5 form) not to make it difficult when mapping to SQL for execution and be counter intuitive.

3.2   Semantics of ConSQL

The semantics of ConSQL queries can be established based on the SQL queries with some massaging and fine tuning. In the current edition of ConSQL, we choose a tuple similarity based semantics for ConSQL in the following way.

Definition 1 (Relaxation). *Given two conditions θ_1 and θ_2 over an attribute A in query Q, θ_2 is called a* relaxation *or a* relaxed condition, *if $e(Q_{\theta_1}) \subseteq e(Q_{\theta_2})$, where e is the evaluation function of query Q_{θ_i} with condition θ_i, and $i \in \{1, 2\}$.*

Distance Function. We assume that there is a type-polymorphic distance function δ^t that can compute the similarity of two identical typed data items. For example, numbers, GPS location based distance, star ratings, and even for complex types such as airports (similarity of LAX and AMS), and so on. For the types δ^t cannot compute the similarity, the distance will be undefined, i.e., ∞.

Definition 2 (Close to Intent). *Let θ_2 be a relaxation of θ_1 is a over an attribute A of type t, and let c_2 and c_1 be the constants respectively. The closeness of a value v to the initial context or the intent is then given by*

$$\chi(v, c_1, c_2) = \begin{cases} 1 & \text{if } v \leq c_1 \\ 1 - \frac{(v - c_1)}{(c_2 - c_1)} & \text{if } c_1 < v \leq c_2 \end{cases}$$

Thus, given two values v_1 and v_2, v_1 is closer to the intent than v_2 if $\chi(v_2, c_1, c_2) \leq \chi(v_1, c_1, c_2)$. Using the definition of closer values, we now can construct the notion of preferred tuples as follows.

Definition 3 (Preferred Tuples). *Let t is a tuple over the scheme $A_1, A_2, \ldots, A_n$. Also let $\theta_i s$ are relaxations corresponding to the initial contexts $\phi_i s$, where $1 \leq i \leq k \leq n$. The (multiplicative) tuple closeness $\tau(t)$ is given by*

$$\tau(t) = \prod_{i=1}^{k} \chi(v_i, c_{1i}, c_{2i})$$

Finally, for two tuples t_1 and t_2, t_1 is preferred over t_2 if $\tau(t_1) \geq \tau(t_2)$ holds.

The tuple closeness ranges between 0 and 1. It must be noted here that preference relation is defined only using the relaxation conditions over the attributes involved, i.e., the remaining attributes do not play any role. Given a top-k response upper limit, we now can easily choose the top k closest responses.

4   Conclusions

In this article, we have defined context as the intent of a query the response to which we try to approximate as tightly as possible. We have leveraged the idea of query relaxation that previously played a major role in designing cooperative query answering systems, and have shown that in chatbot type applications our concept of context can play a significant role in improving conversations and services. The idea presented are at an early stage. The implementation of a query processor is on going while its efficacy has been tested manually by processing the queries by translating them to equivalent SQL queries. Refinement of the model, implementing the polymorphic distance function and context recognition within a text conversation remain as our future research.

References

1. Alsaig, A., Alagar, V.S., Shiri, N.: Contelog: a declarative language for modeling and reasoning with contextual knowledge. Knowl. Based Syst. **207**, 106403 (2020)
2. Chu, W.W., Yang, H., Chiang, K., Minock, M., Chow, G., Larson, C.: Cobase: a scalable and extensible cooperative information system. J. Intell. Inf. Syst. **6**(2/3), 223–259 (1996)
3. Eiter, T., Gottlob, G., Mannila, H.: Disjunctive datalog. ACM Trans. Database Syst. **22**(3), 364–418 (1997)
4. Franconi, E., Tessaris, S.: Relational algebra and calculus with SQL null values. CoRR, abs/2202.10898 (2022)
5. Gaasterland, T., Godfrey, P., Minker, J.: Relaxation as a platform for cooperative answering. J. Intell. Inf. Syst. **1**(3/4), 293–321 (1992)
6. Kalinski, J.: Disjunctive rules, maybe tuples and null values: logic programs with incomplete information. In: ADBIS, 10–13 September, pp. 84–92 (1996)
7. Liu, K., Sunderraman, R.: Indefinite and maybe information in relational databases. ACM Trans. Database Syst. **15**(1), 1–39 (1990)
8. Seipel, D., Weidner, D., Abreu, S.: Intelligent query answering with contextual knowledge for relational databases. In: Queirós, R., Pinto, M., Simões, A., Portela, F., Pereira, M.J. (eds.) SLATE 2021, 1–2 July 2021, Vila do Conde/Póvoa de Varzim, Portugal. OASIcs, vol. 94, pp. 16:1–16:15 (2021)

Classification and Clustering

Detecting Anomalies with Latent*Out*: Novel Scores, Architectures, and Settings

Fabrizio Angiulli, Fabio Fassetti, and Luca Ferragina$^{(\boxtimes)}$

DIMES, University of Calabria, 87036 Rende, (CS), Italy
{f.angiulli,f.fassetti,l.ferragina}@dimes.unical.it

Abstract. Latent*Out* is a recently introduced algorithm for unsupervised anomaly detection which enhances latent space-based neural methods, namely (*Variational*) *Autoencoders*, *GANomaly* and *ANOGan* architectures. The main idea behind it is to exploit both the latent space and the baseline score of these architectures in order to provide a refined anomaly score performing density estimation in the augmented latent-space/baseline-score feature space. In this paper we extend the research on the Latent*Out* methodology in three directions: first, we provide a novel score performing a different kind of density estimation at a reduced computational cost; second, we experiment the combination of Latent*Out* with *GAAL* architectures, a novel type of Generative Adversarial Networks for unsupervised anomaly detection; third, we investigate performances of Latent*Out* acting as a one-class classifier. The experiments show that all the variants of Latent*Out* here introduced improve performances of the baseline methods to which they are applied, both in the unsupervised and in the semi-supervised settings.

Keywords: Anomaly detection · Variational autoencoder · Generative adversarial network · Density estimation

1 Introduction

The task of Anomaly Detection consists in isolating samples in a dataset that are suspected of not being generated by the same distribution as the majority of the data.

Depending on the setting of the dataset, we can distinguish three different families of methods for Anomaly Detection [1,15]. *Supervised methods* consider a dataset, whose items are labeled as normal and abnormal and build a classifier, typically the dataset is highly unbalanced and the anomalies form a rare class. *Semi-supervised methods*, also called one-class classifiers, take in input only examples from the normal class and use them to train the detector. *Unsupervised methods* assign an anomaly score to each object of the input dataset in order to find anomalies in it. There exist several statistical, data mining and machine learning approaches to perform the task of detecting outliers, such as statistical-based [12,16], distance-based [6,7,10,11,25], density-based [13,22], reverse nearest neighbor-based [4,5,19,28], SVM-based [31,33], deep learning-based [14,17], and many others [1,15].

M. Ceci et al. (Eds.): ISMIS 2022, LNAI 13515, pp. 251–261, 2022.
https://doi.org/10.1007/978-3-031-16564-1_24

Among deep learning methods for anomaly detection the ones based on Autoencoders (AE) and Variational Autoencoders (VAE) have shown good performance [3, 14, 20]. The standard application of these architectures to the task of anomaly detection is based on the concept of *reconstruction error*, that is a measure of the difference between the input and the reconstructed data, and relies on the assumption that, since the majority of the data with which they are trained belongs to the normal class, these network are able to reconstruct the inliers better than the outliers.

In [8, 9] the authors state that this approach is too simplistic and highlight the problem that these architectures generalize so well that they can also well reconstruct anomalies [3, 14, 23, 32]; in order to overcome this issue they introduce a novel approach, called Latent*Out*, that is based on the joint use of both the latent space and the reconstruction error. In particular they define two different anomaly scores:

- $\varrho-score$ that is obtained as a k-nearest neighbor estimation on the feature space composed by the latent space combined with the reconstruction error;
- $\zeta-score$ that consists in the difference of the reconstruction error of a certain point with the mean of the reconstruction error of its k nearest neighbor in the latent space.

Moreover they extend the application of Latent*Out* also to other architectures such as GANomaly [2] and *ANOGan* [30].

In this work the Latent*Out* paradigm is expanded in three directions:

- We define a new score that shares with the scores of Latent*Out* the idea of exploiting the latent space, in analogy with them we call it $\gamma-score$. Substantially it is obtained by performing a Gaussian Mixture Model in the space obtained as the Cartesian product of the latent space and the reconstruction error space, thus the computational cost is reduced since it does not requires the computation of the neighborhood. We test $\gamma-score$ against competitors and against the other scores of Latent*Out* and show that it achieves great performances.
- We consider two new architectures, MO-GAAL and SO-GAAL [27] and we modify them in order to make Latent*Out* applicable. We test on these both the original scores and the new one, $\gamma-score$.
- We implement a version of Latent*Out* for the semi-supervised scenario, we adapt the scores to this setting and perform experiments to show the performances of Latent*Out*. In particular, we test the technique exploiting VAE and $GANomaly$ as base architectures since they are easily adaptable to work on semi-supervised scenarios.

The rest of the paper is organized as follows: in Sect. 2 we describe the instruments at the basis of our work and present the contributions in the three subsections, in Sect. 3 we test the introduced methods, finally Sect. 4 concludes the paper.

2   Preliminaries and Method

The majority of the deep learning approach to anomaly detection are based on autoencoders (AE) [17,20,21,26], that are a special type of neural networks that aim at obtaining a reconstruction $\hat{x}$ as close as possible to the input sample x by minimizing the *reconstruction error* $E(x) = \|x - \hat{x}\|_2^2$ after encoding x into a hidden representation in a *latent space*.

A *variational autoencoder* (VAE) is a stochastic generative model that can be seen as a variant of standard AE [24]. The main differences are that a VAE encodes each example as a normal distribution over the latent space instead that as single points, and introduce a regularization term in the loss that maximizes similarity of these distributions with the standard normal distribution. The effect of these operations is that the latent space of a VAE is *continuous*, which means that in this space close points will lead to close decoded representation, thus avoiding the severe overfitting problem affecting standard autoencoders, for which some points of the latent space will give meaningless content once decoded. In the field of anomaly detection VAEs are used, in analogy with standard AE, by defining a *reconstruction probability* [3].

In [8] is proved that it is possible to obtain better results with VAEs by taking into account both the reconstruction error and the latent space. In particular authors propose to consider the enlarged feature space $\mathcal{F} = \mathcal{L} \times \mathcal{E}$, where $\mathcal{L}$ represents the latent space and $\mathcal{E}$ is the reconstruction error space (usually $\mathcal{E} \subseteq \mathbb{R}$) and introduce the first variant of the Latent*Out* algorithm that consists in performing a KNN density estimation in the space $\mathcal{F}$. In particular it is defined the $\varrho-score$ as

$$\varrho-score(x_i) = \frac{1}{k} \sum_{x_j \in \mathrm{N}_k^{\mathcal{F}}(x_i)} \mathrm{d}_{\mathcal{F}}(x_i, x_j),$$

where $\mathrm{N}_k^{\mathcal{F}}(x_i)$ is the set of the k-nearest neighbors of the point x_i according to the distance $\mathrm{d}_{\mathcal{F}}$ that corresponds to the euclidean distance calculated between the images of x_i and x_j on the feature space $\mathcal{F}$.

In [9] a variant of Latent*Out* considering an additional anomaly score, called $\zeta-score$, is presented. This score is related to the difference between the reconstruction error $E(x_i)$ of the point x_i and the mean of the reconstruction errors of its k-nearest neighbors in the latent space, in formula

$$\zeta-score(x_i) = \frac{E(x_i) - \mu\big(\mathrm{N}_k^{\mathcal{L}}(x_i)\big)}{\sigma\big(\mathrm{N}_k^{\mathcal{L}}(x_i)\big)},$$

where $\mathrm{N}_k^{\mathcal{L}}(x_i)$ is the set of the k nearest neighbors in the latest space $\mathcal{L}$ of the image x_i in the same space, and

$$\mu\big(\mathrm{N}_k^{\mathcal{L}}(x_i)\big) = \frac{1}{k} \sum_{x_j \in \mathrm{N}_k^{\mathcal{L}}(x_i)} E(x_j), \quad \sigma^2\big(\mathrm{N}_k^{\mathcal{L}}(x_i)\big) = \frac{1}{k} \sum_{x_j \in \mathrm{N}_k^{\mathcal{L}}(x_i)} \Big(E(x_j) - \mu\big(\mathrm{N}_k^{\mathcal{L}}(x_i)\big)\Big)^2.$$

Next, we present the novel extensions of the Latent*Out* method.

2.1   Definition of LatentOut$_\gamma$

Both $\varrho-score$ and $\zeta-score$ are based on the computation of the distances between the points of the dataset, thus their computation can be slow, especially when the number n of points is huge.

Here we introduce a novel score, called $\gamma-score$ that shares with the other two the use of the space $\mathcal{F}$, but that exploits Gaussian Mixture Models (GMM) to perform density estimation instead of a distance-based method. In practice we reconstruct the unknown density of the data as a mixture of k distributions

$$p\left(z_i|\omega_j,\mu_j,\Sigma_j\right) = \sum_{j=1}^{k} \omega_j g\left(z_i|\mu_j,\Sigma_j\right).$$

where z_i is the image of the point x_i mapped in $\mathcal{F}$ and each $g(\cdot|\mu_j,\Sigma_j)$, $j = 1,\ldots,k$, is a $d+1$-dimensional Gaussian distribution in the feature space $\mathcal{F}$:

$$g\left(z_i|\mu_j,\Sigma_j\right) = \frac{1}{(2\pi)^{(d+1)/2}|\Sigma_j|^{1/2}} \exp\left(-\left(z_i-\mu_j\right)^T \Sigma_j^{-1}\left(z_i-\mu_j\right)\right).$$

The parameters $\omega_j \in \mathbb{R}$, $\mu_j \in \mathbb{R}^{d+1}$, and $\Sigma_j \in \mathbb{R}^{d\times d}$ of the mixture are estimated by using the Expectation-Minimization algorithm. Notice that the Σ_j are diagonal matrices, since co-variances are assumed to be null.

The score relative to LatentOut$_\gamma$ is defined as the value of the density obtained with the parameters ω_j,μ_j,Σ_j that maximize the expectation, in formula

$$\gamma-score\left(x_i\right) = p\left(x_i|\omega_j,\mu_j,\Sigma_j\right).$$

The idea is that, as already pointed out in [8], *outliers tend to lie in the sparsest regions of $\mathcal{F}$* and so the density obtained with the parameters ω_j,μ_j,Σ_j should have lower values on them.

From a computational point of view, the difference is that, given a point x, the calculation of the two previous scores requires to determine all the n distances separating x from all the other points, while the computation of the $\gamma-score$ only requires the evaluation of k $(d+1)$-dimensional gaussian pdfs. Because of this the computational cost goes from $O(nd)$ for the two previous scores to $O(kd)$ for $\gamma-score$.

Moreover, while the previous two scores show some local behavior, in that they only depend on the characteristic of the dataset in the neighborhood of the selected point, the novel score as a more global behavior, in that it approximates the probability density function of the underlying data generating distribution.

2.2   Extension to GAAL Architectures

A *Generative Adversarial Network* (GAN) [18] is a generative model which is composed by two models trained simultaneously: a generator G that aims to capture the distribution of the data in order to reproduce samples as realistic as possible and a discriminator D, that must distinguish the data belonging to the dataset from the ones artificially created by G. AnoGAN [30] with its extensions GAN+ [34] and FastAnoGAN [29], and GANomaly [2] are the first works in which GAN are used for the task of anomaly detection.

Latent*Out* has already been successfully applied to the above mentioned GAN-based architectures. Here we apply Latent*Out* on Single-Objective Generative Adversarial Active Learning (SO-GAAL) [27], a novel adversarial method for anomaly detection based on the mini-max game between a generator that creates potential anomalies and a discriminator that tries to draw a separation boundary between the anomalies and the normal class. We deal also with Multiple-Objective GAAL (MO-GAAL), an extension of SO-GAAL which employs multiple generators with different objectives in order to prevent the generator from falling into the mode collapsing problem.

The GAAL architectures in their standard version are not suitable for being associated with Latent*Out* algorithm, indeed, although they are composed by a latent space $\mathcal{L}$ from which the generator pick the samples in order to create artificial items, they have not one or more layer that maps the input space into $\mathcal{L}$. With the aim of solving this issue, we modify the architecture of SO-GAAL by adding an encoder f_ϕ that receives in input the original data x_i and outputs its latent representation z_i, that in turn is passed to the generator.

The same problem arises for the MO-GAAL architecture, we face it by adding an encoder for each of the M generators $f_\phi^{(1)}, \ldots, f_\phi^{(M)}$ of the network. In this way each point x_i is associated with M latent representations $z_i^{(1)} = f_\phi^{(1)}, \ldots, z_i^{(M)}(x_i)$, where $z_i^{(j)} = f_\phi^{(j)}(x_i)$ for each $j = 1, \ldots, M$, therefore we define as latent transformation of x_i the mean of these points $z_i = \frac{1}{M} \sum_{j=1}^{k} z_i^{(j)}$.

Finally in all the three parts of the GAAL (encoders, generators and discriminator) we add some convolutional layers in order to make them deeper and more suitable for image data.

2.3 Latent*Out* for Semi-supervised Outlier Detection

The semi-supervised setting is characterized by the presence of a training set $T = \{t_1, \ldots, t_n\}$ composed only by normal items and a test set $X = \{x_1, \ldots, x_m\}$ with binary labels $Y = \{y_1, \ldots, y_m\}$ where $y_i = 0$ if x_i is normal and $y_i = 1$ if it is an anomaly.

The application of Latent*Out* to this context instead of to the classical unsupervised setting for which it has been designed, is not automatic, because all the scores must be redefined in order to deal with the fact that the models are trained only on normal data. In particular, given a point x_i in the test set, the semi-supervised versions of both $\varrho-score$ and $\zeta-score$ require the computation of the distances, in the enlarged latent space $\mathcal{F}$, of the distances between x_i and each example t_i of the training set.

$$\varrho-score(x_i) = \frac{1}{k} \sum_{t_j \in N_k^{\mathcal{F}}(x_i)} d_{\mathcal{F}}(x_i, t_j), \quad \zeta-score(x_i) = \frac{E(x_i) - \mu_T(N_k^{\mathcal{L}}(x_i))}{\sigma_T(N_k^{\mathcal{L}}(x_i))},$$

where

$$\mu_T\big(\mathrm{N}_k^{\mathcal{L}}(x_i)\big) = \frac{1}{k} \sum_{t_j \in \mathrm{N}_k^{\mathcal{L}}(x_i)} E(t_j), \quad \sigma_T^2\big(\mathrm{N}_k^{\mathcal{L}}(x_i)\big) = \frac{1}{k} \sum_{t_j \in \mathrm{N}_k^{\mathcal{L}}(x_i)} \Big(E(t_j) - \mu\big(\mathrm{N}_k^{\mathcal{L}}(x_i)\big)\Big)^2.$$

On the other hand, the semi-supervised version of the $\gamma-score$ is computed by first estimating the density p_T on the training data t_i and then evaluating it on the test point

$$\gamma-score\,(x_i) = p_T\,(x_i|\omega_j, \mu_j, \Sigma_j)\,.$$

3   Experimental Results

In this section we report experiments conducted to study the behavior of the proposed techniques. In particular we focus on the following three aspects:

- the impact of the $\gamma-score$ on architectures that have already been introduced in [9] in the original unsupervised setting, and comparison of its performances with the ones of other LatentOut scores and of the relative baseline;
- the application of all LatentOut scores on the new architectures SO-GAAL and MO-GAAL and comparison with baseline method;
- the behavior of LatentOut algorithm in the semi-supervised (one-class) setting in comparison with baseline architectures.

Experimental Settings. In our experiments we consider three standard benchmark datasets, two composed by grayscale images, $MNIST^1$ and $Fashion$-$MNIST^2$, and one composed by three-channels colour images, $CIFAR$-10^3 Both the grayscale datasets consist of 60,000 28×28 pixels images divided in 10 classes, $CIFAR$-10 consists of 60,000 32×32 colour images partitioned in 10 classes.

All these dataset are multi-labelled, thus, in order to make them suitable for anomaly detection, we decide to adopt a *one-vs-all* policy, which means that we consider one class as normal and all the others as anomalous.

In particular, in the unsupervised setting, we consider a dataset composed by all the examples of the normal class and a quantity s of random selected examples from each other class as anomalies.

On the other hand, in the semi-supervised (one-class) setting the training set is composed only by examples from the normal class, while the test set coincides with the with the original test sets of the considered datasets, thus it is composed of examples from both the normal and the anomalous classes.

Latent Out_γ on the unsupervised setting. As a preliminary experiment to test the performances of the new $\gamma-score$, we consider the Fashion-MNIST dataset in a *one-vs-all* unsupervised setting in order to compare it with the previous scores and the baseline architectures in the original experimental environment of [9].

---

[1] http://yann.lecun.com/exdb/mnist/.

[2] https://github.com/zalandoresearch/fashion-mnist.

[3] https://www.cs.toronto.edu/~kriz/cifar.html.

Table 1. AUC for Fashion-MNIST in the *one-vs-all* unsupervised setting ($s = 10$).

DS	Class	GANomaly			VAE		
		Base	Previous best	$\gamma-score$	Base	Previous best	$\gamma-score$
Fashion-MNIST	0	.712±.042	.800±.013	**.820±.010**	.594±.035	**.919±.011**	.914±.011
	1	.972±.006	**.974±.003**	.952±.007	.958±.007	.962±.002	**.980±.003**
	2	.716±.046	.759±.025	**.824±.013**	.534±.024	.900±.031	**.915±.005**
	3	.800±.046	.844±.015	**.866±.005**	.864±.010	.917±.008	**.930±.025**
	4	.762±.063	.806±.030	**.825±.019**	.673±.053	.875±.015	**.928±.003**
	5	.882±.022	.785±.048	**.887±.006**	.887±.017	.921±.018	**.924±.006**
	6	.643±.024	**.746±.016**	.743±.012	.471±.046	**.867±.010**	.820±.020
	7	.966±.007	.923±.011	**.973±.006**	.951±.008	**.977±.002**	.974±.005
	8	.424±.064	.605±.060	**.629±.105**	.540±.072	.911±.003	**.938±.002**
	9	.784±.075	.811±.006	**.921±.027**	.772±.038	.954±.011	**.987±.001**

In Table 1 are reported the results of this experiment, in particular the three columns on the left are relative to GANomaly and the three on the right are relative to the Variational Autoencoder. In this experiment we vary the dimension of the latent space in the interval $[2, 128]$ and report the best performance obtained (as the mean on 5 runs) for each method. For what concerns the $\gamma-score$ the best performance are obtained in the intermediate interval $[8, 16]$, and with values of the parameter k in the interval $[3, 10]$. For each architecture $\gamma-score$ (the third column) obtains the best value of AUC for almost all the normal classes, moreover it is also important to observe that in those few cases in which this does not happen, the best AUC score is obtained by the best of $\varrho-score$ and

Table 2. AUC for MNIST and Fashion-MNIST in the *one-vs-all* unsupervised setting ($s = 10$).

DS	Class	SO-GAAL	Latent*Out*$_{SO-GAAL}$			MO-GAAL	Latent*Out*$_{SO-GAAL}$		
			$\zeta-score$	$\varrho-score$	$\gamma-score$		$\zeta-score$	$\varrho-score$	$\gamma-score$
MNIST	0	.940±.005	.834±.062	**.989±.004**	*.973±.011*	.942±.006	.901±.011	**.982±.006**	.968±.005
	1	.966±.011	.934±.023	**.997±.000**	*.996±.000*	.985±.007	.947±.005	**.998±.000**	.998±.000
	2	.835±.025	.740±.031	**.920±.025**	*.899±.019*	.842±.015	.766±.021	**.912±.008**	.905±.009
	3	.864±.020	.782±.027	*.889±.047*	**.923±.042**	.885±.017	.826±.047	*.878±.018*	**.900±.007**
	4	.900±.008	.874±.020	*.912±.016*	**.917±.011**	.903±.030	.890±.017	*.923±.023*	**.945±.008**
	5	.669±.101	.636±.114	*.909±.017*	**.926±.003**	.731±.006	.659±.035	*.902±.011*	**.915±.003**
	6	.908±.051	.833±.044	**.980±.005**	*.978±.006*	.911±.036	.879±.036	*.971±.002*	**.974±.004**
	7	.872±.028	.854±.020	*.958±.009*	**.980±.007**	.900±.040	.862±.047	*.952±.004*	**.960±.011**
	8	.855±.003	.789±.027	*.876±.013*	**.963±.014**	.824±.032	.802±.038	*.864±.037*	**.881±.010**
	9	.858±.041	.816±.070	*.947±.010*	**.961±.011**	.863±.067	.846±.087	*.950±.004*	**.959±.004**
DS	Class	SO-GAAL	Latent*Out*$_{SO-GAAL}$			MO-GAAL	Latent*Out*$_{SO-GAAL}$		
			$\zeta-score$	$\varrho-score$	$\gamma-score$		$\zeta-score$	$\varrho-score$	$\gamma-score$
Fashion-MNIST	0	.779±.035	.771±.053	**.906±.015**	*.837±.001*	.845±.002	.763±.027	.881±.015	**.886±.009**
	1	.976±.003	.932±.019	**.986±.002**	*.981±.004*	.949±.028	.884±.020	**.983±.003**	.980±.003
	2	.726±.064	.714±.008	**.884±.007**	*.845±.008*	.830±.004	.835±.053	.819±.011	**.856±.008**
	3	.917±.016	.905±.006	*.915±.014*	**.916±.010**	.915±.003	.868±.015	.907±.008	**.911±.007**
	4	.847±.017	.747±.012	**.907±.006**	*.868±.004*	.883±.037	.845±.040	.886±.003	**.898±.010**
	5	.864±.039	*.866±.029*	**.879±.005**	.822±.009	.794±.008	.837±.019	.831±.041	**.845±.016**
	6	.660±.013	.761±.002	**.802±.005**	*.779±.041*	.740±.035	.736±.041	.763±.008	**.770±.016**
	7	**.979±.007**	.960±.013	*.973±.007*	.964±.006	.966±.015	.960±.015	**.973±.005**	.965±.008
	8	.719±.019	.589±.055	**.909±.005**	*.872±.006*	.808±.042	.778±.037	.775±.011	**.826±.004**
	9	.904±.091	.882±.023	**.975±.004**	*.974±.006*	**.984±.006**	.970±.014	.960±.008	.970±.008

$\zeta-score$ (second column) and the performance of the standard architecture (first column) is always lower than the ones of all the LatentOut scores.

Performance on GAAL Architectures. In this section we test LatentOut scores on MO-GAAL and SO-GAAL architectures; in this case the focus is also on $\varrho-score$ and $\zeta-score$, in addition to $\gamma-score$, because the application of both the old scores to these architectures represents a novel contribution. Table 2 shows the results of all the three LatentOut scores and the baseline on MNIST and Fashion-MNIST in an *one-vs-all* unsupervised setting, since the architectures MO-GAAL and SO-GAAL are specific for unsupervised anomaly detection. In this experiment we fix the value of the parameter k for each score, in particular for $\gamma-score$, according to the results obtained in previous section, we set $k = 5$ and for $\varrho-score$ and $\zeta-score$ we follow the indications given in [9] and set $k = 50$ and $k = 200$, respectively. On the other hand, the value of the dimension of the latent space is variable in the interval $[8, 128]$. For both architectures the best values are obtained in the interval $[32, 64]$.

From these results we can conclude that LatentOut is very effective also applied in these architecture, since it always guarantees an improvement over the standard baseline. In particular we can observe that the new $\gamma-score$ is the best score for the majority of the classes, and, in those cases in which this is not true, its performance is almost always very close to the one of the best method.

Table 3. AUC for MNIST, Fashion-MNIST and CIFAR-10 in the *one-vs-all* semi-supervised setting.

Class	MNIST		Fashion-MNIST		CIFAR-10	
	VAE	LatentOut	VAE	LatentOut	VAE	LatentOut
0	.989±.010	**.991±.007**	.711±.007	**.897±.011**	.618±.027	**.625±.033**
1	**.999±.000**	.996±.000	.981±.000	**.982±.005**	.658±.014	**.691±.010**
2	.891±.010	**.957±.006**	.696±.015	**.885±.014**	.474±.022	**.641±.006**
3	.868±.011	**.931±.008**	**.937±.015**	.930±.029	.627±.034	**.628±.033**
4	.932±.021	**.942±.004**	.780±.014	**.912±.003**	.445±.025	**.701±.024**
5	.939±.010	**.953±.001**	**.939±.007**	.936±.004	.554±.025	**.577±.035**
6	.978±.011	**.990±.002**	.563±.017	**.789±.009**	.605±.013	**.728±.039**
7	.954±.011	**.967±.002**	.971±.008	**.981±.016**	.526±.024	**.555±.067**
8	.825±.016	**.948±.019**	.679±.019	**.889±.022**	.577±.004	**.658±.009**
9	.927±.003	**.964±.011**	.848±.029	**.966±.005**	.693±.048	**.697±.057**
Class	MNIST		Fashion-MNIST		CIFAR-10	
	GANomaly	LatentOut	GANomaly	LatentOut	GANomaly	LatentOut
0	.715±.094	**.884±.018**	.775±.012	**.898±.020**	.633±.030	**.716±.015**
1	.986±.052	**.997±.006**	.935±.005	**.972±.003**	.581±.048	**.592±.010**
2	.737±.046	**.792±.030**	.773±.067	**.849±.090**	.628±.001	**.663±.011**
3	.752±.039	**.846±.017**	.779±.019	**.872±.027**	.571±.049	**.575±.017**
4	.835±.017	**.899±.036**	.806±.011	**.846±.017**	.712±.007	**.730±.007**
5	.744±.016	**.808±.016**	.776±.066	**.834±.029**	.539±.022	**.550±.007**
6	.853±.051	**.912±.022**	.604±.007	**.766±.066**	.697±.039	**.712±.003**
7	.764±.097	**.933±.001**	.918±.075	**.968±.020**	.543±.019	**.573±.028**
8	.578±.033	**.796±.047**	.713±.010	**.804±.026**	.580±.031	**.650±.041**
9	**.797±.035**	.781±.004	.895±.017	**.938±.064**	.531±.002	**.613±.031**

Latent*Out* on the Semi-supervised Setting. In the last experiment we test Latent*Out* in the semi-supervised (one-class) setting by considering the architectures VAE and GANomaly as baseline. The results are reported in Table 3; for each dataset and each architecture, on the left column there is the AUC of the baseline and on the right column there is the best AUC obtained by the three scores of Latent*Out*. We vary the dimension of the latent space in the interval $[2, 64]$; the best results are obtained in the interval $[8, 16]$ for Latent*Out*$_{VAE}$, $[16, 32]$ for Latent*Out*$_{GANomaly}$, for $[4, 8]$ for standard VAE and for $[16, 64]$ for standard GANomaly. From these results it is clear that Latent*Out* outperforms both the considered baselines, and the improvement in many cases is huge.

4 Conclusion

In this work we introduce three extensions of the Latent*Out* algorithm: a novel score with a reduced computational cost, a novel architecture and the application to the semi-supervised setting. The experiments show that all the scores of Latent*Out* improve performances of the baseline methods to which they are applied, both in the unsupervised and in the one-class scenarios.

The results in this paper make us believe that the idea behind Latent*Out* of exploiting both the reconstruction error and the latent space can be effective in a wide range of different anomaly detection settings. Because of this, in the future, our main goal is to deal with supervised scenarios in which some anomalies are known in phase of training.

References

1. Aggarwal, C.C.: Outlier Analysis. Springer, Cham (2013). https://doi.org/10.1007/978-3-319-47578-3
2. Akcay, S., Atapour-Abarghouei, A., Breckon, T.P.: GANomaly: semi-supervised anomaly detection via adversarial training. In: Jawahar, C.V., Li, H., Mori, G., Schindler, K. (eds.) ACCV 2018. LNCS, vol. 11363, pp. 622–637. Springer, Cham (2019). https://doi.org/10.1007/978-3-030-20893-6_39
3. An, J., Cho, S.: Variational autoencoder based anomaly detection using reconstruction probability. Technicl report 3, SNU Data Mining Center (2015)
4. Angiulli, F.: Concentration free outlier detection. In: Ceci, M., Hollmén, J., Todorovski, L., Vens, C., Džeroski, S. (eds.) ECML PKDD 2017. LNCS (LNAI), vol. 10534, pp. 3–19. Springer, Cham (2017). https://doi.org/10.1007/978-3-319-71249-9_1
5. Angiulli, F.: CFOF: a concentration free measure for anomaly detection. ACM Trans. Knowl. Discov. Data (TKDD) **14**(1), 4:1–4:53 (2020)
6. Angiulli, F., Basta, S., Pizzuti, C.: Distance-based detection and prediction of outliers. IEEE Trans. Knowl. Data Eng. **2**(18), 145–160 (2006)
7. Angiulli, F., Fassetti, F.: DOLPHIN: an efficient algorithm for mining distance-based outliers in very large datasets. ACM Trans. Knowl. Disc. Data (TKDD) **3**(1), Article 4 (2009)

8. Angiulli, F., Fassetti, F., Ferragina, L.: Improving deep unsupervised anomaly detection by exploiting VAE latent space distribution. In: Appice, A., Tsoumakas, G., Manolopoulos, Y., Matwin, S. (eds.) DS 2020. LNCS (LNAI), vol. 12323, pp. 596–611. Springer, Cham (2020). https://doi.org/10.1007/978-3-030-61527-7_39
9. Angiulli, F., Fassetti, F., Ferragina, L.: Latent *Out*: an unsupervised deep anomaly detection approach exploiting latent space distribution. Mach. Learn. (2022)
10. Angiulli, F., Pizzuti, C.: Fast outlier detection in large high-dimensional data sets. In: Proceedings International Conference on Principles of Data Mining and Knowledge Discovery (PKDD), pp. 15–26 (2002)
11. Angiulli, F., Pizzuti, C.: Outlier mining in large high-dimensional data sets. IEEE Trans. Knowl. Data Eng. 2(17), 203–215 (2005)
12. Barnett, V., Lewis, T.: Outliers in Statistical Data. Wiley, Hoboken (1994)
13. Breunig, M.M., Kriegel, H., Ng, R., Sander, J.: LoF: identifying density-based local outliers. In: Proceedings of International Conference on Management of Data (SIGMOD) (2000)
14. Chalapathy, R., Chawla, S.: Deep learning for anomaly detection: a survey (2019)
15. Chandola, V., Banerjee, A., Kumar, V.: Anomaly detection: a survey. ACM Comput. Surv. 41(3) (2009)
16. Davies, L., Gather, U.: The identification of multiple outliers. J. Am. Stat. Assoc. 88, 782–792 (1993)
17. Goodfellow, I., Bengio, Y., Courville, A.: Deep Learning. MIT Press, Cambridge (2016)
18. Goodfellow, I., et al.: Generative adversarial nets. In: Advances in Neural Information Processing Systems, vol. 27 (2014)
19. Hautamäki, V., Kärkkäinen, I., Fränti, P.: Outlier detection using k-nearest neighbour graph. In: International Conference on Pattern Recognition (ICPR), Cambridge, UK, 23–26 August, pp. 430–433 (2004)
20. Hawkins, S., He, H., Williams, G., Baxter, R.: Outlier detection using replicator neural networks. In: International Conference on Data Warehousing and Knowledge Discovery (DAWAK), pp. 170–180 (2002)
21. Hecht-Nielsen, R.: Replicator neural networks for universal optimal source coding. Science 269(5232), 1860–1863 (1995)
22. Jin, W., Tung, A., Han, J.: Mining top-n local outliers in large databases. In: Proceedings of ACM SIGKDD International Conference on Knowledge Discovery and Data Mining (KDD) (2001)
23. Kawachi, Y., Koizumi, Y., Harada, N.: Complementary set variational autoencoder for supervised anomaly detection. In: IEEE International Conference on Acoustics, Speech and Signal Processing (ICASSP), pp. 2366–2370 (2018)
24. Kingma, D.P., Welling, M.: Auto-encoding variational Bayes (2013)
25. Knorr, E., Ng, R., Tucakov, V.: Distance-based outlier: algorithms and applications. VLDB J. 8(3–4), 237–253 (2000)
26. Kramer, M.A.: Nonlinear principal component analysis using autoassociative neural networks. AIChE J. 37(2), 233–243 (1991)
27. Liu, Y., et al.: Generative adversarial active learning for unsupervised outlier detection. IEEE Trans. Knowl. Data Eng. 32(8), 1517–1528 (2020)
28. Radovanović, M., Nanopoulos, A., Ivanović, M.: Reverse nearest neighbors in unsupervised distance-based outlier detection. IEEE Trans. Knowl. Data Eng. 27(5), 1369–1382 (2015)
29. Schlegl, T., Seeböck, P., Waldstein, S., Langs, G., Schmidt-Erfurth, U.: f-anogan: Fast unsupervised anomaly detection with generative adversarial networks. Med. Image Anal. 54 (2019)

30. Schlegl, T., Seeböck, P., Waldstein, S.M., Schmidt-Erfurth, U., Langs, G.: Unsupervised anomaly detection with generative adversarial networks to guide marker discovery. In: Niethammer, M., et al. (eds.) IPMI 2017. LNCS, vol. 10265, pp. 146–157. Springer, Cham (2017). https://doi.org/10.1007/978-3-319-59050-9_12
31. Schölkopf, B., Platt, J.C., Shawe-Taylor, J., Smola, A.J., Williamson, R.C.: Estimating the support of a high-dimensional distribution. Neural Comput. **13**(7), 1443–1471 (2001)
32. Sun, J., Wang, X., Xiong, N., Shao, J.: Learning sparse representation with variational auto-encoder for anomaly detection. IEEE Access **6**, 33353–33361 (2018)
33. Tax, D.M.J., Duin, R.P.W.: Support vector data description. Mach. Learn. **54**(1), 45–66 (2004)
34. Zenati, H., Foo, C.S., Lecouat, B., Manek, G., Chandrasekhar, V.R.: Efficient GAN-based anomaly detection (2019)

Richness Fallacy

Mieczysław A. Kłopotek[1]([⊠]) [ID] and Robert A. Kłopotek[2] [ID]

[1] Institute of Computer Science, Polish Academy of Sciences, Warsaw, Poland
mieczyslaw.klopotek@ipipan.waw.pl
[2] Faculty of Mathematics and Natural Sciences, School of Exact Sciences,
Cardinal Stefan Wyszyński University in Warsaw, Warsaw, Poland
r.klopotek@uksw.edu.pl

Abstract. The paper points at the grieving problems implied by the richness axiom in the Kleinberg's axiomatic system and suggests resolutions. The richness induces learnability problem in general and leads to conflicts with consistency axiom. As a resolution, learnability constraints and usage of centric consistency or restriction of the domain of considered clusterings to super-ball-clusterings is proposed.

Keywords: Richness axiom · Consistency axiom · Centric consistency axiom · Clustering algorithms · Learnability · Theoretical foundations

1 Introduction

Kleinberg [8] introduced an axiomatic system for distance based clustering functions consisting of three axioms: *scale-invariance* (same clustering should be obtained if all distances between objects are multiplied by the same positive number), *consistency* (same clustering should be obtained, if distances within a cluster are decreased and distances between elements from distinct clusters are increased) and *richness* (for any partition of the dataset into non-empty subsets there should exist a set of distances between the datapoints so that the clustering function delivers this partition).

This set of axioms was controversial from the very beginning. Kleinberg proved himself that the entire axiomatic system is contradictory, and only pairs of these axioms are not contradictory. But also the axiom of consistency turned out to be controversial as it excludes the k-means algorithm from the family of clustering functions. It has been shown in [9] that consistency is contradictory by itself if we consider fixed dimensional Euclidean spaces which are natural domain of k-means application.

It seems to be disastrous for the domain of clustering algorithms if an axiomatic system consisting of "natural axioms" is self-contradictory. It renders the entire domain questionable. Therefore numerous efforts have been made to cure such a situation by proposing different axiom sets or modifying Kleinberg's one, just to mention [1–4, 7, 13–15, 19] etc.

M. Ceci et al. (Eds.): ISMIS 2022, LNAI 13515, pp. 262–271, 2022.
https://doi.org/10.1007/978-3-031-16564-1_25

In this paper, we look at more detail at the richness property and investigate its counterintuitiveness. The richness induces learnability problem in general (see Sect. 2) and leads to conflicts with consistency axiom (see Sect. 4). In the past, we applied embedding into Euclidean space to resolve some problems with consistency axiom, but it does not work for richness (Sect. 3). Therefore, as a resolution, usage of centric consistency (Sect. 4) or restriction of the domain of considered clusterings to super-ball-clusterings (Sect. 5) is proposed.

2   Richness and Learnability

Following [8] let us define: A *partition* Γ of the set of datapoints S into k partitions is the set $\Gamma = \{C_1, \ldots, C_k\}$ such that $C_i \cap C_j = \emptyset$ for $i \neq j$, $C_i \neq \emptyset$ and $S = C_1 \cup C_2 \cup \cdots \cup C_k$. A *clustering function* f is a function assigning a partition Γ to any dataset S with at least two datapoints. A *clustering quality function* Q is a function assigning a real value to a partition.

Richness is easy to achieve. Imagine the following *clustering function*. You order nodes by average distance to other nodes, on tights on squared distance and so on. If no sorting can be achieved, the unsortable points are set into one cluster. Then we create an enumeration of all clusters and map it onto unit line segment. We take the quotient of the lowest distance to the largest distance and state that this quotient mapped to that line segment identifies the optimal clustering of the points. Though the algorithm is simple in principle (and useless also), and meets axioms of richness and scale -invariance, we have a practical problem: As no other limitations are imposed, one has to check up to $\sum_{k=2}^{n} \frac{1}{k!} \sum_{j=1}^{k} (-1)^{k-j} \binom{k}{j} j^n$ possible partitions (Bell number) in order to verify which one of them is the best for a given distance function because there must exist at least one distance function suitable for each of them. This cannot be done in reasonable time even if each check is polynomial (even linear) in the dimensions of the task (n).

Richness is a concept from the realm of clustering, that is un-supervised learning. But if we have an explicit clustering quality function Q, but the clustering function f has to be deduced from it assuming that $Q(f(S))$ is optimal among all possible partitions of S, then we have to do with the typical supervised learning task. Hence we are transferred to the supervised learning domain. The idea of learnability is based on the assumption that the learned concept may be verified against new data. For example, if one learned k-means clustering on a data sample, then with newly incoming elements the clustering should not change (too much). With the richness requirement, if you learnt the model from a sample of size n, then it consists of no more than n clusters. But there exists a sample of size $2n$ such that it defies the learned structure, because it has more than n clusters. That is the clustering structure of the domain is not learnable by definition of richness.

Furthermore, most algorithms of cluster analysis are constructed in an incremental way. But this can be useless if the clustering quality function is designed in a very unfriendly way, for example based on modulo computation. Consider the following clustering quality function $Q_s(\Gamma)$. For each datapoint $i \in S$ we compute its distance d_i to its cluster center under Γ. Let $dmx = \max_{i \in S} d_i$. $q = \sum_{i \in S} round(10000d_i/dmx)$. Then $Q_s(\Gamma) = p - q \mod p$, where p is a prime number (here 3041). This Q_s was applied in order to partition first n points from sample data from Table 1, as illustrated in Table 2. It turns out that the best partition for n points does not give any hint for the best partition for $n + 1$ points therefore each possible partition needs to be investigated in order to find the best one.[1]

Table 1. Data points to be clustered using a ridiculous clustering quality function

Id	x coordinate	y coordinate
1	4.022346	5.142886
2	3.745942	4.646777
3	4.442992	5.164956
4	3.616975	5.188107
5	3.807503	5.010183
6	4.169602	4.874328
7	3.557578	5.248182
8	3.876208	4.507264
9	4.102748	5.073515
10	3.895329	4.878176

Table 2. Partition of the best quality (the lower the value the better) after including n first points from Table 1.

n	Quality	Partition
2	1270	{1, 2}
3	1270	{1, 2} {3}
4	823	{1, 3, 4} {2}
5	315	{1, 4} {2, 3, 5}
6	13	{1, 5} {2, 4, 6} {3}
7	3	{1, 6} {2, 7} {3, 5} {4}
8	2	{1, 2, 4, 5, 6, 8} {3 } {7}
9	1	{1, 2, 4, 5} {3, 8} {6, 9} {7}
10	1	{1, 2, 3, 5, 9} {4, 6} {7, 10} {8}

---

[1] Strict separation [5] mentioned earlier is another kind of a weird cluster quality function, requiring visits to all the partitions.

Under these circumstances let us point at the implications of the so-called learnability theory [6,16] for the richness axiom. On the one hand the hypothesis space is too big for learning a clustering from a sample. On the other hand an exhaustive search in this space is prohibitive so that some theoretical clustering functions do not make practical sense.

Furthermore, if the clustering function can fit any data, we are practically unable to learn any structure of data space from data [10]. And this learning capability is necessary at least in the cases: either when the data may be only representatives of a larger population or the distances are measured with some measurement error (either systematic or random) or both. We speak about a much broader aspect than cluster stability or cluster validity, pointed at by Luxburg [17,18].

In the special case of k-means, the reliable estimation of cluster center position and of the variance in the cluster plays a significant role. But there is no reliability for cluster center if a cluster consists of fewer than 2 elements, and for variance the minimal cardinality is 3.

3   Richness Problems in Euclidean Space

We have shown [9] that some problems with consistency axiom can be resolved by embedding into fixed-dimensional space. But we show here that this is not the case for richness.

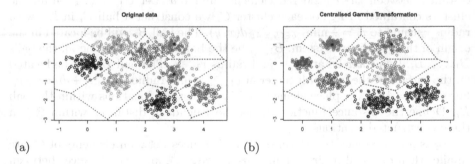

(a)   (b)

Fig. 1. (a) A mixture of 8 normal distributions as clustered by k-means algorithm (Voronoi diagram superimposed). (b) Same data after a centric consistency transformation, clustered by k-means algorithm into 8 groups

While consistency transform turns out to be too restrictive in finite dimensional space [9], the richness is problematic the other way, mainly from the point of view of learnability theory, but also leads to the chaining effect.

Let us ask whether or not it is possible to have richness, that is for any partition there exists always a distance function that the clustering function will return this partition, and yet if we restrict ourselves to $\mathbb{R}^m$, the very same clustering function is not rich any more, or even it is not chaining.

Consider the following clustering function $f()$. If it takes a distance function $d()$ that takes on only two distinct values d_1 and d_2 such that $d_1 < 0.5d_2$ and for any three data points a, b, c if $d(a, b) = d_1, d(b, c) = d_1$ then $d(a, c) = d_1$, it creates clusters of points in such a way that a, b belong to the same cluster iff $d(a, b) = d_1$, and otherwise they belong to distinct clusters. If on the other hand $f()$ takes a distance function not exhibiting this property, it works like k-means. Obviously, function $f()$ is rich, but at the same time, if confined to $\mathbb{R}^m$, if $n > m + 1$ and $k \ll n$, then it is not rich – it is in fact k-rich, and hence not chaining.

Can we get around the problems of all three Kleinberg's axioms in a similar way in $\mathbb{R}^m$? Regrettably,

Theorem 1. *If Γ is a partition of $n > 2$ elements returned by a clustering function f under some distance function d (any distance function allowed by Kleinberg in [8]), and f satisfies consistency, then there exists a distance function d_E embedded in $\mathbb{R}^m$ for the same set of elements such that Γ is the partition of this set under d_E.*

Theorem 1 implies that the constructs of contradiction of Kleinberg axioms are simply transposed from the domain of any distance functions to distance functions in $\mathbb{R}^m$.

Proof. To show the validity of the theorem, we will construct the appropriate distance function d_E by embedding in the $\mathbb{R}^m$. Let $dmax$ be the maximum distance between the considered elements under d. Let $C_1, \ldots, C_k$ be all the clusters contained in Γ. For each cluster C_i we construct a ball B_i in $\mathbb{R}^m$ with radius r_i equal to $r_i = \frac{1}{2} \min_{x,y \in C_i, x \neq y} d(x, y)$. The ball B_1 will be located in the origin of the coordinate system. $B_{1,\ldots,i}$ be the ball in $\mathbb{R}^m$ containing all the balls $B_1, \ldots, B_i$. Its center be at $c_{1,\ldots,i}$ and radius $r_{1,\ldots,i}$. The ball B_i will be located on the surface of the ball with center at $c_{1,\ldots,i-1}$ and radius $r_{1\ldots,i-1} + dmax + r_i$. For each $i = 1, \ldots, k$, place elements of C_i at distinct locations within the ball B_i. Define the distance function d_E as the Euclidean distances within $\mathbb{R}^m$ in these constructed locations.

d_E is a consistency-transform of d, as distances between elements of C_i are smaller than or equal to $2r_i = \min_{x,y \in C_i, x \neq y} d(x, y)$, and the distances between elements of different balls exceed $dmax$.

This means that if f is rich and consistent, it is also rich in $\mathbb{R}^m$.

4   Disturbing Effects of Consistency Axiom on Richness Axiom

Let us recall practical problems with consistency, pointed at by Ben-David [4]: emergence of impression of a different clustering of data (with a different number of clusters) as they show in their Fig. 1. This may imply problems if we apply k-means algorithm with k varying over a range of values to identify most appropriate value of k, e.g. by picking k, for which 90% of variance is variance or there is abrupt break (saturation) in increase of relative variance explained.

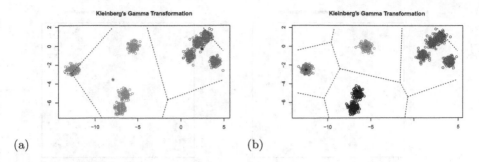

Fig. 2. Data from Fig. 1(a) after Kleinberg's consistency-transformation clustered by k-means algorithm into (a) two and (b) four groups.

The variance explained $Var_E(\Gamma_k)$ by the clustering Γ_k of the dataset S by k-means algorithm is defined as follows. $Var_E(\Gamma_k) = Var(S) - \sum_{c \in \Gamma_k} Var(c)$. The percentage of variance explained $Var_{RE}(\Gamma_k) = Var_E(\Gamma_k)/Var(S) * 100\%$.

We applied this method of choosing k for k-means (R implementation) to a mixture of data points sampled from 8 normal distributions, shown in Fig. 1(a). As visible in Fig. 1(a) and confirmed by the column "Original" of Table 3, representing the percentage of variance explained (Var_{RE}), the k-means algorithm with $k = 8$, separates best the points from various distributions. At about 7–8 clusters we get saturation, 90% explained variance mark is crossed and with more than 8 clusters the relative increase of variance explained $RIV(k) = \frac{Var_{RE}(\Gamma_k) - Var_{RE}(\Gamma_{k-1,})}{Var_{RE}(\Gamma_{k+1,}) - Var_{RE}(\Gamma_{k,})}$, having a pick at $k = 8$, drops abruptly.

Figure 2 illustrates a result of a consistency-transform on the results of the clustering from Fig. 1(a) (outer consistency transform). The "RIV" column next to "Kleinberg" column picks at $k = 2$ and has a smaller pick at $k = 4$ Visually we have two clusters (Fig. 2(a)). We could also classify this data set as having four clusters, as in Fig. 2(b). This renders Kleinberg's consistency axiom counterintuitive, independently of the choice of the particular clustering algorithm - the k-means. Kleinberg's consistency transforms may also create new structures, as illustrated in Fig. 3.

The problems with Kleinberg's consistency transformation gave rise to a proposal of centric consistency transformation [11]. The centric consistency transformation maps points closer to the gravity center of the cluster by some factor λ. The centric consistency impacts the clustering by k-means as follows: Let Γ be an original clustering, $c \in \Gamma_k$ be a cluster to which we apply centric consistency with λ yielding a clustering $\Gamma_{k.c,\lambda}$.

$$Var_E(\Gamma_{k,c,\lambda}) = Var_E(\Gamma_k) - (1 - \lambda^2)Var(c)$$

while the variance of the entire set drops from $Var(S))$ to $Var(S) - (1 - \lambda^2)Var(c)$. The relative variance is then

$$Var_{RE}(\Gamma_{k,c,\lambda}) = \frac{Var_E(\Gamma_k) - (1 - \lambda^2)Var(c)}{Var(S) - (1 - \lambda^2)Var(c)}$$

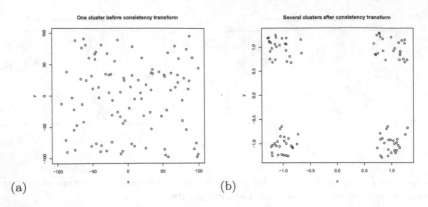

Fig. 3. Problems with consistency in Euclidean space. (a) single cluster before consistency transform. (b) four clusters emerging after consistency transform.

Obviously, $Var_{RE}(\Gamma_{k,c,\lambda}) \geq Var_{RE}(\Gamma_k)$ Consider the quotient

$$\frac{Var_{RE}(\Gamma_{k,c,\lambda}) - Var_{RE}(\Gamma_{k',c,\lambda})}{Var_{RE}(\Gamma_{k",c,\lambda}) - Var_{RE}(\Gamma_{k,c,\lambda})}$$

$$= \frac{Var_E(\Gamma_k) - (1-\lambda^2)Var(c) - Var_E(\Gamma_{k'}) + (1-\lambda^2)Var(c)}{Var_E(\Gamma_{k"}) - (1-\lambda^2)Var(c) - Var_E(\Gamma_k) + (1-\lambda^2)Var(c)}$$

$$= \frac{Var_E(\Gamma_k) - Var_E(\Gamma_{k'})}{Var_E(\Gamma_{k"}) - Var_E(\Gamma_k)}$$

in case $k' < k < k"$ and $c \subseteq c' \in \Gamma_{k'}$ and $c \subseteq c" \in \Gamma_{k"}$. If, however, c is not a subset of any set in $\Gamma_{k'}$ and/or $\Gamma_{k"}$, then the variance reduction in case of k' and/or k" clustering is not that big so that

$$\frac{Var_{RE}(\Gamma_{k,c,\lambda}) - Var_{RE}(\Gamma_{k',c,\lambda})}{Var_{RE}(\Gamma_{k",c,\lambda}) - Var_{RE}(\Gamma_{k,c,\lambda})}$$

$$= \frac{Var_E(\Gamma_k) - (1-\lambda^2)Var(c) - Var_E(\Gamma_{k'}) + (1-\lambda^2)Var(c) + \epsilon'}{Var_E(\Gamma_{k"}) - (1-\lambda^2)Var(c) - \epsilon" - Var_E(\Gamma_k) + (1-\lambda^2)Var(c)}$$

$$= \frac{Var_E(\Gamma_k) - Var_E(\Gamma_{k'}) + \epsilon'}{Var_E(\Gamma_{k"}) - Var_E(\Gamma_k) - \epsilon"} \geq \frac{Var_E(\Gamma_k) - Var_E(\Gamma_{k'})}{Var_E(\Gamma_{k"}) - Var_E(\Gamma_k)}$$

where $\epsilon' \geq 0, \epsilon" \geq 0$.

This means that

Theorem 2. *If the RIV quotient is used for selection of k, then centric consistency preserves or strengthens RIV value for the pick k.*

Richness usage in combination with centric consistency may make sense in general, though counterexamples can be constructed. In [11] we introduced the concept of $k \downarrow$-near-richness. A clustering algorithm possesses this property if it can return any clustering except one with the number of clusters over k. We have shown in [11, Theorem 32] that an axiomatic system consisting of centric consistency, scale-invariance and $k \downarrow$-nearly-richness is not contradictory.

Table 3. Variance explained (in percent) when applying k-means algorithm with $k = 2,\ldots,10$ to data from Figs. 1 (a) (Original), 2 (Kleinberg), and 1(b) (Centric)

k	Original	RIV	Kleinberg	RIV	Centric	RIV
2	49.5	2.36	78.2	**7.89**	50.1	2.36
3	70.7	2.78	88.1	1.25	71.3	2.78
4	78.1	1.43	96	4.38	78.9	1.43
5	83.3	1.20	97.8	1.63	84.2	1.20
6	87.6	1.10	98.9	2.75	88.6	1.10
7	91.6	1.60	99.3	2.00	92.6	1.60
8	94	**8.33**	99.5	2.00	95.1	**8.33**
9	94.3	1.50	99.6		95.4	1.50
10	94.6		99.6		95.6	

5   Super-Ball Separation

Consider a special case of dataset to be clustered.

Definition 1. *Consider data with the following clusters: For each cluster i there exists a radius r_i such that all members of the cluster lie within $r_i > 0$ of any cluster member and at least two cluster elements have positive distance. Furthermore the distance between any convex hull point of cluster i and j amounts to at least $5r_i + 5r_j$. We will call such clusters super-ball-separated and the clustering itself super-ball-clustering.*

The super-ball-separation may be seen as a generalization of nice separation and perfect separation as well as their ball variants see e.g. [12]. It is easily seen that the Algorithm 1 in [12], originally designed to discover perfect-ball-clusterings, can discover the super-ball-clustering also. Algorithm 2 therein can be adapted to confirm/disconfirm that the underlying clustering is a super-ball-clustering. Let us consider only datasets where no two datapoints lie at the same place in space (distances greater than zero).

If we know the number of clusters $k = k_t$ in advance, the Algorithm 1 of [12] (incremental k-means) will always discover the super ball clustering Γ_s correctly. If we choose a too large $k > k_t$, then Algorithm 1 will split some of the clusters into subclusters, however, never creating a cluster from subsets of Γ_s clusters. Hence an iterative procedure starting from a large k ($\leq n$ - the number of cluster elements) and decreasing at each step the number k will finally detect the clustering Γ_s if it exists. Some additional conditions are necessary, like non-zero radii r_i.

If we want to work within the realm of super-ball-clusterings, then obviously Kleinberg's richness cannot be valid for any algorithm. Only half-richness property (all possible clusterings of $k = 1,\ldots,n/2$ clusters) is realistic. Half-richness has no conflict with the scale-invariance property, but it is in conflict with Kleinberg's consistency.

Therefore, the consistency needs to be modified by an additional condition. For any cluster element e let E be a sequence of same cluster elements such that $d(e, E_i) \leq d(e, E_{i+1})$. So let us impose the restriction that for each i $5d(e, E_i) > d(e, E_{i+1})$ after consistency transform and let call this transform 5-consistency. In this way, no super-ball-separation is created inside a cluster. Obviously,

Theorem 3. *Half-richness, 5-consistency and scale-invariance are non-contradictory properties in the ball-separations realm of super-ball-clusterings (that is for algorithms discovering super-ball-clusterings).*

6   Final Remarks

In this paper we have shown that the richness axiom proposed by Kleinberg for clustering is unacceptable in the light of learnability theory and due to its conflicts with Kleinberg's consistency axiom.

We have shown that the problems with this axiom need a resolution not only in terms of restricting the number of clusterings to be considered, but also a change in consistency axiom is needed. We proposed to either use the consistency transformation or by going over to the world of super-ball-separation.

Note that the centric consistency may seem to be quite rigid, but still it turns out to be less restrictive than Kleinberg's consistency as distances between datapoints in different clusters are allowed to get closer.

The super-ball-clustering is very rare case for real world applications and its usefulness is first of all of theoretical nature, that is to show that a sound axiomatic system can be created, contrary go general opinion in literature. Though super-ball-separation is clearly too idealistic, but in practice we encounter situations where in fact the datapoints outside of a strict cluster core are sparse so that an extension of this concept to an approximated one seems to be an interesting area of further research. The sound axiomatic crisp case, investigated here, may be a good starting point for creating sound approximate definitions for clustering function and clustering axioms.

References

1. Ackerman, M., Ben-David, S., Loker, D.: Towards property-based classification of clustering paradigms. In: Lafferty, J., Williams, C., Shawe-Taylor, J., Zemel, R., Culotta, A. (eds.) Advances in Neural Information Processing Systems, vol. 23, pp. 10–18. Curran Associates, Inc. (2010)
2. Ackerman, M., Ben-David, S., Loker, D., Sabato, S.: Clustering oligarchies. In: Proceedings of the Sixteenth International Conference on Artificial Intelligence and Statistics, AISTATS 2013, Scottsdale, AZ, USA, 29 April–1 May 2013, pp. 66–74 (2013)
3. Ben-David, S.: Attempts to axiomatize clustering. In: NIPS Workshop, December 2005

4. Ben-David, S., Ackerman, M.: Measures of clustering quality: a working set of axioms for clustering. In: Koller, D., Schuurmans, D., Bengio, Y., Bottou, L. (eds.) Advances in Neural Information Processing Systems, vol. 21, pp. 121–128. Curran Associates, Inc. (2009)
5. Blum, A.: Thoughts on clustering. In: Essay for the 2009 NIPS Workshop "Clustering: Science or Art?" (2009)
6. Fulop, S.A., Chater, N.: Learnability theory. WIRE Cogn. Sci. 299–306 (2013). https://doi.org/10.1002/wcs.1228
7. Hopcroft, J., Kannan, R.: Computer science theory for the information age (chapter 8.13.2). A Satisfiable Set of Axioms, p. 272ff (2012)
8. Kleinberg, J.: An impossibility theorem for clustering. In: Proceedings of NIPS 2002, pp. 446–453 (2002). http://books.nips.cc/papers/files/nips15/LT17.pdf
9. Kłopotek, M.A., Kłopotek, R.A.: Clustering algorithm consistency in fixed dimensional spaces. In: Helic, D., Leitner, G., Stettinger, M., Felfernig, A., Raś, Z.W. (eds.) ISMIS 2020. LNCS (LNAI), vol. 12117, pp. 352–361. Springer, Cham (2020). https://doi.org/10.1007/978-3-030-59491-6_33
10. Klopotek, M.A.: On the phenomenon of flattening "flexible prediction" concept hierarchy. In: Jorrand, P., Kelemen, J. (eds.) FAIR 1991. LNCS, vol. 535, pp. 99–111. Springer, Heidelberg (1991). https://doi.org/10.1007/3-540-54507-7_9
11. Kłopotek, M.A., Kłopotek, R.A.: Towards continuous consistency axiom. Appl. Intell. 2022 (2022). https://doi.org/10.1007/s10489-022-03710-1
12. Kłopotek, R.A., Kłopotek, M.A.: Solving inconsistencies of the perfect clustering concept. In: Proceedings of PP-RAI'2019 Congress, pp. 273–276 (2019). http://pp-rai.pwr.edu.pl/PPRAI19_proceedings.pdf
13. van Laarhoven, T., Marchiori, E.: Axioms for graph clustering quality functions. J. Mach. Learn. Res. **15**, 193–215 (2014)
14. Meilă, M.: Comparing clusterings: an axiomatic view. In: Proceedings of the 22nd International Conference on Machine Learning. ICML 2005, pp. 577–584. ACM, New York (2005)
15. Strazzeri, F., Sánchez-García, R.J.: Possibility results for graph clustering: a novel consistency axiom (2021). https://arxiv.org/abs/https://arxiv.org/abs/1806.06142
16. Valiant, L.G.: A theory of the learnable. Commun. ACM **27**(11), 1134–1142 (1984)
17. von Luxburg, U.: Clustering stability: an overview. Found. Trends Mach. Learn. **2**(3), 235–274 (2009)
18. von Luxburg, U., Williamson, R.C., Guyon, I.: Clustering: Science or art? Initially, opinion paper for the NIPS Workshop "Clustering: Science or Art" in 2009 (2011)
19. Zadeh, R.B., Ben-David, S.: A uniqueness theorem for clustering. In: Proceedings of the Twenty-Fifth Conference on Uncertainty in Artificial Intelligence. UAI 2009, pp. 639–646. AUAI Press, Arlington (2009)

Adapting Loss Functions to Learning Progress Improves Accuracy of Classification in Neural Networks

Andreas Knoblauch$^{(\boxtimes)}$ ⓘ

KEIM Institute, Albstadt-Sigmaringen University,
Jakobstrasse 6, 72458 Albstadt-Ebingen, Germany
knoblauch@hs-albsig.de

Abstract. Power error loss (PEL) has recently been suggested as a more efficient generalization of binary or categorical cross entropy (BCE/CCE). However, as PEL requires to adapt the exponent q of a power function to training data and learning progress, it has been argued that the observed improvements may be due to implicitly optimizing learning rate. Here we invalidate this argument by optimizing learning rate in each training step. We find that PEL clearly remains superior over BCE/CCE if q is properly decreased during learning. This proves that the dominant mechanism of PEL is better adapting to output error distributions, rather than implicitly manipulating learning rate.

Keywords: Cross entropy · Power error loss · Learning rate · Learning schedule · Random grid path search

1 Introduction

During the last decade deep neural network models have become state of the art for various applications related to classification, including object recognition and detection, image segmentation, speech understanding, autonomous driving, or robot control [5,7,12,14,16,18]. This success can be attributed to an improved understanding of large-scale neural architectures and solving earlier problems like vanishing gradients blocking learning progress [1]. For example, to alleviate these problems various methods have been suggested, like improved activation functions, (dropout) regularization, weight initialization procedures, and optimization methods have been developed [4,6,8,15,17]. We have recently suggested a

This work was supported by the Ministerium für Wirtschaft, Arbeit und Tourismus Baden-Württemberg (VwV Invest BW - Innovation) via the project KICAD (FKZ BW1_0092/02) and by the Deutsches Bundesministerium für Verkehr und digitale Infrastruktur (Modernitätsfonds/mFUND) via the project AI4Infra (FKZ 19F2112C). The author acknowledges support by the state of Baden-Württemberg through bwHPC. The author is also grateful to German Nemirovski, Rober Frank, and Lukas Lorek for valuable discussions and help with the computing infrastructure.

M. Ceci et al. (Eds.): ISMIS 2022, LNAI 13515, pp. 272–282, 2022.
https://doi.org/10.1007/978-3-031-16564-1_26

complementary strategy, focusing on a new family of loss functions that can be adapted to the network model and learning task [9–11].

In fact, neural networks for classification and segmentation typically minimize loss functions based on binary or categorical cross entropy (BCE/CCE). However, it has been pointed out that cross entropy is not always optimal for maximizing likelihood or accuracy [9]. Instead, power error loss (PEL) has been suggested to improve learning performance, both in terms of training time and final accuracy, by means of adapting the exponent q of a power function to training data and learning progress [9,10]. Specifically, PEL is defined to obtain a simple initialization of error signals δ_{nk} for backpropagation by using the power function

$$\delta_{nk}(y_{nk}, t_{nk}) := (1 - 2t_{nk}) \cdot |y_{nk} - t_{nk}|^q \tag{1}$$

assuming a binary or multilabel classification task with target values $t_{nk} \in \{0,1\}$ and sigmoid outputs $y_{nk} = \sigma(a_{nk}) = 1/(1 + e^{-a_{nk}})$, where a_{nk} are dendritic potentials in the output layer. Here the error exponent $q > 0$ is a hyperparameter that may be adapted to the actual distribution of output errors $\epsilon_{nk} := t_{nk} - y_{nk}$. For the special case of fixed $q = 1$, PEL becomes identical to BCE. It has been demonstrated that choosing a fixed $q \neq 1$ can significantly improve performance over both BCE and CCE, depending on the classification model and training data [9]. Moreover, a further boost of learning progress can follow from adapting q during ongoing training [10,11].

However, it has been argued that a comparison of PEL to BCE or CCE would be unfair, as the improvement of PEL may be due to an interaction between the error exponent q and the learning rate of gradient descent: Specifically, (1) implies that for large $q > 1$ or small $q < 1$ error signals become smaller or larger, respectively, than they would be for BCE. As error signals are backpropagated from the output layer to all hidden neurons of the network, and gradients are computed as products of error signals and neuron activities, any scaling of error signals may have a similar effect as scaling the learning rate. So optimizing PEL exponent q may have a similar effect as optimizing learning rate, which can improve learning as well.

The main purpose of this study is therefore to investigate the interaction between optimizing PEL exponent q and optimizing learning rate during backpropagation learning. For this we do a co-optimization of both hyperparameters using a novel optimization method called Random Grid Path Search. If the benefits of PEL would be solely due to implicitly optimizing the learning rate, then there should be no additional improvement when optimizing both learning rate and PEL exponent. By contrast, if there were an additional improvement, this would prove that there is a different mechanism underlying the observed learning improvements after optimizing PEL exponents. As we will see in the following, our experiments favor the latter hypothesis, showing that optimizing PEL typically improves learning beyond optimizing learning rates. Moreover, as previous works on PEL have focused on fixed $q > 1$ in relatively small neural networks, we show that fixed $q < 1$ in large deep neural networks can as well improve learning outcome [11], and further investigate optimal learning schedules of adapting q and learning rate dependent on learning progress.

2   Models and Methods

2.1   Neural Network Models and Data Sets

As experiments in this study involve extensive optimization of hyperparameters, we use a relatively simple Convolutional Neural Network (CNN) with the following layout (see [10] for further details): Input $\rightarrow$ CNN(32) $\rightarrow$ BN(32) $\rightarrow$ ReLU $\rightarrow$ CNN(64) $\rightarrow$ ReLU $\rightarrow$ MaxPool $\rightarrow$ CNN(128) $\rightarrow$ BN $\rightarrow$ ReLU $\rightarrow$ CNN(128) $\rightarrow$ ReLU $\rightarrow$ MaxPool $\rightarrow$ Dropout($p = 0.05$) $\rightarrow$ CNN(256) $\rightarrow$ BN $\rightarrow$ ReLU $\rightarrow$ CNN(256) $\rightarrow$ ReLU $\rightarrow$ MaxPool $\rightarrow$ Dropout($p = 0.1$) $\rightarrow$ FC(1024) $\rightarrow$ ReLU $\rightarrow$ FC(512) $\rightarrow$ Dropout($p = 0.1$) $\rightarrow$ FC(512) $\rightarrow$ σ. Numbers in brackets correspond to channels (CNN/BN) and size (FC). BN and FC correspond to batch normalization and fully connected layers, MaxPool to max-pooling (kernel size 2; stride 2). All CNN-layers employ kernels with size 3, stride 1, and "same" padding. For training we used the CIFAR10 dataset [13]. As optimizer we used ADAM with standard parameters except learning rate [8]. Elsewhere, we have tested our methods also for more advanced setups [9–11], including sequence processing and semantic segmentation using DeepLabV3+ [3].

2.2   Power Error Loss Function (PEL)

In the context of the backpropagation algorithm [2,19], error signals for neurons k and inputs n are generally defined as the partial derivative of the loss function $\mathcal{L}_n$ with respect to the dendritic potentials a_{nk} of neurons k,

$$\delta_{nk} := \frac{\partial \mathcal{L}_n}{\partial a_{nk}} \tag{2}$$

Therefore, the power error loss function (PEL) can be defined by integrating the error signal initialization (1) with respect to the dendritic potentials a_{nk} in the output layer,

$$\mathcal{L}_{nk}^{\text{PEL},(q)}(y, t) := \int \delta_{nk} da_{nk} = (1 - t)F(y) + tF(1 - y) \tag{3}$$

with the antiderivative $F(y) := \int_0^y \frac{y^{q-1}}{1-y} dy$. Proper formulas and Python-based implementations for $F(y)$ are given elsewhere [10].

2.3   Hyperparameter Optimization

For the simple CNN model we performed co-optimization of learning rate and PEL exponent q. As optimal hyperparameters could vary over several orders of magnitude during a learning trial, we defined a large high-resolution grid, with initial learning rate (of first epoch) covering the interval $[10^{-6}; 10^1]$, relative learning rate (multiplied to optimal initial learning rate in subsequent epochs) covering the interval $[10^{-4}; 10^3]$, and the PEL exponent q covering the interval $[\frac{1}{16}; \frac{300}{16}]$. More specifically, the grid sampled the intervals of initial and relative learning rate at 33 (relative) values [1.0, 1.1, 1.2, 1.3, 1.4, 1.5, 1.6, 1.7, 1.8, 1.9, 2.0, 2.2, 2.4, 2.6, 2.8, 3.0, 3.2, 3.4, 3.6, 3.8, 4.0, 4.2, 4.6, 5.0, 5.5, 6.0, 6.5, 7.0, 7.5, 8.0, 8.5, 9.0, 9.5] within each decade, whereas q was sampled uniformly with increment $1/16$. Thus the grid comprised 7×33 values for the learning rates and 300 values for q, giving a total size of $231 \times 300 = 69300$ grid points. For the experiments involving BCE and CCE we optimized only learning rates (1D grid, 231 values). Each grid point was evaluated using 16 learning trials.

Due to the large grid sizes and the necessity to automatically optimize hyperparameters with high resolution, we discarded standard search techniques like grid search or random search, and developed a custom optimization algorithm that we call Random-Grid-Path-Search: This algorithm starts by testing $Z = 4$ randomly chosen grid points (plus additionally the midpoint of the grid) and then selecting the $B = 2$ best grid points for further search. For each of the B best initial grid points, we do a local "grid path search" by iteratively testing the neighboring grid points (within radius $R = 2$) and selecting the best neighbor until a local optimum is found. From the set of $\leq B$ local optima we finally select the best one (often the grid paths converged to a single optimum). As experimental measures of accuracy are noisy (even after averaging over 16 trials), it was necessary to use $R > 1$, to avoid getting stuck in false "noisy" maxima. To further accelerate the search we implemented a "stage-based" hierarchical approach, where we iteratively apply the described scheme to subgrids with increasing resolution. For the PEL experiments we used a 5-stage approach, where the subgrid increments (in both dimensions) where 16, 8, 4, 2, 1 in stages 1,2,3,4,5. With this algorithm we obtained proper (local) optima by evaluating typically only a few hundred of the 69300 grid points. For the BCE/CCE experiments we used the same increments, resulting in the evaluation of typically only some tens of the 231 grid points.

3 Results

Before doing systematic grid search optimizations, we present some reference results (see Fig. 1). For this we tested the CNN model with various (hand-designed) schedules for the PEL exponent q, while learning rates where kept fixed at a common standard value (0.0001). The experiments involving fixed q reproduce basically earlier findings: Best results occur for $q > 1$, whereas $q < 0$ performs worse than BCE and CCE. For fixed q, there are only minor (if any) improvements over BCE and CCE. However, continually adapting q during single learning trials from initially large values $q > 1$ to smaller values $q < 1$ can boost learning, exceeding performance of BCE and CCE, both in terms of learning speed and final outcome. By contrast, for the reverse order (first $q < 0$ and later $q > 1$), learning performance drops. Best results are obtained for a continual adaptation of q (cyan dashed line, corresponding to $q = 60/8 = 7.5$ in epoch

1, $q = 44/8 = 5.5$ in epoch 2, $q = 32/8 = 4$ in epoch 3, $q = 24/8 = 3$ in epoch 4, $q = 12/8 = 1.5$ in epochs 5 and 6, $q = 8/8 = 1$ in epochs 7 and 8, $q = 6/8 = 0.75$ in epochs 9–11, and $q = 4/8 = 0.5$ in the remaining epochs 12–20.

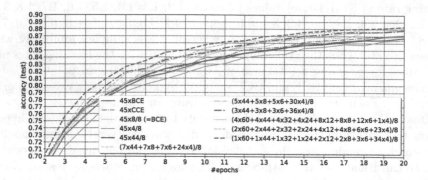

Fig. 1. Accuracy for CIFAR10 data set using a custom DCNN where q is adapted according to predefined time schedules during training as indicated in legends. For example, $45 \times 8/8$ means 45 epochs with $q = 8/8 = 1$ (corresponding to BCE); similarly, $(3 \times 44 + 3 \times 8 + 3 \times 6 + 36 \times 4)/8$ means 3 epochs with $q = 44/8$ followed by 3 epochs with q=8/8 followed by 3 epochs with $q = 6/8$, followed by 36 epochs with $q = 4/8$.

In a first attempt to approach optimal learning we optimized learning rate and q during each of 20 training epochs by using a "greedy" approach. That is, for PEL we conducted a grid search on a 2D grid for learning rates and PEL exponent q. For BCE and CCE we just optimized learning rate. After each training epoch, we selected the best grid parameters, and then started the next training epoch loading the network states of the best networks. Accuracy was estimated based on averaging over 16 training runs for each parameter set. After each epoch, we selected the best parameter set and stored the corresponding network states of all 16 tested networks. These 16 networks were used as initialization for the networks used for grid search of the next training epoch. Note that this corresponds to a "greedy" optimization as we assume that the optimal network maximizing accuracy after epoch i will be the optimal initialization for maximizing accuracy in epoch $i + 1$. We imposed this "greedy" assumption because it is otherwise not viable to do an exhaustive search over all combinations of learning rates and exponents q at any of the 20 epochs.

Figure 2 illustrates the results of the experiments. Figure 2A shows that largest accuracy is consistently obtained for PEL. Best results occur for PEL with small minibatch sizes 16,32,64. Also for most larger minibatch sizes, PEL achieves significantly better results than BCE and CCE. Initial learning phase is relatively noisy, where BCE and CCE may achieve better results, whereas during a later phase PEL typically overtakes BCE and CCE at some point. Larger minibatches generally learn slower as there are fewer weight updates per epoch leading to extended noisy phases.

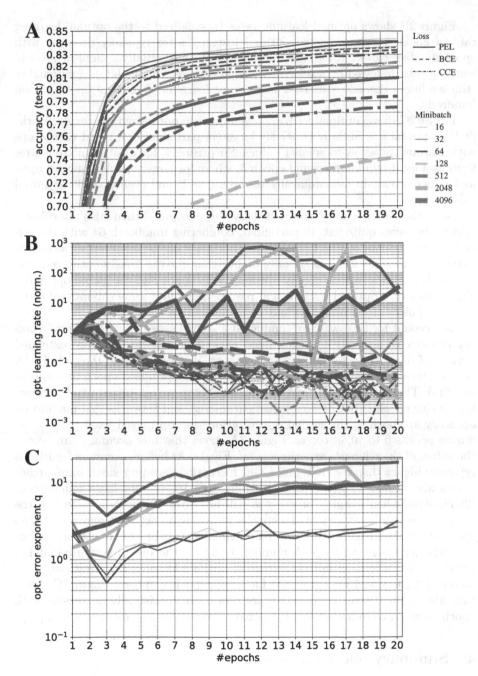

Fig. 2. Results for "greedy" co-optimizing error exponent q and learning rate for each individual training epoch for CIFAR10 data set using a custom DCNN. Data is shown for different loss functions (PEL, BCE, CCE) and minibatch sizes as indicated in the legends. Panels show test accuracy (**A**), optimal learning rate (**B**), and optimal error exponent (**C**).

Figure 2B shows optimal learning rates (normalized to the optimal learning rate in epoch 1). As expected optimal learning rates typically decrease with epoch number. However, for most minibatches there is initially a slight increase in learning rate, followed by a relatively steep decrease. For the larger minibatches (thicker lines) there is a marked increase of learning by factors up to several hundreds.

Figure 2C shows optimal PEL exponent q. As hypothesized in previous works [9,10], optimal q is initially relatively large (in particular $q > 1$ and then drops with learning progress to smaller values. Surprisingly, for most minibatch sizes there is a local minimum around epoch 3, where for larger epoch numbers, optimal q increase again monotonically towards $q > 1$, and even exceed the initial values.

Note also that accuracy grows initially very steep and then, also after 3 or 4 epochs, becomes quite flat. In particular, comparing minibatch 64 with the reference experiments of Fig. 1, shows that accuracy of the reference experiments (with fixed learning rate) is actually higher than for the greedy-optimized networks. This indicates that greedy-optimization cannot find the global optimum. This means that optimal learning will require doing first some suboptimal (non-greedy) choices of learning rate and/or exponents for some epochs.

To account for this, we have conducted a second series of experiments, relaxing the greedy property by using fixed parameter sets during more extended periods of several epochs, and optimizing w.r.t. accuracy only at the end of such epoch intervals. Figure 3 shows results where have optimized epoch intervals of length 5: This means in the first optimization step we optimized fixed parameters during epoch 1 to 5. Selecting best parameters and networks that maximized accuracy at epoch 5, the second optimization step optimized fixed parameters during epochs 6 to 10, et cetera. Figure 3A shows that now accuracy can exceed the values of the reference experiments (cf. Fig. 1). As before, accuracy for PEL is typically higher than for BCE and CCE. Exceptions are very small minibatches (when always $q < 1$; see panel C for minibatch 16) or very large minibatches (2048, 4096, when learning progress is still low after 20 epochs). Note also the steep increases of accuracy at the epochs switching to new parameter sets (i.e., epochs 6, 11, 16). Figure 3B shows that optimal learning rates are now more consistently decreasing with epoch number. Only for the two largest minibatches (2048, 4096) there is an initial increase of optimal learning rate (but only by a factor ≤ 3), followed by a decrease for larger epoch numbers. Figure 3C shows that also optimal error exponents q are now more consistently decreasing with epoch number, as expected and predicted by theoretical considerations in [10].

4   Summary and Conclusions

The main results of this study are the following: First, we have shown that an optimal schedule for adapting error exponents q during learning is to continually decrease q from large values (>1) at the beginning to small values (<1) at the end of training (see Fig. 2). This confirms earlier theoretical considerations

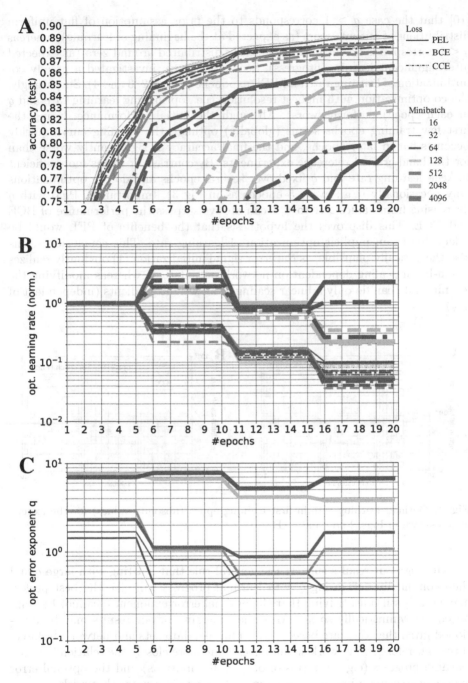

Fig. 3. Results for "greedy" co-optimizing error exponent q and learning rate for intervals of 5 training epochs for CIFAR10 data set using a custom DCNN. Data is shown for different loss functions (PEL, BCE, CCE) and minibatch sizes as indicated in the legends. Panels show test accuracy (**A**), optimal learning rate (**B**; normalized to optimal learning rate of epoch 1; cf., text) and optimal error exponent (**C**).

[10] that the case $q > 1$ corresponds to the prior assumption of flat uniform distributions of output errors (as expected at the beginning of learning), whereas $q < 1$ corresponds to output errors closely distributed around zero (as expected after significant learning progress). Second, we have investigated a greedy co-optimization of learning rate and PEL exponent q (Figs. 2, 3 and 4). Surprisingly, the co-optimization with highest resolution (i.e., optimizing learning rate and q in each epoch) performs worst. Although the initial performance during the first few training epochs seems (globally) optimal, the learning curve quickly becomes flat in later epochs, and the final accuracy is significantly worse than for the hand-crafted reference experiments. By contrast, it seems more efficient to keep learning rate and q fixed over several epochs and do the optimizations only occasionally (e.g., every 5 epochs as in Fig. 3). Importantly, PEL with q decreasing from values > 1 to < 1 can always surpass the performance of BCE and CCE. This disproves the hypothesis that the benefits of PEL would be solely due to an implicit optimization of learning rate. This conclusion seems also theoretically plausible, as the error signal initialization (1) obviously realizes a non-linear scaling dependent on network outputs y_{nk}, whereas modifying the learning rate can do only a linear scaling of all weight gradients (independent of y_{nk}).

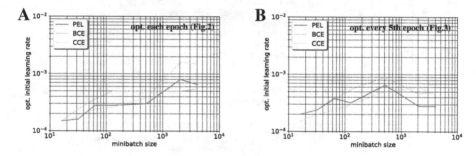

Fig. 4. Optimal learning rate in first training epoch (absolute values) for the experiments shown in Figs. 2 (**A**) and 3 (**B**).

Although our results suggest the general rule that starting with large q and then continually decreasing q with learning progress will give the best performance, it is still unsatisfying that the optimal error exponent q cannot be computed deterministically so far. Also, the generality of our results may be questioned given that they are based only on limited data sets and network architectures. Therefore future work should investigate more closely the relation between learning progress (e.g., in terms of output distributions) and the optimal error exponent q, using a broader spectrum of data sets and network models.

References

1. Bengio, Y., Simard, P., Frasconi, P.: Learning long-term dependencies with gradient descent is difficult. IEEE Trans. Neural Netw. **5**(2), 157–166 (1994)
2. Bishop, C.: Pattern Recognition and Machine Learning. Springer, New York (2006)
3. Chen, L.C., Zhu, Y., Papandreou, G., Schroff, F., Adam, H.: Encoder-decoder with atrous separable convolution for semantic image segmentation. In: Proceedings of the European Conference on Computer Vision (ECCV). arXiv preprint arXiv:1802.02611v2 (2018)
4. Glorot, X., Bengio, Y.: Understanding the difficulty of training deep feedforward neural networks. In: Teh, Y., Titterington, M. (eds.) Proceedings of the 13th International Conference on Artificial Intelligence and Statistics, Proceedings of Machine Learning Research, vol. 9, pp. 249–256. JMLR Workshop and Conference Proceedings, Chia Laguna Resort, Sardinia, Italy (2010)
5. Goodfellow, I., Bengio, Y., Courville, A.: Deep Learning. MIT Press, Cambridge (2016). http://www.deeplearningbook.org
6. He, K., Zhang, X., Ren, S., Sun, J.: Delving deep into rectifiers: surpassing human-level performance on imagenet classification. In: Proceedings of the IEEE International Conference on Computer Vision (ICCV), December 2015
7. He, K., Zhang, X., Ren, S., Sun, J.: Deep residual learning for image recognition. In: 2016 IEEE Conference on Computer Vision and Pattern Recognition, CVPR 2016, Las Vegas, NV, USA, 27–30 June 2016, pp. 770–778. IEEE Computer Society (2016). https://doi.org/10.1109/CVPR.2016.90
8. Kingma, D., Ba, J.: Adam: a method for stochastic optimization. In: Bengio, Y., LeCun, Y. (eds.) 3rd Proceedings of the International Conference on Learning Representations (ICLR). arXiv:1412.6980v9 (2015)
9. Knoblauch, A.: Power function error initialization can improve convergence of back-propagation learning in neural networks for classification. Neural Comput. **33**(8), 2193–2225 (2021)
10. Knoblauch, A.: On the antiderivatives of $x^p/(1-x)$ with an application to optimize loss functions for classification with neural networks. Ann. Math. Artif. Intell. **90**(4), 425–452 (2022)
11. Knoblauch, A., Luniak, P.: Improving learning of neural networks for classification, segmentation, and associative memory. In: Proceedings of the 7th bwHPC Symposium, 8 November 2021, Ulm, Germany (2022, to appear)
12. Krizhevsky, A., Sutskever, I., Hinton, G.: Imagenet classification with deep convolutional neural networks. In: Pereira, F., Burges, C., Bottou, L., Weinberger, K. (eds.) Advances in Neural Information Processing Systems, vol. 25. Curran Associates, Inc. (2012)
13. Krizhevsky, A.: Learning multiple layers of features from tiny images. Technical report, Department of Computer Science, University of Toronto (2009)
14. Ronneberger, O., Fischer, P., Brox, T.: U-Net: convolutional networks for biomedical image segmentation. In: Navab, N., Hornegger, J., Wells, W.M., Frangi, A.F. (eds.) MICCAI 2015. LNCS, vol. 9351, pp. 234–241. Springer, Cham (2015). https://doi.org/10.1007/978-3-319-24574-4_28
15. Ruder, S.: An overview of gradient descent optimization algorithms. arXiv preprint arXiv:1609.04747 (2016)
16. Schmidhuber, J.: Deep learning in neural networks: an overview. Neural Netw. **61**, 85–117 (2015)

17. Srivastava, N., Hinton, G., Krizhevsky, A., Sutskever, I., Salakhutdinov, R.: Dropout: a simple way to prevent neural networks from overfitting. J. Mach. Learn. Res. **15**(56), 1929–1958 (2014)
18. Tan, M., Le, Q.: EfficientNet: rethinking model scaling for convolutional neural networks. In: Chaudhuri, K., Salakhutdinov, R. (eds.) Proceedings of the 36th International Conference on Machine Learning. Proceedings of Machine Learning Research, vol. 97, pp. 6105–6114. PMLR, 09–15 June 2019
19. Werbos, P.J.: Beyond regression: new tools for prediction and analysis in the behavioral sciences. Ph.D. thesis, Harvard University (1974)

Multiscale and Multivariate Time Series Clustering: A New Approach

Jannaï Tokotoko[1]([✉]), Rodrigue Govan[1], Hugues Lemonnier[2], and Nazha Selmaoui-Folcher[1] [ID]

[1] Institute of Exact and Applied Sciences, University of New Caledonia, 98851 Noumea Cedex, France
tokotokojannai@yahoo.fr, nazha.selmaoui@unc.nc
[2] Ifremer, UMR 9220 ENTROPIE (IRD, Univ. Réunion, IFREMER, Univ. Nouvelle-Calédonie, CNRS), BP 32078, 98897 Nouméa Cedex, New Caledonia

Abstract. This paper introduces a new approach to multiscale and multivariate time series clustering based on the *X-MeansTS* method. It is common that the notion of multivariate time series clustering is defined as the grouping of a set of time series, and not the grouping of instances described by a set of common time series. For this second case, the proposed method is the first approach that meets the constraints of this type of clustering formalized in this paper. The quality of a clustering depends strongly on the distance measure used. The choice of the measure can also depend on the domain; in some cases, the clustering of some time series is done by considering more the interval of the measures on the y-axis while considering the shift of the series on the x-axis. The proposed multivariate method remains robust to these shifts.

Keywords: Time series · Clustering · Multiscale · Multivariate

1 Introduction

For the past decades, there has been considerable interest in time series (TS) analysis. Indeed, sensors are ubiquitous and with the motivation to acquire data in various fields of application: environmental monitoring, economy, health, etc., temporal data are more and more available and in mass. Therefore, it is important to develop efficient methods to process this big data. The focus has mainly been on univariate TS (only one temporal variable), but only few methods have been developed for multivariate TS, i.e., problems where we have multiple temporal variables per instance e.g., individuals or objects. Time series analysis are used for various objectives such as subsequence matching, anomaly detection, pattern discovery, clustering, classification, visualization, segmentation, and forecasting. Studies that have started on multivariate TS mostly focused on classification and forecasting methods [4,10]. In the literature, there exists no clustering method allowing to find clusters in multivariate TS. However, there are a variety of clustering methods in univariate TS [2]. Our study focuses on multiscale and

© The Author(s), under exclusive license to Springer Nature Switzerland AG 2022
M. Ceci et al. (Eds.): ISMIS 2022, LNAI 13515, pp. 283–293, 2022.
https://doi.org/10.1007/978-3-031-16564-1_27

multivariate TS clustering. By multiscale TS, we mean time series with their own time scales. However, the multivariate approach proposed in this paper is based on univariate methods. There are two main approaches for univariate TS clustering. The first approach is based on the feature construction. Time series are described by a vector of feature attributes [7] and the instances are clustered using a classic clustering method (*K-Means*, *X-Means*, *DBScan*, ...). The second approach uses similarity measures adapted to time series comparison. Several similarity measures have been proposed in the context of univariate TS clustering. A state of the art can be found in [6]. These measures have been developed and improved in terms of performance, for example the DTW measure (Dynamic Time Warping, [9]) and the SBD measure (Shape Based Distance, [8]). The LCSS (Longest Common SubSequence, [12]) assigns more weight to similar parts of two TS. The ERP distance [3] modifies the edition distance by adding a real penalty related to the number of required operations so that two sequences are similar. All these distance measures compare series considering only the temporal phase shift. They have been combined with basic approaches (*K-Means*) for TS clustering that do not take into account the so-called amplitude drifts. [11] proposes *X-MeansTS*, a new TS clustering approach considering the range of values (on the y-axis) in which the series evolve. The authors use existing clustering methods which use an adapted distance measure (DTW, SDB, ...). From the clusters generated by the existing method, *X-MeansTS* allows to determine an optimal number of clusters based on a new dispersion measure. This measure determines the distance dispersion between the representative and the series per cluster. This measure is based on the entropy and the standard deviation of the values to select the series to retain per cluster. A threshold is defined to determine the maximum value in the dispersion measure computation. In this paper, the new approach proposed uses *X-MeansTS*, on the variables independently and then generates multivariate clusters by ensuring that its clusters respect the dispersion threshold defined per variable. To meet this criterion, the dataset has been reduced to a static data matrix in which the instances are the individuals and the attributes are the cluster representatives corresponding to each of them. With this reduction, it is now possible to use a standard clustering method. The chosen method is *K-Means* where the dispersion measure is computed for each temporal variable describing the instances, for each cluster obtained by *K-Means*. The measure is computed from the series and their representatives. If the measure of a cluster does not verify the criterion for a temporal variable, then *K-Means* is repeated with the instances of this cluster. The operation is repeated hierarchically until the dispersion per variable is lower than the threshold. We tested the method on several multivariate TS datasets from the Time Series Classification repository, provided by the University of California, Riverside (UCR, [1]). In this paper, we propose a new multivariate time series clustering formalism. Then, we will introduce the *X-MeansTS* method used, in the multivariate TS approach for an upstream univariate clustering.

2   Notations and Definitions

Consider a multidimensional space where each instance is described by a set of temporal variables (in the shape of time series). We note $I = \{I_1, I_2, \ldots, I_n\}$ the set of instances and $S = \{s_1, s_2, \ldots, s_m\}$ the set of m temporal variables which can be defined on a common range of time $T = \{t_1, t_2, \ldots, t_p\}$. We will note $I_i = \{s_{i1}, s_{i2}, \ldots, s_{im}\}$ where $s_{ij} = \{s_{ij}(t_1), s_{ij}(t_2), \ldots, s_{ij}(t_p)\}$ a time series of length p with $s_{ij}(t)$ corresponding the signal value related to the s_j variable of the i instance, with $j \in \{1, 2, \ldots, m\}$. We talk about multivariate time series (MTS).

The MTS are multiscale when the set of temporal variables $S = \{s_1, s_2, \ldots, s_m\}$ is defined on a set of time range $\Gamma = \{T^1, T^2, \ldots, T^m\}$. Each time range is associated to a variable and can have a time scale different from another time range i.e., $T^i = \{t_1^i, t_2^i, \ldots, t_{p_i}^i\}$, $T^j = \{t_1^j, t_2^j, \ldots, t_{p_j}^j\}$ and $p_i \neq p_j$.

(Univariate Time Series Clustering). A temporal k-clustering for the variable s_j on a set of instances $I^j = \{s_{1j}, s_{2j}, \ldots, s_{nj}\}$ is a set $C^j = \{C_1^j, C_2^j, \ldots, C_k^j\}$ containing k homogeneous subsets of I^j (in the sense of a distance measure $Dist$ adapted to time series) and each of them having a representative. This representative is also called prototype, can be a series centroid composing each cluster of a variable j, and noted $R_{C_i^j}$. A k-clustering $C^j = \{C_1^j, C_2^j, \ldots, C_k^j\}$ with $\forall i \in \{1, \ldots, k\}$, $C_i^j = \{s_{i_1 j}, s_{i_2 j}, \ldots, s_{i_n j}\}$ (where n_i is a number of instances in cluster C_i^j) verifies the following criteria:

1. $I^j = \cup_{i=1}^k C_i^j$ and $C_i^j \cap C_h^j = \emptyset \ \forall i \neq h$.
2. $Dist(R_{C_i^j}, s) < Dist(R_{C_h^j}, s) \ \forall s \in C_i^j$ and $h \neq i$.

(Multiscale and Multivariate Time Series Clustering (MMTS) and Their Representatives). A k-clustering of MMTS is a set $CM = \{CM_1, \ldots, CM_k\}$ of I, with $CM_i = \{C_i^1, C_i^2, \ldots, C_i^m\}$ where $C_i^j = \{s_{i_1 j}, s_{i_2 j}, \ldots, s_{i_n j}\}$. CM contains k clusters (in the sense of a distance measure $Dist$) and verifies the following criteria:

1. $I = \cup_{i=1}^k CM_i$ and $CM_h \cap CM_i = \emptyset$, $\forall h \neq i$.
2. $\forall s \in C_i^j$, $Dist(R_{C_i^j}, s) < Dist(R_{C_h^j}, s)$ with $h \neq i \ \forall h, i \in \{1, \ldots, k\}$, $\forall j \in \{1, \ldots, m\}$.

(Distance Time Warping Measure, DTW). The distance time warping measure [9] is a distance computed from a path (between the first and last points) presenting an optimal non linear alignment between two time series. Let $q = \{q(t_1), \ldots, q(t_n)\}$ and $c = \{c(t_1), \ldots, c(t_n)\}$, two time series of $T = \{t_1, t_2, \ldots, t_n\}$. In the first place, the DTW measure consists in constructing a matrix M of dimension $n \times n$ where $M(i, j) = (q(i) - c(j))^2$. A warped path $W = \{w_1, \ldots, w_r, \ldots, w_p\}$ where $p \geq n$ is a contiguous sequence of elements from the matrix M and such that the first element is $w_1 = M(1, 1)$, the last element is $w_p = M(n, n)$ and w_r corresponds to $M(i, j)$. The method determines the

optimal alignment by searching into the matrix, the path W minimizing the cumulated (euclidean) distance i.e., $Wo = argmin(\sqrt{\sum_{i=1}^{p} w_i})$.

(Standard Deviation and Cluster Entropy). Let C_i, a cluster of C on T according to a distance measure $Dist$. Let $Dist(C_i) = \{d_{i_1}, \ldots, d_{i_{m_i}}\}$ the set of values of the $Dist$ measure between an instance of C_i and its representative R_{C_i}. Let $\sigma(C_i)$ the standard deviation computed on the distribution of values taken by $Dist(C_i)$, and $E(C_i)$ its entropy measure. $\sigma(C_i) = \sqrt{\sum_{k=1}^{m_i}(d_{i_k} - \overline{d_i})^2}$ where $\overline{d_i}$ is the average of $Dist(C_i)$ and $E(C_i) = -\sum_{k=1}^{m_i} P(d_{i_k}) \times \log(P(d_{i_k}))$.

3   Multiscale and Multivariate Time Series Clustering

The multivariate time series clustering method we propose allows to gather the clusters initially obtained by applying a method, in this case we choose the univariate TS clustering method *X-MeansTS* developed in [11] by considering each variable independently of the others.

For each temporal variable S^i, for $i \in \{1, \ldots, m\}$, we apply the *X-MeansTS* algorithm to obtain refined clusters (C_F^j) based on the dispersion measure $Disp$ defined by: $Disp(C_i) = \frac{\sigma(C_i)}{H(C_i)}$ with C_i a cluster from the set T as introduced in [11]. The *X-MeansTS* algorithm principle appears as these following steps:

1. **(Initial clusters definition)** The time series of a variable are clustered in a minimum number of clusters nb_min_clust. To generate these clusters, we apply the classic *K-MeansTS* algorithm [5] with $k = nb\_min\_clust$ and the distance measure $Dist$ used (which can be DTW, SDB, ...).
2. **(Cluster refinement by applying the dispersion criterion)** The homogeneity criterion is applied to each cluster to only retain the instances verifying the dispersion criterion. The remaining instances are assigned to a reject class. If the number of instances in an initial cluster is lower than nb_min_inst, then this cluster is deleted and its instances are assigned to a reject class.
3. **(Stop criterion application)** If the number of instances in the reject class is greater than nb_min_inst, then we repeat the first steps by taking as new set the reject class. Otherwise, the algorithm stops.

3.1   Principle of the Multivariate Time Series Method

Our multivariate clustering approach takes as input the same parameters from *X-MeansTS* which will be detailed below, with a dispersion threshold which will have to be defined for each variable. The chosen number of clusters for the initial clustering and the number of mininum instances are equally set up for the clusters of all variables. In our multivariate approach, the *X-MeansTS* method is applied to each variable and the input parameters are mostly the same as *X-MeansTS*. These parameters are as follows:

1. nb_min_clust: the number of clusters initially generated by *X-MeansTS*.

2. nb_min_inst: the minimum number of instances accepted per cluster.
3. s_d^j: the minimum threshold of the dispersion measure accepted for the variable j.

We will detail afterwards our multivariate time series clustering approach which generates a number of multivariate clusters automatically determined according to the three criteria previously mentioned. We nevertheless remind that the X-$MeansTS$ algorithm can automatically determine the minimum threshold for each variable, if no threshold is set up.

The Fig. 1 introduces the principle of the multivariate approach. In the first step, X-$MeansTS$ is applied to each variable independently. A matrix M is generated, it includes the set of cluster representatives in column and the instances in row. Each cell value corresponds to the distance between the cluster representative of a variable and the instance series describing the same variable.

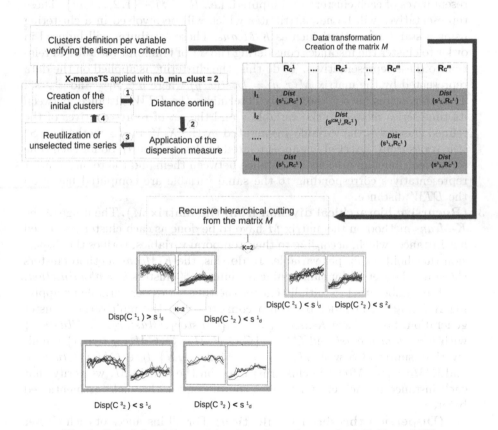

Fig. 1. Principle of the multivariate clustering method X-$MeansMTS$

In this Fig. 1, the example is given with $nb\_min\_clust = 2$. Then, we have a cutting of instances of the matrix, by dichotomy ($k = 2$). The cutting stops when for each multivariate cluster and for each variable, the dispersion distances between the instances and its representatives is lower than the dispersion threshold set up per variable.

The following steps detail the algorithm:

1. **(Cluster definition per variable verifying the dispersion criterion).** For each variable j, the instances of T^j (set of time series of the variable j) are clustered in a minimum number of clusters nb_min_clust. To create these clusters, we apply the classic K-$MeansTS$ algorithm for univariate time series [5] with $k = nb\_min\_clust$ and the distance measure $Dist$ (which can be DTW, SDB, ...). The homogeneity criterion s_d^j (the dispersion measure has to verify the minimum threshold s_d^j) is then applied for each cluster C_i^j of the variable j to keep only the instances verifying this criterion.

2. **(Data transformation).** After X-$MeansTS$ is applied per variable, the representatives of each cluster are computed, i.e., R_{C^j}, $\forall j \in \{1, 2, \ldots, m\}$. These representatives will be new attributes which will be involved in a clustering using a standard method such as K-$Means$. These attributes will be used in order to cluster the instances considering the set of temporal variables in relation to these representatives. To do this, the clustering is applied on the data represented by the matrix $M(x, y) = Dist(x, y)$ where $y \in R_{C_F^h}$ and $x \in s^h$. Thus, for an instance i, the matrix determines the DTW distance, between its time series related to the variable j and the set of representatives of the refined clusters of the variable j generated by the X-$MeansTS$ method on S^j (the series from the set of instances related to the variable j). So for each time series describing an instance, a distance between them and the refined cluster representatives corresponding to the same variable are computed based on the DTW distance.

3. **(Recursive hierarchical division from the matrix M).** The usage of the K-$Means$ method on the matrix M have to be done as each cluster generated has instances which, according to their temporal variables, verifies the dispersion threshold set up per variable. To do this, the K-$Means$ method clusters the static dataset M by maintaining as number of clusters $k = nb\_min\_clust$, i.e., the number used to initially cluster the time series per variable by applying X-$MeansTS$, in order to keep a consistency in the multivariate cluster generation. So we have K-$Means(M) = \{Cluster_1, Cluster_2, \ldots, Cluster_k\}$ with $k = nb\_min\_clust$ and $CM = \{CM_1, CM_2, \ldots, CM_{nb\_min\_clust}\}$ a multivariate subset of S with $\forall I_h \in I, h \in \{1, 2, \ldots, N\}, I_h \in CM_i \cap Cluster_i$, and $|CM_i| \leq N$. After the clustering done on the matrix M, we verify that each instance in each cluster verifies the dispersion threshold as mentioned before.

   (a) **(Dispersion threshold verification).** For all instances of each cluster obtained with K-$Means$, we compute again the dispersion of their time series per variable according to their new representatives. If one of the temporal variables does not verify the dispersion measure, then we hierarchically apply the K-$Means$ method on this subset with $k = nb\_min\_clust$.

(b) **(Stop condition of the hierarchical K-Means).** The previous clustering is now done recursively on the instances of each cluster initially generated by K-Means until the multivariate time series associated with these instances $(CM_i \in CM)$ do not verify the dispersion criteria per variable i.e., $CM_i = \{C_i^1, C_i^2, \ldots, C_i^m\}$, if $Disp(C_i^j) \geq s_d^j$ for $j \in \{1, 2, \ldots, m\}$. If the number of instances is lower than nb_min_inst, then the algorithm stops.

4 Experiments

Experiments will be done on different multivariate datasets, in this case we will use several datasets from the repository (UCR, [1]). We used 7 datasets with various number of instances, dimensions and length series. Each dataset is associated with pre-defined classes.

For these datasets, classes will be used to compute the multivariate TS clustering performance, i.e., the homogeneity and the completeness of the clustering, which the calculations will be detailed.

The following Table 1 introduces the datasets from the UCR repository. We chose these datasets because they are heterogeneous according to the series length and to the numbers of classes and instances that compose these datasets.

Table 1. Multivariate time series datasets description from the UCR repository

Name	# Time series	# Dimensions	Series length	# Classes
ArticularyWordRecognition	275	9	144	25
Heartbeat	204	61	405	2
JapaneseVowels	270	12	29	9
Libras	180	2	45	15
NATOPS	180	24	51	6
PhonemeSpectra	3315	11	217	39
UWaveGestureLibrary	120	3	315	8

4.1 Performance Measures Used

The performance measures used to evaluate our algorithm are the following:

(Cluster homogeneity). Let $L = \{Li | i = 1, \ldots, m\}$ be a set of real labels. We define a_{ij} as the number of instances labeled i assigned to the cluster j. The homogeneity h of a cluster C is defined as:

$$h(C) = \begin{cases} 1 & \text{if } H(L, C) = 0 \\ 1 - \frac{H(L|C)}{H(L)} & \text{otherwise.} \end{cases}$$

Algorithm 1. *X-MeansMTS*

Input:
- I and $S = \{S^1, S^2, \ldots, S^m\}$: multivariate time series
- nb_min_clust: minimum number of clusters at the first step
- $s_d = \{s_d^1, s_d^2, \ldots, s_d^m\}$: dispersion threshold
- nb_min_inst: minimum number of instances

Output:
- C_{FM}: set of clusters

1: **if** firstCall **then**
2:     **for** j in $\{1, 2, \ldots, m\}$ **do**
3:         $C_F^j = X\text{-}MeansTS(S^j, nb\_min\_clust, s_d^j, nbClust, nbMaxIter, recursifCpt)$
4:     **end for**
5: **end if**
6: I: instances $\in S$
7: **if** $|I| > nb\_min\_inst$ **then**
8:     isDispOk = True
9:     Generate the matrix $M(x, y) = Dist(x, y)$ where $y \in R_{C_F^h}$ and $x \in S^h$
10:     K-Means$(M())= \{Cluster_1, Cluster_2, \ldots, Cluster_k\}$ with $k = nb\_min\_clust$
11:     $CM = \{CM_1, CM_2, \ldots, CM_k\}$ k multivariate subset of S with $\forall I_i \in I, I_i \in CM_i \cap Cluster_i$
12:     **for** CM_i in CM **do**
13:         $j = 0$
14:         isDispOk = True
15:         **for** C_i^j in CM_i **do**
16:             **if** $Disp(C_i^j) < s_d^j$ **then**
17:                 isDispOk = False
18:             **end if**
19:             $j = j + 1$
20:         **end for**
21:         **if** isDispOk == False **then**
22:             $S_{temp} = \{S | I \in S \cap CM_i\}$ with I, the instances represented by the multivariate time series
23:             X-MeansMTS$(S_{temp}, nb\_min\_clust, s_d, nb\_min\_inst)$
24:         **else**
25:             $C_{FM} = C_{FM} \cup CM_i$
26:         **end if**
27:     **end for**
28: **end if**
29: **return** C_{FM}

where $H(L|C) = -\sum_{c=1}^{|C|} \sum_{l=1}^{|L|} \frac{a_{lc}}{N} \log \frac{a_{lc}}{\sum_{l=1}^{|L|} a_{lc}}$ and $H(L) = -\sum_{l=1}^{|L|} \frac{\sum_{c=1}^{|C|} a_{lc}}{m} \log \frac{\sum_{c=1}^{|C|} a_{lc}}{m}$

(Cluster completeness). The completeness measures how many similar instances with the same label are in the same cluster: $c = \frac{H(L|C)}{H(L)}$

4.2 Experiment Scenarios

UCR Datasets. Several tests have been made according to combination of input parameter values. The *X-MeansTS* method outperforms the *K-shape* method on complex data, i.e., high dimension datasets or with a high number of classes and with important length of series. As stated, the *X-MeansMTS* method can compute the threshold per variable. To do this, the method clusters the dataset in k groups and provides the minimum, maximum and mean dispersions of these k groups. For the multivariate time series approach tests, we determine different numbers of initial clusters k which are equivalent to the number of initial clusters nb_min_clust parameter considering several values close to the actual number of classes for each dataset, and the tests are also done for nb_min_clust equals to the actual number of classes. So based on a number nb_min_clust and computed dispersions (minimum and median), the multivariate time series method proposed is tested everytime with a number of minimum instances per cluster (nb_min_inst) ranging from 3 to 10 included. Therefore, with 5 different values of nb_min_clust, the three dispersion threshold values computed by the value of nb_min_clust and the eight values of nb_min_inst, for each dataset, the multivariate time series method is applied $5 \times 3 \times 8 = 120$ times with different parameters. A very low threshold values s_d and a very high value of initial cluster nb_min_clust generate clusters with a number of instances that remains very low. For example, the Fig. 2 shows this case, with multivariate groups when the number of minimum instances nb_min_inst is equal to 4.

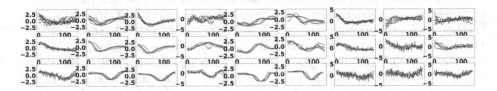

Fig. 2. Example of multivariate time series clusters

The result Table 2 shows mean values of performance obtained with two thresholds (minimum and median) on each dataset. The # Final clusters column shows the mean number of multivariate clusters obtained (with a standard deviation equals to 20). This number is really close to the actual number of classes per dataset, automatically obtained thanks to the *X-MeansMTS* method. In the end, the result Table 2 shows that on complex datasets (e.g., the Phoneme dataset in the Table 1), the *X-MeansMTS* performance remains excellent, which the completeness being 1 indicates. The dispersion measure threshold allows to increase the multivariate cluster homogeneity, when the minimum threshold is low.

Table 2. Mean performance of *X-MeansMTS* method on datasets from UCR repository

Dataset	Final clusters (±20)	V-measure	Homogeneity	Completeness	Name s_d
WordRecognition	141	0,92	0,85	1	Minimum
Heartbeat	56	0,74	0,59	1	Minimum
JapaneseVowels	6	0,48	0,31	1	Minimum
Libras	76	0,88	0,78	1	Minimum
NATOPS	86	0,89	0,8	1	Minimum
PhonemeSpectra	86	0,37	0,23	1	Minimum
GestureLibrary	51	0,88	0,79	1	Minimum
WordRecognition	111	0,81	0,68	1	Median
Heartbeat	26	0,38	0,23	1	Median
JapaneseVowels	6	0,48	0,31	1	Median
Libras	41	0,81	0,68	1	Median
NATOPS	26	0,58	0,41	1	Median
PhonemeSpectra	6	0,34	0,2	1	Median
GestureLibrary	36	0,81	0,68	1	Median

5   Conclusion

In this paper, we proposed a new approach of multiscale and multivariate time series clustering *X-MeansMTS*. The developed approach allows to cluster instances described by a common set of time series. For this purpose, the approach uses the monovariate time series clustering method *X-MeansTS*. This method uses a dispersion measure, applied per cluster to the distances between the series and their representatives and aims to increase the generated cluster homogeneity. One can define the dispersion threshold. In this multivariate approach (*X-MeansMTS*), the application of *X-MeansTS* to each variable generates a static data matrix containing the distances between the instance series and the cluster representatives according to a variable. *X-MeansMTS* keeps the defined threshold per variable by using hierarchical descending clustering on the static data matrix. We tested the method on several datasets.

The automatic search of the threshold of the dispersion measure can be improved, in order to optimize the final homogeneity (of the set of clusters). This measure can be applied to all clustering methods based on distance measures, for example it can be directly adapted to the existing method *K-Means*, for clustering static data. Indeed, the multivariate method generates a static data matrix, on which *K-Means* is applied in a hierarchical way. One of the perspectives of this work is also to supervise the *X-MeansMTS* approach. For this, we can rely on supervised learning approaches based on the exploitation of Gaussian distributions of the data. Indeed, the dispersion measure is adapted to this type of distribution. This measure can for example be integrated in the *LDA* (Linear Discriminant Analysis) supervised approach, which uses the Gaussian distribution of the data to predict. Finally, the objective is to create a supervised classification model that can be interpreted for multiscale and multivariate time series. Indeed, there is currently no method that addresses this issue.

References

1. Bagnall, A.J., et al.: The UEA multivariate time series classification archive (2018). CoRR abs/1811.00075
2. Bar-Joseph, Z., Gerber, G., Gifford, D., Jaakkola, T., Simon, I.: A new approach to analyzing gene expression time series data. In: Proceedings of the Annual International Conference on Computational Molecular Biology, February 2002
3. Chen, L., Ng, R.T.: On the Marriage of Lp-norms and Edit Distance. In: VLDB, pp. 792–803 (2004)
4. Gay, D., Bondu, A., Lemaire, V., Boullé, M., Clérot, F.: Multivariate time series classification: a relational way. In: Song, M., Song, I.-Y., Kotsis, G., Tjoa, A.M., Khalil, I. (eds.) DaWaK 2020. LNCS, vol. 12393, pp. 316–330. Springer, Cham (2020). https://doi.org/10.1007/978-3-030-59065-9_25
5. Huang, X., Ye, Y., Xiong, L., Lau, R.Y., Jiang, N., Wang, S.: Time series k-means: a new k-means type smooth subspace clustering for time series data. Inf. Sci. **367–368**, 1–13 (2016)
6. Jiang, G., Wang, W., Zhang, W.: A novel distance measure for time series: maximum shifting correlation distance. Pattern Recogn. Lett. **117** (2018)
7. Kalpakis, K., Gada, D., Puttagunta, V.: Distance measures for effective clustering of ARIMA time-series. In: ICDM, pp. 273–280 (2001)
8. Meesrikamolkul, W., Niennattrakul, V., Ratanamahatana, C.A.: Shape-based clustering for time series data. In: PaKDD, pp. 530–541 (2012)
9. Müller, M.: Dynamic time warping. In: Müller, M. (ed.) Information Retrieval for Music and Motion, pp. 69–84. Springer, Heidelberg (2007). https://doi.org/10.1007/978-3-540-74048-3_4
10. Ruiz, A.P., Flynn, M., Large, J., Middlehurst, M., Bagnall, A.: The great multivariate time series classification bake off: a review and experimental evaluation of recent algorithmic advances. Data Min. Knowl. Disc. **35**(2), 401–449 (2020). https://doi.org/10.1007/s10618-020-00727-3
11. Tokotoko, J., Selmaoui-Folcher, N., Govan, R., Lemonnier, H.: TSX-means: an optimal K search approach for time series clustering. In: Strauss, C., Kotsis, G., Tjoa, A.M., Khalil, I. (eds.) DEXA 2021. LNCS, vol. 12924, pp. 232–238. Springer, Cham (2021). https://doi.org/10.1007/978-3-030-86475-0_23
12. Vlachos, M., Kollios, G., Gunopulos, D.: Discovering similar multidimensional trajectories. In: ICDE 2002, pp. 673–684 (2002)

Improve Calibration Robustness of Temperature Scaling by Penalizing Output Entropy

Jun Zhang[1,2]([✉]), Wen Yao[2], Xiaoqian Chen[2], and Ling Feng[1]

[1] Tsinghua University, Beijing, China
mcgrady150318@163.com
[2] National Innovation Institute of Defense Technology, Chinese Academy
of Military Science, Beijing, China

Abstract. Data corruption is usually encountered in captured images due to the light condition, weather condition, or the quality of devices. Calibrating deep neural network (DNN) based image classifiers under data corruption is crucial, especially in safety-critical applications. A recent study shows that the widespread post-hoc calibration method temperature scaling (TS) performs poorly under corrupted shifts because it is easy to overfit the validation set. Three issues need to be addressed when improving the calibration robustness of TS: (1) How to measure the data shifts due to corruption? (2) How to reformulate TS with measured metrics? (3) How to improve the robustness of TS? Observing that output entropy increases with data shift intensity caused by corruption, we incorporate an entropy term into TS' optimizing Negative Log-likelihood (NLL) problem (Q1). Since the two terms of loss function are mutually exclusive, we reformulate TS as a multi-objective optimization (MOO) problem (Q2). By solving the MOO problem, a set of scaling parameters can thus be obtained and integrated to improve the calibration robustness of TS (Q3). We propose a novel TS method named MOO-ETS based on the above solutions, consisting of two integration strategies. Experimental results on the corrupted versions of CIFAR-10, CIFAR-100, and TinyImageNet demonstrate that: (1) entropy can measure corrupted data shifts accurately; (2) MOO-ETS can achieve competitive performance under corrupted shifts compared with the state-of-the-art method Deep Ensemble and beat the TS family baselines remarkably.

Keywords: Confidence calibration · Multi-objective optimization · Temperature scaling · Image classification

1 Introduction

A recent research [5] indicates that deep neural network (DNN) models tend to make overconfident predictions, which could cause serious disasters in some

© The Author(s), under exclusive license to Springer Nature Switzerland AG 2022
M. Ceci et al. (Eds.): ISMIS 2022, LNAI 13515, pp. 294–304, 2022.
https://doi.org/10.1007/978-3-031-16564-1_28

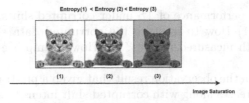

Fig. 1. In real applications, data corruption is frequently encountered in captured images due to the light condition, weather condition, or the quality of devices. Given a pre-trained image classifier, the output entropy increases with image saturation. Inspired by this observation, we incorporate an entropy term to the original TS to improve calibration performance under corrupted shift.

safety-critical scenes such as automatic driving [7], medical image diagnosis [3], etc. Various works [12,14] have aimed at improving the calibration of DNN based image classifiers in recent years. Corruptions are frequently encountered in captured images due to the light condition, weather condition, out of focus, the quality of devices [8], etc., which results in data distribution shifting. Data corruptions have been a big challenge to the robustness of calibration approaches, but rare research works focus on this issue specifically. According to the calibration stage, previous research works can be divided into pre- and post-hoc.

There are two main categories in the pre-hoc class. The first one is the multi-model method that calibrates DNN models by averaging the predictions of multiple models. Bayesian Neural Network (BNN) [1] is a usual method in this category and takes the weights of DNN as random variables and calibrates the models by approximating the posterior distribution. MC Dropout (MCD) [4] can be viewed as a particular type of BNN but is more easy-to-implement. Deep Ensemble (DE) [12] is another multi-model, which trains models by multiple random initialization schemes. In this category, DE is the most competitive approach under corrupted shifts in some benchmarks [14], but the disadvantage is too much memory cost.

The second line in pre-hoc focuses on the regularization in the training phase. MMCE [11] and SB-ECE [10] can improve the calibration performance which is injected into the traditional loss function during the training process. Rare works in this category are concerned about corrupted shifts.

In the category of post-hoc approach, temperature scaling (TS) is a popular method because it is easy to use and can preserve prediction accuracy. Standard TS calibrates the models by the scaling parameter T, minimizing Negative Log-Likelihood (NLL) on the validation set. To improve the expressive power of TS, Ensemble TS [17] calibrates a model by the ensemble of three T with different weights. [14] suggests that TS is easy to overfit on the validation set and calibrate poorly under corrupted shifts. The actual test samples often do not conform to the Independent Identical Distribution (IID) assumption with training data, limiting the effectiveness of TS in real scenarios.

To improve the performance of TS under corrupted shifts, three issues need to be addressed: (1) How to measure the corrupted data shifts? (2) How to reformulate TS with measured metrics? (3) How to improve the robustness of TS?

For Q1, based on the observation result that given a pre-trained classifier, the output entropy increases along with corrupted shift intensity as shown in Fig. 1, we consider using Entropy to measure the shifting and conduct experiments to validate the effectiveness.

For Q2, inspired by the success of label smoothing trick in calibration [15], we borrow the KL divergence between a Uniform distribution and output logits as entropy term, where lower KL divergence value means higher Entropy. Note that NLL and entropy terms are mutually exclusive where one loss increases, the other decreases simultaneously; we reformulate TS as a multi-objective optimization (MOO) problem.

For Q3, by solving the MOO problem via a classical evolutionary algorithm named NSGA-II [2], a set of efficient solutions $TSet$ consists of multiple T can thus be obtained. As shown in Fig. 2, each solution can be viewed as a combination of NLL and Entropy, where lower NLL and KL divergence values mean better fitting on the clean and corrupted dataset, respectively. Based on the obtained $TSet$, we propose a novel method named MOO-ETS to integrate multiple T to boost the expressive power of TS under corrupted shifts.

The contributions of this paper can be summarized as follows: (1) We discover that Entropy can measure corrupted shift accurately. (2) To the best of our knowledge, this is the first work reformulates TS as a MOO problem. (3) The empirical experiments demonstrate that the proposed MOO-ETS can achieve competitive performance under corrupted shifts compared with the state-of-the-art method Deep Ensemble and beat the TS family baselines remarkably.

The remainder of the paper is organized as follows. We introduce the background of this paper in Sect. 2. We propose Multi-Objective Optimized based Ensemble Temperature Scaling (MOO-ETS) method in Sect. 3 and report its performance in Sect. 4. We conclude the paper with a brief discussion of future work in Sect. 5.

2   Background

Given dataset $(x, y) \in D$, the image classification task with k classes can be regarded as learning a mapping function $f(x; \theta)$, where x denotes the input image and θ represents the weight parameters of DNN models. We take o as output logits of model, namely $o = f(x; \theta)$. The i-th element of prediction probability $\hat{y}$ is given by softmax function as follows:

$$\hat{y}_i = \frac{\exp(o_i)}{\sum_{i=1}^{k} \exp(o_i)} \tag{1}$$

2.1   Temperature Scaling

Temperature scaling (TS) is a popular method in the category of post-hoc calibration, which rescales the output logits as follows:

$$\hat{t}(T) = \frac{\exp(o_i/T)}{\sum_{i=1}^{k} \exp(o_i/T)} \tag{2}$$

where $\hat{t}$ denotes the re-scaled outputs, and T denotes the scaling parameter. Standard TS obtains the optimal T by minimizing the NLL loss function based on the validation set D_{val} as follows:

$$\arg\min_{T} nll\_loss[\hat{t}(T), y] \tag{3}$$

where y denotes the target in the dataset, and nll_loss means the NLL loss.

2.2   Expected Calibration Error

Expected calibration error (ECE) is the most commonly used metric to evaluate the calibration of DNN models, which measures the difference in expectation between accuracy and confidence [13]. ECE can be calculated as below:

$$ECE = \sum_{i=1}^{M} \frac{|B_i|}{N} |acc(B_i) - conf(B_i)| \tag{4}$$

where M denotes the size of equally-spaced confidence bins and i represents the bin number. N is the size of dataset. $acc(B_i)$ and $conf(B_i)$ denote the accuracy and confidence of i-th bin.

3   MOO-ETS: Multi-objective Optimization Based Ensemble Temperature Scaling

In this paper, we focus on improving the calibration performance of TS under corrupted shifts. To solve this problem, three critical questions should be addressed: (1) How to measure corrupted shifts? (2) How to reformulate standard TS with measured metrics? (3) How to improve the robustness of TS?

3.1   How to Measure Corrupted Shifts?

Standard TS performs poorly under corrupted shift [14] because it is easy to overfit on the validation set, which has the same distribution as the training set. This is the key factor that reduces the robustness of TS. However, data corruption is frequently encountered in captured images due to the light condition, weather condition, or the quality of devices. The actual test samples often do not conform to the IID assumption with training data, limiting the effectiveness of standard TS in real scenarios.

The first point is making TS measure the corrupted data shift accurately. Based on the observation result that given a pre-train classifier, the output entropy increases with corruption shift intensity as shown in Fig. 1, we consider using output entropy to measure the shifts. We will validate the effectiveness of Entropy in measuring corrupted shifts via empirical experiments.

3.2 How to Reformulate TS with Measured Metrics?

We firstly incorporate an entropy related regularizer into the original loss function of TS. The searching problem can be formulated as follows:

$$\arg\min_{T} nll\_loss(T) + \lambda * ent(T) \tag{5}$$

where ent denotes Entropy related regularizer, which is a function of T, and λ controls the trade-off between the NLL and regularizer. As shown in Fig. 2, given a pre-trained classifier, lower λ denotes the smaller influence of entropy term in the optimization process, when $\lambda = 0$, Eq. 5 is downgraded to the formulation of standard TS. To improve the calibration robustness of TS, we should use many choices of λ to adopt different data shift situations. Observing that multiple λ choices look like the Pareto Front of a MOO problem, we reformulate the searching problem in Eq. 5 as a single-variable (T) two-objective optimization $(nll\_loss$ and $ent)$ problem as follows:

$$\arg\min_{TSet}\{nll\_loss(T), ent(T)\} \tag{6}$$

where $TSet$ denotes the Pareto Front (a set of efficient solutions corresponding to different λ choices). The next key point is choosing Entropy related regularizer. Inspired by the calibration improvement via label smoothing trick [15], we use $KL(U||\hat{y})$ mentioned in this work as entropy regularizer to measure the corrupted shifts, where lower $KL(U||\hat{y})$ values represent higher entropy. The regularizer can penalize lower entropy output by making output logits more uniform. The updated optimization formulation can be written as follows:

$$\arg\min_{TSet}\{nll\_loss(T), KL(U||\hat{t}(T))\} \tag{7}$$

To solve the two-objective optimization problem, we use a classical evolutionary MOO algorithm named NSGA-II [2]. The hyper-parameter choices of NSGA-II are shown in the experimental setting section.

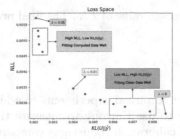

Fig. 2. Each efficient solution T on the Pareto Front denotes a different combination of NLL and entropy, which also corresponds to the choice of λ. Two losses are mutually exclusive, where NLL increases, entropy decreases simultaneously.

As shown in Fig. 2, two loss functions: NLL and Entropy are mutually exclusive; when NLL increases, Entropy decreases simultaneously. Each efficient solution T on the Pareto Front denotes a different combination of NLL and Entropy. T with lower NLL and entropy value represent better fitting on test and corrupted shift settings, respectively.

3.3 How to Improve Robustness of TS?

[17] shows that integrating multiple T can boost the expressive power of TS; we will use $TSet$ from Pareto Front to improve calibration performance under corrupted shifts.

To preserve the performance under test setting when improving the calibration under corrupted shifts, we select Top-K T (denoted as TK) from the $TSet$ by the ascending order of NLL values (which impact the calibration performance under the test setting). The size K is an important hyper-parameter in our methods studied in the experiments. Based on the selective TK, two integration strategies are proposed as follows:

Averaged Prediction. The first strategy denoted MOO-ETS-AP is averaged by the predictions of different T in the TK. The final predictions of models can be written as follows:

$$\hat{t} = \frac{1}{K} \sum_{i=1}^{K} t(T_i) \tag{8}$$

where T_i denotes the i-th T in the TK.

Averaged T. The second strategy denoted MOO-ETS-AT, is averaged by all T in the TK.

$$T_{avg} = \frac{1}{K} \sum_{i=1}^{K} T_i \tag{9}$$

where T_{avg} denotes the average value of TK. The final predictions can be written as $\hat{t} = t(T_{avg})$

4   Experiments

In this section, we conduct empirical experiments to address the following questions: (1) Is it reasonable to use Entropy to measure the corrupted shifts? (2) Does MOO-ETS improve calibration robustness compared with the baselines?

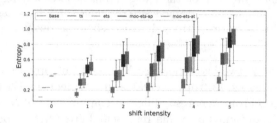

Fig. 3. Entropy on different shift intensities of ResNet-18 CIFAR-10-C task. Each box shows the quartiles' results across all corruption types, while the error bars indicate the minimum and maximum values. Number 0 denotes test setting, and 1–5 denote corrupted shift settings.

4.1   Experimental Setting

Datasets. We evaluate our method on the benchmarks of CIFAR-10, CIFAR-100, and TinyImageNet, which consist of 10, 100, and 200 classes, respectively. For each dataset, we split the original training dataset with 9:1 to obtain the new training and validation set. We call the test set of original dataset as test setting. Corrupted datasets are generated by the corruption functions in [8], which consists of 19 corrupted types for each dataset, and 5 shift intensities for each type. We denote the corrupted datasets as corrupted shift settings.

Baselines. We use ResNet-18 [6], VGG-16 [16] and Densenet-121 [9] as base models to classify images. All the models are trained with 100 epochs. The initial learning rate is 0.1, and it decays by 0.1 in each 20 epochs. We use SGD with momentum (0.9) as our optimizer. The weight decay is 5e−4. Batch size is 500, 500 and 200 for CIFAR-10, CIFAR-100 and TinyImageNet, respectively. We take the base model before calibration, TS [5], Ensemble TS [17] and Deep Ensemble [12] as baselines. They are abbreviated as base, ts, ets and ensemble, respectively. Our proposed MOO-ETS with two integration strategies are denoted by moo-ets-ap and moo-ets-at, respectively.

Metric. ECE is used to evaluate the performance of calibration methods, which is the smaller, the better. We set the bin number as 15 in the experiments.

Hyper-parameters. For the NSGA-II algorithm, we use 200 generations and 15 populations. The size of K in our MOO-ETS is chosen from the set $(1,3,5,7)$.

4.2 Entropy Can Measure Corrupted Shift. (Q1)

We report the overall entropy performance of all TS families with different shift intensities in Fig. 3. As shown in Fig. 3, Our MOO-ETS achieve higher Entropy than other TS methods for each shift intensity. The overall entropy increase with shift intensity for each method, which verifies that output entropy can be utilized to measure the corrupted shifts.

Table 1. Calibration error results of our proposed MOO-ETS and baselines. Test ECE and Shift ECE denote the ECE scores under test and corrupted shift settings, respectively. All results are averaged across all the corruption types and shift intensities by five single runs, and the best scores are highlighted in bold and red color.

Model-dataset	Method										
	Base		ts		ets		ensemble		moo-ets-ap(ours)		moo-ets-at(ours)
	Test ECE	Shift ECE	Test ECE	Shift ECE	Test ECE	Shift ECE	Test ECE	Shift ECE	Test ECE	Shift ECE	Test ECE	Shift ECE
ResNet-18 CIFAR-10	0.0475	0.1849	0.0146	0.1198	0.0144	0.1177	0.0081	0.0993	0.024	0.0721	0.0315	0.0685
ResNet-18 CIFAR-100	0.108	0.2561	0.0271	0.1429	0.0238	0.1368	0.015	0.1169	0.0293	0.0852	0.0377	0.0808
ResNet-18 TinyImageNet	0.2295	0.4641	0.1169	0.2988	0.1948	0.4095	0.027	0.2006	0.0426	0.2103	0.034	0.2011
VGG-16 CIFAR-10	0.0634	0.2364	0.0193	0.1552	0.0224	0.1237	0.015	0.079	0.0334	0.1072	0.0375	0.1039
VGG-16 CIFAR-100	0.1443	0.2894	0.03	0.1179	0.0273	0.1122	0.017	0.1183	0.0913	0.0758	0.1005	0.079
VGG-16 TinyImageNet	0.1851	0.3736	0.0904	0.2485	0.1674	0.3482	0.0406	0.1639	0.0197	0.171	0.0143	0.1647
DenseNet-121 CIFAR-10	0.0453	0.2151	0.014	0.1492	0.0153	0.139	0.012	0.1026	0.0212	0.1165	0.0246	0.1125
DenseNet-121 CIFAR-100	0.1278	0.3117	0.0208	0.1546	0.0316	0.1722	0.0275	0.1075	0.0754	0.0936	0.0885	0.0933
DenseNet-121 TinyImageNet	0.0547	0.3066	0.0694	0.3263	0.0943	0.3608	0.0762	0.1801	0.0295	0.2064	0.0489	0.1862
Average	0.1117	0.2931	0.0447	0.1904	0.0657	0.2133	0.0265	0.1298	0.0407	0.1265	0.0464	0.1211

4.3 MOO-ETS Can Improve Calibration Robustness of TS Family. (Q2)

Overall Performance. We report calibration performance in Table 1. Observed from Table 1, on average of nine image classification tasks, Deep Ensemble achieves the best ECE score under test setting, while our proposed MOO-ETS-AT could be the winner of corrupted shift setting. Compared with TS family baselines: Standard TS and ETS, MOO-ETS-AT can drop ECE scores by 36.4% and 43.2% under corrupted shift settings, respectively, while obtaining similar performance under test setting. Compared with Deep Ensemble, MOO-ETS can perform competitively under corrupted shift settings.

Overall, MOO-ETS can significantly improve calibration performance of TS family methods under corrupted shift setting, even perform competitively compared with the state-of-the-art method Deep Ensemble, while preserving the calibration performance under test setting compared with TS family. The two integrating strategies of MOO-ETS achieve similar calibration performance under test and corrupted shift settings.

Performance on Different Shift Intensities. We report ECE scores on different shift intensities of ResNet-18 CIFAR-10 task in Fig. 4. MOO-ETS can beat TS and ETS for each shift intensity, and the dropped ECE scores increase along with shift intensity. MOO-ETS can also preserve calibration performance compared with the TS family under test setting. For corrupted shift settings, MOO-ETS can perform competitively compared with Deep Ensemble.

Performance on Different Ensemble Size K. We report calibration performance on ensemble size K of ResNet-18 CIFAR-10, CIFAR-100, and TinyImageNet tasks under both settings in Fig. 5. For corrupted shift settings, ECE scores decrease as K increases; we can choose more T from Pareto Front to boost the robustness of TS. For test setting, the optimal K increases with the size of classes, e.g., we can choose 3,5, and 7 for CIFAR-10, CIFAR-100, and TinyImageNet, respectively.

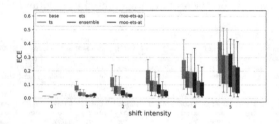

Fig. 4. Calibration error on different shift intensities of ResNet-18 CIFAR-10 task. Each box shows the quartiles of the results across all corruption types, while the error bars indicate the minimum and maximum values. Number 0 denotes test setting, and 1 to 5 denote corrupted shift settings.

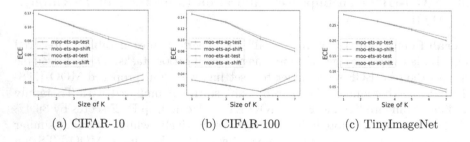

(a) CIFAR-10   (b) CIFAR-100   (c) TinyImageNet

Fig. 5. The analysis of ensemble size K in the ResNet-18 CIFAR-10, CIFAR-100 and TinyImageNet tasks. The suffix 'test' and 'shift' denote the test and corrupted shift setting, respectively.

5 Conclusion and Future Work

We propose a novel method named MOO-ETS by penalizing output entropy to improve the robustness of TS. Empirical experiments on the corrupted version of CIFAR-10, CIFAR-100, and TinyImageNet demonstrate that: (1) entropy can measure corrupted shifts accurately; (2) MOO-ETS can achieve significant improvements under corrupted shift setting while preserving similar performance under test setting compared with TS family baselines; (3) MOO-ETS also performs competitively compared with the state-of-the-art method Deep Ensemble. In the future, we will study how to improve the calibration further when the models have been calibrated in the training phase.

References

1. Blundell, C., Cornebise, J., Kavukcuoglu, K., Wierstra, D.: Weight uncertainty in neural networks. arXiv preprint arXiv:1505.05424 (2015)
2. Deb, K., Pratap, A., Agarwal, S., Meyarivan, T.: A fast and elitist multiobjective genetic algorithm: NSGA-II. IEEE Trans. Evolut. Comput. **6**(2), 182–197 (2002). https://doi.org/10.1109/4235.996017
3. Esteva, A., et al.: Dermatologist-level classification of skin cancer with deep neural networks. Nature **542**(7639), 115–118 (2017)
4. Gal, Y., Ghahramani, Z.: Dropout as a Bayesian approximation: representing model uncertainty in deep learning. In: ICML 2016 Proceedings of the 33rd International Conference on International Conference on Machine Learning - vol. 48, pp. 1050–1059 (2016)
5. Guo, C., Pleiss, G., Sun, Y., Weinberger, K.Q.: On calibration of modern neural networks. In: Proceedings of the 34th International Conference on Machine Learning , vol. 70, pp. 1321–1330 (2017)
6. He, K., Zhang, X., Ren, S., Sun, J.: Deep residual learning for image recognition. In: 2016 IEEE Conference on Computer Vision and Pattern Recognition (CVPR), pp. 770–778 (2016)
7. Hecker, S., Dai, D., Gool, L.V.: Failure prediction for autonomous driving. In: 2018 IEEE Intelligent Vehicles Symposium (IV), pp. 1792–1799 (2018)
8. Hendrycks, D., Dietterich, T.G.: Benchmarking neural network robustness to common corruptions and perturbations. In: International Conference on Learning Representations (2019)
9. Huang, G., Liu, Z., van der Maaten, L., Weinberger, K.Q.: Densely connected convolutional networks. In: Computer Vision and Pattern Recognition (2017)
10. Karandikar, A., et al.: Soft calibration objectives for neural networks. arXiv: Learning (2021)
11. Kumar, A., Sarawagi, S., Jain, U.: Trainable calibration measures for neural networks from Kernel mean embeddings. In: International Conference on Machine Learning, pp. 2805–2814 (2018)
12. Lakshminarayanan, B., Pritzel, A., Blundell, C.: Simple and scalable predictive uncertainty estimation using deep ensembles. In: Advances in Neural Information Processing Systems, vol. 30, pp. 6402–6413 (2017)
13. Naeini, M.P., Cooper, G.F., Hauskrecht, M.: Obtaining well calibrated probabilities using Bayesian binning. In: AAAI 2015 Proceedings of the Twenty-Ninth AAAI Conference on Artificial Intelligence, vol. 2015, pp. 2901–2907 (2015)

14. Ovadia, Y., et al.: Can you trust your model's uncertainty? Evaluating predictive uncertainty under dataset shift. In: Advances in Neural Information Processing Systems, vol. 32, pp. 13969–13980 (2019)
15. Pereyra, G., Tucker, G., Chorowski, J., Łukasz Kaiser, Hinton, G.: Regularizing neural networks by penalizing confident output distributions. In: ICLR (Workshop) (2017)
16. Simonyan, K., Zisserman, A.: Very deep convolutional networks for large-scale image recognition. In: ICLR 2015: International Conference on Learning Representations 2015 (2015)
17. Zhang, J., Kailkhura, B., Han, T.Y.J.: Mix-n-match : ensemble and compositional methods for uncertainty calibration in deep learning. In: International Conference on Machine Learning (2020)

Understanding Negative Calibration from Entropy Perspective

Jun Zhang[1,2](✉), Wen Yao[2], Xiaoqian Chen[2], and Ling Feng[1]

[1] Tsinghua University, Beijing, China
[2] National Innovation Institute of Defense Technology, Chinese Academy of Military Science, Beijing, China
mcgrady150318@163.com

Abstract. Deep neural network (DNN) based image classifiers have successfully applied to various scenes but suffered from severe overconfidence issues, essential to safety-critical applications. In recent years, plenty of research has focused on probabilistic calibration to reduce the risks from overconfident predictions. Recent work presents that it is harder to further calibrate the models calibrated by regularization techniques like label smoothing. We extend the above study to the corrupted dataset setting which is frequently encountered in the image capture processing, such as light condition, weather condition, or the quality of devices. Interestingly, we discover that post-hoc method like temperature scaling (TS) would hurt the calibration performance of original models under corrupted shift of CIFAR-10, CIFAR-100, and TinyImageNet, and we call this phenomenon **Negative Calibration (NC)**. We observe that NC occurs in the decreasing of output entropy when the post-hoc method is applied to the pre-trained model, and we take ResNet-18 CIFAR-10 as an example to understand NC from the perspective of entropy.

Keywords: Negative Calibration · Temperature scaling · Image classification

1 Introduction

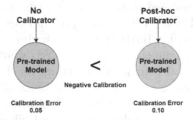

Fig. 1. Negative Calibration: if a pre-trained model achieves a worse calibration error after applying a calibrator.

M. Ceci et al. (Eds.): ISMIS 2022, LNAI 13515, pp. 305–314, 2022.
https://doi.org/10.1007/978-3-031-16564-1_29

Deep neural network (DNN) models have been applied into various domains successfully, such as computer vision [6], natural language processing [2]. However, DNNs suffer from serious overconfidence issues [5], which will cause dangerous consequences, especially in safety-critical applications, such as automatic driving [7], medical image diagnosis [3], etc. In recent years, numerous research works have focused on probabilistic calibration to solve overconfidence issues, improving the reliability of deep learning models. Previous research works can be divided into three classes.

The first one is the average prediction approach which implements calibration by averaging the predicted outputs of multiple models. One famous branch is Bayesian Neural Network (BNN) [1]. Unlike the Frequentist, BNN views each weight of DNN as a random variable, not a fixed value. BNN achieves calibration via sampling from the approximated posterior distribution. As a particular case of BNN, MC Dropout (MCD) [4] obtains the calibrated outputs by sampling the random layer dropout. Deep Ensemble (DE) [10] as a non-Bayesian approach, averages the prediction outputs, which are trained by multiple random initialization schemes. DE is state-of-the-art in this category but needs too much memory cost.

The second type explores the loss functions in the training phase to improve the calibration performance. One line introduces the regularization term which improves calibration performance into the original loss function, e.g., MMCE [9] etc. The other line explores whether the existing loss function can calibrate the model. Lots of works have been proved that focal loss [12] and label smoothing [14] can calibrate the models implicitly.

The third category is the post-hoc approach, of which the most representative work is temperature scaling (TS) [5]. TS calibrates models via scaling the outputs by a learnable parameter T named temperature based on the validation set. Various works have been proposed to improve TS in recent years. Parameterized TS [15] uses a neural network to predict T to improve the expressive power of TS. Ensemble TS [17] calibrates the model by a weighted sum of multiple T.

Interestingly, as reported in [16], there is little improvement room of calibration by the post-hoc methods when the model is trained by specific regularization approach, where the test dataset is sampled from the same distribution as the training dataset. We extend the test scenarios to the corrupted dataset setting, which is frequently encountered in the image capture processing, such as light condition, weather condition, or the quality of devices. Based on the experimental results of CIFAR-10, CIFAR-100, and TinyImageNet, we discover that post-hoc method such as temperature scaling (TS) would hurt the calibration performance of original models under this setting. As shown in Fig. 1, if a pretrained model obtains a worse calibration performance after applying a post-hoc calibrator, we call this phenomenon Negative Calibration (NC).

In this paper, we focus on understanding and eliminating Negative Calibration. [14] shows that penalizing low entropy output can improve calibration implicitly. Inspired by this, we attempt to understand NC from the perspective of entropy. We monitor the average output entropy of the whole dataset in the corrupted image classification benchmarks generated by [8] and discover that NC occurs in the decreasing of output entropy when a post-hoc method is applied to the pre-trained model.

The contributions of this paper can be summarized as follows:

(1) An interesting phenomenon named Negative Calibration is discovered when we extend the work of [16] to the corrupted dataset setting.
(2) We propose to understand Negative Calibration from the perspective of entropy and find that NC occurs when the calibrated output entropy decreases compared with the original model under corrupted dataset setting.

The remainder of the paper is organized as follows. The background of this paper is introduced in Sect. 2. We present Negative Calibration under a corrupted dataset setting and attempt to understand this interesting phenomenon from the perspective of entropy in Sect. 3. We conclude the paper with a brief discussion of future work in Sect. 4.

2   Background

2.1   Notion and Problem Setup

In this paper, we focus on image classification tasks. We let $x \in \mathbb{R}^d$ denote a set of d-dimensional inputs and $y \in \{1, \ldots, k\}$ denote corresponding labels for k-class classification. We assume that a training dataset D consists of N samples $D = \{(x_n, y_n)\}_{n=1}^{N}$.

We use a neural network $f(x; \theta)$ to model $p_\theta(y|x)$ and estimate the parameters θ using the training dataset. We denote the output logits of neural network as $o_n = f(x_n, \theta)$, and probabilistic prediction is $p(y|x_n, \theta) = \sigma_s(o_n)$, where σ_s denotes the softmax function. We define the predicted label as $\hat{y}_n = \arg\max_y p(y|x_n, \theta)$, and the confidence as $p(y = \hat{y}|x, \theta) = \max_k p(y = k|x_n, \theta)$.

2.2   Temperature Scaling

Temperature scaling is a widespread approach of post-hoc calibration, which is easy-to-use and accuracy-preserving. TS scales the output logits by dividing a learnable scaling parameter T as follows:

$$z_n = o_n/T \tag{1}$$

where z_n denotes the scaled output logits. The optimal T is obtained by minimizing the NLL loss function based on the validation set D_{val}.

2.3   Expected Calibration Error

Expected calibration error (ECE) is a classical metric to measure calibration performance, representing the difference in expectation between the accuracy and confidence of model [13]. ECE can be calculated approximately as follows:

$$ECE = \sum_{l=1}^{L} \frac{|B_l|}{N} |acc(B_l) - conf(B_l)| \tag{2}$$

where L denotes the number of equally-spaced confidence bins and $l = 1, \ldots, L$ denotes the bin number l. $acc(B_l)$ and $conf(B_l)$ mean the accuracy and confidence of l-th bin. N is the total number of data samples.

2.4   Loss Functions

We will introduce some proposed loss functions to improve calibration in this section: label smoothing, focal loss, and inverse focal loss.

Label Smoothing (LS). Label smoothing trick can implicitly calibrate DNN [14]. By introducing a label smoothing parameter α, we can re-define the target label as follows:

$$y_{LS} = y(1 - \alpha) + \alpha/k \tag{3}$$

Focal Loss. The core idea in focal loss is that different samples contribute differently to the loss function and should be assigned different weights [11]. The formulation of focal loss can be written as follows:

$$focal\_loss = -(1 - \hat{y})^\gamma \log \hat{y} \tag{4}$$

where γ denotes a weight controlling factor, when $\gamma = 0$, focal loss downgrade to the nll_loss.

Inverse Focal Loss. [16] propose an inverse version of focal loss, to assign larger weights to the samples with larger confidences, which can be defined as follows:

$$inverse\_focal\_loss = -(1 + \hat{y})^\gamma \log \hat{y} \tag{5}$$

The only difference with focal loss is replacing $1 - \hat{y}$ by $1 + \hat{y}$.

3   Negative Calibration Happens When Dataset Shifts

In this section, we extend the experiments of [16] under corrupted dataset setting and observe that TS would hurt the calibration performance of the original model. We call this phenomenon Negative Calibration. Inspired by the conclusion that penalizing output entropy can improve calibration implicitly [14], we attempt to understand NC from the entropy perspective.

Fig. 2. Four examples of corruption types: Gaussian noise, Gaussian blur, fog and JPEG compression, respectively.

3.1 Experimental Settings

Datasets. We use three classical image classification tasks: CIFAR-10, CIFAR-100, and TinyImageNet as evaluation benchmarks, which consist of 10, 100, and 200 classes, respectively. The optimal T in TS is learned from the validation sets of three datasets, sampled 10% percent of the original training set. We show four typical corruption types: noise, blur, weather, and digital in Fig. 2. To extend the experiments of [16] under corrupted dataset setting, we generate CIFAR-10-C, CIFAR-100-C, and TinyImageNet-C by the corruption methods in [8] based on the original test sets of benchmarks. The corrupted dataset includes 19 corruption types with five shift intensities, and each intensity contains 10,000 images. To better distinguish, we call the original and corrupted datasets as **test** and **corruption settings**, respectively.

Models. We choose three frequently used DNN based image classification models: ResNet-18, VGG-16 and Densenet-121 in the experiments. All the models are trained with 100 epochs. The initial learning rate is 0.1, and it decays by 0.1 in each 20 epochs. We use SGD with momentum 0.9 as our optimizer. The weight decay is 5e-4. Batch size is 500, 500 and 200 for CIFAR-10, CIFAR-100 and TinyImageNet, respectively.

Baselines. We use four losses described in 2.4: NLL, Label Smoothing (LS), Focal, and Inverse Focal. The optimal hyper-parameters α in LS and γ in Focal and Inverse Focal are tuned by validation sets of benchmarks. For post-hoc calibration, we use temperature scaling (TS). The results after TS are denoted by the suffix "TS". For label smoothing, the hyper-parameter α is tuned 0.05, 0.05, 0.10 for CIFAR-10, CIFAR-100 and TinyImageNet based on the validation set. For focal and inverse focal, the hyper-parameter γ is tuned 1.0, 1.0, 3.0 for CIFAR-10, CIFAR-100 and TinyImageNet based on the validation set.

Metric. We use ECE to evaluate the calibration performance, and the number of bins in ECE is set 15. The smaller, the better.

3.2 Negative Calibration at First Look

We summarize the calibration performance from three perspectives: overall, shift intensity, and corruption type.

Table 1. Overall performance of ΔECE, ΔEntropy and T for image classification tasks under corruption setting. All the results are averaged across the corruption types and shift intensities under 5 single runs. Row Average denote the mean result of the corresponding column. Negative Calibration is highlighted in underline. Positive and negative values of metrics in the table are highlighted in red and blue, respectively.

Model-Dataset	Method											
	NLL			LS			Focal			Inverse Focal		
	ΔECE(−)	ΔEntropy	T	ΔECE(−)	ΔEntropy	T	ΔECE(−)	ΔEntropy	T	ΔECE(−)	ΔEntropy	T
ResNet-18 CIFAR-10-C	−0.0651	0.219	1.619	0.02	−0.1099	0.903	−0.0087	0.0283	1.05	−0.0567	0.1969	1.564
ResNet-18 CIFAR-100-C	−0.1132	0.5411	1.386	0.038	−0.2518	0.914	−0.1235	0.6009	1.417	−0.1179	0.5957	1.404
ResNet-18 TinyImageNet-C	−0.1653	0.84	1.513	−0.0812	0.5269	1.25	−0.2587	1.5399	2.07	−0.2477	1.4646	2.029
VGG-16 CIFAR-10-C	−0.0811	0.329	1.769	0.0084	−0.043	0.965	−0.1248	0.498	1.932	−0.0779	0.297	1.676
VGG-16 CIFAR-100-C	−0.1715	0.9348	1.574	−0.0033	0.0201	1.006	−0.2049	1.5993	2.123	−0.1922	0.9356	1.743
VGG-16 TinyImageNet-C	−0.1251	0.6773	1.395	−0.0745	0.4811	1.243	−0.194	1.2218	1.839	−0.1743	1.0924	1.839
DenseNet-121 CIFAR-10-C	−0.0659	0.2428	1.672	0.0144	−0.0779	0.931	−0.1155	0.4736	2.033	−0.0007	0.0021	1.003
DenseNet-121 CIFAR-100-C	−0.1571	0.7912	1.563	−0.0206	0.1171	1.043	−0.1961	1.4524	1.964	−0.1197	0.5814	1.4
DenseNet-121 TinyImageNet-C	0.0197	−0.1084	0.957	0.0071	−0.0437	0.985	−0.1193	0.6913	1.303	−0.1024	0.5819	1.25
Average	−0.1027	0.4963	−	−0.0102	0.0688	−	−0.1495	0.9006	−	−0.1211	0.6386	−

Overall Calibration Performance. We report the overall calibration performance results of four losses in Table 1, which is evaluated by nine groups of classification tasks. The ΔECE score denotes the difference between ECE values after and before using TS, where positive values mean that TS would hurt the calibration of original models.

Observed from the ΔECE columns in Table 1, we can find that ECE scores increase when TS is used in certain tasks, such as ResNet-18 CIFAR-10-C and CIFAR-100-C of the LS column (highlighted in red). The phenomenon of hurting calibration by using TS is called Negative Calibration (highlighted underline). As reported in [16], there is little room to improve by TS when the models are trained with specific regularizes, such as label smoothing. Interestingly, TS would hurt the calibration performance when the test scenarios are extended to the corruption setting. Among all the losses, **LS is more likely to happen NC, which occurs in 5 of 9 tasks, while NLL happens once, and the other two losses do not occur yet.**

Calibration Performance on Different Shift Intensities. We report calibration performance of ResNet-18 CIFAR-10-C task on different shift intensities in Fig. 3 (a). As observed from Fig. 3 (a), we can find that ECE scores of all the baselines increase along with shift intensity. ECE scores would decrease for each shift intensity when applying TS into the models trained by NLL, Focal, and Inverse Focal. **However, ECE scores would increase (Negative Calibration) in the context of LS in each shift intensity.**

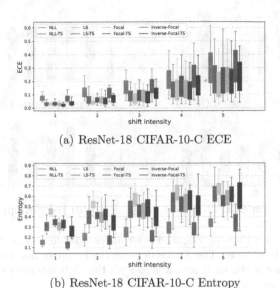

(a) ResNet-18 CIFAR-10-C ECE

(b) ResNet-18 CIFAR-10-C Entropy

Fig. 3. ECE (upper) and Entropy (bottom) performance on different shift intensities of ResNet-18 CIFAR-10-C task. For each method we show the averaged results on the dataset and summarize the results on different shift intensities with a box plot. Each box shows the quartiles across all corruption types while the error bars denote the minimum and maximum values across different corruption types.

Calibration Performance on Different Corruption Types. We report calibration performance of ResNet-18 CIFAR-10-C task on four typical corruption types in the left part of Fig. 4. For each corruption type, we can find that ECE scores increase when TS is applied into the model trained by LS (the third and fourth bars) while decrease for the other three losses. **To sum up, TS would hurt the calibration performance of original models under corruption setting, especially when the models are trained by LS.**

3.3 Understanding Negative Calibration from Entropy Perspective

[14] shows that penalizing low entropy output can improve calibration implicitly, which is equivalent to label smoothing trick. Since NC occurs in the models trained by LS many times, we attempt to understand NC from the perspective of entropy.

Entropy. We re-write the output entropy of a single sample as a function of T in TS as follows:

$$Ent(T) = \sum_1^K -\sigma_s(\frac{\boldsymbol{o}_n}{T}) \log\left[\sigma_s(\frac{\boldsymbol{o}_n}{T})\right] \tag{6}$$

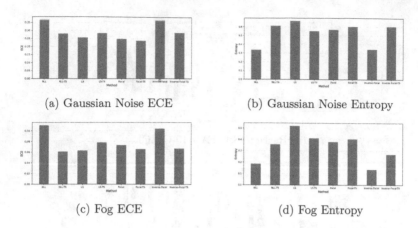

(a) Gaussian Noise ECE   (b) Gaussian Noise Entropy

(c) Fog ECE   (d) Fog Entropy

Fig. 4. ECE (left) and Entropy (right) Performance on different corrupted types of ResNet-18 CIFAR-10-C tasks. The X-axis denotes the baselines and their TS versions, and the Y-axis denotes the ECE scores across all the shift intensities under 5 single runs.

As reported in [5], when $T > 1$, the softmax will be softened by the T which results in the entropy increasing, when $T \to \infty$, the probability approaches $1/K$, the maximum of entropy can be reached. Conversely, when $T < 1$, the entropy decreases as T decreases, the probability is sharpened simultaneously. In the experiments, we report the averaged entropy values across all the samples in the evaluation dataset from three perspectives: overall, shift intensity and corruption type, which are shown in Table 1, Fig. 3 (b) and the right part of Fig. 4, respectively.

Overall Entropy Performance. Observed from the columns of ΔEntropy and T in Table 1, we can find that negative calibration happens when entropy decreases when TS is applied into the pre-trained model (highlighted in blue). Simultaneously, the learned T is less than 1 (highlighted in green). **When the dataset shifts by corruption, the original output probability tends to be sharper. If the learned $T < 1$ is based on the clean validation set, the worse calibration performance would be achieved by TS, namely negative calibration.**

Entropy Performance on Different Shift Intensities. As shown in Fig. 3 (b), the entropy would decrease for each level of shift intensity when applying TS into the ResNet-18 model trained by LS in the CIFAR-10-C task, which is consistent with the overall performance.

Entropy Performance on Different Corruption Types. As reported in the right part of Fig. 4, the entropy decreases for four corruption types: Gaussian

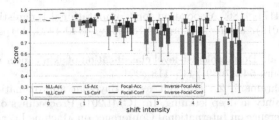

Fig. 5. Accuracy and confidence performance on different shift intensities of ResNet-18 CIFAR-10 task. Each box shows the quartiles across all corruption types while the error bars denote the minimum and maximum values across different corruption types.

noise, Fog, Gaussian blur, and JPEG Compression when TS is applied into the ResNet-18 model trained by LS in the CIFAR-10-C task, which achieves consistent performance with the overall results.

Understanding Negative Calibration via ResNet-18 CIFAR-10 Task. To understand the above conclusion, we show the averaged accuracy and confidence results of ResNet-18 CIFAR-10 task in Fig. 5. Observed from Fig. 5, we can find that accuracy and confidence scores decrease as shift intensity increase, and the dropping extent of accuracy is far more than confidence. Thus the severity of overconfidence increases along with shift intensity. But the averaged confidence is less than averaged accuracy for LS under test setting (Number "0"), which results in the learned $T < 1$. **Negative Calibration happens when test set is corrupted by certain transformations introduced in Fig 2 and the output entropy decreases at the same time.**

4   Conclusions

In this paper, we discover an interesting phenomenon called Negative Calibration when we extend the work of [16] to the corrupted dataset setting. TS hurts the calibration performance of the original pre-trained models, which have been calibrated implicitly by the regularization techniques like label smoothing in the training phrase. We find that Negative Calibration occurs in the situation of entropy decreasing when TS is applied. In the future, we will study the theoretical explanation of Negative Calibration.

References

1. Blundell, C., Cornebise, J., Kavukcuoglu, K., Wierstra, D.: Weight uncertainty in neural networks. arXiv preprint arXiv:1505.05424 (2015)
2. Devlin, J., Chang, M.W., Lee, K., Toutanova, K.: BERT: Pre-training of deep bidirectional transformers for language understanding. In: Proceedings of the 2019 Conference of the North American Chapter of the Association for Computational Linguistics: Human Language Technologies, Volume 1 (Long and Short

Papers), pp. 4171–4186. Association for Computational Linguistics, Minneapolis, Minnesota (2019). https://doi.org/10.18653/v1/N19-1423, https://aclanthology.org/N19-1423

3. Esteva, A., et al.: Dermatologist-level classification of skin cancer with deep neural networks. Nature **542**(7639), 115–118 (2017)
4. Gal, Y., Ghahramani, Z.: Dropout as a Bayesian approximation: representing model uncertainty in deep learning. In: ICML2016 Proceedings of the 33rd International Conference on International Conference on Machine Learning, vol. 48, pp. 1050–1059 (2016)
5. Guo, C., Pleiss, G., Sun, Y., Weinberger, K.Q.: On calibration of modern neural networks. In: Proceedings of the 34th International Conference on Machine Learning, vol. 70, pp. 1321–1330 (2017)
6. He, K., Zhang, X., Ren, S., Sun, J.: Deep residual learning for image recognition. In: 2016 IEEE Conference on Computer Vision and Pattern Recognition (CVPR), pp. 770–778 (2016)
7. Hecker, S., Dai, D., Gool, L.V.: Failure prediction for autonomous driving. In: 2018 IEEE Intelligent Vehicles Symposium (IV), pp. 1792–1799 (2018)
8. Hendrycks, D., Dietterich, T.G.: Benchmarking neural network robustness to common corruptions and perturbations. In: International Conference on Learning Representations (2019)
9. Kumar, A., Sarawagi, S., Jain, U.: Trainable calibration measures for neural networks from kernel mean embeddings. In: International Conference on Machine Learning, pp. 2805–2814 (2018)
10. Lakshminarayanan, B., Pritzel, A., Blundell, C.: Simple and scalable predictive uncertainty estimation using deep ensembles. In: Advances in Neural Information Processing Systems, vol. 30, pp. 6402–6413 (2017)
11. Lin, T.Y., Goyal, P., Girshick, R., He, K., Dollar, P.: Focal loss for dense object detection. IEEE Trans. Pattern Anal. Mach. Intell. **42**(2), 318–327 (2020)
12. Mukhoti, J., Kulharia, V., Sanyal, A., Golodetz, S., Torr, P.H.S., Dokania, P.K.: Calibrating deep neural networks using focal loss. In: Advances in Neural Information Processing Systems, vol. 33, pp. 15288–15299 (2020)
13. Naeini, M.P., Cooper, G.F., Hauskrecht, M.: Obtaining well calibrated probabilities using Bayesian binning. In: AAAI2015 Proceedings of the Twenty-Ninth AAAI Conference on Artificial Intelligence, vol. 2015, pp. 2901–2907 (2015)
14. Pereyra, G., Tucker, G., Chorowski, J., Kaiser, Ł., Hinton, G.: Regularizing neural networks by penalizing confident output distributions. In: ICLR (Workshop) (2017)
15. Tomani, C., Cremers, D., Buettner, F.: Parameterized temperature scaling for boosting the expressive power in post-hoc uncertainty calibration. arXiv: Learning (2021)
16. Wang, D.B., Feng, L., Zhang, M.L.: Rethinking calibration of deep neural networks: do not be afraid of overconfidence. In: NeurIPS (2021)
17. Zhang, J., Kailkhura, B., Han, T.Y.J.: Mix-n-Match : ensemble and compositional methods for uncertainty calibration in deep learning. In: International Conference on Machine Learning (2020)

A New Clustering Preserving Transformation for k-Means Algorithm Output

Mieczysław A. Kłopotek[✉]

Institute of Computer Science, Polish Academy of Sciences, Warsaw, Poland
mieczyslaw.klopotek@ipipan.waw.pl

Abstract. A novel clustering preserving transformation of cluster sets obtained from k-means algorithm is introduced. This transformation may be used to generate new labeled datasets from existent ones. It is more flexible than Kleinberg axiom based consistency transformation because data points in a cluster can be moved away and datapoints between clusters may come closer together.

Keywords: Axioms for clustering · Clustering preserving · k-means · Labelled dataset transformation

1 Introduction

We introduce a novel clustering preserving transformation of cluster sets obtained from k-means algorithm. It may be considered as a contribution towards formulation of clustering axiomatic system as well as a contribution to automated generation of labelled datasets for testing and validation of new algorithms of the k-means family, e.g. for stability studies, susceptibility to minor perturbations, and also as a method for privacy preserving data transformations etc.

As we will refer to the k-means algorithm, let us recall that it is aimed at minimizing the cost function Q (reflecting its quality in that the lower Q the higher the quality) of the form:

$$Q(\Gamma) = \sum_{i=1}^{m} \sum_{j=1}^{k} u_{ij} \|\mathbf{x}_i - \boldsymbol{\mu}_j\|^2 = \sum_{j=1}^{k} \frac{1}{2n_j} \sum_{\mathbf{x}_i \in C_j} \sum_{\mathbf{x}_l \in C_j} \|\mathbf{x}_i - \mathbf{x}_l\|^2 \tag{1}$$

for a dataset $\mathbf{X}$ under some partition Γ of the dataset $\mathbf{X}$ into the predefined number k of clusters (a cluster being a non-empty set, with empty intersections with other clusters), where u_{ij} is an indicator of the membership of data point $\mathbf{x}_i$ in the cluster C_j having the cluster center at $\boldsymbol{\mu}_j$ (which is the cluster's gravity center). [13] modified this definition by dividing the right hand side by m in order to make comparable values for samples and the population, but we will only handle samples of a fixed size so this definition is sufficient for our purposes. A k-means algorithm finding exactly the clustering optimizing Q shall be refereed

M. Ceci et al. (Eds.): ISMIS 2022, LNAI 13515, pp. 315–322, 2022.
https://doi.org/10.1007/978-3-031-16564-1_30

to as k-means-ideal. There exist a number of implementations of algorithms aiming at approximate optimization of the k-means quality criterion. For various versions of k-means algorithm see e.g. [16]. Realistic implementations start from a randomized initial partition and then improve the Q iteratively. Our method of clustering preserving transformation assumes that a clustering by k-means-ideal is available. In practice, one will use some approximation by a realistic k-means. Recall that even stating that the minimum of the function Q was reached, is hard, except for special cases [6].

Kleinberg [5] introduced an axiomatic framework for distance-based clustering functions to which the widely used k-means algorithm belongs.

Two of the Kleinberg's axioms represent transformations under which the clustering should be preserved. This can be viewed as a way to create new labelled data sets for testing of clustering functions. Regrettably, the k-means algorithm (one of the most frequently used clustering algorithms) fails to match requirements of this axiomatic system, It fails on the consistency axiom. Therefore, it is not possible to use of Kleinberg's consistency axiom as a method of generating new labelled data sets.

The consistency axiom of Kleinberg has the following form:

Property 1. *[5] Let Γ be a partition of S, and d and d' two distance functions on S. We say that d' is a Γ-transformation of d ($\Gamma(d) = d'$) if (a) for all $i, j \in S$ belonging to the same cluster of Γ, we have $d'(i, j) \leq d(i, j)$ and (b) for all $i, j \in S$ belonging to different clusters of Γ, we have $d'(i, j) \geq d(i, j)$. The clustering function f has the consistency property if for each distance function d and its Γ-transform d' the following holds: if $f(d) = \Gamma$, then $f(d') = \Gamma$*

The paper [8] suggested a substitute for the consistency axiom, proposing so-called centric consistency.

Definition 1. *[8] Let $\mathcal{E}$ be an embedding of the dataset S with distance function d (induced by this embedding). Let Γ be a partition of this dataset. Let $C \in \Gamma$ and let $\boldsymbol{\mu}_c$ be the gravity center of the cluster C in $\mathcal{E}$. We say that we execute the Γ^* transformation $\Gamma^*(d; \lambda) = d'$ if for some $0 < \lambda \leq 1$ a new embedding $\mathcal{E}'$ is created in which each element x of C, with coordinates $\mathbf{x}$ in $\mathcal{E}$, has coordinates $\mathbf{x}'$ in $\mathcal{E}$ such that $\boldsymbol{x}' = \boldsymbol{\mu}_c + \lambda(\boldsymbol{x} - \boldsymbol{\mu}_c)$, all coordinates of all other data points are identical, and d' is induced by $\mathcal{E}'$. C is then said to be subject of the centric transform.*

The set of possible Γ^*-transformations for a given partition is neither a subset nor superset of the set of possible Kleinberg's Γ-transformation in general.

Property 2. *[8] A clustering method has the property of centric consistency if after a Γ^* transform it returns the same partition.*

Theorem 3. *[7] k-means algorithm satisfies centric consistency property in the following way: if the partition Γ of the set S with distances d is a global minimum of k-means, and $k = 2$, and the partition Γ has been subject to Γ^*-transformation yielding distances d', then Γ is also a global minimum of k-means under distances d'.*

See proof in paper [7]. A more general theorem was introduced in

Theorem 4. *[8, Theorem 34] k-means algorithm satisfies centric consistency in the following way: if the partition Γ is a global minimum of k-means, and the partition Γ has been subject to Γ^*-transform yielding Γ', then Γ' is also a global minimum of k-means.*

For the proof of this theorem see [8, Theorem 34].

The axiom/property 2 may be considered as a rigid one because all data points of a cluster are transformed linearly. Therefore, in this paper, we elaborate another clustering preserving transformation suitable for k-means algorithm so that it can be used with this category of algorithms.

It is worth noting that this transformation differs from Kleinberg's Γ transformation. which required that distances within a cluster are reduced and distances between clusters are increased. With our transformation it is valid to increase distances within a cluster and decrease distances between clusters.

Definition 2. *For a point set P, the centric set transform by factor λ is a transformation assigning each point $\mathbf{x} \in P$ a datapoint $\mathbf{x}'$ such that $\mathbf{x}' = \lambda(\mathbf{x} - \boldsymbol{\mu}(P)) + \boldsymbol{\mu}(P)$ where $\boldsymbol{\mu}(P)$ is the gravity center of the dataset P.*

In the next Sect. 2, we provide a brief overview of various areas where cluster preserving transformations of data are an issue. Then we formulate our proposal of cluster preserving transformation for k-means in Sect. 3 and prove analytically its correctness. In Sect. 4 we illustrate the operation of the transformation with some examples. Section 5 contains some final remarks.

2   Related Work

The most heavily cited (and at the same time contradictory) axiomatic framework for clustering was introduced by Kleinberg [5]. An overview of various efforts to overcome the problems of that framework is provided e.g. in [8].

In this overview of related work let us concentrate therefore on the aspect of cluster-preserving transformations which are often related to diverse practical applications. In [8] also the applicability to the problem of testbed creation for clustering algorithms from the k-means family is addressed.

Roth et al. [15] investigated the issue of preservation of clustering when embedding non-euclidean data into the Euclidean space. They showed that clustering functions, that remain invariant under additive shifts of the pairwise proximities, can be reformulated as clustering problems in Euclidean spaces.

A similar problem was addressed by Kłopotek et al. [9] whereby the issue of interpretation of results of kernel k-means to non-euclidean data was discussed. A cluster-preserving transformation for this specific problem was proposed via increasing of all distances.

Parameswaran and Blough [12] considered the issue of cluster preserving transformations from the point of view of privacy preserving. They designed a Nearest Neighbor Data Substitution (NeNDS), a new data obfuscation technique with strong privacy-preserving properties while maintaining data clusters. Cluster preserving transformations with the property of privacy preserving focusing on the k-means algorithm are investigated by Ramírez and Auñón [14]. Privacy preserving methods for various k-means variants boosted to large scale data are further elaborated in [2]. Keller et al. [4] investigate such transformations for other types of clustering algorithms. A thorough survey of privacy-preserving clustering for big data can be found in [18] by Zhao et al.

Howland and Park [3] proposed models incorporating prior knowledge about the existing structure and developed for them dimension reduction methods independent of the original term-document matrix dimension. Other, more common dimensionality reduction methods for clustering (including PCA and Laplacian embedding) are reviewed by Ding [1].

Larsen et al. [11] reformulate the heavy hitter problem of stream mining in terms of a clustering problem and elaborate algorithms fulfilling the requirement of "cluster preserving clustering".

Zhang et al. [17] developed clustering structure preserving transformations for graph streaming data, when there is a need to sample the graph.

3   A Clustering Preserving Transformation

Theorem 5. *Let a partition $\Gamma_o = \{T, Z_2, \ldots, Z_k\}$ be an optimal partition of a dataset $\mathbf{X}$ under k-means algorithm, that is minimizing $Q(\Gamma)$ from Eq. 1 over all partitions Γ with $card(\Gamma) = card(\Gamma_o)$. Let a subset P of T be subjected to centric set transform yielding $P'(\lambda)$ and $T'(\lambda) = (T-P) \cup P'(\lambda)$. Then partition $\{T', Z_2, \ldots, Z_k\}$ is an optimal partition of $T' \cup Z_2 \cup \cdots \cup Z_k$ under k-means.*

Definition 3. *In the above Theorem 5, $\{T'(\lambda), Z_2, \ldots, Z_k\}$ shall be called Γ^{++} transform of $\{T, Z_2, \ldots, Z_k\}$*

The detailed proof can be found in [10]. Let us just show its outline. Assume the optimal clustering for a given set of objects $\mathbf{X}$ consist of k clusters: T and $Z_2, \ldots, Z_k$. Let T consist of two disjoint subsets P, Y, $T = P \cup Y$ and let us ask the question whether or not centric transform of the set P will affect the optimality of clustering. Then $T'(\lambda) = P'(\lambda) \cup Y$. See Fig. 1 left. The cluster centre of $T'(\lambda)$ will be the same as that of T. We ask if $\{T'(\lambda), Z_2, \ldots, Z_k\}$ is the globally optimal clustering of $T'(\lambda) \cup Z_2 \cup \cdots \cup Z_k$. Assume to the contrary, that there exists a clustering into sets $K_i'(\lambda) = A_i'(\lambda) \cup B_i \cup C_{i,2} \cup \cdots \cup C_{i,k}$, $i = 1, \ldots, k$ where $P = A_1 \cup A_2 \cdots \cup A_k$, and $A_i'(\lambda)$ are the points obtained from A_i when P is subjected to centric set transformation. hence $P'(\lambda) = A_1'(\lambda) \cup \cdots \cup A_k'(\lambda)$, $Y = B_1 \cup \cdots \cup B_k$, $Z_j = C_{1,j} \cup \cdots \cup C_{k,j}$, that, for some $\lambda = \lambda^* \in (0,1)$ has lower clustering quality function value $Q(\{K_1'(\lambda), \ldots, K_k'(\lambda)\})$. Define also the function $h(\lambda) = Q(\{T'(\lambda), Z_2, Z_k\}) - Q(\{K_1'(\lambda), \ldots, K_k'(\lambda)\})$. Due to optimality assumption, $h(1) \leq 0$.

We have shown in [10] that $h(0) \leq 0$. As $h(\lambda)$ is quadratic in λ, we establish in [10] that the coefficient at λ^2 is positive.

As $h(\lambda)$ is a quadratic function in λ, and $h(0) \leq 0$ and $h(1) \leq 0$, then also $h(\lambda) \leq 0$ for any value of λ between 0 and 1. This completes the proof.

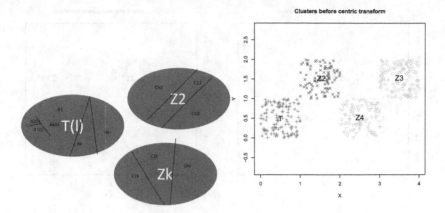

Fig. 1. Left: Illustration of the task. Right: Example clusters before centric transformation

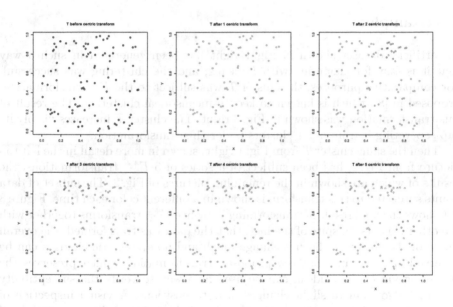

Fig. 2. Example: T cluster before centric transformation, and then after 1, 2, 3, 4 and 5 centric transformations.

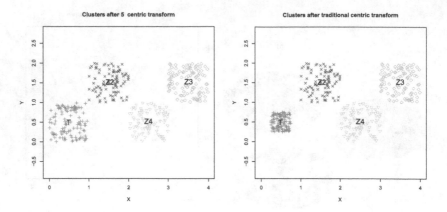

Fig. 3. Left: Example clusters after 5 centric transformations of data from Fig. 1 right. Right: Example clusters after traditional Γ^* transformation of data from Fig. 1 right.

4   Examples

An artificial dataset shown in Fig. 1 right has been generated in such a way that it is easy for clustering with k-means, and the clustering was successful. For comparative purposes, the cluster T was subject to the Γ^* transformation, proposed in [8], which is known to keep k-means-ideal clustering. The result of this transformation was shown in Fig. 3 right. The cluster T increases visibly its gap separating it from other clusters under this transformation.

Then the same cluster T from Fig. 1 right, shown in more detail in the left top picture from Fig. 2, has been subject of a series of 5 Γ^{++} transformations, the results of which are shown in the remaining pictures of Fig. 2. The subset of data points subjected to this transform is shown in a different color each time. Figure 3 left shows the final result of transforming T by five Γ^{++} transforms together with the other clusters. In spite of the fact, that the data was transformed, the overall shape of the cluster T did not change much and no systematic changes can be observed. This means that the proposed transformation, which preserves the clustering via k-means-ideal, can be used as a way to verify the susceptibility of a real algorithm to slight changes in data positions. A visual inspection of Fig. 3 left and right convinces that the Γ^{++} is better than Γ^* as a method of generating new labelled data sets from old ones.

This experiment was repeated 100 times using k-means implemented in R with 100 times restart and neither for Γ^* nor for Γ^{++} any error was reported.

A next experiment was to compare Γ with Γ^{++}. An artificial dataset consisting of 10000 datapoints in 3D uniformly distributed over two squares of same edge, touching each other at a corner point. They were transformed increasing angular distance to the diagonal by factor 1.9. Then either a Γ or Γ^{++} was applied to one of the 2 clusters. Γ consisted in changing angular distance to the diagonal by factor 0.05. Γ^{++} with $\lambda = 0.5$ was applied to a random subset of datapoints of size up to 1/3 of the dataset. Default restart number was used for k-means. No clustering error was observed for Γ^{++} and up to 1% of errors occurred for Γ (mean errors 0.04%, std dev 0.007%).

5 Final Remarks

The new clustering transformation method proposed in this paper not only provides with a new, more flexible test bed for k-means-like algorithms, but also suggests that the rigid Kleinberg's consistency axiom can be relaxed at least for this category of algorithms. Our transformation allows for increase of intra-cluster distances and decrease of inter-cluster distances. This fact may be considered a hint that there is a need to redefine popular formal cluster understanding.

Further research is needed in order to create similar transformations suitable for other clustering algorithms and in this way to find a more general axiomatic system for clustering algorithms.

Though the goal of the proposed cluster preserving transformation was to provide with some new methods for generating testbed data for k-means like algorithms, its applicability in for example some brands of privacy preserving needs further investigation.

References

1. Ding, C.: Dimension reduction techniques for clustering. In: Liu, L., Oezsu, M. (eds.) Encyclopedia of Database Systems. Springer, Boston, MA (2009). https://doi.org/10.1007/978-0-387-39940-9_612

2. Gao, Z.-Q., Zhang, L.-J.: DPHKMS: an efficient hybrid clustering preserving differential privacy in spark. In: Barolli, L., Zhang, M., Wang, X.A. (eds.) EIDWT 2017. LNDECT, vol. 6, pp. 367–377. Springer, Cham (2018). https://doi.org/10.1007/978-3-319-59463-7_37

3. Howland, P., Park, H.: Cluster preserving dimension reduction methods for document classification. In: Berry, M.W., Castellanos, M. (eds.) Survey of Text Mining II, pp. 3–23. Springer, London (2007). https://doi.org/10.1007/978-1-84800-046-9_1

4. Keller, H., Möllering, H., Schneider, T., Yalame, H.: Privacy-preserving clustering. In: Gazdag, S.L., Loebenberger, D., Nüsken, M. (eds.) Crypto Day Matters 32. Gesellschaft für Informatik e.V./FG KRYPTO, Bonn (2021). https://doi.org/10.18420/cdm-2021-32-42

5. Kleinberg, J.: An impossibility theorem for clustering. In: Proceedings of the 15th International Conference on Neural Information Processing Systems (NIPS), pp. 446–453 (2002)

6. Kłopotek, M.A.: An aposteriorical clusterability criterion for k-means++ and simplicity of clustering. SN Comput. Sci. 1(2), 1–38 (2020). https://doi.org/10.1007/s42979-020-0079-8

7. Kłopotek, M.A., Kłopotek, R.A.: In-the-limit clustering axioms. In: Rutkowski, L., Scherer, R., Korytkowski, M., Pedrycz, W., Tadeusiewicz, R., Zurada, J.M. (eds.) ICAISC 2020. LNCS (LNAI), vol. 12416, pp. 199–209. Springer, Cham (2020). https://doi.org/10.1007/978-3-030-61534-5_18 2020, Proceedings, Part II. Lecture Notes in Computer Science, vol. 12416, pp. 199–209. Springer (2020)

8. Kłopotek, M.A., Kłopotek, R.A.: Towards continuous consistency axiom. Appl. Intell. (2022). https://doi.org/10.1007/s10489-022-03710-1

9. Kłopotek, R., Kłopotek, M., Wierzchoń, S.: A feasible k-means Kernel trick under non-Euclidean feature space. Int. J. Appl. Math. Comput. Sci. 30(4), 703–715 (2020). https://doi.org/10.34768/amcs-2020-0052

10. Kłopotek, M.A.: A clustering preserving transformation for k-means algorithm output (2022). https://arxiv.org/abs/2202.10455

11. Larsen, K.G., Nelson, J., Nguyundefinedn, H.L., Thorup, M.: Heavy hitters via cluster-preserving clustering. Commun. ACM **62**(8), 95–100 (2019)

12. Parameswaran, R., Blough, D.M.: A robust data-obfuscation approach for privacy preservation of clustered data. In: Proceedings of the Workshop on Privacy and Security Aspects of Data Mining, pp. 18–25 (2005). https://doi.org/10.1021/om500167r

13. Pollard, D.: Strong consistency of k-means clustering. Ann. Statist. **9**(1), 135–140 (1981)

14. Ramírez, D.H., Auñón, J.M.: Privacy preserving k-means clustering: a secure multi-party computation approach. https://arxiv.org/abs/2009.10453 (2020)

15. Roth, V., Laub, J., Kawanabe, M., Buhmann, J.: Optimal cluster preserving embedding of nonmetric proximity data. IEEE Trans. Pattern Anal. Mach. Intell. **25**(12), 1540–1551 (2003). https://doi.org/10.1109/TPAMI.2003.1251147

16. Wierzchoń, S.T., Kłopotek, M.A.: Modern Clustering Algorithms. Studies in Big Data. springer, Cham (2018). https://doi.org/10.1007/978-3-319-69308-8

17. Zhang, J., Zhu, K., Pei, Y., Fletcher, G., Pechenizkiy, M.: Cluster-preserving sampling from fully-dynamic streaming graphs. Inf. Sci. **482**, 279–300 (2019). https://doi.org/10.1016/j.ins.2019.01.011

18. Zhao, Y., Tarus, S.K., Yang, L.T., Sun, J., Ge, Y., Wang, J.: Privacy-preserving clustering for big data in cyber-physical-social systems: Survey and perspectives. Inf. Sci. **515**, 132–155 (2020). https://doi.org/10.1016/j.ins.2019.10.019

Complex Data

ComplexData

A Transformer-Based Framework for Geomagnetic Activity Prediction

Yasser Abduallah[✉], Jason T. L. Wang, Chunhui Xu, and Haimin Wang

Department of Computer Science and Institute for Space Weather Sciences,
New Jersey Institute of Technology, Newark, NJ 07102, USA
ya54@njit.edu

Abstract. Geomagnetic activities have a crucial impact on Earth, which can affect spacecraft and electrical power grids. Geospace scientists use a geomagnetic index, called the Kp index, to describe the overall level of geomagnetic activity. This index is an important indicator of disturbances in the Earth's magnetic field and is used by the U.S. Space Weather Prediction Center as an alert and warning service for users who may be affected by the disturbances. Early and accurate prediction of the Kp index is essential for preparedness and disaster risk management. In this paper, we present a novel deep learning method, named KpNet, to perform short-term, 1–9 hour ahead, forecasting of the Kp index based on the solar wind parameters taken from the NASA Space Science Data Coordinated Archive. KpNet combines transformer encoder blocks with Bayesian inference, which is capable of quantifying both aleatoric uncertainty (data uncertainty) and epistemic uncertainty (model uncertainty) when making Kp predictions. Experimental results show that KpNet outperforms closely related machine learning methods in terms of the root mean square error and R-squared score. Furthermore, KpNet can provide both data and model uncertainty quantification results, which the existing methods cannot offer. To our knowledge, this is the first time that Bayesian transformers have been used for Kp prediction.

Keywords: Bayesian inference · Deep learning · Geomagnetic index · Uncertainty quantification

1 Introduction

Geomagnetic activities have a crucial impact on Earth, which can affect spacecraft, electrical power grids, observers of the aurora, and navigation systems. They are very complex, and their scales vary in orders of magnitude. Geomagnetic activity modeling and forecasting has therefore been an important subject in space weather research.

Generally, geomagnetic indices are indicators of disturbances in the Earth's magnetic field. One important geomagnetic index, named the Kp index, is used to describe the overall level of geomagnetic activity. Kp values range from 0 to

M. Ceci et al. (Eds.): ISMIS 2022, LNAI 13515, pp. 325–335, 2022.
https://doi.org/10.1007/978-3-031-16564-1_31

9. The U.S. Space Weather Prediction Center (SWPC) at the National Oceanic and Atmospheric Administration (NOAA) classifies the geomagnetic activities into four categories: (i) quiet to unsettled (Kp< 4), (ii) active (Kp= 4), (iii) minor storm (Kp= 5), and (iv) major to severe storm (Kp >5). A storm occurs when there is an exchange of energy from the solar wind to the near-Earth space environment. At any period, if the Kp index indicates a high geomagnetic activity (e.g., a storm), SWPC issues a warning or alert for the affected parties.

Several machine learning-based methods have been developed to forecast the Kp index. Costello [5] designed and implemented a neural network (NN)-based operational system, whose prediction results are posted on the NOAA website.[1] Boberg et al. [3] described a real-time NN algorithm for making three-hour predictions of the Kp index. Shprits et al. [14] explored the relative efficiency of solar wind-based predictions and compared long and short term Kp forecasting. Ji et al. [9] provided a comparison between NN and SVM (support vector machine) algorithms used for Kp forecasting. Tan et al. [16] and later Chakraborty and Morley [4] developed deep learning models for Kp forecasting using long short-term memory (LSTM) networks. In addition to the above Kp prediction methods, several researchers investigated the importance of predictive parameters such as the magnetic component, particle density and velocity, and total magnetic field B to understand the impact of each of the parameters on Kp forecasting performance [20].

In this paper we present a new transformer-based Bayesian deep learning model, named KpNet, for Kp prediction. Our work makes several contributions to the field, listed below.

- KpNet is the first model to utilize transformers to forecast the Kp index for a short-term (i.e., 1–9 hours ahead) period.
- With Bayesian inference, KpNet can quantify both aleatoric and epistemic uncertainties in producing Kp predictions. Aleatoric uncertainty, also known as data uncertainty, measures the noise inherent in data. Epistemic uncertainty, also known as model uncertainty, measures the uncertainty in the parameters of a model [11]. To our knowledge, no previous Kp prediction method can quantify both types of uncertainty.
- KpNet performs better than closely related machine learning methods for short-term Kp prediction in terms of both the root mean square error (RMSE) and R-squared (R^2).

2   Data

2.1   Data Collection

The Kp measurements used in this study are provided by the GFZ German Research Centre for Geosciences.[2] The Kp values in the GFZ site are collected

---

[1] https://www.noaa.gov/.

[2] https://www.gfz-potsdam.de/en/kp-index/.

with a 3-hour cadence where the values range from 0 to 8.33. The solar wind parameters used in this study are taken from the NASA Space Science Data Coordinated Archive.[3] We collect the data in the time period between January 1, 2010 and March 31, 2022. We select the time resolution of the hourly average for the solar wind parameters. Following [16,20], we consider eight solar wind parameters, namely the interplanetary magnetic field (IMF) magnitude average, magnetic field Bx, By, and Bz components, plasma temperature, proton density, plasma speed, and flow pressure. Due to the difference in cadence, where Kp uses a 3-h cadence whereas the solar wind parameters use a 1-h cadence, we process the data by temporally matching the Kp measurements from the GFZ site with the solar wind parameters from the NASA site to create the final dataset.

2.2 Data Labeling

We divide our dataset into two parts: training set and test set. The training set contains the records from January 1, 2010 to December 31, 2021. The test set contains the records from January 1, 2022 to March 31, 2022. The training set and test set are disjoint. The records are labeled as follows. Let t be a given time point and let w be the time window ahead of t, where w ranges from 1 to 9 h for the short-term Kp forecasting studied here. The label of the record at time point t is defined as the Kp value at time point $t + w$ for w-hour-ahead forecasting. Each training record at time point t in the training set has nine values including eight solar wind parameter values at time point t and the label of the training record. Each test record in the test set contains only eight solar wind parameter values; the label of the test record in the test set will be predicted by our KpNet model.

3 Methodology

3.1 The KpNet Architecture

Figure 1 presents the architecture of KpNet, which is created using the keras framework from tensorflow.[4] To enhance the KpNet learning capability and its performance, we add multiple layers to the network. The input of KpNet consists of non-overlapping sequences of records x_{p+1}, x_{p+2}, ..., x_{p+n}, where n is set to 512 in our study. The sequences are passed to a one-dimensional convolution (Conv1D) layer with 64 kernels where the size of each kernel is 1. Conv1D was proven to be well suited for sequential data and was also previously used for geomagnetic index prediction [15]; it learns internal patterns from the input data sequence and passes them to a bidirectional long short-term memory (biLSTM) layer that is configured with 300 neurons. Combining Conv1D and biLSTM layers has shown substantial improvement in performance when dealing with time series as our ablation studies show later.

---

[3] https://nssdc.gsfc.nasa.gov.
[4] https://www.tensorflow.org.

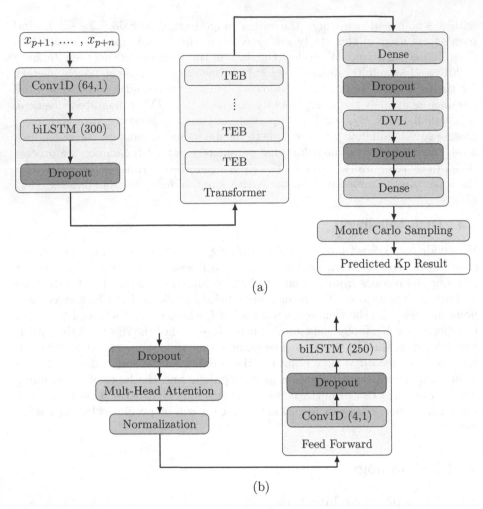

Fig. 1. (a) Architecture of KpNet. (b) Configuration details of a transformer encoder block (TEB) used in KpNet. Each TEB has two dropout layers, a multi-head attention layer, a batch normalization layer, followed by a feed forward network that contains a Conv1D layer, and a bidirectional LSTM (biLSTM) layer.

The biLSTM layer transfers the learned patterns down to a transformer network, which is composed of b transformer encoder blocks (TEBs).[5] Each TEB consists of a multi-head attention layer [18], a batch normalization layer, and a feed-forward network. Generally, transformers for natural language processing (NLP) use layer normalization leading to significant performance gains over batch normalization [18]. However, we use batch normalization here to avoid the effect of outliers in time series which do not exist in NLP word embedding [19].

---

[5] Following [18], we set b to 8 in this study.

The multi-head attention layer provides transformation on the sequential input values to obtain distinct metrics of size h. Here, h is the number of attention heads that is set to 4 and the size of each attention head is also set to 4; the other parameters are left with their default values. The feed-forward network, composed of a Conv1D layer, with 4 kernels where the number of kernels equals the number of attention heads and each kernel size is 1, followed by a biLSTM layer with 250 neurons, helps reduce over-fitting. Notice that each TEB uses a transformer encoder without the decoder because we are dealing with time series instead of language processing and therefore translation is not involved [19].

Furthermore, a dense variational layer (DVL) [17] with 10 neurons is added that uses variational inference [2] to approximate the posterior distribution over the model weights. DVL is similar to a regular dense layer but requires two input functions that define the prior and posterior distributions over the model weights. DVL allows KpNet to represent the weights by a distribution instead of estimated points. In addition, KpNet includes several dense and dropout layers. Each dense layer is strongly connected with its preceding layer where every neuron in the dense layer is connected with every neuron in the preceding layer. Each dropout layer provides a mechanism to randomly drop a percentage of hidden neurons to avoid over-fitting, as explained below.

3.2 Uncertainty Quantification

With the dropout layers, our model's internal structure is slightly different each time the neurons are dropped. This is an important behavior to the Monte Carlo (MC) class of algorithms that depends on random sampling and provides useful information [6]. We use this technique to introduce a distribution interval of predicted values as demonstrated in Sect. 4. Specifically, to quantify the uncertainty with our KpNet model, we use a prior probability, $P(W)$, over the model's weights, W. During training, the eight solar wind parameter values and Kp values, collectively referred to as D, are used to train the model. According to Bayes' theorem,

$$P(W|D) = \frac{P(D|W) \times P(W)}{P(D)}. \tag{1}$$

Computation of the exact posterior probability, $P(W|D)$, is intractable [10], but we can use variational inference [8] to learn the variational distribution over the model's weights parameterized by θ, $q_\theta(W)$, by minimizing the Kullback-Leibler (KL) divergence of $q_\theta(W)$ and $P(W|D)$ [2]. According to [6], a network with dropout layers provides variational approximation. To minimize the KL divergence, we use the dense variational layer (DVL) shown in Fig. 1 and assign the KL weight to $1/N$ where N is the size of the training set [13]. We use the mean squared error (MSE) loss function and the adaptive moment estimation (Adam) optimizer [7] with a learning rate of 0.0001 to train our model. Let $\hat{\theta}$ denote the optimized variational parameter obtained by training the model; we use $q_{\hat{\theta}}(W)$ to denote the optimized weight distribution.

During testing/prediction, our model utilizes the MC dropout sampling technique to produce probabilistic forecasting results, quantifying both aleatoric and epistemic uncertainties. Dropout is used to retrieve K MC samples by processing the test data K times [6].[6] For each of the K MC samples, a set of weights is randomly drawn from $q_{\hat{\theta}}(W)$. For each predicted Kp value, we get a mean and a variance over the K samples. Following [10,12], we decompose the variance into aleatoric and epistemic uncertainties. The aleatoric uncertainty captures the inherent randomness of the predicted result, which comes from the input test data. On the other hand, the epistemic uncertainty comes from the variability of W, which accounts for the uncertainty in the model parameters (weights).

4   Experiments and Results

We conducted a series of experiments to evaluate our proposed KpNet model and compare it with closely related methods. The performance metrics used in the study are the root mean square error (RMSE) and R-squared (R^2) [1]. The smaller RMSE or the larger R^2 a method yields, the more accurate the method is.

4.1   Ablation Studies

In this experiment, we performed ablation tests to analyze and assess the components of the proposed KpNet model. We considered four subnets derived from KpNet: KpNet-C, KpNet-L, KpNet-CL, and KpNet-T. KpNet-C (KpNet-L, KpNet-CL, KpNet-T, respectively) represents the subnet of KpNet in which we remove the Conv1D (64, 1) layer (biLSTM (300) layer, Conv1D (64, 1) and biL-STM (300) layers, transformer network, respectively) while keeping the remaining components of the KpNet framework. For comparison purposes, we turned off the uncertainty quantification mechanism in the five models.

Figure 2 presents the ablation test results of the five models. The $t + wh$, $1 \leq w \leq 9$, on the X-axis represents the w-hour ahead prediction of the Kp index based on the test records in the test set. It can be seen from Fig. 2 that the full network model, KpNet, achieves the best performance among the five models. KpNet-C captures the temporal correlation from the input data but does not learn additional patterns and properties to strengthen the relationship between the test records. KpNet-L captures the properties from the test records but lacks the temporal correlation information to deeply analyze the sequential information in the input data. We also see from Fig. 2 that removing both of the two layers, Conv1D and biLSTM, decreases the performance significantly. Overall, KpNet-T is the worst model, indicating the importance of the transformer network. The results based on RMSE and R^2 are consistent. In subsequent experiments, we used the full network model, KpNet, due to its best performance among the five models.

---

[6] In the study presented here, K is set to 100.

4.2   Comparison with Related Methods

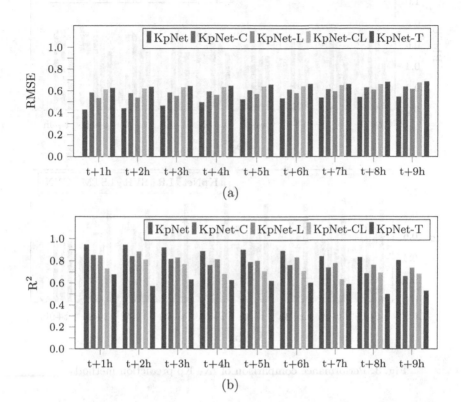

Fig. 2. Results of the ablation tests obtained by removing components of KpNet.

In this experiment, we compared KpNet with four related machine learning methods that have been used in previous Kp forecasting studies, including linear regression (LR) [16], support vector regression (SVR) [9], a long short-term memory (LSTM) network [16], and a convolutional neural network (CNN) that has also been used for geomagnetic index prediction [15]. Since the four methods do not have the ability to quantify both data and model uncertainties, we turned off the uncertainty quantification mechanism in our KpNet model when performing this experiment. Figure 3 presents the results. It can be seen from the figure that KpNet achieves the best performance, producing the most accurate predictions, among the five methods in terms of RMSE and R^2. This finding is consistent with those in the literature [19], which indicate that transformers often perform better than LSTM and CNN models in time series forecasting.

4.3   Uncertainty Quantification Results

Figure 4 shows uncertainty quantification results produced by KpNet on the test set. Due to space limitation, we only present the results obtained from 6-hour

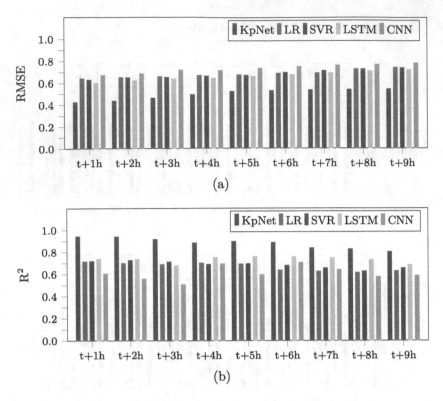

Fig. 3. Performance comparison of five Kp prediction methods.

ahead predictions of the Kp index. In Fig. 4, yellow lines represent observed Kp values (ground truths) while black lines represent predicted Kp values. The light blue region in Fig. 4(a) represents aleatoric uncertainty (data uncertainty). The light gray region in Fig. 4(b) represents epistemic uncertainty (model uncertainty). It can be seen from Fig. 4 that the black lines are reasonably close to the yellow lines, demonstrating the good performance of KpNet, which is consistent with the results in Fig. 3. Figure 4 also shows that the light gray region is much smaller than the light blue region, indicating that the model uncertainties are much smaller than the data uncertainties. Thus, the uncertainty in the predicted result is mainly due to the noise in the input test data. Similar results were obtained from the other w-hour, $1 \leq w \leq 9$, $w \neq 6$, ahead predictions of the Kp index.

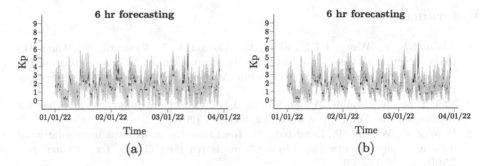

Fig. 4. Uncertainty quantification results produced by KpNet. (a) 6 hr ahead forecasting results with data uncertainty (light blue region). (b) 6 hr ahead forecasting results with model uncertainty (light gray region). Yellow lines represent observed Kp values while black lines represent predicted Kp values. (Color figure online)

5 Conclusion

We presented a novel deep learning model, named KpNet, to perform short-term, 1–9 hour ahead, forecasting of the Kp index based on the solar wind parameters taken from the NASA Space Science Data Coordinated Archive. KpNet combines transformer encoder blocks with Bayesian inference, which is capable of quantifying both aleatoric uncertainty and epistemic uncertainty when making Kp predictions. Our experimental results demonstrated the good performance of KpNet and its superiority over related machine learning methods. These results were based on the data collected in the period between January 1, 2010 and March 31, 2022. The training set contained hourly records from January 1, 2010 to December 31, 2021. The test set contained hourly records from January 1, 2022 to March 31, 2022. To avoid bias in our findings, we performed additional experiments using 10-fold cross validation where the data was divided into 10 approximately equal partitions or folds. The sequential order of the data in each fold was maintained. In each run, one fold was used for testing and the other nine folds together were used for training. There were 10 folds and hence 10 runs where the average performance metric values over the 10 runs were calculated. Results from the 10-fold cross validation were consistent with those reported in the paper. Thus we conclude that the proposed KpNet is a feasible machine learning method for short-term, 1–9 hour, ahead predictions of the Kp index. In the future we plan to extend KpNet to perform long-term Kp forecasting using other data sources such as solar images in addition to solar wind parameters.

Acknowledgments. This work was supported in part by the U.S. National Science Foundation under Grant Nos. AGS–1927578 and AGS–2149748. We acknowledge the use of NASA/GSFC's OMNIWeb and OMNI data.

References

1. Abduallah, Y., Wang, J.T.L., Shen, Y., Alobaid, K.A., Criscuoli, S., Wang, H.: Reconstruction of total solar irradiance by deep learning. In: Proceedings of the International Conference of the Florida Artificial Intelligence Research Society (FLAIRS-34) (2021)
2. Blei, D.M., Kucukelbir, A., McAuliffe, J.D.: Variational inference: a review for statisticians. J. Am. Stat. Assoc. **112**, 859–877 (2016)
3. Boberg, F., Wintoft, P., Lundstedt, H.: Real time Kp predictions from solar wind data using neural networks. Phys. Chem. Earth Part C Sol. Terr. Planet. Sci. **25**(4), 275–280 (2000)
4. Chakraborty, S., Morley, S.K.: Probabilistic prediction of geomagnetic storms and the Kp index. J. Space Weather Space Clim. **10**, 36 (2020). https://doi.org/10.1051/swsc/2020037
5. Costello, K.A.: Moving the Rice MSFM into a real-time forecast mode using solar wind driven forecast modules, Ph.D. dissertation, Rice University (1998)
6. Gal, Y., Ghahramani, Z.: Dropout as a Bayesian approximation: representing model uncertainty in deep learning. In: Proceedings of the 33rd International Conference on Machine Learning, pp. 1050–1059 (2016)
7. Goodfellow, I.J., Bengio, Y., Courville, A.C.: Deep Learning. MIT Press (2016). http://www.deeplearningbook.org
8. Graves, A.: Practical variational inference for neural networks. In: Proceedings of the Annual Conference on Neural Information Processing Systems (2011)
9. Ji, E.Y., Moon, Y.J., Park, J., Lee, J.Y., Lee, D.H.: Comparison of neural network and support vector machine methods for Kp forecasting. J. Geophys. Res. Space Phys. **118**(8), 5109–5117 (2013)
10. Jiang, H., et al.: Tracing Hα fibrils through Bayesian deep learning. Astrophys.J. Suppl. Ser. **256**(20) (2021)
11. Kendall, A., Gal, Y.: What uncertainties do we need in Bayesian deep learning for computer vision? In: Proceedings of the Annual Conference on Neural Information Processing Systems (2017)
12. Kwon, Y., Won, J.H., Kim, B.J., Paik, M.C.: Uncertainty quantification using Bayesian neural networks in classification: application to biomedical image segmentation. Comput. Stat. Data Anal. **142**, 106816 (2020)
13. Ling, Z.H., Dai, L.R.: Minimum Kullback–Leibler divergence parameter generation for HMM-based speech synthesis. IEEE Trans. Audio Speech Lang Process. **20**(5), 1492–1502 (2012)
14. Shprits, Y.Y., Vasile, R., Zhelavskaya, I.S.: Nowcasting and predicting the Kp index using historical values and real-time observations. Space Weather **17**(8), 1219–1229 (2019)
15. Siciliano, F., Consolini, G., Tozzi, R., Gentili, M., Giannattasio, F., De Michelis, P.: Forecasting SYM-H index: a comparison between long short-term memory and convolutional neural networks. Space Weather **19**(2) (2021)
16. Tan, Y., Hu, Q., Wang, Z., Zhong, Q.: Geomagnetic index Kp forecasting with LSTM. Space Weather **16**(4), 406–416 (2018)
17. Tran, D., Dusenberry, M.W., van der Wilk, M., Hafner, D.: Bayesian layers: a module for neural network uncertainty. In: Proceedings of the Annual Conference on Neural Information Processing Systems (2019)
18. Vaswani, A., et al.: Attention is all you need. In: Proceedings of the Annual Conference on Neural Information Processing Systems (2017)

19. Zerveas, G., Jayaraman, S., Patel, D., Bhamidipaty, A., Eickhoff, C.: A transformer-based framework for multivariate time series representation learning. In: Proceedings of the 27th ACM SIGKDD Conference on Knowledge Discovery & Data Mining, pp. 2114–2124 (2021)
20. Zhelavskaya, I.S., Vasile, R., Shprits, Y.Y., Stolle, C., Matzka, J.: Systematic analysis of machine learning and feature selection techniques for prediction of the Kp index. Space Weather 17(10), 1461–1486 (2019)

AS-SIM: An Approach to Action-State Process Model Discovery

Alessio Bottrighi[1], Marco Guazzone[1], Giorgio Leonardi[1], Stefania Montani[1], Manuel Striani[2(✉)], and Paolo Terenziani[1]

[1] DISIT, Università del Piemonte Orientale "A. Avogadro", Alessandria, Italy
{alessio.bottrighi,marco.guazzone,giorgio.leonardi,
stefania.montani,paolo.terenziani}@uniupo.it
[2] Dipartimento di Informatica, Università di Torino, corso Svizzera 185, 10149 Turin, Italy
manuel.striani@unito.it

Abstract. Process model discovery has gained a lot of attention in recent years, to mine a process model from traces of process executions. In our recent work, we have proposed SIM (Semantic Interactive Miner), an innovative process mining tool able to discover the process model in an incremental way: first, a mining module builds an initial process model, called *log-tree*, from the available traces; then, such a model is refined interactively with domain experts, through *merge* and *abstraction* operations. However, in several contexts, traces are richer: they do not record only actions, but also *states* (i.e., values of parameters possibly affected by the actions). A typical example is the medical domain, where traces contain both actions and measurements of patients' parameters. In this paper, we propose AS-SIM (Action-State SIM), the first approach aiming at discovering a comprehensive model, in which two distinct classes of nodes are considered, to capture both actions and states. We focus on the definition and on the discovery of the initial action-state process model (called *action-state log-tree*), while in our future work we will extend SIM's *merge* and *abstraction* operations accordingly.

Keywords: Process mining · Process model discovery · Mining action+state evolution

1 Introduction

Process Mining (PM) [1] describes a family of a-posteriori analysis techniques able to extract non-trivial information from the *event log*, a repository storing the sequences of actions (*traces* henceforth [1]) that have been executed at a given organization. Within PM, *process model discovery* techniques take as an input the traces and build a process model, focusing on its control flow constructs. The mined process model is an oriented graph (whose semantics can typically be modeled as Petri Nets), where nodes represent trace actions and arcs represent the ordering relation between them. Process model discovery is the most relevant and widely used PM sub-field, and has given birth to a large family of algorithms. Such approaches have proved to be successful in several

M. Ceci et al. (Eds.): ISMIS 2022, LNAI 13515, pp. 336–345, 2022.
https://doi.org/10.1007/978-3-031-16564-1_32

applications, such as, e.g., business or production processes. However, in some domains, actions are strictly dependent on the *state* they operate on, and their effect (in terms of the variations they produce on such a state) is not strictly predictable\deterministic. In all such domains, organization logs usually contain also *state* information. As a typical example, in medicine, actions depend on and modify the state of a patient, and patient traces record (possibly timestamped) sequences, where actions are interleaved with state descriptions (consisting of the recording of patient's parameters values), or the information recorded in the hospital information system can be converted into this format (see e.g. [2]). In these cases, mining a model in which only the flow of actions is considered looks reductive, since:

(i)   The model would be incomplete, as it would not consider the state in which actions are performed, and their effects on such a state, and

(ii)   Such pieces of information are indeed available, and can be mined from the traces and exploited to better characterize the model itself.

In this paper, we propose a new line of process mining, that we call *action-state process model discovery*, in which we extend "traditional" process model discovery to consider also states. Though our proposal is general, and could be carried on as an extension of other miners in the literature, for the sake of concreteness in this paper we describe it as an extension of SIM (Semantic Interactive Miner), a mining algorithm we developed in the past [3].

SIM supports experts in an interactive step-by-step procedure for discovering a process model. Interactive process discovery in SIM starts with an algorithm which mines a process model, called *log-tree*, from the log. The log-tree is already a process model, but possibly an overfitting one, since it perfectly corresponds to the input traces. Experts can move progressively to more generalised graph-based process models, through the application of *merge* and *abstraction* operations. Given a set of occurrences of an activity pattern, *merge* operations modify the current model by "putting them together". On the other hand, *abstraction* operations are based on a-priori knowledge about action decompositions, and support the possibility of abstracting component actions into the macro-actions constituted by them. Each process model can then be evaluated by experts both quantitatively and qualitatively. Through a versioning mechanism, experts can also navigate the history of the versions of the model, generate new versions, or backtrack to a previous one, until they "approve" one of the models.

In this paper we present AS-SIM (Action-State SIM), which extends SIM to deal with state mining as well. In particular, in Sect. 2 we present our representation formalism. In Sect. 3, we describe our approach to mine the initial action-state process model. Section 4 is devoted to comparisons and conclusions.

Note that we focus on the definition and on the discovery of the initial action-state process model (called *action-state log-tree*), while in our future work we will extend *merge* and *abstraction* operations accordingly.

2 Action-State Log-Tree: Representation Formalism

In a SIM process model, only actions were considered, and each node in the initial log-tree (or in the graph obtained using merge and abstraction operations) could represent either a single action, or a set of actions to be executed in any order. In AS-SIM we consider input traces containing both actions and state information (about a set of parameters). Therefore, the representation formalism needs to be extended, to collect in the **action-state log-tree** (and, later, in the action-state graph), also state details.

Notation. We denote by $\mathbb{A}$ the domain $\{a_1, \ldots, a_n\}$ of actions, by $\mathbb{P}$ the domain $\{p_1, \ldots, p_m\}$ of parameters, and by $\mathbb{D}_{p_i}$ the domain $\{v_1, \ldots, v_k\}$ of the values that can be assumed by the parameter p_i (where $\mathbb{D}_{p_i}$ is a discrete and finite domain, as discussed in Sect. 3.1). Also, given a set T of traces, we denote by $\mathbb{A}_T$ the set of all the actions appearing in the traces in T.

Definition 1 (state). We define *state* as a nonempty set of pairs $<p_i, v_j>$ (where $p_i \in \mathbb{P}$ and $v_j \in \mathbb{D}_{p_i}$ and p_i appears at most once) appearing together in the trace (i.e., belonging to the same "observation"). Let $\mathbb{S}$ be the domain of possible states.

Notably, by "state" we refer to a set of variables, measured at a given timepoint, that describe the situation of the entity on which the process at hand operates, e.g., a patient if we work in the medical domain. Indeed, in this case, According to [5], "states describe clinical situations in terms of a set of state variables with a clinical sense". In real contexts, some state variables may be unavailable at the time of measurement – so the data collection can be incomplete.

In AS-SIM, we distinguish between two types of nodes for action-state log-trees: action-nodes and state-nodes.

Definition 2 (action-node, state-node). *Action-nodes* represent a pair $<As, Ts>$, and *state-nodes* represent a pair $<Ss, Ts>$, where.

- *As* denotes a (possibly unary) set $\{a_1, \ldots, a_k\}$ of actions ($a_i \in \mathbb{A}$, , $1 \le i \le k$). Actions in the same node are in *any-order* relation.
- *Ss* denotes a (possibly unary) sequence $(s_1, \ldots, s_k)$ of states ($s_i \in S$, $1 \le i \le k$). Notice that, in such a way, each *path* from the root of the action-state log-tree to a given node N denotes a set of possible action-state patterns (called **support patterns** of N henceforth), obtained by following the order represented by the arcs in the path to visit the action-state log-tree, ordering in every possible way the actions in each action-node, and considering the sequence of states in state-nodes. For instance, the path $\{a,b\} \rightarrow (s_1,s_2) \rightarrow \{c\}$, where a,b,c $\in \mathbb{A}$ and $s_1,s_2 \in \mathbb{S}$ represents the support patterns "a b s_1 s_2 c" and "b a s_1 s_2 c".
- *Ts* represents a set of pointers to all and only those traces (called **support traces** henceforth) in the log whose prefixes exactly match the one of the *support patterns* of *As* or *Ss*.

Indeed, action-nodes support a compact representation of sets of actions which can be executed in any order, and state-nodes compact the representation of a sequence of states into a unique node. Two or more actions are in any order only when their order of execution is not relevant to the process goals. Additionally, we introduce a further type of nodes, macro-state-nodes, as a way to incapsulate alternative state-nodes into a unique node.

Definition 3 (macro-state-node). *Macro-state-nodes* represent a set of state-nodes.

3   Mining the Action-State Log-Tree

The mining process takes as an input the event log (i.e., a set of traces), and provides as an output a tree, containing action-nodes and macro-state-nodes (notably, we intend that a macro-state-node may also include a single state-node). In general, input logs can assume different forms. To facilitate the definition of the mining algorithm (Sect. 3.2), we first perform pre-processing operations on the log (Sect. 3.1). Then, we obtain the final action-state log-tree through a post-processing transformation (Sect. 3.3).

3.1   Log Pre-processing

In this pre-processing phase, the values of the parameters in the input traces are discretized, according to a set of discretization functions.

Definition 4 (domain-discretization functions). Given the domain $P = \{p_1, ..., p_m\}$ of parameters, defined (in the input traces) over the domains $\{D_{p_1}, ..., D_{p_m}\}$, we define domain-discretization functions as a set of functions $f_{ddf} = \{f_1, ..., f_m\}$, where, for each $f_i \in f_{ddf}$, f_i is a function that maps values of the domain D_{p_i} into values of a new, finite domain D'_{p_i} (i.e., $f_i: D_{p_i} \rightarrow D'_{p_i}$).

In the approach described in this paper, we assume that domain-discretization functions are provided as an input to AS-SIM. Notably, however, such functions can be either provided by domain experts, or pre-computed on the basis of the (values of the parameters in the) input traces (e.g., considering the values probability distributions).

To clarify the concepts exposed, we refer to the domain of stroke treatment. In particular, the patients admitted to the stroke unit should be stabilized before starting any treatment. The main parameters to monitor to check the patient's state: temperature, glycemia, diastolic and systolic pressure. Further patient's features are considered to define the treatment strategy, such as age and time since the stroke onset. We assume that the domain-discretization function is defined in terms of discretization levels provided by domain experts. An example is reported in Table 1.

Table 1. Discretization levels of patient's parameters.

Parameter	Measure unit	Discretization levels
Temperature (T)	°C	[0–35); [35–37.5); [37.5 and beyond)

(continued)

Table 1. (*continued*)

Parameter	Measure unit	Discretization levels
Glycaemia (G)	mg/dL	[0–50); [50–180); [180 and beyond)
Diastolic pressure (DP)	mmHg	[0–60); [60–120); [120 and beyond)
Systolic pressure (SP)	mmHg	[0–100); [100–185); [185 and beyond)
Age	Years	[0–18); [18–45); [45–80); [80 and beyond)
Time since onset (TSO)	Hours	[0–4.5); [4.5 and beyond)

Moreover, in general, each trace in the log consists of a sequence of elements, where each element may be either an action or a state, with no constraint. In particular, also sequences of states may appear in the traces. To facilitate the construction of state-nodes (see Sect. 2), in the pre-processing phase, we merge sequences of states into a unique element, modeling the sequences of values assumed by the parameters, at each state.

For example, during the patient stabilization phase the monitoring data are collected periodically to check the patient's state, in order to take actions accordingly. This can generate sequences of states in the traces; Fig. 1 shows an excerpt of a trace containing a state sequence.

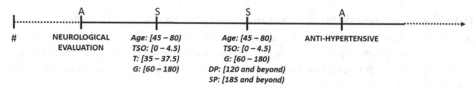

Fig. 1. Excerpt of trace containing sequence of states

In Fig. 1, each vertical line represents an action (identified by the character "A" above the line) or a state (character "S"). After the pre-processing phase, the same trace will be arranged as shown in Fig. 2.

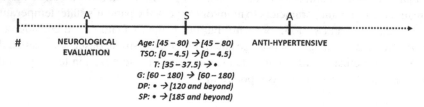

Fig. 2. Trace in Fig. 1 after pre-processing

In Fig. 2, the two consecutive states in Fig. 1 are merged into a unique element. In this merged state, each parameter is represented as a sequence, where not measured data are indicated with the value "•".

3.2 Mining Algorithm

In this section, we present the algorithm to build the action-state log-tree as an extension of the one used in SIM [3]. The pseudocode is shown in Algorithm 1. The function Build-Tree in Algorithm 1 takes as input a variable index, representing a given position in the traces, a node $\langle P,T \rangle$ (either an action-node or a state-node), and two user-defined thresholds α and β (for details, see below). Initially, it is called on the first position in the traces (we assume that all the traces start with a dummy action #) and on the root of the action-state log-tree (which is a dummy node, corresponding to the # action; therefore, initially, index $= 0$, $P = \{\#\}$ and T is the set of all the traces).

Algorithm 1: *Build-Tree* pseudocode.

```
1. Build-Tree(index, ⟨P,T⟩, α, β)
2.  ⟨nextP_S, nextP_A⟩ ← getNext(index+1, T, α)
3.  if (nextP_S ∪ nextP_A) not empty then
4.     nextNodes ← XORvsANY(⟨nextP_S, nextP_A⟩, T, β)
5.     foreach node <P',T'> ∈ nextNodes do
6.        AppendSon(⟨P',T'⟩, ⟨P,T⟩)
7.        Build-Tree(index+|P'|, ⟨P',T'⟩, α, β)
8.     end
9.  end
```

The function *getNext* (see line 2) inspects the traces in T to find all possible next elements (either actions or states). At this stage, "rare" patterns can be ruled out. Specifically, if $P = \{X\}$, and Y is a possible next element, Y will be provided in output by *getNext* only if the edge frequency E_F of the sequence $X > Y$ is above a user-defined threshold α, where:

$$E_F(X > Y) = \frac{|X > Y|}{|T|}$$

being $|X > Y|$ the number of traces in T in which X is immediately followed by Y (i.e., the cardinality of the support traces of Y), and being $|T|$ the cardinality of the support traces of X. Setting $\alpha > 0$ allows to rule out noisy patterns. Note that, if rare/noisy patterns are ruled out, the resulting action-state log-tree is not guaranteed to still have precision $= 1$ and replay-fitness $= 1$ (see [4]); however, in this paper, we will set $\alpha = 0$, and thus consider all the traces in the initial model construction.

On the remaining next states *nextP_S* and next actions *nextP_A*, the function *XORvsANY* identifies possible sets of actions in any-order (see line 4). Notably, (i) we assume that states cannot be in any-order (so they are directly managed as XOR sons of the current node), and (ii) the same action may appear more than once as a son of the current node (e.g., as a "unitary" action-node and\or in one or more "any-order" action-nodes).

To identify any-order sets of actions appearing in *nextP_A*, support traces in T are inspected at positions *index* + 2 (binary any-order), *index* + 3 (ternary any-order) and so on, until no "wider" any-order can be determined. For the sake of simplicity, let us consider the case of binary any-orders: *XORvsANY* calculates the dependency frequency

$A \rightarrow B$ between every action pair $\langle A,B \rangle$ in $nextP_A \times nextP_A$ by considering sequences of two actions A (at position $index + 1$) and B (at position $index + 2$) following P in the traces T as follows:

$$A \rightarrow B = \frac{1}{2}\left(\frac{|A > B|}{\sum_{C \in A_T} |A > C|} + \frac{|A > B|}{\sum_{D \in A_T} |D > B|} \right)$$

where, always considering the traces in T, $|A > B|$ is the number of traces in which A is immediately followed by B, $|A > C|$ is the number of traces in which A is immediately followed by some action $C \in A_T$, and $|D > B|$ is the number of traces in which B is immediately preceded by some action $D \in A_T$. If both the dependency frequencies of $A \rightarrow B$ and $B \rightarrow A$ are above the given (user-defined) threshold β, this means that A and B occur frequently in any-order in the same traces. Thus, *XOR vs ANY* identifies an any-order relation between A and B, and creates a new node $A\&B$ with the associated support traces.

The output *nextNodes* of the function *XORvsANY* is a set of nodes $\langle P', T' \rangle$. Each node is appended to the action-state log-tree (function *AppendSon*; see line 6), and *Build-Tree* is recursively applied to each node (with the first parameter *index* properly set according to the cardinality of P'; see line 7).

Finally, a-posteriori, we create a dummy node $, and connect all the leaves to it.

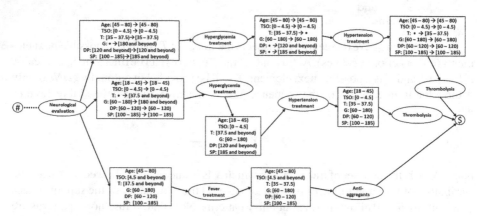

Fig. 3. Action-state log-tree for patients' stabilization in stroke disease

Figure 3 shows the action-state log tree, based on traces from the stroke domain focusing on the patients' stabilization phase described in Sect. 3.1. Here, the states are represented by rectangles, while the actions by ellipses. In the process in Fig. 3, after the neurological evaluation, different stabilization strategies are performed (including Hyperglycemia, Fever or Hypertension treatments) on the basis of the patients' states, before adopting the most appropriate treatment (Thrombolysis or Anti-aggregants).

3.3 Model Post-processing: Generating Macro-State-Nodes

The action-state log-tree, as modeled so far, represents in a compact way the input traces. However, an additional simplification can be provided. Given the fact that, in many real situations, the number of state parameters may be high, as well as the number of values they may assume (even after discretization), states deriving from different traces are rarely identical, so that the action-state log-tree "spans" in many different branches whenever states appear. We propose, as a first "compacting" step, to automatically merge all the state-nodes in the action-state log-tree which are preceded by the same action-node and are followed by the same action-nodes, into a unique macro-state-node, representing their union.

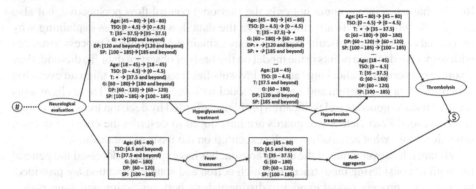

Fig. 4. Action-state log-tree in Fig. 3 after post-processing

Figure 4 shows the action-state log tree in Fig. 3, after the post-processing step. In particular, the two upper branches are composed by the same sequence of actions, therefore the post-processing algorithm merges the two state-nodes between "Neurological evaluation" and "Hyperglycemia treatment"; the state-nodes between "Hyperglycemia treatment" and "Hypertension treatment" and the state-nodes between "Hypertension treatment" and "Thrombolysis".

A possible interpretation of this generalization is that the same treatment strategy can be applied for patients having slightly different conditions.

Notably, besides making models more compact and "readable", the introduction of macro-state-nodes facilitates further processing stages (operating on states), as discussed in future work (see Sect. 4).

3.4 Properties of the Action-State Log-Tree Model

The action-state log-tree is a tree, and has precision = 1 (i.e., each path in the action-state log-tree corresponds to at least one trace in the log), and replay-fitness = 1 (i.e., each trace in the log can be replayed in the action-state log-tree with no errors) (see [4]).

4 Comparisons, Future Work and Conclusions

In this paper, we have proposed the first process model discovery approach mining and providing an explicit representation of the flow of both actions and states from input traces. Indeed, some approaches have started to face such an important issue. In the area of medical data mining, Kamisalic et al. [5] have considered the case of traces representing sequences of patient visits. Each visit assesses the state of a patient, and is followed by a prescription (indicating the therapies to be followed until the next visit). From this input, they mine a graph, where nodes represent patient' states, and arcs are labeled by the therapies leading from the input to the output states. With respect to our approach, therefore, they focus on states and on their transitions, not on the overall action-state process model.

In the PM literature, it is worth mentioning the work by De Leoni et al. (see, e.g., [6, 7]), that takes into account not only the (actions) control flow perspective, but also the data flow one. Specifically, they analyze the data flow to find rules explaining why individual cases take a particular path, i.e., they explain the behavior of process instances with respect to decision points in the model on the basis of observed data. To this end, they resort to conformance checking (another PM sub-field) techniques to align an event log containing data information and a process model with decision points. These alignments are then used to generate a classification problem, afforded by decision trees. The output is an extended Petri Net, where guards are introduced to describe the effect of data on transitions, i.e., what actions have to be executed on the basis of data values.

Notably, however, none of the above approaches has explicitly addressed the general problem of considering input traces in which action and state information are provided, to discover a process model explicitly distinguishing between action and state nodes. Such a generalized model is, in our opinion, extremely important in all domains where the effects of actions on states have to be analysed. We therefore believe that our approach may become the starting point of a new stream of research in the area of process model discovery.

In the future, we plan to extend the work along several lines.

First of all, we will conduct an extensive experimental evaluation. Though the proposed methodology is general, we will consider the medical context, where we have been operating since a long time. In particular, we will resort to the public available patient data in the Mimic database [8].

Other major evolutions regard the discovery process. The action-state log-tree is already a process model, but possibly an overfitting one, since it perfectly corresponds to the input traces (see Sect. 3.4). Experts may want to move progressively to more generalised graph-based process models. In SIM, considering only actions, such generalizations could be obtained through the application of a wide class of *merge* and *abstraction* operations. In the future, we aim at extending AS-SIM along these directions, and more specifically we will work on the:

- Definition of *intra-state generalization* operations. Such operations will apply to macro-state-nodes, to discover hidden pieces of information (e.g., to abstract parameter trends from sequences of states) and to use them in order to simplify the state representation, making it more "compact";
- Definition of *merge* and *abstraction* operations, generalizing and complementing the current operations in SIM in order to consider also state-nodes.

References

1. van der Aalst, W.M.P.: Process Mining - Data Science in Action. 2nd edn. Springer, Heidelberg (2016). https://doi.org/10.1007/978-3-662-49851-4
2. Wang, S., McDermott, M.B.A., Chauhan, G., Ghassemi, M., Hughes, M.C., Naumann, T.: MIMIC-Extract: a data extraction, preprocessing, and representation pipeline for MIMIC-III. In: Ghassemi, M. (ed.) ACM CHIL 2020: ACM Conference on Health, Inference, and Learning, Toronto, Ontario, Canada, 2–4 Apr 2020, pp. 222–235. ACM (2020). https://doi.org/10.1145/3368555.3384469
3. Bottrighi, A., Canensi, L., Leonardi, G., Montani, S., Terenziani, P.: Interactive mining and retrieval from process traces. Expert Syst. Appl. **110**, 62–79 (2018). https://doi.org/10.1016/j.eswa.2018.05.041
4. Buijs, J.C.A.M., van Dongen, B.F., van der Aalst, W.M.P.: On the role of fitness, precision, generalization and simplicity in process discovery. In: Meersman, R., et al. (eds.) On the Move to Meaningful Internet Systems: OTM 2012. Lecture Notes in Computer Science, vol. 7565, pp. 305–322. Springer, Heidelberg (2012). https://doi.org/10.1007/978-3-642-33606-5_19
5. Kamisalic, A., Riano, D., Welzer, T.: Formalization and acquisition of temporal knowledge for decision support in medical processes. Comput. Methods Programs Biomed. **158**, 207–228 (2018). https://doi.org/10.1016/j.cmpb.2018.02.012
6. de Leoni, M., van der Aalst, W.M.P.: Data-aware process mining: discovering decisions in processes using alignments. In: Shin, S.Y., Maldonado, J.C. (eds.) Proceedings of the 28th Annual ACM Symposium on Applied Computing, SAC 2013, Coimbra, Portugal, 18–22 Mar 2013, pp. 1454–1461. ACM (2013). https://doi.org/10.1145/2480362.2480633
7. de Leoni, M., Felli, P., Montali, M.: A holistic approach for soundness verification of decision-aware process models. In: Trujillo, J.C., et al. (eds.) Conceptual Modeling. Lecture Notes in Computer Science, vol. 11157, pp. 219–235. Springer, Cham (2018). https://doi.org/10.1007/978-3-030-00847-5_17
8. Johnson, A.E.W., et al.: MIMIC-III, a freely accessible critical care database. Sci. Data **3**, 160035 (2016)

Combining Active Learning and Fast DNN Ensembles for Process Deviance Discovery

Francesco Folino(✉) , Gianluigi Folino , Massimo Guarascio ,
and Luigi Pontieri

ICAR-CNR, Via P. Bucci 8/9C, Rende (CS), Italy
{francesco.folino,gianluigi.folino,massimo.guarascio,
luigi.pontieri}@icar.cnr.it

Abstract. Detecting deviant traces in business process logs is a crucial task in modern organizations due to the detrimental effect of certain deviant behaviors (e.g., attacks, frauds, faults). Training a Deviance Detection Model (DDM) only over labeled traces with supervised learning methods unfits real-life contexts where a small fraction of the traces are labeled. Thus, we here propose an Active-Learning-based approach to discovering a deep DDM ensemble that exploits a temporal ensembling method to train and fuse multiple DDMs sharing the same DNN architecture, devised in a way ensuring rapid convergence in relatively few training epochs. Experts' supervision is required only on small numbers of unlabelled traces exhibiting high values of (epistemic) prediction uncertainty, estimated in an ensemble-driven fashion. Tests on real data confirmed the approach's effectiveness, even compared to the results obtained by state-of-the-art supervised methods in the ideal case where all the data are labeled.

Keywords: Process deviance discovery · Deep ensembles · Active learning

1 Introduction

(Process) *Deviance mining* [4] refers to the problem of detecting and analyzing traces exhibiting "deviant" behaviors in a process log. Most existing solutions (e.g., [1,3, 8,10]) reuse Machine Learning (ML) methods to train a *Deviance Detection Model* (*DDM*), i.e., a classifier discriminating deviant traces from normal ones, on propositional trace representations derived by extracting relevant behavioral patterns from the traces and using them as data features. Notably, such a pattern-based approach allows for dealing with interpretable data representations, which will turn helpful when trying to explain/justify (possibly with the help of post-hoc explanation methods) the predictions that the discovered DDM makes on new traces. In particular, [1,10] showed the advantage of mixing up different pattern families for the sake of higher expressivity, while [3] proposed to train an ensemble of DDMs via a multi-view learning strategy, where each DDM is induced from a distinct pattern-based view of the training data. Starting from empirical evidence of the superiority of deep models w.r.t. shallow ML ones in predictive process mining tasks, the approach of [3] was recently enhanced in [5] by leveraging Deep Neural Networks (DNNs) to implement the base DDMs of the ensemble and different (possibly trainable) functions for combining them.

ⓒ The Author(s), under exclusive license to Springer Nature Switzerland AG 2022
M. Ceci et al. (Eds.): ISMIS 2022, LNAI 13515, pp. 346–356, 2022.
https://doi.org/10.1007/978-3-031-16564-1_33

Unfortunately, as observed in [3], in many real-life deviance mining scenarios, a limited number of log traces have a known deviance-class label, since labeling a process instance as either deviant or normal is a time-consuming and/or monetary costly task. This opens the opportunity of exploiting an *Active Learning* (*AL*) strategy [13], where human experts are put into the learning process and repeatedly asked to inspect and label a small number of unlabelled log traces (usually chosen on the basis of prediction uncertainty), hoping these additional labeled examples will help find a better DDM.

However, adapting AL strategies to discover deep DDMs poses some critical issues. First, as the class-membership probabilities returned by such classifiers are not calibrated (and often overconfident), estimating prediction uncertainty directly on these probabilities may perform worse than random sampling [13]. On the other hand, though good uncertainty estimates can be obtained with Bayesian DNNs or DNN ensembles, such models are far more costly to train and apply [13], and definitely less suitable for interactive AL settings. Moreover, as typical deep models are data-hungry, they may incur under/over-fitting issues when trained on relatively small amounts of examples.

In the light of the considerations above, we propose an approach to the DDM discovery problem that relies on learning an ensemble of deep DDMs from both labeled and unlabeled traces. Some major technical features of our proposal are: **(F1)** All the base DDMs are devised to ingest propositional trace representations produced by employing only two (interpretable and compact enough) of the pattern families used in previous work [1,3,5,10] (see Sect. 2) to curb the levels of sparsity and heterogeneity of the representations. **(F2)** For the sake of efficiency and feasibility in an AL-oriented setting, we discover an approximated DDM ensemble via a fast temporal ensemble-learning strategy [6] (see Sect. 3) that finds diverse deep DDMs in as many training epochs as those needed to optimize a single DNN of the same structure and use non-trainable functions to fuse the base DDMs' predictions; to ensure repeated rapid convergence to good local optima, we employ a cyclic learning-rate schedule and a DNN architecture with both dropout and residual-like components (see Sect. 4). **(F3)** The unlabeled instances to pass to the expert are chosen according to their prediction uncertainty, estimated based on how they were probabilistically classified by the base DDMs (see Sect. 3).

Experiments conducted on real-life log data (Sect. 5) showed that this approach yields compelling accuracy results at the cost of reasonable efforts by the experts, even if compared to the results that (computationally more costly) state-of-the-art methods obtain in the ideal, unlikely scenario where all the log traces are labeled.

2 Background, Problem Setting and Other Related Work

Process Traces and Deviance-Detection-Model Discovery Problem. As usually done in Deviance Mining, we assume that for every execution instance of the process under analysis, a distinguished *trace* is stored, consisting of a (temporally ordered) sequence of *events*, plus several instance-level attributes that did not vary during process executions. Each event is a tuple keeping information on the process activity that was executed correspondingly to the event and possibly further information (e.g., the executor, activity parameters, and performance measures).

The ultimate task our work is meant to support consists in deciding whether a process instance is deviant or not, based on its associated trace. Given a set D^L of traces, labelled as either *normal* or *deviant*, the "deviance mining" problem [1,3,8,10] amounts to inducing a *Deviance Detection Model (DDM)* from D^L, to classify novel traces. This basic problem setting is extended by assuming that another collection D^U of example traces, without associated ground-truth class labels, is supplied. To exploit this complementary source of information, we propose to adopt an active-learning formulation of the DDM discovery problem where the learning system is iteratively provided with small samples of traces, drawn from D^U and labeled manually by an expert.

Behavioral Patterns. Many previous approaches to the discovery of a DDM [1,3,8,10] rely on training classification models against some propositional encoding of the log traces for the sake of interpretability. To this end, the event list of each trace is turned into a list of symbols (possibly, the sole activity labels appearing in the trace) before extracting a set of relevant sequential patterns from these discrete sequences and then using them as features to build up a fixed-length encoding of the traces. Different kinds of patterns were used to this end so far: *individual activity* [14] (IA), *discriminative pattern* [8] (DP), and several pattern types borrowed from bio-informatics (e.g., tandem/maximal repeats and their alphabet-abstracted versions).

Specifically, IA patterns descend from regarding every activity label a as a distinct feature for any trace τ, storing how many times a occurs in τ. DP patterns [8] are frequent, closed, possibly non-contiguous sub-sequences of activity labels, capturing behaviors that repeatedly occur in either a trace or multiple ones. DP patterns can be found efficiently using the CLIPER algorithm of [8], which computes frequent candidates via an a-priori-like computation and then returns the most discriminative ones based on the Fischer score. Each DP pattern is eventually used as a non-negative integer feature for any trace τ, storing how frequently the pattern occurs in τ.

Other Related Work. Deep learning (DL) solutions were exploited recently for several Process Mining tasks going beyond deviance detection, such as predicting process outcomes, next events, and remaining execution time (in a Predictive Process Monitoring setting). For these forecasting tasks, sequence-oriented DL architectures leveraging recurrent or 1-D convolutional layers were shown quite effective, [9]. Compelling accuracy results were also obtained with multi-view learning schemes [2,11]. However, the higher expressivity of such DL models, compared to feed-forward neural networks applied to pattern-based trace encodings are likely to pose some higher difficulty in obtaining reliable and comprehensible explanations for the predictions returned.

3   Snapshot-Ensemble-Based Learning Approach

As mentioned above, in our approach, a deviance-detection model (DDM) is discovered from and applied to a propositional trace representation, for the sake of interpretability and fast convergence to non-overfitting solutions. To this end, we only resort to the families of *individual activity* patterns and *discriminative patterns* (see Sect. 2), since: (i) the combination of these two pattern families was shown in [3,5] to outperform other pattern combinations in the discovery of DDMs, and (ii) this combination allows to obtain more compact (and easier to understand) descriptions of deviant behaviors.

Using these patterns as non-negative integer features (counting up the occurrences of the respective patterns), every log trace τ is converted into a fixed-length tuple $\vec{\tau} = x_1, \ldots, x_N$, where each x_i is either one of such pattern-related features evaluated on τ or one of the data attributes (represented via one-hot encoding, if categorical) originally stored in τ. Denoting as $X_1, \ldots, X_N$ the domains these features are defined upon, we will refer to the universe of such data instances as $\mathcal{U} = X_1 \times X_2 \times \ldots \times X_N$.

Several such propositional data instances are assumed to be available, which are logically partitioned into two subsets: a set D^L of instances equipped with a deviance-class label in $\{normal, deviant\}$ and a set D^U of unlabelled instances.

Our approach consists in exploiting these data to discover an ensemble of *Deep Neural Networks* (*DNNs*), which takes the form defined below.

Definition 1 (DDM Ensemble). *A DDM Ensemble model $\mathcal{M}$ over the instance universe $\mathcal{U}$ is a tuple of the form $\langle M_1, \ldots, M_k, \phi \rangle$ for some $k \in \mathbb{N} \setminus \{1\}$, such that: (i) for each $i \in [1..k]$, $M_i : \mathcal{U} \to [0,1]$ is a DNN model that can map any instance $\vec{\tau} \in \mathcal{U}$ to a "deviance score" $M_i(\vec{\tau})$ providing an estimate for the probability that the process instance from which $\vec{\tau}$ was derived is deviant, and (ii) $\phi : [0,1]^k \to [0,1]$ is an aggregation operator that, for any $\vec{\tau} \in \mathcal{U}$, allows for merging the predictions returned by $M_1, \ldots, M_k$ for $\vec{\tau}$ into an overall prediction $\phi(\{M_1(\vec{\tau}), \ldots, M_k(\vec{\tau})\})$. Hereinafter, we will call $M_1, \ldots, M_k$ the base models of $\mathcal{M}$ and ϕ the combiner of $\mathcal{M}$.* $\square$

Specifically, in the current implementation of our framework, the base models in a DDM ensemble share the same DNN architecture (see Sect. 4), including dropout layers and residual-like connections that allow for fast training convergence and higher robustness to over/under-fitting the labeled examples when the latter are relatively few.

The proposed ensemble-based AL (Active Learning) approach is sketched in Algorithm 1. Three alternative options are available for selecting the combiner strategy, which correspond to pick up the maximum (MAX), average (AVG), and median (MEDIAN) of the predictions of the base models, respectively. These simpler but faster ensemble combination functions fit better real AL settings than the trainable combiners in [3,5].

A first version of the DDM ensemble is built by looking only at the given set in D^L of labeled data instances by leveraging a fast ("snapshot-based") approximated ensemble-learning procedure, denoted in the algorithm as train_snapshot. To improve this DDM ensemble, limited feedback is acquired from experts, who are asked to iteratively analyze and label a small number (controlled by parameter b, named hereinafter the *expert budget*) of instances selected from the given set D^U of unlabelled instances. The choice is based on prediction uncertainty scores (higher scores are preferred) in the attempt to provide the training procedure with novel labeled examples that are as informative as possible —details on the uncertainty scores are given later on.

Procedure train_snapshot. The training procedure follows the fast *snapshot ensemble* strategy in [6], which allows for ensembling multiple neural networks at the same cost as training a single one. Specifically, given a training set D_{TR}^L enriched with the ground-truth-labeled instances in X (and a validation set D_{VAL}^L), we train a DNN of the form described in Sect. 4 in the required number e of epochs, by making it converge to k local minima along its optimization path: the model parameters obtained for each minimum are saved in the ensemble as a distinct base DDM. To ensure repeated rapid convergence, a cyclic learning rate schedule is combined with a classic SGD scheme.

Algorithm 1. Lifecycle of a DDM ensemble model in our approach.[a]

Require: labelled data $D^L \subseteq \mathcal{U}$, unlabelled data $D^U \subseteq \mathcal{U}$, no. $m \in \mathbb{N} \cup \{\infty\}$ of AL steps, per-step budget $b \in \mathbb{N}$, no. k of base models, combination strategy $comb \in \{\texttt{MAX}, \texttt{AVG}, \texttt{MEDIAN}\}$, number $e \in \mathbb{N}$ of training epochs, initial learning rate lr, validation percentage $val\%$;

Ensure: a DDM ensemble $\mathcal{M} = \langle M_1, \ldots, M_k, \phi \rangle$ s.t. ϕ implements the aggregation operator specified via $comb$.

1: split D^L into training set D^L_{TR} and validation set D^L_{VAL} s.t. $|D^L_{VAL}| = |D^L| \times val\%/100$;

2: $\mathcal{M} := \texttt{train\_snapshot}(D^L_{TR}, D^L_{VAL}, k, comb, epochs)$ {induce a DDM ensemble model M of the form $M = \langle M_1, \ldots, M_k, \phi_{comb} \rangle$}

3: **for** $i = 1..m$ **do**

4:     **wait** for condition $(|D^U| \geq b$ **and** *an expert is available for labelling*$) = $ **true**;

5:     select $X \subseteq D^U$ such that $|X| = b$ **and** $\left(\text{H}^{\mathcal{M}}_{ep}(x) \geq \text{H}^{\mathcal{M}}_{ep}(x')\right) \forall (x, x') \in X \times D^U \setminus X$; {see paragraph *Uncertainty estimation* for the definition of $\text{H}^{\mathcal{M}}_{ep}(\cdot)$}

6:     $D^U := D^U \setminus X$; $D^L_{TR} := D^L_{TR} \cup X$;

7:     $\mathcal{M}' := \texttt{train\_snapshot}(D^L_{TR}, D^L_{VAL}, k, comb, epochs)$;

8:     **if** $\mathcal{M}'$ performs better than $\mathcal{M}$ over D^L_{VAL} {the two models are compared based on F1 scores (see Sect. 5, paragraph *Evaluation metrics*)) in current implementation} **then**

9:         $\mathcal{M} := \mathcal{M}'$;

10:    **end if**

11: **end for**

---

[a] Dataset D^U can possibly change over time through the addition of novel unlabeled traces; such a stream-like scenario is handled by setting $m = \infty$ and letting the current version of $\mathcal{M}$ can be applied at any moment, without necessarily waiting for the end of the algorithm.

Precisely, the first base model M_1 of the ensemble is randomly initialized and trained for $\lfloor e/k \rfloor$ epochs by using a variable learning rate η: η is initially set to lr and then progressively lowered, from one epoch to the next, according to a shifted-cosine schedule [6]. Each base model M_i is eventually chosen as the version of the DNN (among those obtained at the end of different training epochs) achieving the highest F1 score on D^L_{VAL}. Then, till $i \leq k$ a further base DDM M_i is trained similarly, after initializing it with the same model weights as M_{i-1} and setting again η to lr.

Uncertainty Scores. Let $\mathcal{M} = \langle M_1, \ldots, M_k, \phi \rangle$ be a DDM ensemble (such that ϕ implements the AVG operator) and $x \in \mathcal{U}$ be any data instance representing some trace τ. Let $p^{(q)}_{dev}(x) = M_q(x)$ be the prediction returned, for x by the q-th model in the ensemble, and $p^{\mathcal{M}}_{dev}(x) = \phi(\{M_i(x) \mid i \in [1..k]\})$ be the prediction returned for x by the ensemble as a whole. The total uncertainty affecting the ensemble prediction for x can be quantified through the entropy measurement $H^{\mathcal{M}}_{tot}(x) = -p^{\mathcal{M}}_{dev}(x) \times \log p^{\mathcal{M}}_{dev} - (1 - p^{\mathcal{M}}_{dev}) \times \log (1 - p^{\mathcal{M}}_{dev})$. This value sums up two components: an *aleatoric* (or irreducible) component $H^{\mathcal{M}}_{al}(x)$ and an *epistemic* one $H^{\mathcal{M}}_{ep}(x)$, the latter of which could be possibly reduced by using further training examples in the learning process. Aleatoric uncertainty can be estimated as $H^{\mathcal{M}}_{al} \approx \frac{1}{k} \sum_{q=1}^{k} H^q(x)$, where H^q is the entropy related to the prediction of each base model M_q —i.e. $H^q = -p^{(q)}_{dev}(x) \times \log p^{(q)}_{dev}(x) - \left(1 - p^{(q)}_{dev}\right) \times \log \left(1 - p^{(q)}_{dev}\right)$. Thus, the epistemic uncertainty involved in using the ensemble $\mathcal{M}$ to classify x can be quantified as the difference between the total and aleatoric entropies: $H^{\mathcal{M}}_{ep} \approx H^{\mathcal{M}}_{tot} - H^{\mathcal{M}}_{al}$.

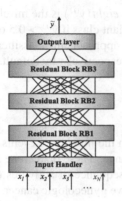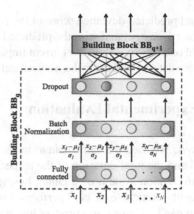

Fig. 1. Base DNN Architecture (left), and details on the Building Block sub-net (right)

It is the latter uncertainty measurement that Algorithm 1 adopts when choosing which unlabelled data instances are to be passed to the expert and then used as additional manually-labeled examples. It is worth noting that this strategy is related to the Bayesian Active Learning by Disagreement (BALD) criterion [13], which prefers unlabelled instances receiving more diverging predictions. Though the models in our DDM ensemble are not trained in a truly independent way, the empirical study in Sect. 5 shows our way of estimating epistemic uncertainties to work effectively in practice.

4 Details: Base DNN Architecture and Loss Function

As mentioned above, the base classifiers in the proposed ensemble-based DDM model share the same architecture, depicted on the left-hand side of Fig. 1.

The architecture consists of the following stack of components: (i) an *input* layer furnishing the propositional pattern-based encoding of any input trace; (ii) three instances of a *Residual Block* sub-net (denoted in the figure as RB_1, ..., RB_3), which consist each of two instances of a *Building-block* sub-net linked one another by a skip connection as shown in the right-hand side of Fig. 1; (iii) an output layer consisting of a single neuron equipped with a *sigmoid* activation function, which eventually returns a deviance score in [0,1]. Each building block is composed in its turn of three components: *(i)* a fully-connected layer including 128 neurons with *tanh* activation functions, *(ii)* a batch-normalization layer and *(iii)* a dropout layer with a dropout rate of 0.25.

Loss Function. A weighted variant of the *Mean Absolute Error* (MAE) is employed as the loss function for training the above-described base DDM's architecture, in order to deal with the case of unbalanced classes, which frequently occurs in real deviance detection settings. The loss, which is meant to pay more attention to the (rarer) training instances of the minority deviant class is defined as follows: $\frac{1}{n}\sum_{i=1}^{n}|y^{(i)} - \tilde{y}^{(i)}| \cdot weight(y^{(i)})$, where n is the number of instances in the training set, $y^{(i)}$ and $\tilde{y}^{(i)}$ are the real and predicted deviance score of the i-th instance, and $weight(y^{(i)})$ is the misclassification cost associated with the predicted class (i.e. the deviant class if $\bar{y}^{(i)} > 0.5$ or the normal one otherwise). In the current implementation of our approach, misclassification costs are set to 1 and 2 for the normal and deviant classes, respectively, by default.

5   Experimental Evaluation

Datasets. We used the same data as in [3, 10], enriched with the number of events contained and the total duration of the respective process instance. These data come from the information system of a Dutch hospital and store the sequence of activities (e.g., treatments, clinical tests, etc.) performed on patients who have gynecologic cancer. This dataset can be viewed as a process log, where each trace stores the history of a distinct patient, for a total of 1,142 traces, 150,291 events, and 624 distinct activities. Using the two alternative classification criteria in [3, 10] we produced two datasets, namely BPI_{dM13} and BPI_{dM16}, where each trace was labeled as either *deviant* (label = 1) iff the diagnosis code was "M13" and "M16", respectively or *normal* (label = 0) otherwise. Both datasets feature an unbalanced class distribution, for there are 310 (resp., 216) deviant traces in BPI_{dM13} (resp., BPI_{dM16}) vs 832 (resp., 926) normal traces. To avoid favorable biases, both datasets were pre-processed by removing all the attributes (in particular, those linked to diagnoses) that leak information on class labels.

Parameter Setting. We ran Algorithm 1 using different ensemble combination modes and making m (number of AL steps) range from 0 to 8. The other parameters were fixed as follows: $lr = 0.001$ (initial learning rate), $e = 32$ (number of training epochs), $b = 20$ (per-step budget), $k = 5$ (number of base models), $val\% = 10$ (validation set percentage). We so simulated a scenario where an expert is available, for a maximum of 8 days, to label $b = 20$ traces per day, for a total of $b_T = b \times m = 160$ traces at most. As usual, the expert was modeled by an oracle-like procedure consisting in unveiling and using the actual ground-truth labels of the selected b tuples. Let us denote the configurations $comb = \text{MAX}$, $comb = \text{AVG}$, and $comb = \text{MEDIAN}$ tested for the algorithms as *ensemble_max*, *ensemble_avg* and *ensemble_median*, respectively.

From each dataset, we randomly selected a 20% sample D_{TEST} of the instances and kept it apart to evaluate the models discovered by Algorithm 1 and its competitors. The remaining instances were randomly partitioned into two subsets of the same size, which were used to instantiate the parameters D^L and D^U of the algorithm, with the latter playing as a set of instances with unknown (actually, hidden) class labels.

Evaluation Metrics. Three metrics were used to evaluate models' predictions: (i) *AUC* (i.e. the area under the ROC curve); (ii) *G-mean* [7] (i.e. the geometric mean of True-Positive and True-Negative rates); and (iii) *F1* score (i.e., the harmonic mean of *Precision* and *Recall*) for the deviant class. The latter two metrics are more appropriate for imbalanced-class settings like ours. Hereinafter, we will use F1 as the metric of choice.

Test Results. Results in Table 1 back the effectiveness of our AL strategy in improving an ensemble DDM model obtained by using only the labeled traces originally available in the training set ($m = 0$). Indeed, it is easily seen that 8 AL iterations allow us to

Table 1. Results obtained, with different ensemble combination options, after different numbers of AL iterations (i.e., different settings of parameter m in Algorithm 1). The extreme setting $m = 0$ corresponds to use only the labeled data available in the training set, with no actual AL iteration. Column $\Delta F1\%$ reports the relative increases of $F1$ scores achieved with respect to the use case $m = 0$.

m	Model	BPI_{dM13}				BPI_{dM16}			
		AUC	G-$Mean$	$F1$	$\Delta F1\%$	AUC	G-$Mean$	$F1$	$\Delta F1\%$
0	ensemble_max	0.818	0.716	0.581	0%	0.862	0.729	0.490	0%
	ensemble_avg	0.803	0.689	0.548	0%	0.877	0.791	0.579	0%
	ensemble_median	0.807	0.687	0.544	0%	0.878	0.791	0.579	0%
1	ensemble_max	0.813	0.723	0.586	1%	0.877	0.759	0.526	7%
	ensemble_avg	0.823	0.752	0.619	13%	0.877	0.791	0.579	0%
	ensemble_median	0.817	0.747	0.614	13%	0.872	0.799	0.602	4%
2	ensemble_max	0.813	0.723	0.586	1%	0.884	0.769	0.548	12%
	ensemble_avg	0.823	0.752	0.619	13%	0.888	0.795	0.621	7%
	ensemble_median	0.817	0.747	0.614	13%	0.883	0.805	0.629	9%
3	ensemble_max	0.854	0.745	0.617	6%	0.884	0.769	0.548	12%
	ensemble_avg	0.823	0.752	0.619	13%	0.888	0.795	0.621	7%
	ensemble_median	0.817	0.747	0.614	13%	0.883	0.805	0.629	9%
4	ensemble_max	0.854	0.745	0.617	6%	0,900	0.814	0.636	30%
	ensemble_avg	0.823	0.752	0.619	13%	0.899	0.810	0.660	14%
	ensemble_median	0.817	0.747	0.614	13%	0.902	0.815	0.653	13%
5	ensemble_max	0.854	0.745	0.617	6%	0.900	0.814	0.636	30%
	ensemble_avg	0.823	0.752	0.619	13%	0.899	0.810	0.660	14%
	ensemble_median	0.817	0.747	0.614	13%	0.886	0.810	0.660	14%
6	ensemble_max	0.854	0.745	0.617	6%	0.900	0.814	0.636	30%
	ensemble_avg	0.823	0.752	0.619	13%	0.899	0.810	0.660	14%
	ensemble_median	0.817	0.747	0.614	13%	0.886	0.810	0.660	14%
7	ensemble_max	0.854	0.745	0.617	6%	0.887	0.830	0.643	31%
	ensemble_avg	0.838	0.757	0.634	16%	0.888	0.837	0.673	16%
	ensemble_median	0.844	0.743	0.619	14%	0.899	0.849	0.686	18%
8	ensemble_max	0.854	0.745	0.617	6%	0.887	0.830	0.643	31%
	ensemble_avg	0.838	0.757	0.634	16%	0.887	0.832	0.680	17%
	ensemble_median	0.847	0.763	0.643	18%	0.899	0.849	0.686	18%

achieve a tangible gain on all metrics and datasets. Specifically, when setting $m = 8$ an improvement of about 5%, 11%, and 18% (resp., 2%, 7%, and 18%) is gained, w.r.t. to non AL-enhanced setting $m = 0$, on dataset BPI_{dM13} (resp., BPI_{dM16}) in terms of AUC, G-$Mean$, and $F1$, respectively, when using the ensemble_median combination mode.

Comparable achievements are obtained, on both datasets, with option *ensemble_avg*, whereas *ensemble_max* performs slightly worse on BPI$_{dM13}$.

Interesting insight can be drawn from Fig. 2, showing the ratio $\Delta F1_m\%/\Delta F1_8\%$, for $m = 0 \ldots 8$. This ratio can be regarded as the performance gap (in terms of F1) between the version of the DDM ensemble obtained after m AL iterations and the final version of it obtained after 8 AL iterations. From this figure, it emerges that just after 3 (resp., 4) iterations on BPI$_{dM13}$ (resp., BPI$_{dM16}$), i.e., when just nearly 38% (resp., 50%) of the total expert budget, say $b_T = b \times 8 = 160$, has been consumed, all of our DDM ensembles have almost caught up with their respective fully-grown versions.

Table 2 compares the results obtained by our best-performing DDM ensemble, here renamed *Ours*, with those of the following DDM discovery methods: (i) a non ensemble-based *baseline* method that consists in training (just one instance of) the base DDM architecture of Sect. 4 against the labeled set D^L available at both the middle ($m = 4$) and last iteration ($m = 8$) of the AL procedure encoded by Algorithm 1; (ii) three state-of-the-art multi-view ensembling approaches defined in [3,5], namely method *HO-DDM-mine* [3] and two variants of method *MVDE* [5], evaluated in the ideal fully-supervised scenario where the class labels of all the instances in $D^L \cup D^U$ are exploited to train the DDM models —while our approach only uses such information for the instances originally stored in D^L and the few ones that it iteratively selects.

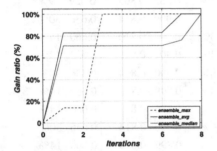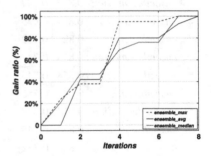

Fig. 2. Ratio between the F1's gain obtained by our approach after $m \in [0..8]$ AL iterations and that obtained after 8 AL iterations on BPI$_{dM13}$ (left) and BPI$_{dM16}$ (right).

Table 2. Comparing our best-performing DDM ensemble to the baseline and existing ensemble-based methods (evaluated in the ideal fully-labeled data scenario).

Model	Setting	BPI$_{dM13}$			BPI$_{dM16}$		
		AUC	G-Mean	F1	AUC	G-Mean	F1
Ours (best config. with $m = 4$)	Active-learning	0.823	0.752	0.619	0.899	0.810	0.660
Ours (best config. with $m = 8$)		0.847	0.763	0.643	0.899	0.849	0.686
Baseline ($m = 4$)		0.831	0.728	0.606	0.861	0.802	0.660
Baseline ($m = 8$)		0.800	0.757	0.630	0.870	0.820	0.667
HO-DDM-mine [3]	Fully-supervised	0.841	0.741	0.633	0.878	0.778	0.643
MVDE-Stack [5]		0.878	0.801	0.675	0.874	0.815	0.608
MVDE-Max [5]		0.864	0.654	0.566	0.907	0.668	0.545

When evaluated after 4 and 8 AL iterations, our approach outperforms the baseline in all metrics and across all datasets —if abstracting from the fact that, after 4 iterations on BPI_{dM13} (resp., BPI_{dM16}), the AUC (resp., F1) score of *Baseline* is better than (resp., comparable with) that of our method. Anyway, as our approach does not entail extra computational costs w.r.t. baseline, it looks preferable to the latter.

The (unfair) comparison between our approach and competitors *HO-DDM-mine*, *MVDE-Stack* and *MVDE-Max* allows us to make a further important observation: although these competitors (which need to train different base models on many different pattern-based views of the log data) are far more costly than our approach, the latter performs not drastically worse than them on dataset BPI_{dM13}, and even better (at least in terms of F1) on BPI_{dM16}. In particular, on BPI_{dM13}, it performs better than *MVDE-Max*, pretty similarly to *HO-DDM-mine* and only worse than *MVDE-Stack*, which, however, exploits a trainable (and hence more expensive) stacking-based function to merge the base models' predictions, differently from our approach. We can hence conclude that our approach ensures a suitable trade-off between effectiveness and efficiency in settings where the DDM ensemble needs to be updated frequently over time.

6 Conclusion and Future Work

The experimental results presented above make us confident in the practical usefulness of the proposed approach in real DDM discovery settings where a relatively small fraction of the log traces have an associated deviance-class label, and human experts are available to spend a limited amount of time providing additional supervision.

As future work, we plan to test the approach on more logs (possibly in streaming-data scenarios), while using hyperparameter optimization methods and different DNN architectures. Moreover, we will investigate combining our AL scheme with semi-supervised-learning methods and interpretable DNN models like that proposed in [12].

References

1. Bose, R.P.J.C., van der Aalst, W.M.P.: Discovering signature patterns from event logs. In: IEEE Symposium on Computational Intelligence and Data Mining (CIDM 2013), pp. 111–118 (2013)
2. Cuzzocrea, A., Folino, F., Guarascio, M., Pontieri, L.: A multi-view multi-dimensional ensemble learning approach to mining business process deviances. In: 2016 International Joint Conference on Neural Networks (IJCNN), pp. 3809–3816 (2016)
3. Cuzzocrea, A., Folino, F., Guarascio, M., Pontieri, L.: A robust and versatile multi-view learning framework for the detection of deviant business process instances. Int. J. Coop. Inf. Syst. 25(04), 1740003 (2016)
4. Folino, F., Pontieri, L.: Business process deviance mining. In: Encyclopedia of Big Data Technologies. Springer (2019). https://doi.org/10.1007/978-3-319-77525-8
5. Folino, F., Folino, G., Guarascio, M., Pontieri, L.: A multi-view ensemble of deep models for the detection of deviant process instances. In: ECML-PKDD Workshop, pp. 249–262 (2020)
6. Huang, G., Li, Y., Pleiss, G., Liu, Z., Hopcroft, J.E., Weinberger, K.Q.: Snapshot ensembles: Train 1, get m for free. arXiv preprint arXiv:1704.00109 (2017)

7. Kubat, M., Holte, R., Matwin, S.: Learning when negative examples abound. In: Proceedings of 9th European Conference on Machine Learning (ECML 1997), pp. 146–153 (1997)

8. Lo, D., Cheng, H., Han, J., Khoo, S.C., Sun, C.: Classification of software behaviors for failure detection: a discriminative pattern mining approach. In: Proceedings of 15th International Conference on Knowledge Discovery and Data Mining (KDD 2009), pp. 557–566 (2009)

9. Neu, D.A., Lahann, J., Fettke, P.: A systematic literature review on state-of-the-art deep learning methods for process prediction. Artif. Intell. Rev. 55(2), 801–827 (2022)

10. Nguyen, H., Dumas, M., La Rosa, M., Maggi, F.M., Suriadi, S.: Mining business process deviance: a quest for accuracy. In: Meersman, R., et al. (eds.) OTM 2014. LNCS, vol. 8841, pp. 436–445. Springer, Heidelberg (2014). https://doi.org/10.1007/978-3-662-45563-0_25

11. Pasquadibisceglie, V., Appice, A., Castellano, G., Malerba, D.: A multi-view deep learning approach for predictive business process monitoring. IEEE Trans. Serv. Comput. (Early Access) (2021). https://doi.org/10.1109/TSC.2021.3051771

12. Pasquadibisceglie, V., Castellano, G., Appice, A., Malerba, D.: Fox: a neuro-fuzzy model for process outcome prediction and explanation. In: 2021 3rd International Conference on Process Mining (ICPM), pp. 112–119 (2021)

13. Ren, P., et al.: A survey of deep active learning. ACM Comput. Surv. (CSUR) 54(9), 1–40 (2021)

14. Suriadi, S., Wynn, M.T., Ouyang, C., ter Hofstede, A.H.M., van Dijk, N.J.: Understanding process behaviours in a large insurance company in Australia: a case study. In: Proceedings of the 25th International Conference on Advanced Information Systems Engineering (CAiSE 2013), pp. 449–464 (2013)

Temporal Graph-Based CNNs (TG-CNNs) for Online Course Dropout Prediction

Zoe Hancox(✉) and Samuel D. Relton

The University of Leeds, Leeds, UK
{Z.L.Hancox,S.D.Relton}@leeds.ac.uk

Abstract. Due to the global pandemic, the use of online courses is increasing significantly; yet the rate of student dropout from online courses is rising. The Accessible Culture & Training Massive Open Online Course (ACT MOOC) dataset is comprised of a temporal sequence of student actions and subsequent dropout information. We introduce a novel approach based upon temporal graphs, which uses the sequence of (and time between) events to predict dropout. The dataset consists of 7,047 users, with a dropout rate of 57.7%. The Temporal Graph-Based Convolutional Neural Network (TG-CNN) models developed in this study are compared against baseline models and existing models in the literature. Performance is assessed using the AUC, accuracy, precision, recall, and F1 score. Our novel TG-CNN model achieves an AUC score of **0.797**, which improves upon previous literature: JODIE 0.756, TGN + MeTA 0.794, TGN 0.777, and CoPE 0.762. Our model offers a novel and intuitive formulation of this problem, with state-of-the-art performance.

Keywords: Temporal graphs · Dropout prediction · Neural networks

1 Introduction

Massive Open Online Courses (MOOCs) allow people to study and learn a wide range of material wherever and whenever they choose [8]. Despite this, student retention with MOOCs is low and course dropout is high [15]. The COVID-19 pandemic has caused a rise in the number of students partaking in online courses. Simultaneously, MOOC dropout rates are increasing as educational resources are forced to move to online and an epidemic of screen fatigue sets in [13]. Additionally, it has been noted that dropout rates are higher from MOOCs compared to in-person and off-line courses [13]. Predicting user dropout based on clickstream data could enable identification of behaviour patterns prior to dropout, to target interventions designed to encourage course completion [5,7].

Graph networks (GNs) are rising in popularity in machine learning (ML) [12, 16]. GNs capture object interactions, and are used to represent social networks and recommender systems, for example where nodes may represent people and

© The Author(s), under exclusive license to Springer Nature Switzerland AG 2022
M. Ceci et al. (Eds.): ISMIS 2022, LNAI 13515, pp. 357–367, 2022.
https://doi.org/10.1007/978-3-031-16564-1_34

edges depict messages from one person to another. Convolutions applied over graph structures have been shown to learn effectively in various tasks [1,12].

The model described in this paper utilises temporal GNs, a three-dimensional Convolutional Neural Network (3D CNN) and Long Short-Term Memory (LSTM) units, to evaluate clickstream data (user actions) making use of the elapsed time between events to improve predictive performance of student dropout.

The 3D CNN component has the ability to capture short-term temporal patterns of user actions, whilst the LSTM can cover longer-term associations. Our formulation of this problem as a graph classification task diverges from previous work in this area, where node and edge level classification are typically used. The approach outlined in this research utilises temporal graphs and 3D CNNs to represent, and learn from, structured temporal data. This generates two novel contributions:

1. Previous MOOC articles have focused on classifying the nodes/edges within a graph, as opposed to our formulation which classifies entire graph structures.
2. LSTMs assume a constant elapsed time between sequence elements, an issue which has received some attention in the literature [4,15]. Our TG-CNN approach offers an alternative formulation of this problem (including variable time dilation), which can model more complex temporal links.

2   Methods

2.1   Dataset and Modelling Approach

The Accessible Culture & Training MOOC (ACT MOOC) dataset includes timestamped actions and dropout labels for 7,047 users[1]. There are 97 potential clickstream actions a user can take which are timestamped. Timestamps are counted in seconds from the first interaction a user makes with the online course. In total over 411,749 interactions are captured, with the highest number of actions taken by one user totalling 505. Dropout occurred in 57.69% of users.

Our approach to this dropout prediction task is to turn this sequence of events into a temporal multigraph and formulate dropout prediction as a binary graph classification task, where each student has their own temporal graph to be classified. In particular, the $n = 97$ possible actions form the nodes of this graph and the temporal edges capture the elapsed time between actions. This can be stored in a 3D tensor $G(i, j, k) = t_k$ where $i, j \in \{1, \ldots, n\}$ are nodes in the graph, and t_k is the elapsed time for the kth edge in the temporal graph. An example with 4 possible actions is shown in Fig. 1.

In practice, we actually store $G(i, j, k) = \exp(-\gamma t_k)$, where $\gamma > 0$ is a trainable parameter of the model. This has two benefits:

---

[1] Stanford Network Analysis Project - https://snap.stanford.edu/data/act-mooc.html.

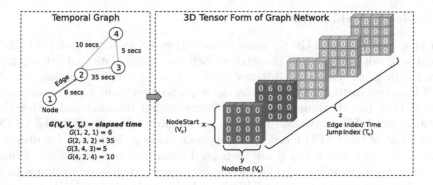

Fig. 1. Graph network visualisation showing connections between actions completed by a user in both graph form and tensor form. This example has only 4 possible actions, so is much smaller than the $97 \times 97 \times 100$ tensor that is used in this project.

1. Actions taken in quick succession or simultaneously have a value close to 1 and actions with a greater temporal gap are closer to 0. Events that are not related have value zero. This allows the temporal graph to be stored as a sparse 3-tensor, saving significant memory in the representation of the data.
2. Elapsed time can be rescaled to avoid extreme values in the neural network and potential under/overflow when using half-precision arithmetic.

The classification of these temporal graphs is performed using a neural network architecture based upon 3D convolutions. In particular, each convolution covers $m = 2, 3, 4, \ldots$ consecutive timesteps and each filter, of size $n \times n \times m$, is applied across the temporal dimension of the 3-tensor. The filters utilise information about the sequence of actions taken and the elapsed time between actions and the convolution operation (with stride 1) collapses this 3-tensor of size $n \times n \times K$ into a vector of length $K - m + 1$. The output of the convolutions is a sequence of feature vectors capturing short patterns (accounting for elapsed time); we pass these to an LSTM which allows for longer temporal patterns, built from these sub-patterns, to be captured.

For this particular task we use the most recent 100 actions of each user to reduce computational burden, front-padding any 3-tensors representing sequences of length less than 100 to ensure the most recent actions are always at the end of the 3-tensor. We also experiment with using a secondary pipeline of filters with stride 2, referred to as a "2-stream architecture", and concatenate the two before the final FCLs of the neural network.

Torch version 1.7.0, Tensorflow 2.8.0, NumPy 1.19.2, Pandas 1.2.4, Scikit-Learn 0.23.1, and CUDA 10.2.89 were used on a desktop with a NVIDIA RTX 3090 (Table 1), and the N8 Bede machine based at Durham University: an IBM Power 9 system with NVIDIA V100 GPUs (Table 2).

2.2 Model Architecture

Our proposed Temporal Graph-based Convolutional Neural Network (TG-CNN) model is shown in Fig. 2. The model described can handle data that is irregularly sampled in time. The input 3-tensor of size $97 \times 97 \times 100$ is fed into the 3D CNN layer, extracting information on the sequence of actions and elapsed times. This is then passed through a Batch Normalisation function and a Rectified Linear unit (ReLU) activation function, before proceeding through the LSTM. The output of the LSTM has dropout applied, passing the hidden features into a FCL. Dropout and a ReLU are then used again before a final FCL. Binary cross entropy logits loss is used for the binary target. Adam optimisation with L2 regularization is used to smooth oscillations during training. Our implementation also utilises a learning rate (LR) scheduler, multiplying the LR by 0.9 with an exponential decay each 10,000 steps. Early stopping is used with a patience of 50, to checkpoint the model when the validation loss decreases, interrupting execution when the model gets stuck in a local minima.

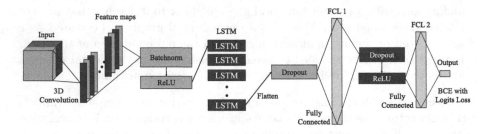

Fig. 2. Model architecture of proposed TG-CNN.

The γ variable controlling time dilation is a trainable parameter within the model. Longer gaps between actions could mean the action was less relevant, which could be modelled by increasing γ. An ablation study was performed to test the model with the γ variable, without the LSTM layer, without the exponential scaling parameter and with elastic net (L1 + L2) regularisation.

Additionally, we investigate the impact of adding a second stream (2-stream) to the network, where there is a second row 3D-CNN and LSTM using filters with a stride of 2, whereas the original 1-stream has a stride of 1. The output of the two independent streams are concatenated after the FCLs.

This model has interpretability potential, as the CNN features could be viewed directly and visualised to show the filters learnt from the data, whereas with LSTMs this is more difficult to comprehend.

2.3 Model Evaluation

This work adopts a 80/10/10 train/validation/test split for the TG-CNN models. Previous models using similar data have primarily focused on area under the receiver operating characteristic curve (AUC), so hyperparameters were optimised for best AUC score on the validation set and test set results are reported.

To optimise the model based the on validation set AUC value, we conducted a random hyperparameter search by sampling the number of epochs [25, 50, 75, 100], LR [0.1, 0.01, 0.05, 0.001, 0.005, 0.0001], number of filters [32, 64, 128], filter size [4, 16, 32, 64], number of LSTM hidden cells [16], L2 regularisation (L2 reg) parameter [1e−1, 1e−2, 1e−3, 1e−4, 1e−5, 5e−2, 5e−3, 5e−4], FCL size [128,256,512,1028,2056], and dropout rate [0.2, 0.3, 0.4, 0.5]. For the 2-stream models, the two streams each had filters of the same size but with different strides. This results in 230,400 possible combinations of hyperparameter values, necessitating the use of random search instead of a grid search.

We also fit two baseline models to the dataset for comparison, an LSTM (BL-LSTM) and Recurrent Neural Network (BL-RNN) architecture - a single RNN layer (LSTM or RNN respectively) followed by two FCLs. Five-fold cross-validation was used to optimise the AUC over the hyperparameter combinations for these baseline models. The BL-LSTM and BL-RNN models were tuned by optimising the LR, the number of epochs, the hidden units in the RNNs, and the number of hidden neurons in the FCLs.

3   Results

We tested the TG-CNN models with 1,430 hyperparameter samples from the list in Sect. 2.3. The best performing hyperparameters and performance metrics for these models are shown in Table 1, metrics are averaged over ten runs to show robustness. BL-LSTM and BL-RNN were each fitted using 5-fold cross-validation with 576 different hyperparameter combinations. The best performing BL-LSTM achieved an AUC of 0.783, 79.26% accuracy, 0.800 precision, 0.852 recall and an F1-score of 0.824 within 20 epochs, using a LR of 0.01, 128 hidden LSTM neurons and FCL sizes of 32 and 16. The best performing BL-RNN achieved an AUC of 0.778, 78.86% accuracy, 0.801 precision, 0.844 recall and an F1-score of 0.819 within 20 epochs, with a LR of 0.001, 32 hidden RNN neurons and FCL sizes of 64 and 32.

Table 2 shows the best variants of each model in the ablation study, and compares to existing models in the literature (which also use the model with maximal AUC for a fair comparison). The best performing variant overall was the TG-CNN with fixed time dilation $\gamma = 1$ (AUC 0.797), closely followed by the 2-stream version (AUC 0.796). The average results after 10 re-runs led to the 2-stream model achieving the best performance (Table 1). Eliminating the LSTM component led to significantly poorer results (AUC 0.705). These results show the importance of both the multi-streams and LSTM component in achieving good performance with this approach.

4   Discussion

Table 2 shows the predictive performance of models with the best AUC score using the ACT MOOC dataset. Our novel TG-CNN approach has state-of-the-

Table 1. Hyperparameter values and test set metrics for the best performing variants of the TG-CNN architecture (mean ± standard deviation from 10 runs).

Parameter	Variable γ	1-stream	2-stream	No LSTM	No exp	Elastic net
Epochs	50	75	52	25	100	100
LR	0.05	0.05	0.0005	0.0001	0.001	0.001
# Filters	64	64	128	64	32	32
Filter size	16	32	4	4	64	32
RNN cells	128	64	32	N/A	16	16
L2 Reg	1e−3	1e−2	5e−4	N/A	N/A	N/A
FCL size	1028	512	2056	512	1024	1024
Dropout	0.5	0.5	0.3	0.5	0.3	0.5
Test AUC	0.662 ± 0.08	0.748 ± 0.02	**0.763 ± 0.01**	0.705 ± 0.02	0.710 ± 0.02	0.711 ± 0.02
Accuracy	0.703 ± 0.06	0.771 ± 0.02	**0.775 ± 0.01**	0.583 ± 0.01	0.690 ± 0.02	0.702 ± 0.02
Precision	0.687 ± 0.06	0.764 ± 0.02	**0.773 ± 0.02**	0.581 ± 0.01	0.677 ± 0.02	0.688 ± 0.02
Recall	0.923 ± 0.05	0.883 ± 0.02	0.859 ± 0.04	**1.00 ± 0.00**	0.871 ± 0.02	0.886 ± 0.02
F1-Score	0.782 ± 0.03	**0.817 ± 0.01**	0.811 ± 0.02	0.735 ± 0.01	0.761 ± 0.01	0.774 ± 0.02

Table 2. Best AUC results of user dropout prediction using the ACT MOOC dataset, from our results (left columns) and from the results in the literature (right columns).

TG-CNN and Baseline models	AUC	Literature models	AUC
TG-CNN 1-stream	**0.797**	TGN + MeTA [16]	0.794
TG-CNN 2-stream	0.796	TGN + TNS [17]	0.791
BL-LSTM	0.783	TGN [12]	0.777
BL-RNN	0.779	CoPE [21]	0.762
TG-CNN 1-stream $\gamma = 4.819$	0.760	JODIE [7]	0.756
TG-CNN with Elastic Net	0.758	TGAT + TNS [17]	0.755
TG-CNN without LSTM	0.750	NPPCTNE [23]	0.745
TG-CNN without the Exponential	0.744	TGAT [16]	0.743

art performance on this task, despite being more intuitive and conceptually simpler than some of the other approaches in previous literature.

The ablation study demonstrated that the LSTM layer and the exponential function enabled the model to learn more effectively, this is potentially due to the LSTM enabling long-term memory alongside the filters learnt from the CNN. The γ variable converged to an average value of 4.819 in the best performing model, which suggests that actions taken closer together are more important to predicting dropout than actions further apart. The cut-off caused by $\hat{t} = \exp(-4.819 \times t)$ is sharper than when $\gamma = 1$, therefore when $\gamma = 4.819$ and the elapsed time t is more than $47\,\mathrm{s}$ $\hat{t}$ will round to 0.

Other advantages of the TG-CNN approach include the constant tensor size, allowing for optimisation of the underlying linear algebra operations, and the ability to extract temporal features in parallel using 3D convolutions, as opposed to RNN-based architectures that require sequential processing through time. The 3-tensor structure enables the filters to be extracted back into a intuitive graph structure. This could serve as a visual tool to show which sequences of events and temporal patterns lead to dropout.

5   Related Work in MOOC Dropout

We searched the IEEE database using the terms "MOOC AND predict*". 95 papers were found, 24 of these were analysed from their title and abstract, and we found that 4 used CNNs [11,18,20,22]. Only 1 paper utilised GNs [15], where the problem was formulated as a node/edge prediction task over time, to which they applied a novel data augmentation approach to existing models. This differs from our formulation of this problem as a graph-level classification task.

Learner behaviour feature matrices, weighted by importance, have been used alongside CNNs to predict dropout from clickstream data and improve predictive accuracy compared to basic models [8,11,18,22]. In [20], the authors use CNNs alongside Squeeze-and-Excitation Networks (SE-Net) and a Gated Recurrent unit (GRU). The GRU enables maintenance of the time series relationship between the clickstream data and the SE-Net helps with automatic feature extraction, this resulted in an accuracy above 90% on their dataset. Interestingly, Edmond Meku Fotso et al. found simple RNNs provided better accuracy compared to LSTMs and GRUs [4]. Standard ML algorithms and ensemble methods including Support Vector Machines, Logistic Regression, Multi-layered Perceptrons, and Decision Trees have also been applied to this task [5,6,8].

Moving away from clickstream data, video views and quiz behaviour have been identified as significant factors contributing to dropout prediction [4]. In other work, course information and the type of interaction (solving problems v.s. watching videos etc.) were found to be important in an analysis based upon GRUs with attention weightings [10].

The JODIE model (see Table 2) utilises RNNs to learn and update embeddings that represent individual interactions between users and actions [7]. The actions and users each have their own RNNs to generate separate static and dynamic embeddings. The embeddings dynamically change over time, capturing the temporal aspect in a statically sized graph. These two RNNs are used together for the user embeddings to update the item embedding and vice versa. The JODIE model alters the embeddings significantly after longer periods of time, implicitly assuming that actions taken closer together have smaller impact.

6   Related Work in Graph Learning

Searching the Web of Science and IEEE databases using the terms "Convolution AND (3d OR three\$dimension*) AND (time OR temporal) AND graph AND predict* AND network\$" returned 18 papers. Of these, there were 5 relevant papers using temporal GNs [1–3,9,19], though they were focused on node and edge detection. To the best of our knowledge, this is the first work to develop temporal graph-based 3D CNN models for graph-level classification.

At the time of writing (12th May 2022), Kumar et al. had 200 citations of their paper [7]. To observe if any other researchers had used the ACT MOOC dataset processed by Kumar et al. (7,047 users), we screened these 200 papers and found 11 papers performing dropout/node prediction tasks.

Four of these utilised RNN components in their model architectures to process time, e.g. [14]. The others used graph models, all based on a node/edge classification formulation of the task. Wang et al. used temporal graph networks (TGNs) with dual message passing mechanisms (TGN + MeTA), to augment data and retain semantics for edge-level prediction and node classification [16]. These messaging passing techniques involve memory translation and cross-level propagation, to adapt the model with temporal and topological features to ignore noise more effectively. This increases the previously obtained AUC scores by 1.7% [12], with no cost to efficiency and reducing the overfitting that occurs due to noisy data. Other models tweak neighbourhood propagation techniques using temporal information e.g. [17,23]. Zhang et al. use ordinary differential equations and GNs to observe model changes over time and for information propagation [21]. By contrast, our TG-CNN approach makes use of a novel 3-tensor structure, storing the temporal graphs in a sparse and intuitive format, which is easily amenable to feature extraction using convolutions for graph classification.

7   Study Limitations and Future Work

The ACT-MOOC dataset in this project provides clickstream events as numerical labels. Events/clickstreams descriptions are not provided. Therefore, reasoning for dropout cannot be explained.

Limits on the amount of computation time available meant we were unable to perform a full grid-search of the TG-CNN hyperparameters, and it is likely that a more optimal configuration could be found. Nevertheless, this approach improves upon previous work and can be adapted to a range of different graph classification tasks. In future work we aim to incorporate attention mechanisms into this approach, to enable further trust in the model and explain why certain predictions may have occurred [10].

The variant of the model including the time dilation factor γ as a trainable parameter performed the poorest (Table 1). The reason behind this is unclear, though the additional complexity modelled by the time dilation will increase the difficulty of the underlying optimisation problem; it is possible that a more optimal hyperparameter set could be found to improve this performance.

8   Conclusions

We propose a novel model for the classification of temporal graphs, using student online course dropout data to develop and test the method. Our approach provides a unique formulation of this problem, compared to previous formulations of node/edge prediction tasks. This method improves upon current state-of-the-art models in terms of AUC score, and our approach has a number of other benefits in terms of memory utilisation and parallel processing compared with other approaches. In future work we aim to extend this approach further, incorporating attention mechanisms to improve explainability of the model output.

Acknowledgements. This work made use of the facilities of the N8 Centre of Excellence in Computationally Intensive Research provided and funded by the N8 research partnership and EPSRC (Grant No. EP/T022167/1). ZH is supported through funding by the EPSRC (Grant No. EP/S024336/1).

References

1. Chen, J., Liao, S., Hou, J., Wang, K., Wen, J.: GST-GCN: a geographic-semantic-temporal graph convolutional network for context-aware traffic flow prediction on graph sequences. In: 2020 IEEE International Conference on Systems, Man, and Cybernetics (SMC), vol. 2020, pp. 1604–1609. IEEE, October 2020. https://doi.org/10.1109/SMC42975.2020.9282828, https://ieeexplore.ieee.org/document/9282828/

2. Cheng, D., Wang, X., Zhang, Y., Zhang, L.: Graph neural network for fraud detection via spatial-temporal attention. IEEE Trans. Knowl. Data Eng. **14**(8), 1 (2020). https://doi.org/10.1109/TKDE.2020.3025588, https://ieeexplore.ieee.org/document/9204584/

3. Doosti, B., Naha, S., Mirbagheri, M., Crandall, D.: HOPE-Net: a graph-based model for hand-object pose estimation. In: Proceedings of the IEEE Computer Society Conference on Computer Vision and Pattern Recognition, pp. 6607–6616, March 2020. https://arxiv.org/abs/2004.00060v1

4. Fotso, J.E.M., Batchakui, B., Nkambou, R., Okereke, G.: Algorithms for the development of deep learning models for classification and prediction of behaviour in MOOCS. In: Proceedings of 2020 IEEE Learning With MOOCS, LWMOOCS 2020, pp. 180–184 (2020). https://doi.org/10.1109/LWMOOCS50143.2020.9234363

5. Haiyang, L., Wang, Z., Benachour, P., Tubman, P.: A time series classification method for behaviour-based dropout prediction. In: Proceedings - IEEE 18th International Conference on Advanced Learning Technologies, ICALT 2018, pp. 191–195 (2018). https://doi.org/10.1109/ICALT.2018.00052

6. Hong, B., Wei, Z., Yang, Y.: Discovering learning behavior patterns to predict dropout in MOOC. In: 2017 12th International Conference on Computer Science and Education (ICCSE), pp. 700–704. No. ICCSE. IEEE, August 2017. https://doi.org/10.1109/ICCSE.2017.8085583, http://ieeexplore.ieee.org/document/8085583/

7. Kumar, S., Zhang, X., Leskovec, J.: Predicting dynamic embedding trajectory in temporal interaction networks. In: Proceedings of the 25th ACM SIGKDD International Conference on Knowledge Discovery & Data Mining, pp. 1269–1278. ACM, New York, USA, July 2019. https://doi.org/10.1145/3292500.3330895, https://dl.acm.org/doi/10.1145/3292500.3330895

8. Liu, K., Tatinati, S., Khong, A.W.: A weighted feature extraction technique based on temporal accumulation of learner behavior features for early prediction of dropouts. In: Proceedings of 2020 IEEE International Conference on Teaching, Assessment, and Learning for Engineering, TALE 2020, pp. 295–302 (2020). https://doi.org/10.1109/TALE48869.2020.9368317

9. Pham Van, L.L., Tran, Q.B., Pham, T.L., Tran, Q.L.: Node-aware convolution in graph neural networks for predicting molecular properties. In: Proceedings - 2020 12th International Conference on Knowledge and Systems Engineering, KSE 2020, pp. 120–125, November 2020. https://doi.org/10.1109/KSE50997.2020.9287744

10. Pulikottil, S.C., Gupta, M.: ONet - a temporal meta embedding network for MOOC dropout prediction. In: Proceedings - 2020 IEEE International Conference on Big Data, Big Data 2020, pp. 5209–5217 (2020). https://doi.org/10.1109/BigData50022.2020.9378001

11. Ren, Y., Huang, S., Zhou, Y.: Deep learning and integrated learning for predicting student's withdrawal behavior in MOOC. In: Proceedings - 2021 2nd International Conference on Education, Knowledge and Information Management, ICEKIM 2021, pp. 81–84 (2021). https://doi.org/10.1109/ICEKIM52309.2021.00026

12. Rossi, E., Chamberlain, B., Frasca, F., Eynard, D., Monti, F., Bronstein, M.: Temporal Graph Networks for Deep Learning on Dynamic Graphs, pp. 1–16, June 2020. arXiv:abs/2006.10637

13. Sun, Z., Harit, A., Yu, J., Cristea, A.I., Shi, L.: A brief survey of deep learning approaches for learning analytics on MOOCs, vol. 12677 LNCS. Springer International Publishing (2021). https://doi.org/10.1007/978-3-030-80421-3_4

14. Wang, A.Z., Ying, R., Li, P., Rao, N., Subbian, K., Leskovec, J.: Bipartite dynamic representations for abuse detection. In: Proceedings of the ACM SIGKDD International Conference on Knowledge Discovery and Data Mining, pp. 3638–3648 (2021). https://doi.org/10.1145/3447548.3467141

15. Wang, L., Wang, H.: Learning behavior analysis and dropout rate prediction based on MOOCs data. In: Proceedings - 10th International Conference on Information Technology in Medicine and Education, ITME 2019, pp. 419–423 (2019). https://doi.org/10.1109/ITME.2019.00100

16. Wang, Y., Cai, Y., Liang, Y., Ding, H.: Adaptive Data Augmentation on Temporal Graphs. NeurIPS (NeurIPS), pp. 1–13 (2021)

17. Wang, Y., Cai, Y., Liang, Y., Ding, H., Wang, C., Hooi, B.: Time-aware neighbor sampling for temporal graph networks. arXiv preprint pp. 1–12 (2021). http://arxiv.org/abs/2112.09845

18. Wen, Y., Tian, Y., Wen, B., Zhou, Q., Cai, G., Liu, S.: Consideration of the local correlation of learning behaviors to predict dropouts from MOOCs. Tsinghua Sci. Technol. **25**(3), 336–347 (2020). https://doi.org/10.26599/TST.2019.9010013

19. Wu, Z., Wang, M., Wang, J., Zhang, W., Fang, M., Xu, T.: DeepWORD: a GCN-based approach for owner-member relationship detection in autonomous driving. Institute of Electrical and Electronics Engineers (IEEE), March 2021. https://arxiv.org/abs/2103.16099v2

20. Zhang, Y., Chang, L., Liu, T.: MOOCs dropout prediction based on hybrid deep neural network. In: Proceedings - 2020 International Conference on Cyber-Enabled Distributed Computing and Knowledge Discovery, CyberC 2020, pp. 197–203 (2020). https://doi.org/10.1109/CyberC49757.2020.00039

21. Zhang, Y., Xiong, Y., Li, D., Shan, C., Ren, K., Zhu, Y.: CoPE: modeling continuous propagation and evolution on interaction graph. In: International Conference on Information and Knowledge Management, Proceedings, pp. 2627–2636 (2021). https://doi.org/10.1145/3459637.3482419

22. Zheng, Y., Gao, Z., Wang, Y., Fu, Q.: MOOC dropout prediction using FWTS-CNN model based on fused feature weighting and time series. IEEE Access **8**, 225324–225335 (2020). https://doi.org/10.1109/ACCESS.2020.3045157

23. Zhou, Y., Luo, S., Pan, L., Liu, L., Song, D.: Continuous temporal network embedding by modeling neighborhood propagation process. Knowl.-Based Syst. **239**, 107998 (2022). https://doi.org/10.1016/j.knosys.2021.107998

Graph Convolutional Networks Using Node Addition and Edge Reweighting

Wen-Yu Lee[✉]

GREE, Inc., Tokyo, Japan
wenyu.lee@gree.net

Abstract. Graph convolutional networks (GCNs) provide a promising way to explore datasets that have graph structures in nature. The presence of corrupted or incomplete graphs, however, dramatically decreases the performance of GCNs. To improve the performance, recent works on GCNs reweighted edges or added missing edges on the given graphs. On top of that, this paper further explores the domain of node addition. This paper presents a simple but effective extension of GCNs by combining node addition and edge reweighting. Node addition adds new nodes and edges as communication centers to the original graphs. By doing so, nodes can share information together for efficient inference and noise reduction. Moreover, edge reweighting re-distributes the weights of edges, and even removes noisy edges considering local structures of graphs for performance improvement. Based on four publicly available datasets, the experimental results demonstrate that the proposed approach can achieve better performance than four state-of-the-art approaches.

Keywords: Graph convolutional network · Semi-supervised learning · Node addition · Edge reweighting

1 Introduction

For a long time, convolutional neural networks (CNNs) have been widely used in various applications, such as image classification [4,14], image retrieval [17, 19], semantic segmentation [5,9], and clothing recommendation [10,24]. Classical CNNs focus on the problems where a data instance can be represented in a regular grid structure [1], *e.g.*, an image. With a regular structure, filters can directly be applied to extract effective features for model generation. However, many problems involve irregular structures in nature, and these datasets are commonly modeled as irregular graph structures, such as social relation analysis. As a result, generalized CNNs have been rapidly developed for irregular graph structures from single-relational and even multi-relational data instances, see, *e.g.*, [1–3,8].

Earlier, Bruna *et al.* extended the classical convolution operator based on the spectral representation of graphs for generalized CNNs [1]. Extending the work in [1], Defferrard *et al.* proposed a computationally efficient method to perform convolution operations on graphs [2]. Subsequently, Kipf and Welling considered the classical graph-based semi-supervised learning (SSL) problems, where

© The Author(s), under exclusive license to Springer Nature Switzerland AG 2022
M. Ceci et al. (Eds.): ISMIS 2022, LNAI 13515, pp. 368–377, 2022.
https://doi.org/10.1007/978-3-031-16564-1_35

the objective is to predict labels for unlabeled nodes based on labeled nodes, see *e.g.*, [25, 26]. They then developed graph convolutional networks (GCNs) for graph-based SSL problems [8]. Further, Veličković *et al.* added self-attention layers to reweight edges of graphs [20]. Jiang *et al.* then showed performance improvement by combining Kipf and Welling's GCNs and their edge reweighting method [6]. Recently, Rong *et al.* studied edge removal and showed the effectiveness on preventing over-smoothing [21]. Yu *et al.* presented graph-revised convolutional networks (GRCNs), which are capable of adding new edges and reweighting edges [23].

While most works resorted to edge-based refinement, this paper further explores the domain of node addition. Node addition allows nodes with similar features to share information together and reduce noisy information. This paper further considers edge reweighting so as to determine proper weights for edges adjacent to the added nodes, and remove noisy edges. Overall, this paper presents a simple but effective extension of GCNs in [8] by node addition and edge reweighting, for graph-based SSL problems. Compared to [6, 20], node addition considers the addition of new nodes and new edges to original graphs, while the two works focused on reweighting the edges existing in the original graphs. In contrast to [23], the proposed approach further considers the addition of new nodes and the removal of noisy edges. Overall, this paper has three main contributions as follows:

- To the best of our knowledge, this paper presents the first work on adding new nodes for graph-based CNNs on SSL problems.
- This paper presents a new method to reweight edges and even to remove noisy edges of a given graph. Besides, the method benefits node addition by determining proper weights for edges adjacent to the new nodes.
- We conduct experiments on four datasets to validate the effectiveness of the proposed approach on node classification.

The remainder of this paper is organized as follows. Section 2 reviews the GCN method for graph-based SSL problems. Section 3 details the proposed approach. Section 4 evaluates the performance of the proposed approach. Section 5 concludes this paper.

2 Background

This section briefly reviews the SSL of using the GCN method in [8]. Let $G(V, E)$ be a graph with nodes V and edges $E \subseteq V \times V$. An adjacency matrix $A \in \mathbb{R}^{|V| \times |V|}$ provides a representation of whether pairs of nodes in G contain edges connecting them, where $|V|$ is the number of nodes of graph G. Typically, $A_{ij} = 1$ if the nodes i and j are adjacent, and $A_{ij} = 0$ otherwise. We are given a feature matrix $X = (x_1, x_2, ..., x_n) \in \mathbb{R}^{n \times p}$ for n instances, where x_i is the feature vector of instance i, and p is the dimension of a feature vector. For SSL, each node of graph G is associated with the feature vector of an instance, and thus, $|V| = n$. The adjacency matrix, A, is associated with the relationship between pairs of the instances, *e.g.*, similarity. We let $Y \in \mathbb{R}^{n \times c}$ be a label matrix and L be the set of nodes with labels, where $Y_{ij} = 1$ if $i \in L$ and the label of x_i is j, and $Y_{ij} = 0$ otherwise.

Given X, A, Y, and L, as the inputs for SSL, an r-layer GCN method performs layer-wise propagation as follows.

$$H^{(u+1)} = \sigma(\tilde{D}^{-\frac{1}{2}}\tilde{A}\tilde{D}^{-\frac{1}{2}}H^{(u)}W^{(u)}), \tag{1}$$

where $H^{(0)} = X$, $u = 0, 1, ..., (r-1)$ is a layer index, $\tilde{A} = A + I$ means to add a self-loop of every node, $I \in \mathbb{R}^{n \times n}$ is the identify matrix, $\tilde{D} \in \mathbb{R}^{n \times n}$ is a diagonal matrix with $\tilde{D}_{ii} = \sum_{j=1}^{n} \tilde{A}_{ij}$, $W^{(u)}$ is the weight matrix that is going to be learned for the u-th layer, and $\sigma(.)$ is an activation function, *e.g.*, ReLU(.). It is worth mentioning that $\tilde{D}^{-\frac{1}{2}}\tilde{A}\tilde{D}^{-\frac{1}{2}}$ is a symmetric normalization for $\tilde{A}$. If we let $\hat{A} = \tilde{D}^{-\frac{1}{2}}\tilde{A}\tilde{D}^{-\frac{1}{2}}$, then $\hat{A}_{ij}$ can be viewed as the weight of edge that connects the nodes associating to x_i and x_j of graph G. Note that the output of the propagation is defined as,

$$Z = \text{softmax}(H^{(r)}W^{(r)}), \tag{2}$$

where $Z \in \mathbb{R}^{n \times c}$. Row Z_i refers to the prediction of the node associating to x_i for the c classes. Finally, the GCN method defines the loss function as,

$$\zeta_{\text{pred}} = -\sum_{i \in L}\sum_{j=1}^{c}(Y_{ij}\ln Z_{ij}), \tag{3}$$

to measure how good or bad the model does.

3   Proposed Approach

Figure 1 outlines the proposed approach. We first group the given nodes into several clusters, followed by adding a new node for each cluster. In each time we add a new node into a cluster, we also connect the new node to the nodes in the same cluster by adding new edges. More details about the node addition will be presented in Sect. 3.1. After adding the new nodes and edges, we will obtain a new graph. To combine edge reweighting with the GCN method, we add a parameter that will be learned for each node of the new graph, and then modify the GCN method partially. More details about edge reweighting will be presented in Sect. 3.2. Subsequently, the resultant network is used to produce a classification result. To make the performance less sensitive to clustering results, we will repeat the above process several times, and take the average of their classification results as the final results.

3.1   Node Addition

In real-world applications, inputs for SSL might be incomplete or contain noisy data. The goal of node addition is to reduce the influence of the imperfect inputs.

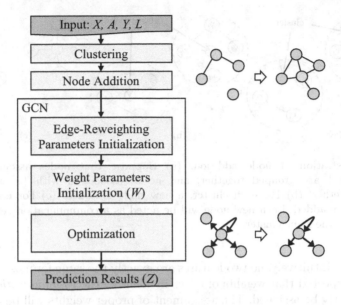

Fig. 1. Overall flow of the proposed approach.

More specifically, the motivation of node addition is twofold: (1) it is expected that nodes with similar features can share information together and also reduce noise by averaging, and (2) adding edges for isolated nodes or even ordinary nodes is capable of increasing the performance of SSL. Considering the two, we intend to provide a *communication center* that connects nodes with similar features, and helps information sharing.

For node addition, we group nodes of graph G into clusters, based on the features associated with the nodes, so that nodes with similar features can be assigned to the same cluster. For each cluster, we then add a node in it, where the feature of the new node is set to be the average of the features of the original nodes in the cluster. That is, the new node is placed at the center of the cluster. In each time we add a new node for a cluster, we also add an edge between the new node and each of the original nodes in the cluster. Figure 2 illustrates the idea, where seven nodes that are associated with seven instances, say $\{x_1, x_2, ..., x_7\}$, respectively.

So far, we have built a *communication center* by adding a new node and some edges, for each cluster of nodes with similar features. Note that the number of nodes added and the number of edges added in graph G are equal to the number of clusters and the total number of nodes in graph G, respectively. Because node addition changes graph G, we will use the superscript, $'$, for the notations associated with the new graph. For example, $G'(V', E')$ is the new graph, $X' \in \mathbb{R}^{n' \times p}$ is the new feature matrix, and $A' \in \mathbb{R}^{n' \times n'}$ is the new adjacency matrix, where n' is the number of nodes in graph G', and obviously $(n' - n)$ is the number of nodes added into G. Note that entries of A' are either

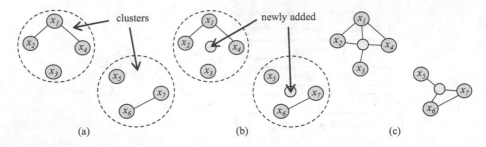

Fig. 2. Illustration of node addition. (a) Suppose that nodes associated with $\{x_1, x_2, x_3, x_4\}$ are grouped together, and nodes associated with $\{x_5, x_6, x_7\}$ are grouped together. (b) For each cluster, a new node is added. (c) For each cluster, new edges are added. Each new node will be acted as a communication center of the nodes within the same cluster.

zeros or one. Intuitively, no two features are exactly the same for most cases, and thus it is expected that weights of edges incident to the *communication centers* could properly be assigned. The assignment of proper weights will be covered in the scope of our edge reweighting method. Later in Sect. 3.2, we will generate a weighted adjacency matrix from A', and then do edge reweighting.

For practical implementation, we use k-means clustering, and we apply a simple heuristic as follows. We randomly pick m ($m = 5$ in this paper) distinct numbers from 0 to $n/10$ as the candidates of the k-value, where we set $n/10$ as a bound because we do not intend to generate too many small-size clusters. After clustering, there can be a few nodes far from the center of each cluster. Mostly, it is beneficial to refine the clustering result. We thus calculate the average distance between each point in a cluster to the cluster center. For each cluster, we then remove nodes from the cluster if their distances to the cluster center are greater than the average distance of the cluster. Finally, we update the cluster center based on the remaining nodes of each cluster. Note that we do not remove nodes from the graph. We only refine the clustering result. In addition, node addition places the new nodes on the updated centers, and adds the new edges only for the remaining nodes.

For each of the m candidates of the k-value, we will go through the GCN optimization flow shown in Fig. 1. Eventually, we can get m different results. Because it is difficult to find a proper k-value for a given graph, we take the average of the m results for the prediction result.

3.2 Edge Reweighting

This section reviews layer-wise propagation first, and details the implementation of edge reweighting afterward. Based on graph G', Eq. (1) can directly be rewritten as,

$$H'^{(u+1)} = \sigma(\tilde{D}'^{-\frac{1}{2}} \tilde{A}' \tilde{D}'^{-\frac{1}{2}} H'^{(u)} W'^{(u)}). \tag{4}$$

In Eq. (4), we can see that the operation, $\hat{A}' = \tilde{D}'^{-\frac{1}{2}}\tilde{A}'\tilde{D}'^{-\frac{1}{2}}$, normalizes $\tilde{A}'$ in a symmetric way. More specifically, the operation assigns a weight for each edge considering the degrees of nodes in graph G'. Generally, the operation is used when edges of graph G' are undirected, $i.e.$, A' is symmetric. $\hat{A}'$ will be symmetric if A' is symmetric. If edges of graph G' are directed, the operation can be replaced with $\hat{A}' = D'^{-1}\tilde{A}'$, so that each row in $\hat{A}'$ sums to one. That is, weights of edges pointing from each node sums to one. In many cases, $\hat{A}'$ will not be symmetric if edges of graph G' are directed.

Edge reweighting will view the given graph, $i.e.$, G', as a directed graph. Typically, an undirected graph can be converted into a directed graph by replacing the undirected edge between each pair of nodes with two directed edges in opposite direction. Practically, if edges of the initial graph, $i.e.$, graph G, are directed, we will use two directed edges to connect a new node to each of the original nodes in the same cluster for the stage of node addition. Then for Eq. (4), we will replace $\tilde{D}'^{-\frac{1}{2}}\tilde{A}'\tilde{D}'^{-\frac{1}{2}}$ with $D'^{-1}\tilde{A}'$. Note that if edges of the initial graph are undirected, we will not make any change for Eq. (4). It is worth mentioning that our edge reweighting method has two advantages. Firstly, we can consider only the edges pointing from (or to) a node at each time, and thus drastically reduce the complexity of reweighting. Secondly, our method can not only be applied to undirected graphs, but also directed graphs. Note that no matter whether edges of graph G (and thus, graph G') are directed or undirected, the methods introduced below can be applied and are exactly the same.

Given graph G', we create a vector $b = (b_1, b_2, ..., b_{n'})^T \in \mathbb{R}^{n' \times 1}$, as parameters to be learned. Each parameter is assigned to exactly a node of graph G'. The parameter of a node will be added to the weight of every edge pointing from the node. Formally, we generate an adjacency matrix, say $B \in \mathbb{R}^{n' \times n'}$, where

$$B_{ij} = \max(\hat{A}'_{ij} + b_i, 0), \forall i, (i,j) \in E'. \tag{5}$$

Note that b_i can be negative, and $\max(.)$ forces negative values to be zero. If B_{ij} equals zero, there is no edge pointing from node i to node j. That is, some edges can be removed if b_i is negative. We then do normalization by $\hat{B} = D'^{-1}B$ so that weights of edges points from each node sums to one. If b_i is positive and b_i is much greater than any of $\hat{A}'_{ij}$ with $(i,j) \in E'$, normalization can make all of the values of $\hat{B}_{ij}$ with $(i,j) \in E'$ be almost the same. That is, parameters $\{b_1, b_2, ..., b_{n'}\}$ can be used to reduce the difference of edge weights or remove some edges of graph G'.

The resultant layer-wise propagation is as follows:

$$H'^{(u+1)} = \sigma(\hat{B}H'^{(u)}W'^{(u)}). \tag{6}$$

Similar to [6, 12], we then define the loss function used to optimize the parameters as,

$$\zeta_{\text{graph}} = (1 - \lambda) \sum_{i,j=1}^{n'} \left(\|x_i - x_j\|_2^2 \hat{B}_{ij} \right) + \lambda \sum_{i,j=1}^{n'} \hat{B}_{ij}^2, \tag{7}$$

Table 1. Statistics of four publicly available datasets.

Dataset	#Nodes	#Features	#Edges	#Classes
PubMed	19,717	500	44,324	3
CS	18,333	6,805	81,894	15
Computers	13,752	767	245,861	10
Photo	7,650	745	119,081	8

Table 2. Comparison the accuracy (%) of the GCN, GAT, GLCN, GRCN, and our models, on the datasets.

	PubMed	CS	Computers	Photo
GCN	73.14 ± 3.53	90.95 ± 0.29	80.75 ± 1.96	89.56 ± 1.89
GAT	74.22 ± 3.66	89.73 ± 0.37	NA*	NA*
GLCN	73.10 ± 3.44	90.02 ± 0.30	79.15 ± 1.67	89.27 ± 1.35
GRCN	73.18 ± 3.06	90.85 ± 0.35	81.22 ± 2.01	89.71 ± 1.54
Ours (GCN)	74.38 ± 3.30	91.85 ± 0.45	82.39 ± 2.09	90.52 ± 1.44
Ours (GRCN)	74.28 ± 3.18	92.48 ± 0.34	82.63 ± 1.89	90.87 ± 1.32

* The results of the GAT model on the datasets of Computers and Photo are not available (NA) due to a known issue for the sparse version [16]. The dense version on any of the two, however, requires more memory than is available in our environment, *i.e.*, 252GB.

where the former encourages nodes with larger distance in features to have smaller weights, the latter tries to remove noisy edges, and $1 \geq \lambda \geq 0$ is a constant used to control the relative importance between the two terms. Finally, the loss function of our approach is set to be $\zeta_{\text{pred}} + \beta\zeta_{\text{graph}}$, where $\beta \geq 0$ is also a constant used to control the relative importance. Empirically, λ and β are set to be 0.9 and 0.1, respectively.

4 Experiments

We implemented the proposed approach based on PyTorch [13] and scikit-learn [15]. For comparative studies, we evaluated the performance of (1) the GCN model [8], (2) the GAT model [20], (3) the GLCN model [6], (4) the GRCN model [23], (5) our extension on the GCN model, and (6) our extension on the GRCN model. Note that all of the models used the same optimizer (*i.e.*, Adam [7]), learning rates, weight decays, and the number of hidden units, based on the settings of GRCN. All of them were also based on PyTorch. It is worth mentioning that the GLCN model was implemented by ourselves, because we did

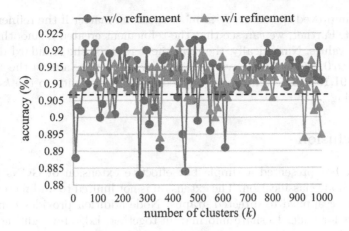

Fig. 3. Case study of refining clustering results or not to our extension on GCNs on the Photo dataset. Note that the dotted line shows the accuracy of a baseline, where no clustering is used (*i.e.*, $k = 0$, for k-means clustering).

not find its PyTorch implementation. For evaluation, we conducted experiments based on four publicly available datasets from [11] and [18]. The statistics of the datasets are shown in Table 1.

In Table 1, column "#Nodes" lists the number of nodes, "#Features" the dimension of each feature vector, "#Edges" the number of edges, and "#Classes" the number of classes for classification. The preparation of the datasets is the same as [22], where 20 instances of each class were used for training data. Overall, there were 500 and 1,000 instances used for validation data and testing data, respectively. There were 20 and 30 instances of each class used for training data and validation data, respectively. For testing data, the classes and instances were first removed if the number of instances of a class is smaller than 50. The remaining data were used as the testing data.

Table 2 shows the experimental results. For each dataset, we reported the average results over five runs with random splits on training, validation, and testing data (the data numbers were kept the same). As can be seen, our extensions achieved better performance than the other models. Based on the results, we can see that adding edges on existing graphs (*i.e.*, GRCN and Ours) is helpful to achieve better performance than the others in most cases. Consider our extensions on GCNs and GRCNs. We noticed that combining the edges added by GRCNs and the edges by our approach mostly improved the performance. However, some edges might become noisy for accuracy, see, the result on the PubMed dataset.

Remind that in Sect. 3.1, we not only refine the clustering results, but also take the average of several optimized results, so as to make our performance less sensitive to the quality and the k-values of k-means clustering. Figure 3 shows the result of a case study that compares our extension on GCNs of refining clustering results or not on the Photo dataset. As can be seen, node addition by clustering improved the accuracy in most cases, no matter if the refinement was used or not. Further, we can see that the refinement created a smoothing effect for most k-values. Numerically, the refinement reduced the standard deviations from 0.008 to 0.005, while the average accuracy values were almost the same, *i.e.*, 91.11% to 91.10%. The refinement did reduce the sensitivity to the k-values of the clustering.

5   Conclusion

This paper has presented a simple but effective extension of GCNs for semi-supervised node classification. The extension is not limited to undirected graphs; it can also be applied to directed graphs. Node addition provides communication centers for nodes to share information together. Edge reweighting not only reweights edges, but also removes noisy edges for high performance. Future works include the node removal and dynamic graph modification for GCNs.

References

1. Bruna, J., Zaremba, W., Szlam, A., LeCun, Y.: Spectral networks and deep locally connected networks on graphs. In: Proceedings of International Conference on Learning Representations (2014)
2. Defferrard, M., Bresson, X., Vandergheynst, P.: Convolutional neural networks on graphs with fast localized spectral filtering. In: Proceedings of International Conference on Neural Information Processing Systems, pp. 3844–3852 (2016)
3. Duvenaud, D., et al.: Convolutional networks on graphs for learning molecular fingerprints. In: Proceedings of International Conference on Neural Information Processing Systems, pp. 2224–2232 (2015)
4. He, K., Zhang, X., Ren, S., Sun, J.: Deep residual learning for image recognition. In: Proceedings of the IEEE Conference on Computer Vision and Pattern Recognition, pp. 770–778 (2016)
5. Jégou, S., Drozdzal, M., Vázquez, D., Romero, A., Bengio, Y.: The one hundred layers tiramisu: fully convolutional DenseNets for semantic segmentation. In: Proceedings of International Workshop on Computer Vision in Vehicle Technology (2017)
6. Jiang, B., Zhang, Z., Lin, D., Tang, J., Luo, B.: Semi-supervised learning with graph learning-convolutional networks. In: Proceedings of IEEE/CVF Conference on Computer Vision and Pattern Recognition, pp. 11313–11320 (2019)
7. Kingma, D.P., Ba, J.: Adam: a method for stochastic optimization. In: arXiv preprint arXiv:1412.6980 (2014)
8. Kipf, T.N., Welling, M.: Semi-supervised classification with graph convolutional networks. In: Proceedings of International Conference on Learning Representations (2017)
9. Liu, J., Zhou, Q., Qiang, Y., Kang, B., Wu, X., Zheng, B.: FDDWNet: a lightweight convolutional neural network for real-time semantic segmentation. In: Proceedings of IEEE International Conference on Acoustics, Speech and Signal Processing, pp. 2373–2377 (2020)

10. Manandhar, D., Yap, K.H., Bastan, M., Heng, Z.: Brand-aware fashion clothing search using CNN feature encoding and re-ranking. In: Proceedings of the IEEE International Symposium on Circuits and Systems, pp. 1–5 (2018)
11. Namata, G., London, B., Getoor, L., Huang, B.: Query-driven active surveying for collective classification. In: Proceedings of International Workshop on Mining and Learning with Graphs (2012)
12. Nie, F., Wang, X., Huang, H.: Clustering and projected clustering with adaptive neighbors. In: Proceedings of the ACM SIGKDD International Conference on Knowledge Discovery and Data Mining, pp. 977–986 (2014)
13. Paszke, A., et al.: Automatic differentiation in pytorch. In: Proceedings of NIPS Workshop on Autodiff (2017)
14. Pedersen, M., Christiansen, H., Azawi, N.H.: Efficient and precise classification of CT scannings of renal tumors using convolutional neural networks. In: Proceedings of Foundations of Intelligent Systems: 25th International Symposium, pp. 440–447 (2020)
15. Pedregosa, F., et al.: Scikit-learn: machine learning in python. J. Mach. Learn. Res. **12**(85), 2825–2830 (2011)
16. AssertionError in assert not torch.isnan(h_prime).any(). https://github.com/Diego999/pyGAT/issues/11 (2018)
17. Radenović, F., Tolias, G., Chum, O.: Fine-tuning CNN image retrieval with no human annotation. IEEE Trans. Pattern Anal. Mach. Intell. **41**(7), 1655–1668 (2019)
18. Shchur, O., Mumme, M., Bojchevski, A., Günnemann, S.: Pitfalls of graph neural network evaluation. In: arXiv preprint arXiv:1811.05868 (2018)
19. Valem, L.P., Pedronette, D.C.G.: A denoising convolutional neural network for self-supervised rank effectiveness estimation on image retrieval. In: Proceedings of the International Conference on Multimedia Retrieval, pp. 294–302 (2021)
20. Veličković, P., Cucurull, G., Casanova, A., Romero, A., Liò, P., Bengio, Y.: Graph attention networks. In: Proceedings of International Conference on Learning Representations (2018)
21. Rong, Y., Huang, W., Xu, T., Huang, J.: DropEdge: towards deep graph convolutional networks on node classification. In: Proceedings of International Conference on Learning Representations (2020)
22. Yang, Z., Cohen, W.W., Salakhutdinov, R.: Revisiting semi-supervised learning with graph embeddings. In: arXiv preprint arXiv:1603.08861 (2016)
23. Yu, D., Zhang, R., Jiang, Z., Wu, Y., Yang, Y.: Graph-revised convolutional network. In: Proceedings of European Conference on Machine Learning and Principles and Practice of Knowledge Discovery in Databases (2020)
24. Yu, W., Zhang, H., He, X., Chen, X., Xiong, L., Qin, Z.: Aesthetic-based clothing recommendation. In: Proceedings of the World Wide Web Conference, pp. 649–658 (2018)
25. Zhou, D., Bousquet, Q., Lal, T.N., Weston, J., Schölkopf, B.: Learning with local and global consistency. In: Proceedings of NIPS Foundation Advances in Neural Information Processing Systems, pp. 321–328 (2003)
26. Zhu, X., Ghahramani, Z., Lafferty, J.: Semi-supervised learning using Gaussian fields and harmonic functions. In: Proceedings of International Conference on Machine Learning, pp. 912–919 (2003)

Audio Super-Resolution via Vision Transformer

Simona Nisticò[ID], Luigi Palopoli[ID], and Adele Pia Romano[✉][ID]

DIMES Department, University of Calabria, Rende, Italy
{simona.nistico,luigi.palopoli,adelepia.romano}@dimes.unical.it

Abstract. Audio super-resolution refers to techniques that improve the quality of audio signals, usually by exploiting bandwidth extension methods, whereby audio enhancement is obtained by expanding the phase and the spectrogram of the input audio traces. These techniques are therefore much significant for all those cases where audio traces miss relevant parts of the audible spectrum. In many cases, the given input signal contains the low-band frequencies (the easiest to capture with low-quality recording instruments) whereas the high-band must be generated. In this paper, we illustrate a system for bandwidth extension that works on musical tracks and generates the high-band frequencies starting from the low-band ones. The system, called *ViT Super-resolution* (ViT-SR), features an architecture based on a Generative Adversarial Network and Vision Transformer model. Some experiments, which are accounted for in the paper, serve the purpose to prove the effectiveness of the presented approach. In particular, our purpose was to demonstrate that it is possible to faithfully reconstruct the high-band signal of an audio file having only its low-band spectrum available as the input, therewith including the usually difficult to synthetically generate harmonics associated with the input track which significantly contribute to the final perceived sound quality.

Keywords: Audio super-resolution · Music enhancement · Generative adversarial networks · Transformers · Vision transformer

1 Introduction

Bandwidth Extension (hereafter, BWE) is a process intended to improve audio tracks quality by means of the extension of their signal frequency range and, in [24], audio enhancement is claimed to have undergone significant recent breakthroughs in performances thanks to the exploitation of deep learning techniques [14,16].

It is common knowledge that audio signals encoding musical tracks are best appreciated if the full band of frequencies are available on tracks. In particular, a good low frequency response and the presence of high frequencies (including sound harmonics) are universally understood to be elements of high quality

© The Author(s), under exclusive license to Springer Nature Switzerland AG 2022
M. Ceci et al. (Eds.): ISMIS 2022, LNAI 13515, pp. 378–387, 2022.
https://doi.org/10.1007/978-3-031-16564-1_36

audio. In this context, the reference [23] offers an overview of the BWE techniques and an example of its application. Unfortunately, quite often, either signal sampling rate used to record musical tracks are rather low or a part of band frequency is absent because of some phenomenon, such as those determined by the use of inadequate or outdated recording technologies (for instance, this is the case of very old musical records). In this context, BWE technique application can be exploited in order to recover missing frequencies to enrich the sound and eventually obtain an improvement of musical tracks quality.

In this general context, bandwidth extension models have been recently proposed in the literature whose purpose is that of generating high-band frequency musical signals from low-frequency ones [2,3,17].

It is important to point out that only the BWE part of the audio enhancement process is dealt with here, while it is assumed that other procedures useful to ameliorate audio track quality (like, e.g., noise reduction) were already run on the input audio files.

Thus, in this paper, the *ViT-SR* system is presented that realizes an audio super-resolution model to be applied to music signals. In particular, we propose a neural system that generates high-band frequencies of a musical track from the low-band frequencies assuming that the former one are missing or largely corrupted in the input audio file. To this end, we shall both describe the details of our system architecture and provide evidence that demonstrate the good performances that our system is capable to attain.

Our approach to the problem of musical track enhancement relies on processing the audio signals in the frequency domain, whereby quality enhancement is obtained via a careful extension of phase and spectrogram data. In order to expand the spectrogram content, we have implemented a Generative Adversarial Network (GAN) that encapsulates a Vision Transformer (ViT) which allows us to enhance the musical track quality by exploiting an image super-resolution technique.

The rest of the paper is organized as follows. In the next section, we shall describe the system structure and related implementation details. After that, in Sect. 3, we shall present the results of the experimental campaign we have conducted. In the last section, we shall summarize the paper contribution and briefly discuss future research.

2   System Structure

This section illustrates the structure of our system and its functionality. The starting point of our work has been the architecture shown in Fig. 1, that was proposed in *Super-Resolution for Music Signals Using Generative Adversarial Networks* [4]. The general architecture of our system basically coincides with that of [4], with the important difference that the structure of the generator and discriminator networks we employ are new and different from the ones adopted in [4]. In particular, our idea is to exploit the embedding capability of transformers architecture to extract features optimally for super-resolution.

Moreover, the range of frequencies we altogether deal with is the entire audible spectrum 20 Hz to 20000 Hz, whereas in the paper [4] a restricted set of 257 spectral points are considered. To grossly illustrate the architecture, given an audio signal X, a Short-Time Fourier Transform (STFT) with *Hamming* window function [22] is applied on it to map the signal onto the frequency domain. From the frequency-domain representation, the dB-scale spectrogram and the phase are then computed. Subsequentially, the phase spectrum is extended by means of a flip operation, and, moreover, using a GAN, the spectrogram is further extended, as explained below in more detail. At the end of the extension process, the two parts of the signal are recombined together and, finally, the signal is reconstructed by Inverse Short-Time Fourier Transform (ISTFT).

The Generative Adversarial Network Architecture. The main component of the ViT-SR architecture is the GAN that extends the dB-scale spectrum. The general behaviour of the two subnetworks is as follows:

- the generator network is responsible to generate the high-band spectrogram (from 11 kHz to 22 kHz) starting from the low-band portion (0 Hz to 11 kHz) given in input;
- then, in order to implement the adversarial game, the discriminator network has to distinguish if its input data is real or synthetic.

All that given, and as already stated, the main contribution of this work relies in redesigning both the generator and the discriminator architectures in order to attain rather good functional performances when considered in the light of some well-known measures we shall illustrate in the following.

The generator and the discriminator architectures are shown in Fig. 2.

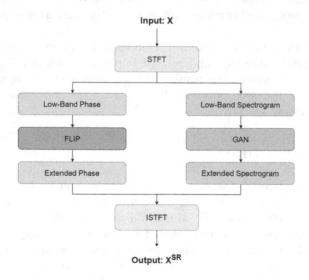

Fig. 1. ViT-SR architecture modified from [4]

The innovation introduced with ViT-SR is the use of a transformer architecture for realizing the BWE task. In fact, the transformer architectures have proven effective in a number of fields and their use on music tracks is also not new [9]. In the cited reference, the attention mechanism is readjusted for the music tracks data type. However, the structure we employ and the data flow it realizes is, to the best of our knowledge, new. In particular, we have chosen the Vision Transformer [13], that is the first transformer developed to analyze images rather than sequential data. Thus, while starting from sequential data (the music signals), we actually make the transformer work on images. This transformer architecture works on squared-images and, first of all, it breaks it into patches. Then, the patches are linearized and processed by the transformer encoder. An overview of the ViT model resulting from these modifications is depicted in Fig. 3. The structure of the transformer encoder is shown on the right of Fig. 3: the block "*Multi-Head Attention*" is the layer that implements the attention function of the transformer. In our architecture, we have modified the standard ViT architecture: in particular, after going through the encoder layers, data go through a *MultiLayer Perceptron Head* (MLP Head) [13] that returns the labels. In fact, the ViT was proposed as a classifier, but we use it to map data into a latent space, so we have removed the so called *MLP Head*.

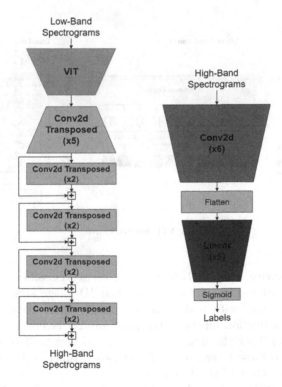

Fig. 2. Implemented Network. (left) Generative Network. (right) Discriminative Network.

Furthermore, we do not consider the feature in position zero, because also this is relative to the classification task.

As far as the generative network is concerned, we have implemented it as a *simil-autoencoder*, because we have a latent space to create a bottleneck in the dataflow, but the output data have a different shape as compared to the input (as it happens, instead, when using autoencoders). The ViT is used as the encoder module, so it is the first module in the generative network architecture: we use the *huggingface.co* [1] implementation with the following settings: 32 as patch size, 288 as hidden size and 8 as hidden layers. The other parameters are set to default values.

After that, there are five transpose 2D convolution layers that serve the purpose of bringing the dimensions from the latent space to the output space. The kernel size is set to 5 for all these layers except for the last, where the kernel size is 4. In the first layer we have set a padding value equal to 1, while in the others it is 2 and, furthermore, the stride value is set to 2 for all layers.

The other layers are associated with skip connections (indicated in Fig. 2 with arrows), an essential component of current state-of-the-art deep neural networks [11,21], to refine the data. In particular, we have inserted one skip connection every two layers. In this case, each layer has the same parameters: 3 as kernel size and 1 as padding value.

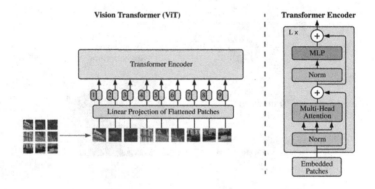

Fig. 3. ViT architecture

The discriminative network behaves as a binary classifier. It is composed by six 2D convolutional layers that reduce the 2D-image size, a flatten layer, and some linear layers with a decreasing number of neurons. In ouptut, we use the *sigmoid* activation function. For the Conv2D layers, we set kernel size to 5 and stride to 2 for the first three and kernel size to 3 to the last three. Moreover, between these layers, a *GeLU* activation function is used, that is the same activation function of the ViT model.

As the loss function, we have decided to use the *Binary Cross-Entropy* loss for the discriminator, because it is the typical loss for binary classification task, while for the generative network we have used the *Mean Squared Error* loss, which is well-suited to quantify the reconstruction error. Another important aspect to consider is the optimizer: for the discriminative network we have chosen *Adam* [12], very popular for image classification, and for generative network we have chosen *AdamW* [18], that is a variant of *Adam* which adopts the *weight decay* and was recommended by *huggingface.co* for working with vision transformers [8].

3   Experiments

Dataset. For the experimental campaign the *Free Music Archive* (FMA) dataset has been used. It was presented in [5] as an opensource dataset for *Music Information Retrieval* (MIR) tasks: according to the authors, there were not sufficiently large datasets of musical pieces, which made it more difficult to make progress in the analysis of this type of data and MIR tasks.

FMA consists of a large and diversified collection of musical tracks grouped by subgenres. For our tests, we have consider 70 of these groups. For each subgenre, we have used 70% of the data as the training set, 15% as the validation set, and 15% as the test set. Eventually, we have obtained a training set, a validation set, and a test set including, respectively, 2664, 572 and 571 musical tracks.

Metrics. To evaluate the performance of the GAN we have used two metrics: the *Signal-to-Noise Ratio* (SNR) [10] and the *Log-Spectral Distance* (LSD) [6], that are two common measures for evaluating audio quality. Both SNR and LSD are used to compare two spectrograms. To illustrate, the former measure provides information on the quantity of useful signals as compared to the noise present in an audio files. It can be expressed, as showed in Eq. (1), as the ratio of the power of a signal P_s (meaningful input) to the power of background noise P_n (meaningless or unwanted input).

$$SNR = \frac{P_s}{P_n} \ , \ 0 \leq SNR < \infty \tag{1}$$

Instead, the LSD is a measure to evaluate the difference between two log-scale spectrograms and provides information about how similar are the frequency contents and can be formulated as follows in Eq. (2), where $P(\omega)$ and $\hat{P}(\omega)$ represent two power spectrogram.

$$LSD = \sqrt{\frac{1}{2\pi} \int_{-\pi}^{\pi} \left[10 \frac{P(\omega)}{\hat{P}(\omega)} \right]^2 d\omega} \tag{2}$$

For a technique to attain good performances, it should be associated with the largest possible SNR values and the smallest possible LSD values.

Experiments Setup. In the FMA dataset, musical tracks have a sampling rate equal 44100 Hz. By exploiting the *librosa* library [19], we have computed the

STFT with 4096 as the window length and, then, the phase and dB-spectrogram have been generated. Both these components are matrices with 2049 rows. To apply our model, we have considered the first 1025 matrix rows as low-band and the following 1024 rows as high-band (the portion of the spectrum our system is asked to generate). We have trained our GAN for 100 epochs and saved the best weights according to the performances obtained over the validation set.

To evaluate the model, we have then carried out some experiments.

First of all, we have verified if there is a qualitative improvement obtained by adding the high frequencies to signals only containing the lower frequency band. To this end, we have computed SNR and LSD in order to compare the real spectrogram and synthetic spectrogram, to gauge how similar these two are and, moreover, we have compared the real spectrogram and the spectrogram with only the low-frequencies (reference measure values) in order to contrast the similarity measure values returned in the two cases.

The results of this first phase of experimentation are showed in Fig. 4 and 5. From the figures, it is possible to see that with ViT-SR extension, we obtained a clearly larger SNR value and a smaller LSD w.r.t. the reference measure values, which means that the application of the technique improved the quality of the artificially depleted signal by generating useful and faithful information.

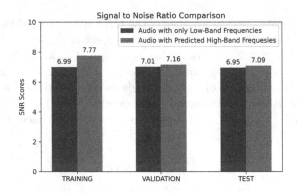

Fig. 4. SNR on Training Set, Validation Set and Test Set

The goal of the second part of experimentation was to compare our GAN ViT-SR model to other image super-resolution techniques presented in the literature. Indeed, some few systems were recently proposed in this context [4,15,25,26]. For all those systems, we were not in the conditions to conduct a direct comparison since neither their codes nor the datasets they used were available to us and, moreover, sometimes also the reference frequency ranges differ. Therefore, we opted for conducting a (admittedly, limited) comparison of the indirect kind and, to this purpose, we selected the system presented in the paper "Audio Super Resolution using Neural Networks" [15] as the approach presented there is the most similar to our own. Thus, in order to get at least a rough idea of

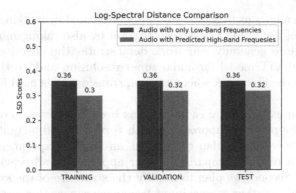

Fig. 5. LSD on Training Set, Validation Set and Test Set

the relative performances of our system as opposed to that presented in [15], we indirectly compared them on the basis of the performances that the systems separately demonstrated w.r.t. the Cubic Spline interpolation [20] method, a conventional signal processing baseline for audio super-resolution.

Table 1. Comparisons on SNR and LSD

	Metric	Cubic-spline interp	Proposed method	Improvement
Audio-SR-NN [15]	LSD	3.5	3.4	2.9%
	SNR	29.4	30.1	0.7 dB
ViT-SR	LSD	0.73	0.32	**56.2%**
	SNR	4.86	7.09	**2.23 dB**

The results obtained in this experiment are shown in the Table 1 (where the name of our "competitor" system is indicated as "Audio-SR-NN"). The table displays the values of both the SNR and LSD metrics. The values indicate that, for both metrics, our system seems to outperform Audio-SR-NN (considered the conditions charaterizing the test, this conclusion remains however merely indicative).

4   Conclusions

In this paper, we have presented a neural architecture for audio enhancement. The enhamcement is obtained by extending the input signal, which is assumed to miss the high half of the frequency spectrum, by generating those missing frequency signals in a faithful way. To do so, we exploit a GAN architecture that incorporates a ViT model. This architecture realizes a ViT Super-resolution approach which allows us to attain our application goal by working on the visual representation of the audio tracks spectrogram by also taking into account the signal phase. More generally, our work demostrates the effectiveness of GAN architecture and ViT model for audio super-resolution and, to the best of our knowledge, it is the first work where a vision transformer is used for this type of task.

Moreover, an experimental campaign has been conducted in order to assess the quality of the proposed approach, which turned out to be quite promising.

As for future work, we plan to conduct an ampler experimental campaign in order to get a reliable comparison of our approach to all previous proposals of similar kind. Second, we plan to modify the structure of the system so as to make it working on rather narrow input frequency ranges (for instance, the fundamental mid-range – with frequencies going 400 Hz to 1200 Hz) while remaining capable to faithfully reconstruct all missing frequency signals, which would make it suitable for audio restoration applications. Also, we plan to develop experiments devoted to assess the effectiveness of our approach when used to ameliorate a present but rather poor high-band in the audio signal.

For what concerns phase information, we will consider other solutions as the adoption of a GAN network and the adoption of Griffin-Lim algorithm [7], which do not use phase information for audio reconstruction.

References

1. https://huggingface.co/docs/transformers/index
2. Andreev, P., Alanov, A., Ivanov, O., Vetrov, D.: HiFi++: a unified framework for neural vocoding, bandwidth extension and speech enhancement. arXiv preprint arXiv:2203.13086 (2022)
3. Chen, X., Yang, J.: Speech bandwidth extension based on Wasserstein generative adversarial network. In: 2021 IEEE 21st International Conference on Communication Technology (ICCT), pp. 1356–1362. IEEE (2021)
4. Dai, J., Zhang, Y., Xie, P., Xu, X.: Super-resolution for music signals using generative adversarial networks. In: 2021 IEEE 4th International Conference on Big Data and Artificial Intelligence (BDAI), pp. 1–5. IEEE (2021)
5. Defferrard, M., Benzi, K., Vandergheynst, P., Bresson, X.: FMA: a dataset for music analysis. arXiv preprint arXiv:1612.01840 (2016)
6. Erell, A., Weintraub, M.: Estimation using log-spectral-distance criterion for noise-robust speech recognition. In: International Conference on Acoustics, Speech, and Signal Processing, pp. 853–856. IEEE (1990)
7. Griffin, D., Lim, J.: Signal estimation from modified short-time Fourier transform. IEEE Trans. Acoust. Speech Signal Process. **32**(2), 236–243 (1984)
8. Guo, M.H., et al.: Attention mechanisms in computer vision: a survey. arXiv preprint arXiv:2111.07624 (2021)
9. Huang, C.Z.A., et al.: Music transformer. arXiv preprint arXiv:1809.04281 (2018)
10. Johnson, D.H.: Signal-to-noise ratio. Scholarpedia **1**(12), 2088 (2006)
11. Kim, J., Englebienne, G., Truong, K.P., Evers, V.: Deep temporal models using identity skip-connections for speech emotion recognition. In: Proceedings of the 25th ACM International Conference on Multimedia, pp. 1006–1013 (2017)

12. Kingma, D.P., Ba, J.: Adam: a method for stochastic optimization. arXiv preprint arXiv:1412.6980 (2014)
13. Kolesnikov, A., et al.: An image is worth 16 × 16 words: transformers for image recognition at scale (2021)
14. Kuleshov, V., Enam, S.Z., Ermon, S.: Audio super resolution using neural networks. arXiv preprint arXiv:1708.00853 (2017)
15. Kuleshov, V., Enam, S.Z., Ermon, S.: Audio super resolution using neural networks. arXiv preprint arXiv:1708.00853 (2017)
16. Li, K., Lee, C.H.: A deep neural network approach to speech bandwidth expansion. In: 2015 IEEE International Conference on Acoustics, Speech and Signal Processing (ICASSP), pp. 4395–4399. IEEE (2015)
17. Liu, Y.: Recovery of lossy compressed music based on CNN super-resolution and GAN. In: 2021 IEEE 3rd International Conference on Frontiers Technology of Information and Computer (ICFTIC), pp. 623–629. IEEE (2021)
18. Loshchilov, I., Hutter, F.: Decoupled weight decay regularization. arXiv preprint arXiv:1711.05101 (2017)
19. McFee, B., et al.: librosa: audio and music signal analysis in python. In: Proceedings of the 14th Python in Science Conference, vol. 8, pp. 18–25. Citeseer (2015)
20. McKinley, S., Levine, M.: Cubic spline interpolation. Coll. Redwoods **45**(1), 1049–1060 (1998)
21. Oyedotun, O.K., Al Ismaeil, K., Aouada, D.: Why is everyone training very deep neural network with skip connections? IEEE Trans. Neural Netw. Learn. Syst., 1–15 (2022)
22. Podder, P., Khan, T.Z., Khan, M.H., Rahman, M.M.: Comparative performance analysis of hamming, hanning and blackman window. Int. J. Comput. Appl. **96**(18), 1–7 (2014)
23. Smaragdis, P., Raj, B.: Example-driven bandwidth expansion. In: 2007 IEEE Workshop on Applications of Signal Processing to Audio and Acoustics, pp. 135–138. IEEE (2007)
24. Su, J., Wang, Y., Finkelstein, A., Jin, Z.: Bandwidth extension is all you need. In: ICASSP 2021–2021 IEEE International Conference on Acoustics, Speech and Signal Processing (ICASSP), pp. 696–700. IEEE (2021)
25. Wang, H., Wang, D.: Time-frequency loss for CNN based speech super-resolution. In: ICASSP 2020–2020 IEEE International Conference on Acoustics, Speech and Signal Processing (ICASSP), pp. 861–865. IEEE (2020)
26. Wang, H., Wang, D.: Towards robust speech super-resolution. IEEE/ACM Trans. Audio Speech Lang. Process. **29**, 2058–2066 (2021)

Similarity Embedded Temporal Transformers: Enhancing Stock Predictions with Historically Similar Trends

Kenniy Olorunnimbe[✉] and Herna Viktor

School of Electrical Engineering and Computer Science, University of Ottawa, Ottawa, Canada
{molor068,hviktor}@uottawa.ca

Abstract. Price prediction is essential in financial market research, as it is often used as a primary component for trading strategy or portfolio management specialisations. As these strategies rely on more than one future prediction point, the accuracy of a multi-horizon forecast is very important. Classical models, such as autoregressive integrated moving average (ARIMA), are not very good at multi-horizon forecasting. Also, current approaches employing deep learning do not usually factor in the heteroscedasticity of financial market time series. We introduce the *similarity embedded temporal transformer* (SeTT) algorithm by extending the state-of-the-art temporal transformer architecture with time series forecasting and statistical principles. We employ similarity vectors generated from historical trends across different financial instruments that are used to adjust the weight of the temporal transformer model during the training process. We conducted independent experiments across two time frames with volatile extrapolation periods using 20 companies from the Dow Jones Industrial Average. By focusing on the historical windows that are most similar to the current window in the self-attention tuning process, SeTT outperformed both the classical financial models and the baseline temporal transformer model in terms of predictive performance.

Keywords: Deep learning · Financial price prediction · Temporal transformer · Stock market volatility · Multi-horizon forecast

1 Introduction

Making future predictions on the prices of financial instruments such as stocks, bonds, and futures using historical time series data is an important area of research and represents an integral component of automated trading and portfolio management systems [10,15]. Recently, temporal transformers, a sequential deep learning approach, have been shown to outperform classical financial models [3,15].

© The Author(s), under exclusive license to Springer Nature Switzerland AG 2022
M. Ceci et al. (Eds.): ISMIS 2022, LNAI 13515, pp. 388–398, 2022.
https://doi.org/10.1007/978-3-031-16564-1_37

However, a drawback of temporal transformers is that they do not consider financial market data. That is, current temporal transformer architectures do not fully address the non-constant nature of the volatility of data across different periods of time. They also do not utilise the fact that the volatility of a specific day is strongly related to the previous day [13].

The paper is organised as follows. Section 2 introduces background work, and Sect. 3 details our novel similarity-embedded temporal transformer architectures. Our experimental evaluation is presented in Sect. 4, and Sect. 5 concludes the paper.

2   Background

The transformer is a state-of-the-art architecture in sequential deep learning [14] that simplifies learning complexities by eschewing the ordered input dependencies using positional encoding. The architecture consists of an encoder and a decoder that learn the relationship between input and output sequences using multi-head attention, consisting of multiple *attention mechanisms* [17]. An attention mechanism consists of learnable weighted combinations of all elements in two sequences, allowing every position in one sequence to connect to every position in the other sequence. The transformer has attention mechanisms both between and within the encoder and decoder sequences, called self-attention or intra-attention [17]. The multi-head attention allows the decoder to search for the most relevant part of the encoder sequence based on the weights learned in the training process.

Although the transformer architecture uses positional encoding and self-attention to draw relationship data points in a sequence, it remains insensitive to its immediate locality, which may be problematic in time series data. Moreover, the default representation of the attention mechanism assumes homoscedasticity, that is, constant mean and variance in the attention search space [17]. The lack of locality bias is an important drawback of the transformer model for temporal data, with long-distance feature sequences purported to have equal weight as a local feature sequence [7]. To address the issue, Lim et al. [8] incorporated *locality enhancements* into a temporal transformer model. However, their work employs the default attention search mechanism by assuming that the historical sequence is independent and identically distributed (IID). Financial time series data are non-IID; the data distributions are not identical across the different historical horizons [13]. In this work, we address this shortcoming of earlier work by introducing our novel SeTT algorithm that makes the temporal transformer aware of the distribution of the domain-specific financial data sequences, as discussed next.

3   SeTT Architecture

A financial time series is modelled as finite historical targets $y_{t, t \in \{1, ..., T\}}$ available to the model, where y_t is the target at the current point in time, $y_{\tau, \tau \in \{1, ..., H\}}$

is the forecast horizon, and $y_l \leftarrow y_{i,i\in\{T-l+1,...,T\}}$ are the targets of the most recent time series of size l, that is, the current window. Our similarity embedded temporal transformer (SeTT) architecture introduces the concept of a sliding window to learn similar time series.

Our SeTT algorithm consists of four steps. (1) A current window and multiple historical windows are generated from the time series data. (2) The historical windows are individually compared with the current window. These comparisons return vectors of $1s$ if similar or $0s$ if dissimilar, each of size l. (3) The individual vectors from the comparison are combined into a similarity vector (4) The similarity vector is embedded into the temporal transformer to mute the dissimilar time series during training. Algorithm 1 depicts the pseudocode of our SeTT algorithm.

Algorithm 1. similarity embedded temporal transformer (SeTT)

Require:

1: $X_{t,t\in\{1,...,T\}}$ ▷ time series features of size T

2: $y_{t,t\in\{1,...,T\}}$ ▷ time series targets (Δprice) of size T

3: l ▷ current window size; $l < T$

Ensure: $\hat{y} \sim y_{\tau,\tau\in\{1,...,H\}}$ ▷ vector of H future quantile forecasts

4:

5:   $SimVec \leftarrow$ GENERATESIMVEC(y,T,l)

6:   $Q \leftarrow W_q X, K \leftarrow W_k X, V \leftarrow W_q X$ ▷ initialise attention

7:   $\mathcal{L}(y,W) = L_q(y,\hat{y},q)$ ▷ update W_q, W_k, W_v by minimising quantile loss

8: **repeat**

9:      A(Q, K, V) $\leftarrow$ softmax $\left(\frac{Q\{K \cdot SimVec\}^T}{\sqrt{d_k}} \right) V \cdot SimVec$

10:      $\hat{y} \leftarrow L_q(y,\hat{y},q)$

11: **until** stopping condition is met

12:

13: **procedure** GENERATESIMVEC(y,T,l)

14:      $y_l \leftarrow y_{\{T-l,...,T\}}$ ▷ current window

15:      $SimVec \leftarrow [[\underbrace{1}_{T\ times}]]$ ▷ initialise with T-sized vector of $1s$

16:      **for all** s in $0...(T-l+1)$ **do**   ▷ index of sliding historical windows

17:          $y_s \leftarrow y_{\{s,...,s+l\}}$

18:          $c \leftarrow$ COMPAREWIN(y_l, y_s)

19:          $\hat{y}_s \leftarrow [\underbrace{1}_{s\ times}, \underbrace{c}_{l\ times}, \underbrace{1}_{T-s\ times}]$ ▷ pad both sides of l-sized vector with $1s$

20:          $SimVec.append(\hat{y}_s)$

21:      **return** $matProd(SimVec)$ ▷ column-wise matrix multiplication

22:

23: **procedure** COMPAREWIN(y_l, y_s)

24:      **if** $S(y_l, y_s) > \alpha$ **then return** 1   ▷ $\alpha = 0.05$; S: similarity-test function

25:      **else return** 0

Historical Windows. The historical windows are foundational to our algorithm and are used to compute the similarity vector. At current time y_t, they are all the available windows, $y_M \leftarrow y_{i,i\in\{1,...,T-l\}}$, of length l not in the current window y_l. These windows are generated by sliding a vector of size l across the historical time frame of length M. The historical windows are generated from the time series targets in the first variant of our SeTT algorithm.

It is common to see frequent *runs*, that is, an uninterrupted sequence of positive or negative stock price changes [13]. We incorporated this into the second

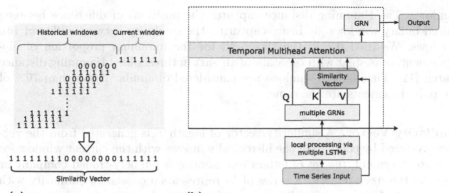

(a) Similarity vector of current and historical windows for each stock symbol.   **(b)** SeTT architecture using matrix of similarity vectors.

Fig. 1. Model architecture and illustration of similarity vectors.

variant called run-similarity embedded temporal transformer (r-SeTT) that uses the binary representation of the time series targets (Δprice) within the time horizon; that is,

$$y_{ri} = \mathrm{RUN}(y_i) = \begin{cases} 1 & \text{if } y_i \geq 0 \\ 0 & \text{otherwise} \end{cases} \tag{1}$$

where y_i is a price change, and y_{ri} is the binary representation of the price change. The historical run windows at y_t are all the available positive (1) or negative (0) price change windows of length l, $y_{rM} \leftarrow y_{ri, i \in \{1,...,T-l\}}$, not in the current run window $y_{rl} \leftarrow y_{ri, i \in \{T-l+1,...,T\}}$.

Comparison Between Windows. After the historical windows are generated, they are individually compared to the current window. To test for similarity on the SeTT algorithm, we alternatively employed Cramér–von Mises (CM), Kolmogorov–Smirnov (KS) and Epps–Singleton (ES) tests of fit. These are a class of non-parametric tests for the null hypthesis (H_0) that two independent samples have the same probability [1,2].

The null hypothesis H_0 in all tests is that two samples are drawn from the same distribution. The p-value is computed based on the values in the distribution, and the null hypothesis is rejected if the p-value is greater than 5% ($\alpha = 0.05$). Otherwise, it is accepted. A consistent sequence of similar historical windows is determined from the historical windows to obtain windows similar to y_l as a vector of 1s, whereas dissimilar windows are represented as a vector of 0s.

Because the input to the r-SeTT's similarity vector routine is a binary representation, we used the *Hamming distance* to measure the similarity between windows. The Hamming distance captures the positions of difference between paired binary windows [6], hence capturing the similarity between different run windows. We used 0.7 as the threshold for the minimum proportion of disagreement, consistent with the state-of-the-art in thresholded Hamming distance search [11]. That is, two windows are considered dissimilar if less than 70% of the paired elements are the same.

Similarity Vector. A similarity vector of length T is generated from the vectors produced by comparing the historical windows with the current window for each stock symbol, the set of letters representing a publicly-traded company or asset on the stock market. A series of 1s represents a consistent similarity with the current window, potentially across different time horizons; 0s represents otherwise. The current window is also part of the similarity vector instantiated as a series of 1s. The vectors are then collapsed in a column-wise matrix multiplication to a vector of the same length as T, as illustrated in Fig. 1a. A matrix consisting of individual vectors for each stock symbol is incorporated into the model architecture.

Model Training. As shown in Fig. 1b, the attention mechanism employs the similarity vector when searching through the temporal attention network. The same similarity vector is used for the entirety of the training period, making it a constant complexity $O(1)$ addition to a regular transformer architecture. Recall that the transformer architecture searches the attention mechanism for the most similar encoder sequence in the attention network, assuming that the input data is homoscedastic, with constant mean and variance [17]. In this research, we ensured that the model only searched through historical windows in the attention network that are identical to the current window during training and inference by muting dissimilar sequences using the similarity vector.

The temporal transformer models are designed to forecast multiple horizons and are trained using past observation combined with past and future known patterns (i.e., date). In contrast to one-step-ahead prediction, *multi-horizon forecast* [18] involves making price prediction multiple steps in advance, that is, a trading week (5 days) prediction. This approach is more accurate and efficient than recursive one-step-ahead forecasts when the input involves dynamic and static historical and future attributes, such as financial time series [8,18].

The future value of a time series is an unknown random variable over a forecast distribution [4]. Instead of estimating the absolute value, we estimated the probability over a range, known as a *quantile forecast* [4]. A quantile forecast is robust when it is difficult to make absolute predictions as it does not assume future distribution and instead provides probabilistic forecasts that we take action on [18]. This is also useful for optimising risks associated with a financial decision, as it provides the best-case and the worst-case scenario of the prediction target over a probability range [8]. In the training process, the predictions are made across all quantiles of interest. The total loss on multiple forecast horizons is minimised using *quantile loss* as the loss function.

$$L_q(y, \hat{y}, q) = \begin{cases} q(y - \hat{y}) & \text{if } y - \hat{y} \geq 0 \\ (1 - q)(\hat{y} - y) & \text{if } y - \hat{y} < 0 \end{cases} = max(q(y - \hat{y}), (1 - q)(\hat{y} - y))$$

(2)

For comparison with other baseline models, we used 50% (p50) and 90% (p90) quantiles to be consistent with previous work [7,8]. Also, the 50% quantile is the target forecast, as it is the median value of the predictive distribution.

4 Experimental Evaluation

Data. We used time series stock market data and financial data to evaluate our model. We leverage the stock market symbols for the companies listed in the Dow Jones Industrial Average (DJIA).

Table 1. Data and attributes.

Industry		Symbol	Weight(%)
TECH	Information technology	AAPL, CSCO, INTC, MSFT	10.54
HEALTH	Managed health care	UNH	7.88
HOME	Home improvement	HD	6.65
FOOD	Soft drink, food industry	KO, MCD	5.82
PHARM	Pharmaceutical industry	JNJ, MRK	4.82
FIN	Financial services	V	4.34
AERO	Aerospace and defense	BA	4.01
CONST	Construction and mining	CAT	3.73
CONG	Conglomerate	MMM	3.38
ENT	Broadcasting and entertainment	DIS	3.18
APPAREL	Apparel	NKE	2.93
RET	Retailing	WMT	2.69
GOODS	Fast-moving consumer goods	PG	2.61
OIL	Petroleum industry	CVX	2.07
TELE	Telecommunication	VZ	0.97

(a) Industry and stock symbols of 20 of the 30 DJIA companies, showing their collective industry weight in the DJIA index

Attribute	Abbreviation	Description	Type
Open, High, Low & Close prices	–	Daily open, high, low & Close prices	–
Volume	–	Volume of shares traded in a day	–
Shares outstanding	–	All share available for trade	–
Earning per share (EPS)	–	Profit for each outstanding share	–
1-day log price change	p_logd	Log. change in close price over 1 day	Target
1-day log volume change	v_logd	Log. change in volume over 1 day	Observed numeric
Shares turnover	s_turnover	All share available for trade	Observed numeric
Price variation	p_variatn	% change in open & close prices	Observed numeric
Price-Earning (P/E) ratio	per	Share price relative to EPS	Observed numeric
Normalised High, Low Close	norm_h, norm_l, norm_c	High & Close prices normalised with Open price	Observed numeric
Year, Month, Weekday	Year, month, weekday	Categorical representations of Year, Month & Weekday	Known categorical
Industry	Industry	Market sector of company	Static categorical

(b) Features used as input to the model.

DJIA is a stock market index, weighted by price, consisting of 30 prominent U.S. companies across 20 industries. We required both market and corporate results data across all time frames for our experiments. Thus, we only used 20 stocks (Table 1a) for which we could obtain the complete data set, of the 30 companies available in the index, across 15 industries with a total weight of 65.02% of the overall index. Historical market and fundamental data were obtained from SimFin, an online financial data resource.

We used the attributes in Table 1b as input to the models, consisting of the market and fundamental data across all the data sets. The close, low, and high market prices are normalised with the open price to enable the model to learn from these features more efficiently, as their scale can vary quite wildly over different periods [16]. For prediction targets, we used the *differenced* price representation; that is, we computed the logarithmic change in price over 1 day [4]. This makes it stationary, a necessary step to ensure the time series is independent of time and useable for modelling purpose [13]. We further confirmed the stationarity by running the Augmented Dickey–Fuller (ADF) test [13].

Evaluation Period. We constructed independent data sets consisting of information about the 20 companies shown in Table 1a across two independent volatile extrapolation periods. To determine the periods, we used the volatility index (VIX)[1], a real-time index that measures stock market volatility based on relative strength with the S&P 500 index. A VIX of 0–12 is considered low, 13–19 is normal, and above 20 is high. Higher values can also be expected during times of financial issues, as evident in the COVID-19-induced extreme volatility at the start of 2020.

We evaluated data within the periods July 2019 to July 2020 and July 2018 to July 2020, both of which included a volatile extrapolation period. All but the last 5 trading days were used as training data within each period. After the training regime, the model was tested on the last 5 days that constitute the test set.

Size of the Current Window. The model consistently performs best at a window size of 16 for most of the stocks, as shown in Table 2. The same experiments were conducted on the other time frames with similar observations across p50 and p90 metrics.

Table 2. Comparison between the similarity functions during volatile period 2019-07-01 to 2020-07-01. cm = Cramér–von Mises test, es = Epps–Singleton test, ks = Kolmogorov–Smirnov test.

	p50			p90		
	cm	es	ks	cm	es	ks
AAPL	**0.183**	0.260	0.236	0.233	**0.185**	0.315
BA	0.923	**0.812**	0.977	1.039	**0.973**	1.003
CAT	0.303	**0.218**	0.341	0.497	**0.387**	0.568
CSCO	0.377	**0.29**	0.346	0.479	**0.374**	0.404
CVX	0.379	0.357	**0.294**	0.503	**0.347**	0.464
DIS	0.218	0.241	**0.192**	0.229	0.321	**0.165**
HD	0.379	**0.361**	0.393	0.405	**0.342**	0.456
INTC	0.331	0.362	**0.329**	0.233	0.247	**0.164**
JNJ	0.276	0.249	**0.222**	0.299	**0.16**	0.346
KO	0.327	**0.266**	0.349	0.314	**0.301**	0.301
MCD	0.283	**0.205**	0.323	0.378	**0.266**	0.446
MMM	0.316	**0.217**	0.336	0.257	**0.21**	0.226
MRK	0.350	**0.307**	0.401	0.430	**0.275**	0.579
MSFT	0.248	**0.175**	0.204	0.277	0.221	**0.172**
NKE	**0.375**	0.378	0.564	0.290	**0.236**	0.618
PG	0.316	0.357	**0.303**	0.341	0.396	**0.283**
UNH	**0.337**	0.383	0.468	0.459	**0.391**	0.689
V	0.253	**0.21**	0.240	0.262	0.274	**0.222**
VZ	0.494	**0.452**	0.583	0.571	**0.496**	0.720
WMT	**0.238**	0.260	0.320	0.129	0.111	**0.103**

4.1 Experimental Results and Discussion

In this section, we present the results of our experiments across both periods. We compared the result with the temporal fusion transformer (TFT) [8] model as a baseline. For simplicity and consistency with the original TFT work, we focus on 0.5 and 0.9 for the quantile metrics. We also evaluated with mean absolute scaled error (MASE) across the full prediction horizon. We used MASE because it scales the predictions independently with the actual target values and is resistant to extreme outliers [5].

For completeness, we compared the transformer models with autoregressive integrated moving average (ARIMA) and generalized autoregressive conditional heteroskedasticity (GARCH), two models commonly used to predict time series data in the financial domain [4]. One approach when using GARCH for prediction

---

[1] investopedia.com/terms/v/vix.asp.

is to fit it on the residual of an ARIMA model, where the GARCH mean-output is the volatility estimate, and it is added to the ARIMA output [9].

Table 3. Two years training history with volatile extrapolation period 2018-07-01 to 2020-07-01.

	p50					p90				
	ARIMA	ARIMA-GARCH	SeTT	r-SeTT	TFT	ARIMA	ARIMA-GARCH	SeTT	r-SeTT	TFT
AAPL	0.71 (+236.3%)	0.689 (+226.2%)	0.26 (+23.2%)	0.248 (+17.4%)	**0.211**	0.72 (+313.2%)	0.613 (+251.5%)	0.185 (+5.9%)	0.209 (+20.0%)	**0.174**
BA	2.268 (+179.3%)	2.267 (+179.3%)	**0.812**	0.911 (+12.2%)	0.823 (+1.4%)	2.606 (+167.9%)	2.606 (+167.9%)	**0.973**	1.064 (+9.4%)	1.009 (+3.8%)
CAT	0.768 (+252.4%)	0.764 (+250.5%)	**0.218**	0.229 (+5.3%)	0.261 (+19.8%)	1.026 (+165.3%)	1.005 (+159.9%)	**0.387**	0.394 (+1.9%)	0.429 (+11.0%)
CSCO	0.582 (+100.5%)	0.571 (+96.8%)	**0.29**	0.396 (+36.4%)	0.361 (+24.4%)	0.787 (+110.7%)	0.734 (+96.4%)	**0.374**	0.399 (+6.9%)	0.415 (+11.0%)
CVX	0.662 (+85.4%)	0.663 (+85.5%)	**0.357**	0.436 (+22.1%)	0.417 (+16.7%)	0.678 (+95.5%)	0.679 (+95.9%)	**0.347**	0.507 (+46.1%)	0.457 (+31.8%)
DIS	0.73 (+202.8%)	0.728 (+202.0%)	**0.241**	0.274 (+13.7%)	0.328 (+36.0%)	0.749 (+133.4%)	0.755 (+135.1%)	**0.321**	0.373 (+16.0%)	0.453 (+41.1%)
HD	0.696 (+105.2%)	0.701 (+106.8%)	0.361 (+6.3%)	**0.339**	0.344 (+1.5%)	0.758 (+153.4%)	0.742 (+148.2%)	0.342 (+14.2%)	**0.299**	0.32 (+7.1%)
INTC	0.831 (+200.4%)	0.839 (+203.3%)	0.362 (+30.8%)	0.372 (+34.5%)	**0.277**	0.777 (+464.0%)	0.753 (+446.1%)	0.247 (+79.3%)	0.166 (+20.2%)	**0.138**
JNJ	0.321 (+29.0%)	0.324 (+30.3%)	**0.249**	0.318 (+27.8%)	0.356 (+43.1%)	0.309 (+93.0%)	0.299 (+86.8%)	**0.16**	0.354 (+121.4%)	0.366 (+128.8%)
KO	0.639 (+153.5%)	0.62 (+146.0%)	0.266 (+5.6%)	0.3 (+19.0%)	**0.252**	0.673 (+123.6%)	0.629 (+108.9%)	**0.301**	0.35 (+16.1%)	0.312 (+3.5%)
MCD	0.543 (+165.1%)	0.546 (+166.7%)	0.205 (+0.3%)	0.233 (+13.6%)	**0.205**	0.537 (+102.2%)	0.511 (+92.1%)	**0.266**	0.267 (+0.4%)	0.286 (+7.6%)
MMM	0.561 (+158.6%)	0.565 (+160.7%)	**0.217**	0.402 (+85.4%)	0.269 (+24.2%)	0.728 (+246.5%)	0.749 (+256.9%)	**0.21**	0.26 (+23.6%)	0.231 (+10.0%)
MRK	0.59 (+92.1%)	0.549 (+78.9%)	**0.307**	0.380 (+26.7%)	0.43 (+40.0%)	0.804 (+192.6%)	0.71 (+158.2%)	**0.275**	0.487 (+77.1%)	0.496 (+80.3%)
MSFT	0.589 (+240.2%)	0.571 (+229.5%)	0.175 (+0.8%)	**0.173**	0.194 (+12.1%)	0.76 (+243.9%)	0.717 (+224.4%)	**0.221**	0.264 (+19.5%)	0.244 (+10.4%)
NKE	1.447 (+282.9%)	1.436 (+279.9%)	**0.378**	0.409 (+8.2%)	0.401 (+6.0%)	1.23 (+421.1%)	1.174 (+397.3%)	**0.236**	0.348 (+47.6%)	0.327 (+38.4%)
PG	0.688 (+140.9%)	0.67 (+134.6%)	0.357 (+25.0%)	0.295 (+3.4%)	**0.286**	0.838 (+222.6%)	0.796 (+206.4%)	0.396 (+52.6%)	**0.26**	0.288 (+10.9%)
UNH	0.723 (+88.7%)	0.693 (+80.9%)	**0.383**	0.457 (+19.3%)	0.402 (+5.0%)	0.961 (+145.7%)	0.891 (+127.9%)	**0.391**	0.566 (+44.8%)	0.489 (+25.0%)
V	0.5 (+140.8%)	0.446 (+114.6%)	0.21 (+1.1%)	**0.208**	0.243 (+16.9%)	0.544 (+98.0%)	0.417 (+51.7%)	**0.274**	0.286 (+4.2%)	0.329 (+20.0%)
VZ	0.678 (+50.0%)	0.674 (+49.2%)	**0.452**	0.489 (+8.2%)	0.503 (+11.3%)	0.763 (+53.8%)	0.743 (+49.7%)	**0.496**	0.527 (+6.1%)	0.603 (+21.4%)
WMT	0.306 (+17.8%)	0.316 (+21.7%)	**0.26**	0.386 (+48.5%)	0.26 (+0.1%)	0.257 (+357.6%)	0.227 (+304.3%)	0.111 (+97.9%)	0.119 (+111.5%)	**0.056**

The quantile loss comparisons in Tables 3 and 4 show that we reduced the evaluation error during both volatile periods. All the deep learning models performed better than the ARIMA and ARIMA-GARCH models across both periods; our models made noticeable improvements to the baseline TFT model with performance improvements greater than 10% in most cases. The greatest improvement we noticed was 136%.

Table 4. One year training history with volatile extrapolation period 2019-07-01 to 2020-07-01.

	p50					p90				
	ARIMA	ARIMA-GARCH	SeTT	r-SeTT	TFT	ARIMA	ARIMA-GARCH	SeTT	r-SeTT	TFT
AAPL	0.66 (+286.7%)	0.646 (+278.6%)	0.229 (+34.2%)	**0.171**	0.316 (+85.1%)	0.616 (+374.5%)	0.547 (+321.7%)	0.283 (+118.3%)	**0.13**	0.256 (+96.9%)
BA	2.246 (+210.4%)	2.235 (+208.8%)	0.754 (+4.1%)	0.735 (+1.6%)	**0.724**	2.602 (+173.1%)	2.605 (+173.5%)	0.96 (+0.8%)	0.953 (+0.0%)	**0.953**
CAT	0.768 (+200.1%)	0.77 (+200.8%)	0.299 (+16.9%)	**0.256**	0.277 (+8.2%)	1.026 (+141.5%)	1.035 (+143.8%)	0.522 (+22.9%)	**0.425**	0.483 (+13.6%)
CSCO	0.581 (+85.0%)	0.578 (+84.2%)	0.333 (+6.2%)	**0.314**	0.432 (+37.6%)	0.781 (+190.5%)	0.768 (+185.6%)	**0.269**	0.321 (+19.4%)	0.349 (+29.9%)
CVX	0.793 (+188.9%)	0.788 (+186.0%)	**0.275**	0.313 (+14.0%)	0.29 (+5.6%)	0.694 (+142.4%)	0.71 (+148.0%)	0.292 (+2.1%)	**0.286**	0.303 (+5.8%)
DIS	0.741 (+353.0%)	0.736 (+350.1%)	0.193 (+17.8%)	**0.164**	0.256 (+56.5%)	0.755 (+199.3%)	0.769 (+205.0%)	0.297 (+17.6%)	**0.252**	0.361 (+43.3%)
HD	0.702 (+174.3%)	0.705 (+175.4%)	**0.256**	0.322 (+26.0%)	0.261 (+2.1%)	0.759 (+164.3%)	0.751 (+161.4%)	0.316 (+10.0%)	0.306 (+6.6%)	**0.287**
INTC	0.87 (+199.6%)	0.884 (+204.5%)	**0.29**	0.331 (+13.8%)	0.308 (+5.9%)	0.81 (+386.7%)	0.767 (+361.3%)	0.177 (+6.6%)	0.213 (+28.3%)	**0.166**
JNJ	0.4 (+65.7%)	0.402 (+66.7%)	**0.241**	0.243 (+0.7%)	0.299 (+23.7%)	0.401 (+101.7%)	0.394 (+98.0%)	0.291 (+46.2%)	**0.199**	0.28 (+40.8%)
KO	0.544 (+159.6%)	0.477 (+127.7%)	0.237 (+12.9%)	0.266 (+27.0%)	**0.21**	0.617 (+129.6%)	0.461 (+71.5%)	0.285 (+6.2%)	0.305 (+13.7%)	**0.269**
MCD	0.613 (+259.7%)	0.617 (+262.2%)	0.187 (+9.8%)	0.180 (+10.7%)	**0.17**	0.598 (+160.9%)	0.585 (+155.3%)	0.286 (+24.9%)	**0.229**	0.237 (+3.3%)
MMM	0.561 (+278.5%)	0.567 (+282.8%)	0.178 (+19.9%)	0.163 (+9.8%)	**0.148**	0.728 (+293.2%)	0.76 (+310.5%)	0.191 (+3.0%)	0.222 (+19.8%)	**0.185**
MRK	0.563 (+101.4%)	0.568 (+103.1%)	0.331 (+18.2%)	**0.28**	0.36 (+28.8%)	0.772 (+167.5%)	0.783 (+171.3%)	0.404 (+50.4%)	**0.289**	0.404 (+39.8%)
MSFT	0.535 (+254.1%)	0.517 (+242.1%)	0.205 (+35.6%)	**0.151**	0.16 (+6.1%)	0.689 (+213.3%)	0.647 (+194.2%)	0.281 (+27.7%)	0.25 (+13.9%)	**0.22**
NKE	1.447 (+351.7%)	1.432 (+346.8%)	0.335 (+4.6%)	**0.32**	0.321 (+0.3%)	1.23 (+609.9%)	1.152 (+564.4%)	0.297 (+71.5%)	**0.173**	0.275 (+58.4%)
PG	0.735 (+158.4%)	0.723 (+154.4%)	**0.284**	0.371 (+30.4%)	0.308 (+8.5%)	0.963 (+164.0%)	0.936 (+156.7%)	**0.365**	0.391 (+7.2%)	0.393 (+7.6%)
UNH	0.681 (+172.3%)	0.63 (+151.7%)	0.357 (+42.6%)	**0.25**	0.329 (+31.4%)	0.904 (+272.9%)	0.783 (+223.2%)	0.468 (+93.3%)	**0.242**	0.422 (+74.3%)
V	0.47 (+210.7%)	0.422 (+179.0%)	0.16 (+5.5%)	0.158 (+4.5%)	**0.151**	0.505 (+181.3%)	0.393 (+118.9%)	**0.179**	0.19 (+6.0%)	0.2 (+11.4%)
VZ	0.647 (+106.6%)	0.643 (+105.5%)	0.333 (+6.3%)	0.363 (+15.8%)	**0.313**	0.722 (+85.5%)	0.704 (+81.1%)	0.435 (+11.8%)	0.391 (+0.6%)	**0.389**
WMT	0.308 (+20.5%)	0.309 (+20.8%)	0.305 (+19.5%)	**0.256**	0.26 (+1.6%)	0.256 (+401.3%)	0.254 (+396.5%)	0.137 (+168.6%)	**0.051**	0.121 (+136.0%)

We observed very similar performance with the MASE metric, as presented in Table 5. Our algorithms always outperformed the ARIMA and GARCH models and usually outperformed the baseline TFT with significant improvements.

Table 5. Comparison using MASE across both volatile extrapolation periods.

	ARIMA	ARIMA-GARCH	SeTT	r-SeTT	TFT	ARIMA	ARIMA-GARCH	SeTT	r-SeTT	TFT
AAPL	0.613 (+236.3%)	0.595 (+226.2%)	0.225 (+23.2%)	0.214 (+17.4%)	**0.182**	0.498 (+286.7%)	0.488 (+278.6%)	0.173 (+34.2%)	**0.129**	0.238 (+85.1%)
BA	1.465 (+179.3%)	1.465 (+179.3%)	**0.525**	0.589 (+12.2%)	0.532 (+1.4%)	1.081 (+210.4%)	1.076 (+208.8%)	0.363 (+4.1%)	0.354 (+1.6%)	**0.348**
CAT	0.631 (+252.4%)	0.628 (+250.5%)	**0.179**	0.189 (+5.3%)	0.214 (+19.8%)	0.548 (+200.1%)	0.55 (+200.8%)	0.214 (+16.9%)	**0.183**	0.198 (+8.2%)
CSCO	0.535 (+100.5%)	0.525 (+96.8%)	**0.267**	0.364 (+36.4%)	0.332 (+24.4%)	0.428 (+85.0%)	0.426 (+84.2%)	0.245 (+6.2%)	**0.231**	0.318 (+37.6%)
CVX	0.587 (+85.4%)	0.588 (+85.5%)	**0.317**	0.387 (+22.1%)	0.37 (+16.7%)	0.514 (+188.9%)	0.51 (+186.9%)	**0.178**	0.203 (+14.0%)	0.188 (+5.6%)
DIS	0.714 (+202.8%)	0.713 (+202.0%)	**0.236**	0.268 (+13.7%)	0.321 (+36.0%)	0.546 (+353.0%)	0.543 (+350.1%)	0.142 (+17.8%)	**0.121**	0.189 (+56.5%)
HD	0.754 (+105.2%)	0.759 (+106.8%)	0.39 (+6.3%)	**0.367**	0.373 (+1.5%)	0.586 (+174.3%)	0.588 (+175.4%)	**0.214**	0.269 (+26.0%)	0.218 (+2.1%)
INTC	0.65 (+200.4%)	0.656 (+203.3%)	0.283 (+30.8%)	0.291 (+34.5%)	**0.216**	0.589 (+199.6%)	0.599 (+204.5%)	**0.197**	0.224 (+13.8%)	0.208 (+5.9%)
JNJ	0.433 (+29.0%)	0.438 (+30.3%)	**0.336**	0.43 (+27.8%)	0.481 (+43.1%)	0.443 (+65.7%)	0.445 (+66.7%)	**0.267**	0.269 (+0.7%)	0.33 (+23.7%)
KO	0.853 (+153.5%)	0.827 (+146.0%)	0.355 (+5.6%)	0.4 (+19.0%)	**0.336**	0.562 (+159.6%)	0.493 (+127.7%)	0.244 (−12.9%)	0.275 (+27.0%)	**0.216**
MCD	0.661 (+165.1%)	0.665 (+166.7%)	0.25 (+0.3%)	0.283 (+13.6%)	**0.249**	0.549 (+259.7%)	0.553 (+262.2%)	0.168 (+9.8%)	0.169 (+10.7%)	**0.153**
MMM	0.582 (+158.6%)	0.587 (+160.7%)	**0.225**	0.417 (+85.4%)	0.28 (+24.2%)	0.476 (+278.5%)	0.482 (+282.8%)	0.151 (+19.9%)	0.138 (+9.8%)	**0.126**
MRK	0.699 (+92.1%)	0.651 (+78.9%)	**0.364**	0.461 (+26.7%)	0.509 (+40.0%)	0.512 (+101.4%)	0.516 (+103.1%)	0.301 (+18.2%)	**0.254**	0.327 (+28.8%)
MSFT	0.548 (+240.2%)	0.531 (+229.5%)	0.162 (+0.8%)	**0.161**	0.181 (+12.1%)	0.423 (+254.1%)	0.409 (+242.1%)	0.162 (+35.6%)	**0.119**	0.127 (+6.1%)
NKE	1.408 (+282.9%)	1.397 (+279.9%)	**0.368**	0.398 (+8.2%)	0.39 (+6.0%)	1.264 (+351.7%)	1.25 (+346.8%)	0.293 (+4.6%)	**0.28**	0.281 (+0.3%)
PG	0.87 (+140.9%)	0.847 (+134.6%)	0.451 (+25.0%)	0.373 (+3.4%)	**0.361**	0.754 (+158.4%)	0.742 (+154.4%)	**0.292**	0.38 (+30.4%)	0.316 (+8.5%)
UNH	0.665 (+88.7%)	0.637 (+80.9%)	**0.352**	0.42 (+19.3%)	0.37 (+5.0%)	0.486 (+172.3%)	0.449 (+151.7%)	0.255 (+42.6%)	**0.179**	0.235 (+31.4%)
V	0.494 (+140.8%)	0.44 (+114.6%)	0.208 (+1.1%)	**0.205**	0.24 (+16.9%)	0.371 (+210.7%)	0.333 (+179.0%)	0.126 (+5.5%)	0.125 (+4.5%)	**0.119**
VZ	0.964 (+50.0%)	0.959 (+49.2%)	**0.643**	0.696 (+8.2%)	0.715 (+11.3%)	0.813 (+106.6%)	0.809 (+105.5%)	0.418 (+6.3%)	0.456 (+15.8%)	**0.394**
WMT	0.437 (+17.8%)	0.451 (+21.7%)	**0.371**	0.55 (+48.5%)	0.371 (+0.1%)	0.381 (+20.5%)	0.382 (+20.8%)	0.378 (+19.5%)	**0.316**	0.322 (+1.6%)
	(a)2018-07-01 to 2020-07-01					(b)2019-07-01 to 2020-07-01				

We tested for the distinction between the error distribution of the different models using the non-parametric Mann–Whitney–Wilcoxon test [12]. Table 6 compares the minimum return error of our models with all three baseline models. The tests suggest that the null hypothesis is rejected, and our models have significantly improved predictive performance.

Table 6. Comparison of error distribution between SeTT/r-SeTT and the baseline models using a non-parametric Mann–Whitney test.

	Metric	2018–2020	2019–2020
(r-)SeTT/TFT	p50	0.013617	0.044054
(r-)SeTT/ARIMA	p50	0.000002	0.000002
(r-)SeTT/ARIMA-GARCH	p50	0.000002	0.000002
(r-)SeTT/TFT	p90	0.000851	0.008308
(r-)SeTT/ARIMA	p90	0.000002	0.000002
(r-)SeTT/ARIMA-GARCH	p90	0.000002	0.000002
(r-)SeTT/TFT	Mase	0.013617	0.063723
(r-)SeTT/ARIMA	Mase	0.000002	0.000002
(r-)SeTT/ARIMA-GARCH	Mase	0.000002	0.000002

4.2 Discussion

The results presented in Sect. 4.1 above show that, by extending the transformer model with our similarity vector, we achieved statistically significant reductions in evaluation error during volatile extrapolation periods. Without the similarity vector, the model could focus on an incorrect time frame during the sample extrapolation regime, as Fig. 2a shows, and the cost of that misplaced focus is a loss of predictive performance. However, Fig. 2b shows that, with the similarity vector, the model could focus on the time frame that has the most similar run to the current window, and the model makes better extrapolations as a result.

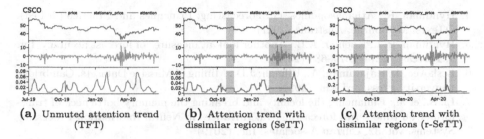

(a) Unmuted attention trend (TFT)   **(b)** Attention trend with dissimilar regions (SeTT)   **(c)** Attention trend with dissimilar regions (r-SeTT)

Fig. 2. Attention on historical window for CSCO (Cisco Systems, Inc.) - 2019-07-01 to 2020-07-01.

Intuitively, when considering the graph for SeTT in Fig. 2b, we notice that the stationary price trend within the masked time frame between February and April 2020 is dissimilar to the current window of the extrapolation period in July 2020. This shows that it is appropriate for that period to have less weight in the forecast. When the dissimilar period is not masked, the model pays more attention to that period, affecting the eventual predictive performance by over 30% across the different evaluation metrics.

5   Conclusions

In this work, we introduced the novel SeTT architectures based on sliding windows in the context of financial time series. A similarity matrix, consisting of symbol-specific similarity vectors that compare the current window with historical windows, was generated based on tested statistical methods. We demonstrated significantly improved extrapolation results in terms of error metrics compared with baselines and the state of the art.

In this research, we only focused on each company in isolation from its industry or other related companies when generating the similarity vectors. Our future work will extend this approach to consider a lookup window that includes the industry or general stock market, hence taking advantage of the similarities that might exist between historical trends of related companies.

References

1. Goerg, S.J., Kaiser, J.: Nonparametric testing of distributions - the Epps-Singleton two-sample test using the empirical characteristic function. Stata J. **9**(3), 454–465 (2009)
2. Hollander, M., Wolfe, D.A., Chicken, E.: Nonparametric Statistical Methods. Wiley, 3rd edn. (2013)
3. Hu, X.: Stock price prediction based on temporal fusion transformer. In: 2021 3rd International Conference on Machine Learning, Big Data and Business Intelligence (MLBDBI), pp. 60–66 (2021)

4. Hyndman, R., Athanasopoulos, G.: Forecasting: Principles and Practice (3rd ed). OTexts: Melbourne, Australia (2021)
5. Hyndman, R.J., Koehler, A.B.: Another look at measures of forecast accuracy. Int. J. Forecast. **22**(4), 679–688 (2006)
6. Leskovec, J., Rajaraman, A., Ullman, J.D.: Mining of Massive Datasets. Cambridge University Press, 3rd edn. (2020)
7. Li, S., et al.: Enhancing the locality and breaking the memory bottleneck of transformer on time series forecasting. In: Advances in Neural Information Processing Systems, vol. 32. Curran Associates, Inc. (2019)
8. Lim, B., Arik, S.O., Loeff, N., Pfister, T.: Temporal fusion transformers for interpretable multi-horizon time series forecasting. Int. J. Forecast. **37**(4), 1748–1764 (2021)
9. Mustapa, F.H., Ismail, M.T.: Modelling and forecasting S&P 500 stock prices using hybrid arima-garch model. J. Phys. Conf. Ser. **1366**(1), 012130 (2019)
10. Olorunnimbe, K., Viktor, H.L.: Deep learning in the stock market - a systematic survey of practice, backtesting and applications. Artif. Intell. Rev. (2022). https://doi.org/10.1007/s10462-022-10226-0
11. Ong, E.J., Bober, M.: Improved hamming distance search using variable length hashing. In: IEEE Conference on Computer Vision and Pattern Recognition (CVPR). IEEE (2016)
12. Paquet, E., Soleymani, F.: QuantumLeap: hybrid quantum neural network for financial predictions. Expert Syst. Appl. **195**, 116583 (2022)
13. Prado, M.L.D.: Advances in Financial Machine Learning. Wiley, 1st edn. (2018)
14. de Santana Correia, A., Colombini, E.L.: Attention, please! a survey of neural attention models in deep learning. Artif. Intell. Rev. (2022). https://doi.org/10.1007/s10462-022-10148-x
15. Sezer, O.B., Gudelek, M.U., Ozbayoglu, A.M.: Financial time series forecasting with deep learning: a systematic literature review: 2005–2019. Appl. Soft Comput. **90**, 106181 (2020)
16. Soleymani, F., Paquet, E.: Financial portfolio optimization with online deep reinforcement learning and restricted stacked autoencoder-DeepBreath. Expert Syst. Appl. **156**, 113456 (2020)
17. Vaswani, A., et al.: Attention is all you need. In: Advances in Neural Information Processing Systems, vol. 2017-December (2017)
18. Wen, R., Torkkola, K., Narayanaswamy, B., Madeka, D.: A multi-horizon quantile recurrent forecaster. In: NIPS'17: Proceedings of the 31st International Conference on Neural Information Processing Systems. Curran Associates Inc. (2017)

Investigating Noise Interference on Speech Towards Applying the Lombard Effect Automatically

Gražina Korvel[1]([✉]), Krzysztof Kąkol[2], Povilas Treigys[1], and Bożena Kostek[3]

[1] Institute of Data Science and Digital Technologies, Vilnius University, Vilnius, Lithuania
grazina.korvel@mif.vu.lt
[2] PGS Software, Wrocław, Poland
[3] Audio Acoustics Laboratory, Faculty of Electronics, Telecommunications and Informatics,
Gdansk University of Technology, Gdansk, Poland

Abstract. The aim of this study is two-fold. First, we perform a series of experiments to examine the interference of different noises on speech processing. For that purpose, we concentrate on the Lombard effect, an involuntary tendency to raise speech level in the presence of background noise. Then, we apply this knowledge to detecting speech with the Lombard effect. This is for preparing a dataset for training a machine learning-based system for automatic speech conversion, mimicking a human way to make speech more intelligible in the presence of noise, i.e., to create Lombard speech. Several spectral descriptors are analyzed in the context of Lombard speech and various types of noise. In conclusion, pub-like and babble noises are most similar when comparing Spectral Entropy, Spectral RollOff, and Spectral Brightness. The larger values of these spectral descriptors, the more the speech-in-noise signal is degraded. To quantify the effect of noise on speech, containing the Lombard effect, an average formant track error is calculated as an objective image quality metric. For image quality assessment Structural SIMilarity (SSIM) index is employed.

Keywords: Lombard effect · Noise background · Structural SIMilarity (SSIM) index

1 Introduction

A number of approaches to robust speech processing are seen in the literature and practical solutions. Despite this, when we refer to the recognition of real-life speech in noise, and especially when noise profiling is a necessary step to process the speech signal correctly, the progress in this area is below expectation. This study builds on the idea of incorporating the Lombard effect (LE) into speech in adverse environments. The Lombard phenomenon, named after the French otolaryngologist Étienne Lombard, occurs in speech production in the presence of noise [1]. He observed that when patients were exposed to loud noise during a conversation, they involuntary raised their voice level and speech became more intelligible. So, to build a human-centric system with

© The Author(s), under exclusive license to Springer Nature Switzerland AG 2022
M. Ceci et al. (Eds.): ISMIS 2022, LNAI 13515, pp. 399–407, 2022.
https://doi.org/10.1007/978-3-031-16564-1_38

ambient intelligence to generate speech with LE for better intelligibility, first, we need to learn about noise interference on speech characteristics. Second, to enable the system to generate Lombard speech when noise is detected and correctly labeled, the interference sound recognition model should be trained on speech with this phenomenon present in it. Moreover, it is evident that by applying a deep model, a large amount of data with the Lombard effect is needed.

Since the discovery of LE, this phenomenon has been extensively studied by a wide range of specialists to find solutions to improve the performance of automatic speech recognition systems in noisy environments [2] or increase speech intelligibility by converting the speaking style from normal to Lombard speech [3, 4]. The idea is that LE may be applied to speech synthesizers, allowing them to adapt to noisy conditions [5–7]. It should be noted that text-to-speech systems adapt to the noise condition during the training process. In contrast, noise profiling still needs to be examined, though some research has already been carried out in this direction with promising results [8, 9].

As already mentioned, our long-term goal is to build a human-centric interface for ambient intelligence to generate speech with the Lombard effect, which could perform automatic adaptation during noise inference. This research investigates the effect of noise interference on Lombard speech. We need such analysis to determine whether speech available on the Internet is with LE or not because we want to use them for the deep network training. So, we investigate to what extent and how the clean speech with LE differs from Lombard speech in noise. For this purpose, specifically, rapidly changing areas of speech such as voiced/unvoiced transitions are examined. To indicate such changes in spectral energy, frequency tracks should be estimated. Both the location and the number of peaks are important in this context. The study deals with various additive noises and different SNR (Signal-to-Noise) levels.

2   Estimation of Frequency Tracks

We conduct the speech analysis that is based on the signal intensity at each time-frequency point. The process of determining frequency tracks in a speech signal is shifted to finding them in a spectrogram, a visual representation of the distribution of signal acoustic energy across frequencies and over time. The darkness of the energy bands is used to estimate the signal intensity. The spectrogram creation process consists of the calculation of the discrete Fourier transform of each short-time frame of speech signal:

$$X_l(k) = \sum_{n=0}^{N-1} x_l(n)w(n)e^{\frac{-2\pi jkn}{N}} \tag{1}$$

where $X_l(k)$ are Fourier transform coefficients ($k = 0, \ldots, N_{FT} - 1$, N_{FT} is the number of Fourier transform coefficients), $x_l(n)$ – the samples of lth short-time frame of signal ($l = 0, \ldots, L - 1$, and L denotes the number of short-time frames), N – the length of the signal, $w(n) = 0, 54 - 0.46\mathrm{con}(2\pi n/N - 1)$ is the Hamming window function, and j is the imaginary unit.

The obtained values are then collected together, and a spectrogram image is built up. A graphical representation of the spectrogram obtained is given in Fig. 1, where both the clean Lombard speech fragment and the same speech fragment corrupted by nonstationary street noise at 0 dB SNR are displayed. For the purpose of this analysis, the spectrogram representation is generated using Hamming windows of size 512. This window size gives smoothed Fourier spectrum. At the same time, the frequency resolution is sufficient for frequency tracking. An overlap of 256 is used to avoid losing part of the information due to the window operation.

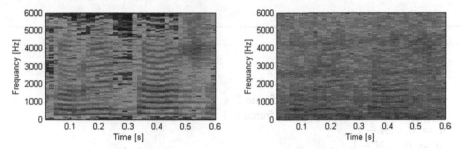

Fig. 1. The spectrogram of clean (the left side) and the noisy (the right side) speech signal.

Various tracking methods and their modifications were proposed [10, 11]. In this research, we used a classical algorithm proposed by McAulay and Quartieri (McA-Q) [12]. The detection of frequency tracks is performed in spectrograms. First, all local maxima of the spectrogram are detected in each short-time frame l. These maxima are called peaks. The estimated peaks, i.e., the amplitudes and their frequencies, are then passed to the tracking algorithm, whose aim is to remove partial trajectories. According to the McA-Q algorithm, frame-to-frame peak matching is performed. The process of matching each spectrum peak in frame l to the peaks in frame $l + 1$, is presented by the following pseudo-code, shown in the algorithmic form.

The result of applying the tracker to the Lombard speech signal is shown in Fig. 2.

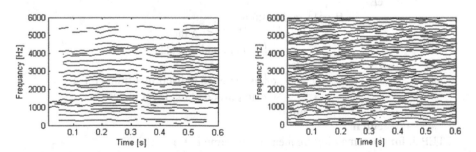

Fig. 2. The estimated frequency tracks of clean (the left side of the figure) and the noisy (the right side) Lombard speech signal.

As we see from the pseudo-code given above, matching each spectrum peak in frame l to the peaks in frame $l + 1$ consists of 3 main steps. In the first step, for each frequency ω_n^l in frame l a search is done for a frequency ω_m^{l+1} in frame $l + 1$, which is the nearest to this frequency and whose absolute distance is less than the threshold (i.e., Δ). In the second step, it is checked, if the frequency ω_m^{l+1} has no better match to unmatched frequencies of frame l. If this condition is satisfied, then the frequencies are matched, and their amplitudes are interpolated between the frames. Otherwise, the adjacent remaining lower frequency ω_{m-1}^{l+1} (if such exists) is tested. In the last step, for the remaining frequencies in frame $l + 1$, for which no matches were made, frequencies are created in frame l with zero amplitude, and the match is made.

Algorithm

INPUT:

ω_n^l – the frequency on frame l

ω_m^{l+1} – the frequency on frame $l + 1$

N – the total number of peaks in frame l

M – the total number of peaks in frame $l + 1$

$n = 0, ..., N - 1$

$m = 0, ..., M - 1$

$p = 0, ..., M - 1$

$p \neq m$

for each frequency in frame l **do**

    STEP 1. **if** $|\omega_n^l - \omega_m^{l+1}| \geq \Delta$ **then**

                ω_n^l is matched to itself in a frame $l + 1$

                the amplitude of ω_n^l is set to zero

       else

          if $(|\omega_n^l - \omega_m^{l+1}| < |\omega_n^l - \omega_p^{l+1}| < \Delta)$ **then**

             ω_m^{l+1} is declared to be a candidate to ω_n^l

          end if

       end if

    STEP 2. **if** $(|\omega_m^{l+1} - \omega_n^l| < |\omega_m^{l+1} - \omega_{p+1}^l|$, where $p > i)$ **then**

                ω_n^l is matched to ω_m^{l+1}

       else

          if ω_{m-1}^{l+1} exists **then**

             if $|\omega_n^l - \omega_{m-1}^{l+1}| < \Delta$ **then**

                ω_n^l is matched to ω_{m-1}^{l+1}

             else

                ω_n^l is matched to itself in a frame $l + 1$

                the amplitude of ω_n^l is set to zero

             end if

       end if

    STEP 3. **for** the remaining frequencies in frame $l + 1$

             frequencies are created in frame l with zero amplitude

             the match is made

The comments on the algorithm:

- If the frequencies are matched, they are eliminated from further consideration.
- Δ denotes a matching interval [12]

Figure 2 shows the frequency tracks of the clean speech segment and the same speech segment corrupted by nonstationary street noise at 0 dB SNR. A ratio of 0 dB indicates the signal level is the same as the noise level; therefore, degradation of formant tracks of noisy speech is visible.

3 Image Comparison Technique

To quantify the effect of noise on speech, containing the Lombard effect, an average formant track error is calculated as an objective image quality metric. For image quality assessment Structural SIMilarity (SSIM) index is calculated. The SSIM index was developed by Wang et al. [13] to evaluate the quality of two images based on the perspective of image formation, i.e., the image luminance, contrast, and structural similarity. The above-mentioned advantages of this method make it sensitive to changes in the image, which is very important in our study. It should also be noted that SSIM is widely used as the quality indicator of the images being compared [14, 15].

Let x and y be two nonnegative image signals. The structural SSIM index is calculated by the following formula [13]:

$$S(x, y) = [l(x, y)]^{\alpha} \cdot [c(x, y)]^{\beta} \cdot [s(x, y)]^{\gamma} \qquad (2)$$

where $\alpha > 0$, $\beta > 0$ and $\gamma > 0$ are weights (in this research parametrized as $\alpha = \beta = \gamma = 1$), $l(x, y)$ is the luminance comparison function, $c(x, y)$ is the contrast comparison function, $s(x, y)$ is the structure comparison function. The functions are given by:

$$l(x, y) = \frac{2\mu_x\mu_y + C_1}{\mu_x^2 + \mu_y^2 + C_1} \qquad (3)$$

$$c(x, y) = \frac{2\sigma_x\sigma_y + C_2}{\sigma_x^2 + \sigma_y^2 + C_2} \qquad (4)$$

$$s(x, y) = \frac{\sigma_{xy} + C_3}{\sigma_x\sigma_y + C_3} \qquad (5)$$

where μ_x and μ_y, σ_x and σ_y, and σ_{xy} are the local means, standard deviations, and cross-covariance of the images being compared, respectively. The constants C_1, C_2, and C_3 are used to avoid instability [13]. The overall similarity measure SSIM is in the range -1 to 1. A value of 1 indicates an ideal agreement between two images, while a value of -1 indicates the given images are very different. In this research, using the SSIM index, we calculated the difference between the image of estimated frequency tracks of the clean speech signal and that of the noisy speech signal.

4 Experimental Setup

Eight speakers (four males and four females) were separately asked to utter fifteen sentences. The speakers were untrained healthy native students of the Gdansk University of Technology. Each speaker was asked to repeat each sentence twice under a different acoustic treatment (in a room with and without an acoustically treated interior that suppresses reverberation). To obtain the Lombard effect while speaking, closed headphones played back the interfering noise were used. The recordings were split into smaller segments, the length of which was 1 s. As a result, 2719 recordings of the acoustically treated room and 3109 ones of the room without acoustic treatment were used in the experiment. This is to balance stratification sampling. Moreover, in this research, we employ four real-life noise recordings, including babble speech (i.e., a mix of many talkers), city streets, rain, and pub that were added to audio data. These recordings were taken from the YouTube platform. The sampling rate of speech and noise signal has been adjusted to 16 kHz before the test.

5 Experimental Results

The experiment is designed to measure the influence of noise interference on the frequency tracks of Lombard speech. The effect of different types of noise was investigated at varying levels of SNR, from -10 dB to 40 dB (i.e., from high to slightly distorted speech). The investigation carried out concerned both acoustically treated and untreated rooms, however, the results were quite similar. As the untreated room conditions are closer to a typical real-life scenario, therefore these results are shown in Table 1. The SSIM index values indicating the correspondence between the shape of a speech signal with LE and its noisy version on different SNR conditions are contained in Table 1. The graphical representation of the results obtained is given in Fig. 3.

The best results were obtained for babble speech noise, followed by recordings mixed with pub noise. Results for city street and rain noises are very similar. For city street noise, there is a slightly higher estimate at the 10 dB condition towards a better result. Further, we analyzed the spectrum of noise signals. The following spectral envelope shape parameters were extracted: Spectral Entropy, Spectral RollOff, and Spectral Brightness. The normalized values are given in Table 2.

When comparing the spectrum-based values (see Table 2) of the noise signal analyzed, we can observe that the spectral entropy, which gives a measure of spectrum irregularity [16], reflects the unpredictability of these signals. This may have led to lower values of the SSIM index values for these noises. Also, the amount of high-frequency information, which is reflected by Spectral Brightness and RollOff, has a direct impact on the SSIM index presented in Table 1.

The investigations carried out also contain the first attempt to automatic noise profile based on recordings contained in the MODALITY multimodal corpus of English speech recordings [17]. Based on the frequency characteristics, the classification model was built. For that purpose, Naïve Bayes [18] algorithm was employed. The following target classes were used: airport, babble speech, car noise, exhibition, restaurant, street noise, subway, train, and pink noise. The test was performed only for a 2-s frame, and the window was moved by 2 s. An example of when the recording was classified as "street," which is the correct classification, is given in Fig. 4.

Table 1. The SSIM index values for recordings (a room without acoustic treatment)

		−10 dB	0 dB	10 dB	20 dB	30 dB	40 dB
Pub noise	Mean	0.412	0.428	0.514	0.634	0.767	0.862
	STD	0.001	0.001	0.002	0.003	0.003	0.002
City street noise	Mean	0.3733	0.3789	0.4198	0.5379	0.6985	0.8188
	STD	0.0007	0.0007	0.0013	0.0025	0.0033	0.0030
Babble speech noise	Mean	0.5214	0.5618	0.6610	0.7656	0.8547	0.9146
	STD	0.0015	0.0019	0.0026	0.0027	0.0021	0.0016
Rain noise	Mean	0.3672	0.3701	0.3843	0.5214	0.6959	0.8202
	STD	0.0006	0.0007	0.0009	0.0026	0.0035	0.0031

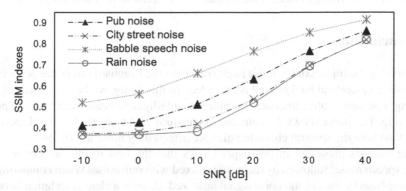

Fig. 3. The SSIM index values for recordings of a room without acoustic treatment.

Table 2. The normalized spectral characteristics of the noise signals

	Spectral entropy	Spectral RollOff	Spectral brightness
Pub noise	0.88	0.51	0.34
City street noise	0.97	0.84	0.84
Babble speech noise	0.82	0.47	0.17
Rain noise	1.00	1.00	1.00

In the context of noise profiling, the model's usefulness is measured by evaluating its stability, not the correctness of classification. It can be seen that this process of classification fluctuates while the averaging mode is stable (dashed and solid lines in Fig. 4).

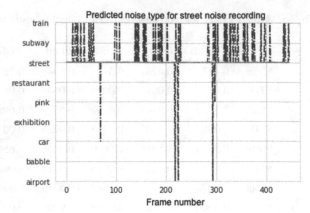

Fig. 4. Classification results on the real-world recordings − the solid line represents the classification in the averaging mode, while the dashed line represents the momentary classification.

6 Conclusions

By analyzing the impact of the noise interference on the Lombard effect, we believe that the existing theoretical background is extended. In this study, we have pointed out that building a human-centric interface for ambient intelligence is an extension of speech processing. The paper shows the outcome of the study that analyzes Lombard speech to understand how the spectral characteristics are affected by noise interference.

The analysis presented in this paper shows that the best results are obtained for babble speech noise, followed by recordings mixed with pub noise. When comparing the spectrum-based values of the noise signal analyzed, there is a clear correlation between the obtained SMM indexes and the obtained spectral characteristics. The greater the Spectral Brightness, RollOff, and Entropy of the interference noise signal, the more the speech signal is degraded.

The influence of noise interference was tested on Lombard speech through an experiment in an acoustically treated room. In real-life, different kinds of noise can be intermingled. The model was not tested against such a combination of noises. The first attempt to profiling noise automatically revealed that LE is applicable in this case. However, this issue needs to be further investigated, which we intend to do in the future.

It is envisioned that the results of this analysis allow for developing a method of monitoring and enhancing speech automatically in the presence of noise. The ultimate goal is to prepare a system capable of synthetically generating Lombard speech through noise profiling.

Acknowledgments. This research is funded by the European Social Fund under the No 09.3.3-LMT-K-712 "Development of Competences of Scientists, other Researchers and Students through Practical Research Activities" measure.

References

1. Lombard, E.: Le signe de l'elevation de la voix. Ann. Mal. de L'Oreille et du Larynx, 101–119 (1911). Zollinger, S.A., Brumm, H.: The lombard effect. Current Biol. **21**(16), 614–615 (2011)
2. Uma Maheswari, S., Shahina, A., Nayeemulla Khan, A.: Understanding Lombard speech: a review of compensation techniques towards improving speech based recognition systems. Artif. Intell. Rev. **54**(4), 2495–2523 (2021)
3. Li, G., Hu, R., Zhang, R., Wang, X.: A mapping model of spectral tilt in normal-to-Lombard speech conversion for intelligibility enhancement. Multimed. Tools Appl. **79**(27), 19471–19491 (2020)
4. Kakol, K., Korvel, G., Kostek, B.: Improving objective speech quality indicators in noise conditions. In: Data Science: New Issues, Challenges and Applications, pp. 199–218. Springer, Cham (2020). https://doi.org/10.1007/978-3-030-39250-5_11
5. Bollepalli, B., Juvela, L., Airaksinen, M., Valentini-Botinhao, C., Alku, P.: Normal-to-Lombard adaptation of speech synthesis using long short-term memory recurrent neural networks. Speech Commun. **110**, 64–75 (2019)
6. Paul, D., Shifas, M.P., Pantazis, Y., Stylianou, Y.: Enhancing speech intelligibility in text-to-speech synthesis using speaking style conversion. arXiv preprint arXiv:2008.05809 (2020)
7. Korvel, G., Kakol, K., Kurasova, O., Kostek, B.: Evaluation of Lombard speech models in the context of speech in noise enhancement. IEEE Access **8**, 155156–155170 (2020)
8. Novitasari, S., Sakti, S., Nakamura, S.: Dynamically adaptive machine speech chain inference for tts in noisy environment: listen and speak louder. Proc. Interspeech **2021**, 4124–4128 (2021)
9. Yue, F., Deng, Y., He, L., Ko, T., Zhang, Y.: Exploring machine speech chain for domain adaptation. In: ICASSP 2022–2022 IEEE International Conference on Acoustics, Speech and Signal Processing (ICASSP), pp. 6757–6761 (2022)
10. Lampert, T.A., O'Keefe, S.E.: On the detection of tracks in spectrogram images. Pattern Recogn. **46**(5), 1396–1408 (2013)
11. Bhattacharjee, M., Prasanna, S.M., Guha, P.: Speech/music classification using features from spectral peaks. IEEE/ACM Trans. Audio, Speech Lang. Process. **28**, 1549–1559 (2020)
12. McAulay, R., Quatieri, T.: Speech analysis/synthesis based on a sinu-soidal representation. IEEE Trans. Acoust. Speech Signal Process. **34**(4), 744–754 (1986)
13. Wang, Z., Bovik, A.C., Sheikh, H.R., Simoncelli, E.P.: Image quality assessment: from error visibility to structural similarity. IEEE Trans. Image Process. **13**(4), 600–612 (2004)
14. Peng, J., et al.: Implementation of the structural SIMilarity (SSIM) index as a quantita- tive evaluation tool for dose distribution error detection. Med. Phys. **47**(4), 1907–1919 (2020)
15. Zini, S., Bianco, S., Schettini, R.: Deep residual autoencoder for blind universal JPEG restoration. IEEE Access **8**, 63283–63294 (2020)
16. Wei, Y., Zeng, Y., Li, C.: Single-channel speech enhancement based on subband spectral entropy. J. Audio Eng. Soc. **66**(3), 100–113 (2018)
17. Czyzewski, A., Kostek, B., Bratoszewski, P., Kotus, J., Szykulski, M.: An audio-visual corpus for multimodal automatic speech recognition. J. Intell. Inf. Syst. **49**(2), 167–192 (2017). https://doi.org/10.1007/s10844-016-0438-z
18. Barber, D.: Bayesian Reasoning and Machine Learning. Cambridge University Press (2012). ISBN 978-0-521-51814-7

Medical Applications

Towards Polynomial Adaptive Local Explanations for Healthcare Classifiers

Jamie Duell[1]([⊠])(iD), Xiuyi Fan[2], and Monika Seisenberger[1](iD)

[1] School of Mathematics and Computer Science, Swansea University, Swansea, Wales
{853435,m.seisenberger}@swansea.ac.uk
[2] School of Computer Science and Engineering, Nanyang Technological University,
Singapore, Singapore
xyfan@ntu.edu.sg

Abstract. Local explanations aim to provide transparency for individual instances and their associated predictions. The need for local explanations is prominent for high-risk domains such as finance, law and health care. We propose a new
model-agnostic framework for local explanations "Polynomial Adaptive Local
Explanations (PALE)", to combat the lack of transparency of predictions through
adaptive local models. We aim to explore explanations of predictions by assessing
the impact of instantaneous rate of change in each feature and the association with
the resulting prediction of the local model. PALE optimises a complex black-box
model and the local explanation models for each instance, providing two forms of
explanations, one provided by a localised derivative of an adapting polynomial,
thus emphasising instance specificity, and the latter a core interpretable logistic
regression model.

1 Introduction

The use of eXplainable Artificial Intelligence (XAI) methods enable clarity for the communication of black-box model predictions enabling a person's rights for a *right for
explanation*' in Europe's General Data Protection Regulation (GDPR) [14]. As of 2016,
there exist variations of XAI surrogate models that explore different approaches to
localised explanations, though the premise of XAI greatly predeceased the recent influx
[5]. Perturbation methods have seen success and wide application in the medical domain
[3,9,13,17], popular examples being Local Interpretable Model-Agnostic Explanations
(LIME) [11], SHapley Additive exPlanations (SHAP) [8] and Scoped Rules (Anchors)
[12], where SHAP explores feature summary through additive marginal contribution
evaluation and Anchors and LIME explore local surrogate models from a set of readily
interpretable models e.g. linear regression.

In this work, we specifically aim to approach local explanations for tabular data
with Electronic Health Records (EHR) being a fundamental asset to population and
individual health research. In exploration of clinical care, it has been a standing point
that local explanations hold high importance to promote trust [16]. For example, being
in the area of individual health, explanations would naturally need to contain patient

This work is supported by the UKRI AIMLAC CDT, funded by grant EP/S023992/1.

M. Ceci et al. (Eds.): ISMIS 2022, LNAI 13515, pp. 411–420, 2022.
https://doi.org/10.1007/978-3-031-16564-1_39

specificity, to deal with a case-by-case basis of care. We've seen the development of tools utilising pre-existing XAI methods in addendum to data exploration and analytic techniques [6].

With the lack of consistency across explanations [4] they can prove to be untrustworthy. In order to better adapt local explanations on an instance level we need to provide optimized scale-ability, highlighting patient specificity. In an attempt to create clear, efficient and patient specific local explanations, we propose the Polynomial Adaptive Local Explanations (PALE) framework, an end-to-end model aiming to mutually optimize both a complex model and each local explanation with a focus on tabular data. This should enable the transparency of patient predictions in a local domain, by producing explanations on how each patient and each feature can impact the outcome through local surrogate models that adapt to patient specific cases, as such in this work we

1. Produce an end-to-end framework that optimises both the complex model and the local model for each instance;
2. Produce explanations based on the derived scaling polynomial models to understand uni-variate feature impact for local instances;
3. Produce explanations based on a logistic regression model to understand uni-variate feature impact for local instances;
4. Compare local explanations and local explanation performance across the different XAI methods.

2   Related Work

Exploration of local surrogate model explanations saw an effective rise posterior to the efforts of LIME. LIME is a model-agnostic method with a primary focus on local explanation where a local linear model is used on a perturbed set around the instance $\mathbf{x}$. An explanation $\mathcal{E}$ for local point $\mathbf{x}$ is defined

$$\mathcal{E}(\mathbf{x}^{(j)}) = \arg\min_{g \in G} L(f, g, \pi_{\mathbf{x}^{(j)}}) + \Omega(g),$$

where we have a local linear model g from a set of linear models G, aiming to minimise the error of the local linear model, where perturbations around instance $\mathbf{x}^{(j)}$ are subject to a neighbourhood π, where the fidelity of the local model is measured against the complex model f, through L. The Ω term is used to reduce the complexity of the local model g. Perturbations are created around the mean of the data set within one standard deviation following a Gaussian distribution. See [11] for details.

There are various branches of LIME, to which end, the original framework has been adapted and extended in various cases. The authors of deterministic-LIME (DLIME) [18] extend the LIME framework by producing an adaptive neighbourhood using k-nearest neighbours and hierarchical clustering in an attempt to provide consistent explanations. In [20] the authors introduce Stabilized-LIME (S-LIME) which also surrounds the improvement of perturbation points for better local explainability, stability in the former DLIME and S-LIME are measured using the Jaccard similarity coefficient. [10] introduces local explanations and example-based local explanations, where weighting is carried out using random forests for supervised neighbourhood selection.

In [1] the authors propose a ensemble approach to LIME, namely LimeOut in order to reduce the reliance of sensitive features, in order to achieve this the authors replicate a similar idea to drop out techniques that are used in neural networks, aiming to maintain model performance. The authors of [19] introduce Bayesian LIME (BayLIME), in efforts to obtain consistency in explanations and maintain model robustness through integration of prior knowledge and the adaptation of Bayesian reasoning.

Extrapolating to local model fits [15] introduces Tree-LIME, this replacing the local linear model with a decision tree based approach for local interpretability. The authors of [2] draw more comparable intentions, as the authors aimed to fit a quadratic model to extend the LIME local model, the intent to analyse the performance improvement against the linear model. Therefore, the development of this inspired the intent for creating a framework with instance specific explainability to any polynomial degree that fits best for a given case. Feature attribution methods have explored specific feature-types, where we see focus on continuous features, enhancing the idea for the selective perturbation strategy [7].

3 Method

3.1 PALE Framework

We propose a complete framework to optimise the complex model f over all data X, therefore, $f(X)$ denotes our black-box model, where we minimise the residual loss $\mathcal{L}_f$ of the complex model. Our model uses the same neighbourhood setting that is used in the LIME framework. We optimise the local explainer loss for each jth instance, where $X = [\ldots, \mathbf{x}^{(j)}, \ldots]$. We search for the optimal local models $g_m \in G$, where G is a set of polynomial models, for an instance in the local neighbourhood $\pi_{\mathbf{x}^{(j)}}$. Local model error is minimised through $\mathcal{L}_{g_m^{(j)}}$, where the optimal m polynomial degree for each instance is obtained. The framework aims to produce local explanations over classification problems, therefore we assume the complex model f to be some classifier.

Adaptive Model. Introducing PALE, the generated surrogate data set $\mathcal{Z}^{(j)}$ is weighted by some neighbourhood $\pi_{\mathbf{x}^{(j)}}$, for an instance of interest $\mathbf{x}^{(j)}$. The surrogate set can be represented by $\{z', \mathbf{y}\} = \mathcal{Z}^{(j)}$, where an instance $\mathbf{z}'_s$ in the surrogate set is defined by $\mathbf{z}'_s \in \mathbb{R}^{1 \times N}$, the surrogate data is given by $z' \in \mathbb{R}^{M \times N}$ and labels $\mathbf{y} \in [0, 1]$. We let $f(\mathbf{z}'_s)$ for each instance $\mathbf{z}'_s$ be the labels of the surrogate set using the prediction probability as the target for $g_m(z')$.

We first aim to have a scaling polynomial fit for instance adaptation in order to both provide better localised model performance as well as to provide insight into feature attribution and the affect of feature alteration in the local domain. This is carried out through the optimisation of the objective function $\mathcal{L}(X; \Phi, \cdot, \Psi)$, to obtain the optimal parameter set for both the local and complex model. We optimise our complete objective function in one function to avoid inconsistencies in local explanations, ensuring we obtain the same random seed for perturbation strategies and data split.

$$\mathcal{L}(X; \Phi, \cdot, \Psi) = \underbrace{\mathcal{L}_f(X; \cdot)}_{\text{Complex model loss}} + \lambda_p(\cdot)$$

$$+ \underbrace{\sum_{j=1}^{M} \mathcal{L}_{g_m^{(j)}}(\mathcal{Z}^{(j)}; \Phi) + \lambda_p(\Phi)}_{\text{Explainer loss}} + \underbrace{\sum_{j=1}^{M} \mathcal{L}_{u^{(j)}}(\mathcal{Z}'^{(j)}; \Psi) + \lambda_p(\Psi)}_{\text{Logistic Loss}}$$

$\mathcal{L}_{g_m^{(j)}}$ is used to minimize the loss where we use the root mean-squared error ($RMSE$) to determine localised model performance of some surrogate set, namely $\mathcal{Z}^{(j)}$ in the neighbourhood $\pi_{\mathbf{x}}^{(j)}$ determining error in each model to the mth degree polynomial for a prediction $g_m(\mathbf{z}'_s)$ for the instance of the surrogate set, and the fidelity to the labels y_s assigned by $f(\mathbf{z}'_s)$. We carry this out for the number of instances in each surrogate set and minimise the loss, we do this for every instance $\mathbf{x}^{(j)}$.

$\mathcal{L}_u^{(j)}$ defines the loss function for the logistic regression function $u^{(j)}$ in the surrogate set $\mathcal{Z}'^{(j)}$, by default this is given by the uni-variate binary cross-entropy loss function for each feature z'_i of each instance z'_s in the surrogate set with respect to the true label for the instance in the surrogate set y'_s. The regularization parameter $\lambda_p(\Phi)$ of our local model in given example to be λ_2 ridge regression, in an attempt to avoid over fitting of the local models whilst keeping all features as non-zero weights.

We let $\lambda_p(\cdot)$ be a placeholder for the parameters regularized in the complex model to obtain a best fit e.g. coefficients in regression. Both Φ and Ψ concretely represent coefficients of the local regression model and logistic regression model. From the integration of this objective function, we can then obtain the optimal set of parameters returned for the ideal polynomial for the local model and using a select complex model and associated loss function, we then obtain the matrices of optimal coefficients Φ' and Ψ', where each row is a vector corresponding to coefficients for the $\mathbf{x}^{(j)}$ instance to be explained.

$$\{\Phi', \cdot', \Psi'\} := arg\,min_{\{\Phi, \cdot, \Psi\}}[\mathcal{L}(X; \Phi, \cdot, \Psi)]. \qquad (1)$$

Once obtaining the optimal fit for the local model, we extract the best performing m^{th} degree polynomial as the model to explain. After determining the best fit for each instance and extracting the coefficient matrix Φ', we do this to obtain the optimal local models $g_m^{(j)}$ for each $\mathbf{x}^{(j)}$.

Adaptive Local Explanations. With models obtained from Eq. 1 we generate explanations, we produce an ordered absolute value where the associative value corresponds to the feature importance ranked by its value $|\frac{\partial g_m^{(j)}}{\partial x_i}|$ for each feature i to gauge a descending order of feature importance. Generalising to a scaling polynomial fit, we can observe the partial derivative for the mth polynomial degree, such that for each feature x_i we observe the affect of change, where every other feature is kept static $\mathbf{x}_{/i}^{(j)}$, therefore,

$$g_m^{(j)}(x_i + \Delta x_i, \mathbf{x}_{/i}^{(j)}) = g_m^{(j)}(x_i, \mathbf{x}_{/i}^{(j)}) + (\Delta x_i) \cdot \frac{\partial g_m^{(j)}}{\partial x_i}(x_i, \mathbf{x}_{/i}^{(j)}),$$

as such we obtain a complete set of polynomial model partial derivative based explanations over the given data set X. We refer to this set of polynomial explanations as $\mathcal{E}_p(X)$, where each row corresponds to a instance $\mathbf{x}^{(j)}$, and each column corresponds to the features,

$$
\mathcal{E}_p(X) = \begin{bmatrix} \frac{\partial g_m^{(1)}}{\partial x_1} & \frac{\partial g_m^{(1)}}{\partial x_2} & \cdots & \frac{\partial g_m^{(1)}}{\partial x_k} \\ \frac{\partial g_m^{(2)}}{\partial x_1} & \frac{\partial g_m^{(2)}}{\partial x_2} & \cdots & \frac{\partial g_m^{(2)}}{\partial x_k} \\ \vdots & \vdots & \ddots & \vdots \\ \frac{\partial g_m^{(r)}}{\partial x_1} & \frac{\partial g_m^{(r)}}{\partial x_2} & \cdots & \frac{\partial g_m^{(r)}}{\partial x_k} \end{bmatrix}.
\tag{2}
$$

Precision. We introduce a form of local precision, this is a user defined level of precision which is in the range [0,1]. The term γ, is a flexible user influenced term that binds whether an instance explanation is returned, to a given precision of local fidelity where a returned explanation given the value for $\gamma = 1$ would determine $|(f(\mathbf{x}^{(j)}) - g_m^{(j)}(\mathbf{x}^{(j)}))| = 0$. This meaning that the prediction of the local model g accurately represents the point of interest predicted from our complex model f, meaning $g_m^{(j)}(\mathbf{x}^{(j)}) = f(\mathbf{x}^{(j)})$. This is determined through a term given the complex and local model for an instance of interest and a measure of precision γ, such that,

$$
Precision(g_m^{(j)}, f, \mathbf{x}^{(j)}; \mathcal{T}, \gamma) = |(f(\mathbf{x}^{(j)}) - g_m^{(j)}(\mathbf{x}^{(j)}))|,
$$
$$
\text{s.t. } Precision \leq 1 - \gamma.
$$

We also allow the user to select a target value, $\mathcal{T} \in \{0, 1\}$ (*1 by default in the binary case*), this will allow for the partial derivative of the local regression to be associated with some user defined $\mathcal{T}$ for an explanation. If the local model does not meet the precision requirements, the instance explanation will not be returned. Therefore, the purpose of this in the applied case is to return only locally precise explanations.

3.2 Logistic Explanation

In addition to the prior, we provide explanations with respect to the odds ratios (OR), through uni-variate logistic regression analysis on each feature in the perturbed set $\mathbf{z}_i'$. We introduce the logistic model as the function $u^{(j)}$, where $u^{(j)}$ is the local logistic regression model over a surrogate set for instance $\mathbf{x}^{(j)}$. The localised model is a uni-variate model to explore individual feature importance. To achieve this, we introduce a secondary surrogate set $\mathcal{Z}'$ where, $\{z', \mathbf{y}'\} = \mathcal{Z}'$. A feature vector is denoted by $\mathbf{z}_i' \in \mathbb{R}^{m \times 1}$ and associated label is a binary case $\mathbf{y}' \in \{0, 1\}$, therefore,

$$
u^{(j)}(z_i') = P(\mathbf{y}' \mid z_i') = \frac{1}{1 + (exp(-(\Psi_i \times z_i')))}.
\tag{3}
$$

We introduce a modified version of OR to center odds at the value 0 for ease of interpretation, the logistic explanation $\mathcal{E}_l$ where Ψ_i is the returned log odds, can be represented by,

$$
\mathcal{E}_l(\mathbf{x}_i^{(j)}) = exp(\Psi_i) - 1.
\tag{4}
$$

4 Comparative Methods

We introduce a comparisons of explanations returned by XAI methods. We include SHAP, a linear model, higher degree polynomials and logistic explanations as the XAI methods.

Jaccard Index. We can explore the Jaccard similarity index for v features, for this paper we explore $v = 5$. The Jaccard index can be defined by $J(A, B) = \frac{|A \cap B|}{|A \cup B|}$, to compare returned sets of feature names between two XAI methods.

Pearson Correlation Coefficient. We also compare the Pearson r correlation coefficient for the sets of explanations, given the absolute values returned from the XAI methods.

Logistic Comparison. We can use the shift in odds ratio in either $\mathbb{R}^+$ or $\mathbb{R}^-$ of non-absolute value explanations, for each feature i of an instance, to determine similarity between the derived explanation and odds ratio explanation. We determine the ratio of shared explanation shift $LogCompare$ for any $\mathbf{x}^{(j)}$ over N features as,

$$LogCompare(\mathbf{x}^{(j)}) = \begin{cases} \frac{1}{N} \sum_{i=1}^{N} \mathbb{1}_{[x_i]}, & \text{if } sgn\left(\frac{\partial g_m^{(j)}}{\partial x_i}\right) = sgn\left(\mathcal{E}_l(\mathbf{x}_i^{(j)})\right), \\ 0, & \text{otherwise.} \end{cases} \quad (5)$$

5 Results

Data for this study uses artificial data from the Simulacrum[1], a synthetic data set developed by Health Data Insight CiC derived from anonymous cancer data provided by the National Cancer Registration and Analysis Service (NCRAS)[2], which is part of Public Health England. We extract a subset of lung cancer patients from the Simulacrum to demonstrate the proposed method. We focus on binary classification problems for the demonstration of this framework. The binary classes we aim to predict are $< 6\,months$ and $> 6\,months$ survival time.

We use an XGBoost model with a 70% train and 30% testing data split as our complex model to demonstrate the explanatory model. The model performance is evaluated using the Root Mean Squared Error (RMSE), obtaining the following,

Class	Precision	Recall	F1-Score
$< 6\,months$	0.97	0.97	0.97
$> 6\,months$	0.98	0.98	0.98

Posterior to this, we determine a local patient instance of interest to explain.

---

[1] https://simulacrum.healthdatainsight.org.uk/.

[2] http://www.ncin.org.uk/about_ncin/.

- Age 66, Sex 0, Morph 8140, Weight 85.90, Height 1.67, Dose Administration 8, Chemo Radiation 0.0, Regimen Outcome Description 0.0, Admin Route 1.0, Regimen Time Delay 0.0, Regimen Stopped Early 1.0, Cycle Number 1.0, Grade 1.0, Cancer Plan 0.0, Cancer Registration Code 301.0, T Best 4.0, N Best 2.0, M Best 0.0, Laterality 2.0, CNS 1.0, ACE 9.0, Performance 0.0, Clinical Trial 2.0.
Prediction: > 6 *Months,*
Actual: > 6 *Months.*

We explore how higher degree polynomial functions can inform feature attribution on a local level. We use the partial derivative for the 2nd and 3rd degree polynomials, to determine how each feature i interacts with the output for our local model.

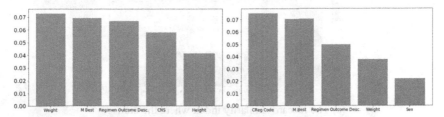

(a) Derivation of the quadratic polynomial term - Simulacrum patient instance.

(b) Derivation of the cubic polynomial term - Simulacrum patient instance.

Fig. 1. The explanation determines how an instantaneous increase in each feature value x_i influences the local polynomial function $g_m^{(j)}$, where we have $g_2^{(j)}$ for Fig. 1a, and $g_3^{(j)}$ for Fig. 1b.

Evaluating the explanations for the first 5 feature, we observe that the quadratic derivative determine *Weight, M Best* and the *Regimen Outcome Description* to have a high attribution in the local model. Conversely, when observing the 3rd degree polynomial, we see *Cancer Registration code* followed by *M Best* and *Regimen Outcome Description* as the highest attribution in the local model (Fig. 1).

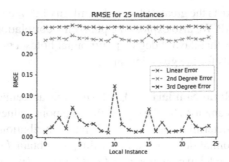

Fig. 2. RMSE measurements for a subset of 25 Simulacrum patient instances. We can observe how the increase in polynomial degree improves the local model accuracy.

We explore the performance of each model to the mth degree polynomial, looking at the $RMSE$ returned for the local model $g_m^{(j)}$ for 25 instances $\mathbf{x}^{(j)} : j = \{1, 2, ..., 25\}$. From this, we determine that an increase in polynomial degree has significant impact on the local model performance over each surrogate set $\mathcal{Z}^{(j)}$ (Fig. 2).

XAI Models - Similarity Measures. For the comparison of XAI models, we determine the Jaccard similarity index between the sets of g_m and the response given by SHAP. Although the PALE framework extracts the ideal polynomial degree and produces an explanation for each instance, we instead manually extract explanations for each degree and compare the similarities amongst each degree polynomial and SHAP (Fig. 3).

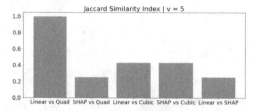

(a) The Jaccard similarity index where the number of returned features returned is 5.

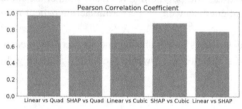

(b) Pearson correlation coefficient between XAI methods.

Fig. 3. A comparison of explanations given by the linear model, quadratic model, cubic model and the SHAP model for a patient instance.

We observe there exists the greatest Jaccard similarity between that of the 3rd degree polynomial fit and SHAP. We also explore the pearson r correlation coefficient between each model and identify that the 3rd degree polynomial holds a greater correlation with SHAP than other models for the given instance.

Interpretable Odds Ratio Similarity. Exploring the agreement between both the quadratic and cubic explanations for the signed floating point values, as opposed to absolute values, so we can determine the amount of shared attribution between the logistic model and local polynomial derivations. From this, we obtain $LogCompare(\mathbf{x}^{(j)}) = 0.48$ for the quadratic model explanation and $LogCompare(\mathbf{x}^{(j)}) = 0.65$ for the cubic model explanation. Therefore, we observe in the given case, the cubic explanation has a greater similarity in explanation with the logistic model than that of the quadratic model.

6 Conclusion and Future Work

We use a similar classification problem as seen in [4,7], where under similar predictions surrounding survival we see great influence from the likes of *M Best*, *Weight*, amongst other features. Therefore, we observe the selection of important features hold a degree of accuracy with clinical knowledge of cancer survival. The contribution of this work is an end-to-end framework that optimizes both the local and complex model to provide an explanation of how change to a feature will influence the outcome of the model prediction in the local setting. We emphasise the need for patient specificity, thus we produce an adaptive framework at the local level through adaptive polynomials.

We identify that the uni-variate approach shows single feature interaction with the local model, and although predictions are reliant on the kernel and localised feature perturbations which can lead to explanation instability, with ongoing research being focused in this area for the extension of LIME, we instead focus on improving the interpretable local model by adapting explanations to each local instance to increase local specificity. Extending upon this, the interpretable comparison with the logistic regression model poses questions towards the disagreement of explanations, to further analyse this, we will consider statistical significance against the explanations given. We acknowledge the problem of potential polynomial overfitting despite regularization. Further research will be carried out in order to approach the addressed issues and expand upon the framework. We also aim to explore this method beyond EHRs.

References

1. Bhargava, V., Couceiro, M., Napoli, A.: LimeOut: an ensemble approach to improve process fairness. In: Koprinska, I., et al. (eds.) ECML PKDD 2020. CCIS, vol. 1323, pp. 475–491. Springer, Cham (2020). https://doi.org/10.1007/978-3-030-65965-3_32
2. Bramhall, S., Horn, H., Tieu, M., Lohia, N.: Qlime-a quadratic local interpretable model-agnostic explanation approach. In: SMU Data Science Review, No. 1, Article 4. vol. 3 (2020)
3. Dindorf, C., et al.: Classification and automated interpretation of spinal posture data using a pathology-independent classifier and explainable artificial intelligence (XAI). Sensors (Basel) 21(18), 6323 (2021)
4. Duell, J., Fan, X., Burnett, B., Aarts, G., Zhou, S.: A comparison of explanations given by explainable artificial intelligence methods on analysing electronic health records. In: 2021 IEEE EMBS International Conference on Biomedical and Health Informatics (BHI) (IEEE BHI 2021), Athens, Greece, July 2021
5. Främling, K.: Decision theory meets explainable AI. In: Calvaresi, D., Najjar, A., Winikoff, M., Främling, K. (eds.) EXTRAAMAS 2020. LNCS (LNAI), vol. 12175, pp. 57–74. Springer, Cham (2020). https://doi.org/10.1007/978-3-030-51924-7_4
6. Kapcia, M., Eshkiki, H., Duell, J., Fan, X., Zhou, S., Mora, B.: ExMed: an AI tool for experimenting explainable AI techniques on medical data analytics. In: 2021 IEEE 33rd International Conference on Tools with Artificial Intelligence (ICTAI), pp. 841–845 (2021)
7. Kovvuri, V.R.R., Liu, S., Seisenberger, M., Müller, B., Fan, X.: On understanding the influence of controllable factors with a feature attribution algorithm: a medical case study. arXiv:2203.12701 (2022)
8. Lundberg, S.M., Lee, S.: A unified approach to interpreting model predictions. In: Advances in NeurIPS 30: Annual Conference on NeurIPS, pp. 4765–4774 (2017)

9. Peng, J., et al.: An explainable artificial intelligence framework for the deterioration risk prediction of hepatitis patients. J. Med. Syst. **45**(5), 1–9 (2021). https://doi.org/10.1007/s10916-021-01736-5

10. Plumb, G., Molitor, D., Talwalkar, A.: Model agnostic supervised local explanations. In: Proceedings of the 32nd International Conference on Neural Information Processing Systems, NIPS 2018, pp. 2520–2529. Curran Associates Inc., Red Hook, NY, USA (2018)

11. Ribeiro, M., Singh, S., Guestrin, C.: "Why Should I Trust You?" explaining the predictions of any classifier. arXiv:1602.04938 (2016)

12. Ribeiro, M., Singh, S., Guestrin, C.: Anchors: High-precision model-agnostic explanations. In: Proceedings of the Thirty-Second AAAI Conference on Artificial Intelligence, pp. 1527–1535. AAAI Press (2018)

13. Sarp, S., Kuzlu, M., Wilson, E., Cali, U., Guler, O.: The enlightening role of explainable artificial intelligence in chronic wound classification. Electronics **10**(12), 1406 (2021)

14. Selbst, A.D., Powles, J.: Meaningful information and the right to explanation. Int. Data Priv. Law **7**(4), 233–242 (2017)

15. Shi, S., Zhang, X., Li, H., Fan, W.: Explaining the predictions of any image classifier via decision trees. ArXiv abs/1911.01058 (2019)

16. Tonekaboni, S., Joshi, S., McCradden, M., Goldenberg, A.: What clinicians want: contextualizing explainable machine learning for clinical end use. In: MLHC (2019)

17. Yoo, T.K., et al.: Explainable machine learning approach as a tool to understand factors used to select the refractive surgery technique on the expert level. Transl. Vis. Sci. Technol. **9**(2), 8 (2020)

18. Zafar, M.R., Khan, N.: Deterministic local interpretable model-agnostic explanations for stable explainability. Mach. Learn. Knowl. Extr. **3**(3), 525–541 (2021)

19. Zhao, X., Huang, W., Huang, X., Robu, V., Flynn, D.: BayLIME: bayesian local interpretable model-agnostic explanations. In: de Campos, C., Maathuis, M.H. (eds.) Proceedings of the Thirty-Seventh Conference on Uncertainty in Artificial Intelligence. Proceedings of Machine Learning Research, vol. 161, pp. 887–896. PMLR, 27–30 July 2021

20. Zhou, Z., Hooker, G., Wang, F.: S-LIME: stabilized-LIME for model explanation. In: Proceedings of the 27th ACM SIGKDD Conference on Knowledge Discovery & Data Mining, KDD 2021, pp. 2429–2438. Association for Computing Machinery, New York (2021)

Towards Tailored Intervention in Medicine Using Patients' Segmentation

Petr Berka[1,2](✉) ⓘ, Maciej Pondel[3] ⓘ, David Chudán[1] ⓘ,
and Agnieszka Siennicka[4] ⓘ

[1] Prague University of Economics and Business, Prague, Czech Republic
berka@vse.cz
[2] University of Finance and Administration, Prague, Czech Republic
[3] Wroclaw University of Economics and Business, Wroclaw, Poland
[4] Wroclaw Medical University, Wroclaw, Poland

Abstract. In the patient treatment process, many aspects should be addressed – among all the patients' psychological profiles. Individual approaches to each patient can be unrealizable due to various limitations in a public healthcare system. We propose the patients' segmentation method based on the standardized Coping Inventory for Stressful Situations (CISS) instrument and clustering algorithm that finds similarities of patients and enables the application of tailored interventions directed and adjusted to specifics of identified segments of patients. We carried out two clustering experiments using k-means clustering with automated parameter tuning. In the first experiment we clustered all examples, in the second experiment we clustered examples that remain after outlier removal. When removing outliers we obtained better clustering results in terms of Davies-Bouldin index.

Keywords: Coping inventory for stressful situations · Clustering · Outlier detection

1 Introduction

Patient treatment is a complex process consisting of proper diagnosis, prescription of medications, application of medical procedures, building patients' awareness of their disease, knowledge transfer about what allowed and forbidden is, and motivation of patients to follow all recommendations. Physicians cover areas related to diagnosis, medical procedures and medications. Psychologists should provide patients with relevant education and encouragement to strengthen treatment results.

Patients profiling is a step towards to deliver tailored interventions. Tailored interventions have been proposed as an appropriate way to improve medication adherence. With the key points of delivering interventions according to a patient's specific barriers to medication adherence, tailored interventions involve an integrated process of identifying nonadherent patients, detecting barriers to medication adherence, and delivering potential solutions according to the patients' barriers to medication adherence [8]. Having characteristics describing selected element of patient's psychological profile a medical

© The Author(s), under exclusive license to Springer Nature Switzerland AG 2022
M. Ceci et al. (Eds.): ISMIS 2022, LNAI 13515, pp. 421–429, 2022.
https://doi.org/10.1007/978-3-031-16564-1_40

unit can divide patients into segments – groups of similar patients and prepare educational materials adjusted to those segments significantly simplifies tailored interventions or makes it possible.

The goal of this study is to group patients according to CISS (coping inventory for stressful situations) methodology in order to build patients' profiles where patients education can be delivered in a personalized way. The rest of the paper is structured as follows: Sect. 2 describes the CISS methodology used to evaluate stress coping strategies, Sect. 3 presents the two clustering experiments, Sect. 4 compares and discusses the clustering results and Sect. 5 concludes the paper.

2   Stress Coping Strategies

Coping with stress and anxiety are permanent aspects of everyday life. Coping strategies play a major role in an individual's physical and psychological well-being when we are confronted with negative or stressful life events. Coping can be defined as "...continuously changing behavioral or cognitive efforts to meet inner and/or outer demands which compel subjective limits of the person or exceed her/his self-resources" [7]. Although several psychometric assessment tools were developed to assess behavioral strategies accepted by people to cope with stressful conditions, Coping Inventory for Stressful Situations (CISS) is superior to other assessment tools. CISS evaluates behavioral strategies towards stressful conditions using three compound criteria: task-oriented coping, emotion-oriented coping, and avoidance coping [4]. Task-oriented coping is a primary control style that is adaptive when situations are appraised as changeable; the focus is maintained, and emotions are controlled. Emotion-oriented coping is a secondary control style adaptive when situations are appraised as unchangeable. The aim is to reduce stress, but over the long term it can increase stress and produce negative outcomes like anxiety, and depression. Avoidance coping is adaptive in the short-term for uncontrollable problems; just do not let the problem bother you and focus on something more interesting. This criterion can be further split into two subparts: avoidant-distracted coping and avoidant-social coping. Evaluation of coping strategies is based on self-assessment using a questionnaire consisting of almost 50 questions, each answered in the scale 1 to 5. The total scores for each of the three criteria are computed as sum of answers of corresponding questions. The higher the value the more a person is engaged in corresponding coping strategies.

Stress coping strategies may influence outcomes of patient treatment. Task-oriented approach, which allows patients to concentrate on lifestyle interventions, and possibly suppress emotional exhaustion, may improve the outcomes of heart failure treatments [5]. This coping style is believed to be the most efficacious, however, only when the stressful situation is controllable. Emotion-oriented coping involves efforts that are self-oriented. Their aim is to reduce emotional tension through emotional responses (e.g. self- blame), self-preoccupation, and fantasizing (daydreaming reactions). This coping style has usually been associated with negative outcomes; however, it can be beneficial in situations that remain highly uncontrollable. Avoidance-oriented coping involves activities and cognitive changes aimed at avoiding the stressful situation. Interventions that target these stress coping strategies may help patients take a more active role in their heart failure management and may improve psychological and cardiac outcomes [6]. Deep knowledge of the patients' profiles basing on stress coping strategies help to tailor the interventions, to deliver treatment and recommendation adjusted to patient's individual stress coping strategy.

3   Clustering CISS Data

3.1   Used CISS Data

The data used in our study have been collected in several cardiological centers in Poland. The sample consists of data about 159 elderly females and 598 elderly males. The available data contain not only the values of CISS scores but also the information about gender, age and BMI of the persons. Table 1 shows basic characteristics of the data: average values for all numeric variables for females and males separately, and in the whole data. The major difference between genders is in the values of emotion-oriented CISS criterion; this difference is reported also in [4]. Table 2 shows the correlation between the CISS criteria, also these values correspond to results of other authors [2]. We used these data to run and compare two clustering experiments; only the values of Task-oriented, Emotion-oriented, and Avoidant scores have been used. We used the RapidMiner system (https://rapidminer.com), a widely used data science platform in our experiments.

Table 1. Basic characteristics of the data.

Gender	Age	BMI	Task-oriented	Emotion-oriented	Avoidant
Female	65,8	27,7	55,2	43,2	45,4
Male	63,2	28,5	54,6	40,8	44,6
Total	63,8	28,3	54,7	41,3	44,8

Table 2. Correlation between CISS criteria.

	Task-oriented	Emotion-oriented	Avoidant
Task-oriented	1	0,099	0,237
Emotion-oriented	0,099	1	0,336
Avoidant	0,237	0,336	1

3.2   Clustering Experiment 1

The first clustering experiment was a standard application of k-means clustering algorithm on the whole data, i.e. clustering of 757 examples. Using the automated parameter tuning option, we let the system to change the number of clusters K in the range 2 to 10 (10 was chosen not to end-up with too many clusters), and the numeric distance/similarity measure to take any value from the list of 12 different functions. This results in more than 100 particular runs of the clustering algorithm. It seems, that the parameter which mostly affects the quality of clustering is the number of clusters. The clustering quality (in terms of Davies-Bouldin index) improves with increasing number of clusters. Davies-Bouldin index [3] evaluates the quality of clustering considering the intra-cluster distance (that should be low) and inter-cluster distance (that should be high). The lower the value of the Davies-Bouldin index, the better the clustering. We obtained the value 1.089 for 10 clusters as the best result. The cluster centroids, together with the number of examples in the clusters are shown in Table 3. Since k-means algorithm computes centroids as means we show also the standard deviations in the table.

Table 3. Cluster centroids for the first experiment

Cluster-ID	Count	Task-oriented	Emotion-oriented	Avoidant
0	127	57,54 ± 4,69	41,81 ± 3,97	39,73 ± 4,06
1	65	62,59 ± 5,65	56,55 ± 4,71	54,46 ± 6,69
2	36	50,47 ± 5,11	29,50 ± 6,78	29,75 ± 4,84
3	61	49,31 ± 5,71	55,80 ± 5,94	39,72 ± 5,24
4	52	66,92 ± 4,23	29,31 ± 6,40	35,50 ± 5,43
5	137	50,69 ± 3,90	47,73 ± 3,98	49,85 ± 4,02
6	81	63,37 ± 5,19	42,59 ± 3,58	52,59 ± 4,19
7	14	30,00 ± 6,65	21,86 ± 5,04	30,64 ± 7,11
8	92	42,95 ± 4,21	33,79 ± 5,97	45,94 ± 5,21
9	92	57,71 ± 4,93	31,13 ± 4,75	49,12 ± 4,61

When comparing the centroids with average values which are 54, 74 for task-oriented coping, 41, 27 for emotion-oriented coping, and 44, 80 for avoidant coping we can for instance see, that cluster 4 consists of people who focus on task-oriented coping and almost ignore emotion-oriented and avoidant coping, people from cluster 3 focus mainly on emotion-oriented coping, people from cluster 1 pay attention to all three forms of coping while people from cluster 7 have no coping strategies at all.

3.3 Clustering Experiment 2

In the second experiment we first removed the outliers and clustered the remaining examples. The idea behind is that after removing outliers, that can be treated separately, we will have better (more homogeneous) clustering results.

We used the local outlier factor (LOF) method to identify the outliers [1]. The method is based on a concept of local density. Values of LOF approximately equal to 1 indicate similar density to neighbors, values smaller than 1 indicate higher density, and values greater than 1 indicate lower density. If an example has lower density than its neighbors, it is an outlier. So a threshold specified for LOF values that are equal or greater than one can be used to decide which examples should be considered as outliers. Table 4 shows the effect of changing the threshold on our clustering experiment. It also shows the values of Davies-Bouldin index for the best clustering.

Table 4. Outlier detection and removal

LOF threshold	No. Outliers	No. Clustered examples	Davies-Bouldin index
1	678	79	0,746
1.25	205	552	1,047
1.5	63	694	1,077
1.75	19	738	1,081
2	6	751	1,083
2.25	3	754	1,085
2.5	2	755	1,085

As can be seen from Table 4, setting the threshold closer to 1 will remove more examples as outliers, and when clustering the remaining data, the quality of clusters will improve. Again, as in experiment 1, we used the automated parameter tuning for number of clusters K in the range 2 to 10 and all numeric distance/similarity measures. The optimal number of clusters was again 10 in all particular runs.

Table 5 shows details of clustering results (number of examples and centroids for every cluster) for outlier detection and removal based on the LOF threshold 1.75. We choose this threshold as a trade-off between the number of removed examples and the quality of clustering. Again, as in the first experiment, we can interpret different clusters. There is a group of people who have low coping strategies in general (cluster 6), people with high coping strategies for all three factors (cluster 1), people that focus mainly on tsk-oriented coping (clusters 0, 1, 5, 7), or people that focus mainly on emotion-oriented coping (cluster 4).

Table 5. Cluster centroids for the second experiment

Cluster-ID	Count	Task-oriented	Emotion-oriented	Avoidant
0	101	60,68 ± 4,46	42,04 ± 4,05	39,57 ± 4,41
1	63	63,30 ± 5,31	55,52 ± 4,60	53,16 ± 5,86
2	29	55,59 ± 6,02	25,79 ± 5,50	28,90 ± 4,50

(continued)

Table 5. (*continued*)

Cluster-ID	Count	Task-oriented	Emotion-oriented	Avoidant
3	81	46,48 ± 4,69	40,05 ± 4,21	37,52 ± 5,01
4	53	50,34 ± 4,99	56,23 ± 5,01	40,81 ± 4,50
5	86	61,22 ± 4,14	40,80 ± 4,19	53,19 ± 3,94
6	41	37,78 ± 5,11	27,59 ± 5,82	38,24 ± 6,09
7	61	64,97 ± 4,44	28,20 ± 4,39	45,59 ± 5,85
8	89	51,71 ± 4,58	32,79 ± 4,46	47,87 ± 4,22
9	134	50,86 ± 4,00	48,33 ± 3,76	49,86 ± 4,00

4   Clustering Comparison and Discussion

Both clustering experiments resulted in grouping people according to their attitude to coping with stress and anxiety according to three dimensions, task-oriented, emotion-oriented and avoidant coping. We can see clear difference in coping strategies between people assigned to different clusters.

When comparing the two experiments we made an interesting observation worth further investigation. Although the difference between Davies-Bouldin index for the reported results for the first and the second experiment is not very big (1.089 for the first experiment, 1.081 for the second experiment) the clustering itself is quite different. This can be seen not only from the Tables 4 and 5 and Figs. 1 and 2 but also from the Table 6 that summarizes the cluster membership of the examples obtained by both

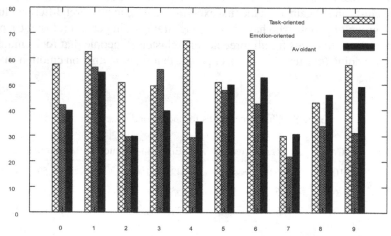

Fig. 1. Plot of centroid values of Task-oriented, Emotion-oriented, and Avoidant criteria for the first experiment.

experiments. Let us stress, that Table 6 shows only those examples that were clustered in both experiments, i.e. examples that were not removed as outliers. The rows in the table correspond to the results of the first experiment, columns correspond to the results of the second experiment. Some clusters are almost mapped one to one between the experiments (e.g. cluster 1 from the first experiment contains almost the same examples as cluster 1 from the second experiment). But some clusters from the first experiment are split into several clusters in the second experiment (e.g. examples from cluster 8 from the first experiment are almost uniformly divided into clusters 3, 6 and 8 in the second experiment).

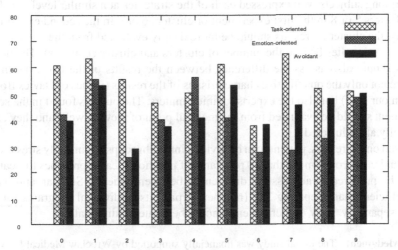

Fig. 2. Plot of centroid values of Task-oriented, Emotion-oriented, and Avoidant criteria for the second experiment.

Table 6. Cross-tabulation of clustering results.

	cluster_0	cluster_1	cluster_2	cluster_3	cluster_4	cluster_5	cluster_6	cluster_7	cluster_8	cluster_9	Sum
cluster_0	89			26	1	2			6	3	127
cluster_1		58								3	61
cluster_2			22	9			4		1		36
cluster_3				7	52						59
cluster_4	8		7					35			50
cluster_5				3			4		5	124	136
cluster_6	4	5				65		1		2	77
cluster_7								11			11
cluster_8				36			26		28	2	92
cluster_9						15		25	49		89
Sum	101	63	29	81	53	86	41	61	89	134	738

5 Conclusions

The performed clustering experiments brought some insights into how people cope with stress and anxiety. We were able to identify groups of persons with similar coping strategies. In the context of coping with the stressor in the form of disease (e.g. chronic heart disease), the most favorable coping strategy is the task-oriented subtype, whereas avoidant strategy is the worst. Emotion oriented coping can be problematic as this may create a risk of depression as a comorbidity of the somatic illness. Therefore, it can be concluded that people with the most favorable configuration of the coping strategy can be found in clusters 0, 1, 4 and 6 for the first experiment and in clusters 0, 1, 5 and 7 for the second experiment. Consequently, the lack of a cluster with the domination of the avoidance-based strategy is a very promising result. It is difficult to predict coping with stress among subjects who expressed each of the strategies at a similar level (eg. Those from clusters 5 or 8 in the first experiment or clusters 6 or 9 in the second experiment). Anyway, the clustering results should be more deeply evaluated from the medical point of view (this applies both to the number of clusters and cluster centroids). The domain experts should also assess the difference between the results of the two experiments. Because not only the quantitative characteristics of the results (values of Davies Bouldin index in our case) but also the experts' opinion maters. The outliers found in the second experiment should be evaluated from the medical point of view as well (and they can be eventually also clustered).

Based on the results, patients can be divided into subgroups of similarly stress coping individuals. 10 groups mean that 10 programs of psychological approaches to treatment should be prepared to address the differences between patients. Such an approach is more efficient than preparing 757 (number of patients) individual programs for each patient separately. Only outlier patients should be treated individually.

Acknowledgment. The initial study was financially supported by Wroclaw Medical University (activity of Laboratory for Applied Research on Cardiovascular System, ST–722). The current analyses were performed within the consortium HeartBIT 4.0 – Application of innovative Medical Data Science technologies for heart diseases funded from the European Union's Horizon 2020 program, grant agreement number 857446.

References

1. Breunig, M.M., Kreigel, H.P., Ng, R.T., Sander, J.: LOF: Identifying density-based local outliers. In: Proceedings of the ACM SIGMOD International Conference on Management of Data, SIGMOD, pp. 93–104 (2000)
2. Choi, Y., et al.: Psychometric properties of the coping inventory for stressful situations in Korean adults. Psychiatry Invest. **14**(4), 427–433 (2017)
3. Davies, D.L., Bouldin, D.W.: A cluster separation measure. IEEE Trans. Pattern Anal. Mach. Intell. **PAMI-1**(2), 224–227 (1979)
4. Endler, N.S., Parker, J.D.: Multidimensional assessment of coping: a critical evaluation. J. Pers. Soc. Psychol. **58**(5), 844–854 (1990)
5. Janowski, K., Kurpas, D., Kusz, J., Mroczek, B., Jedynak, T.: Emotional control, styles of coping with stress and acceptance of illness among patients suffering from chronic somatic diseases. Stress. Health **30**(1), 34–42 (2014)

6. Klein, D.M., Turvey, C.L., Pies, C.J.: Relationship of coping styles with quality of life and depressive symptoms in older heart failure patients. J. Aging Health **19**(1), 22–38 (2007)
7. Bittles, A.H., Parsons, P.A. (eds.): Stress. Studies in Biology, Economy and Society, Palgrave Macmillan UK, London (1996). https://doi.org/10.1007/978-1-349-14163-0
8. Xu, H.Y., Yu, Y.J., Zhang, Q.H., Hu, H.Y., Li, M.: Tailored interventions to improve medication adherence for cardiovascular diseases. Front. Pharmacol. **11**, 510339 (2020)

Application of Association Rules to Classify IBD Patients

Agnieszka Dardzinska[1] [iD] and Anna Kasperczuk[2][✉] [iD]

[1] Department of Mechanics and Applied Computer Science, Bialystok University of Technology,
Wiejska 45c, 15-351 Bialystok, Poland
[2] Department of Biocybernetics and Biomedical Engineering,
Bialystok University of Technology, ul. Wiejska 45c, 15-351 Bialystok, Poland
a.kasperczuk@pb.edu.pl

Abstract. The characterization of inflammatory bowel diseases (IBD) is not always easy. In particular cases of titled lesions, it is difficult for physicians to differentiate between UC and CD. There-fore, it is important to search for new differentiating symptoms. The obtained information will deepen the current knowledge about UC and CD. For this reason, finding an optimal classifier to support further analysis of medical data is a decisive factor for correct patient assignment. Built system using CBA method with very high efficiency is able to predict which group of diseases a new, undiagnosed patient belongs to (sensitivity 94.1%, specificity 85.9%). The built model can be an excellent method to support physicians in decision making.

Keywords: IBD · Association rules · Classification · CBA method

1 Introduction

Methods developed within the framework of knowledge mining from databases, which consist in modeling, i.e. searching for dependencies in data, became an answer to the growing amount of data collected in repositories. The collected data is analyzed and the knowledge gained is used to obtain new information about phenomena. Data mining is the process of complex search for hidden, unknown and potentially useful information from data. Data mining is a key phase of the knowledge discovery process in databases. Its goal is to choose an appropriate algorithm and method for finding patterns and relationships in the dataset under study [1–6].

The discussed phenomenon is also seen in the medical field [1, 7]. Among other things, the use of new measurement procedures and laboratory methods has contributed to the development of statistical research in medicine. An equally important aspect for the development of this field was more accurate maintenance of patients' medical records. At first, physicians performed basic data correlation analysis based on their experience, patient observations, and test results. Simple statistical observations were transformed with the development of technology. With the availability of databases to

M. Ceci et al. (Eds.): ISMIS 2022, LNAI 13515, pp. 430–437, 2022.
https://doi.org/10.1007/978-3-031-16564-1_41

store large amounts of information, it was possible to electronize medical records. It was data mining that made it possible to discover relationships that would have been difficult to see without the use of data.

1.1 Classification Rules

An important method of mining in the area of medical data is classification [3, 4, 8]. In medical systems, it is used primarily to find rules for classifying individual diseases. Then, thanks to the found rules, it is possible to automatically diagnose the next patients. Classification consists in finding mappings of data into a set of predefined classes. Based on the information from the database, a model is built. The database is divided into two sets: training set and test set. The training set is used to build the model while the test set is used to test the model. The model is designed to classify new objects in the database. Classification is a data mining method with teacher (supervision). The classification process can be divided into several steps [2, 8–10]:

1. Building a model (classifier),
2. Model testing,
3. Prediction of unknown values.

The classification performance is evaluated using metrics [1]. In the field of machine learning, specifically in a statistical classification problem, a confusion matrix, also known as an error matrix, is built from which metrics are computed (Table 1) [1, 2],

Table 1. Confusion matrix

Observed events	Expected events	
	Class 1	Class 2
Class 1	TP	FN
Class 2	FP	TN

The following measures are often used in the literature [1]:

1. Sensitivity – (TPR) probability of correct classification provided that the case is positive:

$$TPR = TP/(TP + FN) \tag{1}$$

2. Specificity (TNR) probability of a correct classification, provided that the case is negative:

$$TNR = TN/(TN + FP) = 1 - FPR \tag{2}$$

3. False alarm rate – (FPR) is the rate of instances falsely classified as a given class:

$$FPR = FP/(FP + TN) = 1 - TNR \qquad (3)$$

4. False Discovery Rate – Type I error control rate (FDR). The FDR is designed to control the expected proportion of "discoveries" that are false (incorrect rejections):

$$FDR = FP/(FP + TP) \qquad (4)$$

1.2 Association Rules

Association rules can be compared to "if-then" constructs (if-then), which have found their way into programming languages. The most popular application of association rules is the basket analysis. It involves discovering patterns of customer behavior, that is, finding groups of products that are purchased together. Although association rules have many applications, it is on the example of the aforementioned analysis that a formal description of association rules was created. Association analysis is also used in medicine [2].

First, let us define a set of binary attributes $I = \{i1, i2, ..., in\}$ where $n \in N$. Any set $X \subseteq I$ will be called a "commodity set", while a set of commodities of power k will be called the "k-set of goods". Let the transaction base be the set of executed transactions $D = \{t1, t2, ..., tm\}$, where $m \in N$. A transaction ti, where $i \in N \wedge i \leq m$, can be represented as a vector taking binary values. If commodity ik appears in the transaction then ti $[k] = 1$, where $k \in N \wedge k \leq n$, if commodity ik is not in the transaction then ti $[k] = 0$ (ti $[k]$ denotes the k-th element of vector ti). Each such row represents one transaction.

An association rule is written in the form [2]: $A \Rightarrow B$.

where $A \subset I, B \subset I, A \neq \emptyset, B \neq \emptyset$, and $A \cap B = \emptyset$. The set A is called the set of conditioned values, while the set B is called the set of conditioned values.

Rule measures [2]:

1. **Support:**
   This is a parameter that determines how often an object appears in the dataset, or more precisely, what percentage of all association rules a particular rule represents. The formula represents the ratio of the number of transactions (containing entirely goods from the set A and B) in the data set (transaction database) to the number of all transactions.

$$support(A \Rightarrow B) = \frac{|P_{A \cup B}|}{|P|} \qquad (5)$$

2. **Confidence:**
   This is a parameter that describes how many percent of association rules starting with a particular set of conditional values constitute a given rule. In other words, it is the ratio of the number of occurrences of transactions that contain all goods from set A and B to the number of transactions that contain only goods from set A.

$$confidence(A \Rightarrow B) = \frac{|P_{A \cup B}|}{|P_A|} \qquad (6)$$

3. **Lift:**

The lift measure is a parameter that describes the correlation between the predecessor and successor of an association rule. If the events that are represented by the predecessor and successor of the rule are independent, then the value of the increment parameter is 1.

$$lift(A \Rightarrow B) = \frac{confidence(A \Rightarrow B)}{support(B)} \tag{7}$$

4. **Conviction:**

This parameter was created as an alternative to certainty, which is not as accurate in terms of the direction of association. This measure compares the probability that A would occur in a transaction without B if they were dependent on the actual frequency of occurrence of A without B.

$$conviction(A \Rightarrow B) = \frac{1 - support(B)}{1 - confidence(A \Rightarrow B)} \tag{8}$$

1.3 Classification Using Association Rules Association Rules

The best-known association rule-based classification model is CBA (Classification Based on Association) [11]. The CBA algorithm uses a special type of association rules for classification. These are called class association rules (CAR). These rules have a constraint that the successor of a rule is a single item that is associated with a class label. The CBA algorithm can be divided into 3 steps [12].

- Association Rule Extraction (CAR),
- Pruning and ordering the extracted rules,
- Classifying new objects.

The first stage involves modifying the Apriori algorithm. To reduce the size of the considered CAR sets, pruning based on the pessimistic rule error rate can be applied. According to research published in a paper, the precision of the classifier does not depend on pruning, so implementations of the regular Apriori algorithm can also be used. Regularly mined rules are not used for classification, as the number of rules can be very large, even in small datasets. The generation of conflicting rules, as well as the lack of a default rule, can also be an obstacle. The CBA algorithm dealt with this problem by using data range pruning, which reduced the number of extracted rules.

Two versions of the CBA algorithm can be distinguished, which differ in their running times. The M1 version is slower, while the M2 version, due to data access limitation, runs faster. Rules in pruning are ranked in order of strength:

- A rule is ranked higher if its confidence is greater than the confidence of the next rule,
- If a tie occurs when ranking against the trust value, then the support value is taken into account, then the rule with the higher support value is ranked higher,
- If a tie occurred in both of these cases then the ranking is determined by the time of rule extraction, the higher ranking place is occupied by the rule extracted earlier.

Rules are processed in ranking order. In the case of processing each rule, all matching transactions are removed. If a rule does not cover at least one of the instances, it is also pruned. In the CBA algorithm, data pruning is combined with the cleaning of default rules, i.e. rules added at the end of the rule set with the majority class in the uncovered transactions on the left side and in the empty left side. This allows classification, even in the case where the query does not match any of the of the rules. If the default rule reduces the total number of errors in the learning set then the algorithm cleans all rules below it [13].

1.4 Inflammatory Bowel Diseases

Inflammatory bowel diseases are of interest to researchers worldwide. The term refers to chronic and recurrent diseases of the gastrointestinal tract. This group in-cludes the two conditions of ulcerative colitis (UC) and Crohn's disease (CD). When the clinical picture is relatively diverse, they can be distinguished with greater or lesser certainty. Despite intensive and prolonged research, in many cases the diagnosis is ambiguous, so that methods for a rapid and effective diagnostic method are still being sought [14, 15].

Inflammatory lesions in UC most commonly occur in the colon, but the course of the disease may be variable. In some cases, it is initially mild, with increasing prob-lems within a few weeks. Often, however, the disease begins suddenly. In such cases, due to the lack of results of conservative treatment, surgical treatment is implemented at an early stage of the disease. Often the disease has a pattern of alternating increases in problems and going into remission, after which it flares up again. Patients may have problems throughout their lives [16–18]. For the second condition of CD, when only the intestine is affected it is particularly difficult to distinguish from UC. There are two forms of the disease. The first involves disease involvement of the entire length of the colon, while the second involves segmental lesions throughout the colon [19–22].

2 Material and Methods

The study concerned the analysis of the data collected from patients with IBD. We obtained data about patients of the Department of Gastroenterology and Internal Diseases of Bialystok Clinical Hospital (Poland). The mean age in the CD group (N= 66) was 34.4 years (SD= 14.3), while in the UC group (N= 86) it was 40.8 years (SD= 17.7). Based on the medical data of the patients such as lab results, basic interview results, we want to build a model to classify the patients into two groups: CD and UC. In this work we use CBA method to build classification model. In this study, a classification model was used that was created based on documentation of modules such as caret, arules, arulesCBA, and arulesViz [23]. Additionally, in this paper, we compare the performance of CBA with a classifier using a support vector machine (SVM) to compare the methods and select the most efficient algorithm for our problem. We use Weka 3.8.5 softwere and RStudio 1.3.

3 Result

In line with the research methodology for each classifier, a table of errors was built, respectively for the CBA and SVM classifier – Table 2. Based on the values included in the matrix, the measures assessing the quality of the binary classification were calculated at a later stage. For the mentioned CBA, the sensitivity was 94.2% and specificity was 85.9%. For the mentioned SVM classifier, the sensitivity was 93.1% and specificity was 84.9% (Table 2).

Table 2. Classification results

	TPR	TNR	FPR	FDR
CBA	94.1%	85.9%	14.1%	10.0%
SVM	93.1%	84.9%	15.1%	11.0%

The results obtained by both methods indicate very good classification. The CBA method gave slightly better results of correct classification compared to SVM.

We present some of the strongest rules generated using the CBA algorithm:

1. $Creatinine < 0, 69 AND$ bloodinstool $=''$ $no''AND$ smoking $=''$ $no''ANDPLT <$ $524, 5 \Rightarrow UC$
2. $Creatinine < 0, 69 AND$ bloodinstool $=''$ $no''AND$ smoking $=''$ $no''ANDPLT \geq$ $524, 5 \Rightarrow CD$
3. $Creatinine < 0, 69 AND$ bloodinstool $=''$ $yes''AND$ smoking $=''$ $no'' \Rightarrow UC$
4. $Creatinine < 0, 69 AND$ bloodinstool $=''$ $yes''AND$ smoking $=''$ $yes'' \Rightarrow CD$

Rule 1 and 2: if creatinine is maintained at the level mentioned in the rule, there is no blood in the feces and the patient is not a smoker, then a PLT level greater than or equal to $524.5 \times 10 \char`^ 3$ / uL will be characteristic of CD, and lower than that level will indicate a UC.

Rule 3 and 4: If the patient's creatinine level is kept below 0.69 mg / dL, there is no blood in the feces, the person becomes a smoker, the patient will be diagnosed with CD, but if patients is not smoker will be diagnosed with UC.

4 Conclusion

The conducted analyses of real medical data highlighted that the application of data mining algorithms gives measurable results. This is evidenced by the high predictive quality of the built model (TPR = 94.1%; TNR = 85.9%). The built model, due to its excellent classification and predictive abilities, can be a recommendation system for the physician when making a diagnosis. The results of the work have been presented to clinicians, who see the possibility of testing them in a hospital setting.

Based on the results, it can be concluded that the combination of two data mining methods gives better results when classifying medical data.

More accurate classification allows to reduce the cost of further diagnostics,

- Saving valuable time of doctors and patients,
- Reducing the time from noticing symptoms to diagnosis,
- Reducing the stress associated with suspected disease,
- Reducing the number of unnecessary invasive procedures when a disease is false-ly suspected.

However, looking at the time in which the classification was done, it can be concluded that the hybrid method is slower than the classification done with the J48 classifier. Although the time difference between the hybrid classifier and traditional classifiers was only 1.7 s, in the case of a huge database this difference could be much larger, which could affect the functionality of the method.

It would be important to test the algorithm on a larger dataset to get more confident results.

Acknowledgements. This research was funded by Ministry of Science and Higher Education in Poland, grant number W/WM-IIB/3/2021 and WZ/WM-IIM/3/2020.

References

1. Han, J., Kamber, M.: Data Mining: Concepts and Techniques, Morgan Kaufmann Publishers, Second Edition, 21–27 (2006)
2. Dardzinska, A.: Action Rules Mining. Springer, Heidelberg (2013). https://doi.org/10.1007/978-3-642-35650-6
3. Dardzinska, A., Kasperczuk, A.: Decision-making process in colon disease and Crohn's disease treatment. Acta Mech. Autom. **12**(3), 227–231 (2018)
4. Dardzinska, A., Romaniuk, A.: Mining of frequent action rules, machine intelligence and big data in industry. In: 6th International Conference on Pattern Recognition and Machine Intelligence, pp. 87–95 (2016)
5. Dzhuguryan, T., Deja, A.: Sustainable waste management for a city multifloor manufacturing cluster: a framework for designing a smart supply chain. Sustainability **13**(3), 1540 (2021)
6. Dzhuguryan, T., Deja, A., Wisnicki, B., Jozwiak, Z.: The design of sustainable city multi-floor manufacturing processes under uncertainty in supply chains. Sustainability **12**(22) (2020)
7. Gürdal, O., Dardzinska, A.: A new approach to clinical medicine by action rules. Int. J. Develop. Res. **7**(1), 11032–11039 (2017)
8. Ras, Z., Dardzinska, A.: From data to classification rules and action. Int. J. Intell. Syst. Wiley **26**(6), 572–590 (2011)
9. Kasperczuk, A., Dardzinska, A.: Comparative evaluation of the different data mining techniques used for the medical database. Acta Mech. Autom. **10**(3), 233–238 (2016)
10. Zdrodowska, M., Dardzinska, A.: Classification and action rules in identification and self-care assessment problems. Technol. Health Care **30**(1), 257–269 (2022)
11. Jiří, F., Tomáš, K.: Classification based on associations (CBA) - a performance analysis. In: CEUR Workshop Proceedings, vol. 2204, pp. 1–9 (2018)

12. Vanhoof, K., Depaire, B.: Structure of association rule classifiers: a Review. In: 2010 International Conference on Intelligent Systems and Knowledge Engineering (ISKE) (2010)
13. Hahsler, M., Johnson, I., Kliegr, T., Kuchar, J.: Associative classification in R: arc, arulesCBA, and rCBA. R J. **11**(2), 254–267 (2019)
14. Dolapcioglu, C., Soylu, A., Kendir, T., Ince, A.T., Dolapcioglu, H., Purisa, S.: Coagulation parameters in inflammatory bowel disease. Int. J. Clin. Exp. Med. **7**, 1442–1448 (2014)
15. Giuffrida, P., Corazza, G.R., Di Sabatino, A.: Old and new lymphocyte players in inflammatory Bowel disease. Dig. Dis. Sci. **63**(2), 277–288 (2017). https://doi.org/10.1007/s10620-017-4892-4
16. Gren, S.T., Grip, O.: Role of monocytes and intestinal macrophages in Crohn's disease and ulcerative colitis. Inflamm. Bowel Dis. **22**, 1992–1998 (2016)
17. Cappello, M., Morreale, G.C.: The role of laboratory tests in Crohn's disease. Clin. Med. Insights Gastroenterol. **9**, 51–62 (2016)
18. Daniluk, J., Daniluk, U., Reszec, J., Rusak, M., Dabrowska, M., Dabrowski, A.: Protective effect of cigarette smoke on the course of dextran sulfate sodium-induced colitis is accompanied by lymphocyte subpopulation changes in the blood and colon. Int. J. Colorectal Dis. **32**(11), 1551–1559 (2017). https://doi.org/10.1007/s00384-017-2882-9
19. Priyamvada, S., Gomes, R., Gill, R.K., Seksena, S., Alrefai, W.A., Dudeja, P.K.: Mechanisms underlying dysregulation of electrolyte absorption in IBD associated diarrhea. Inflamm. Bowel Dis. **21**, 2926–2935 (2015)
20. Sarfati, M., Wakahara, K., Chapuy, L., Delespesse, G.: Mutual interaction of Basophils and T cells in chronic inflammatory diseases. Front. Immunol. **6**, 399 (2015)
21. Schieffer, K.M., Bruffy, S.M., Rauscher, R., Koltun, W.A., Yochum, G.S., Gallagher, C.G.: Reduced total serum bilirubin levels are associated with ulcerative colitis. PLoS ONE **12**, e0179267 (2017)
22. Zho, G.X., Liu, Z.J.: Potential roles of neutrophils in regulating intestinal mucosal inflammation of inflammatory bowel disease. J. Dig. Dis. **18**(9), 495–503 (2017)
23. Johnson, I.: Package 'arulesCBA' (2020). https://doi.org/10.32614/RJ-2019-048

Unsupervised Learning Based Rule Generating System with Temporal Features Extractions Tuned for Tinnitus Retraining Therapy

Xin Zhang[1]([✉]), Xinyan Shi[1], and Pamela Thompson[2]

[1] University of North Carolina at Pembroke, Pembroke, NC 28372, USA
xin.zhang@uncp.edu
[2] Catawba College, Salisbury, NC 28144, USA

Abstract. The paper studies the effect of Tinnitus Retraining Therapy using an experimental approach. Each patient was assigned to either the control group where a treatment involves sound masking tests, or the experimental group in which a treatment involves counseling and/or sound masking tests. One additional feature with these medical records is that the treatment varies from one to another due to different frequencies of the visits. The irregular sampling rate and the sparsity in the treatment records present challenges to data analysis using temporal sound feature extraction design. The authors proposed an unsupervised learning system to uncover the relationship among treatment patterns, symptom improvements, and other factors in the data with association rules based on specially designed temporal features and sound descriptors.

Keywords: Association rules · Sparse data · Sound descriptors · Tinnitus retraining therapy

1 Introduction

Tinnitus, an extremely common condition, is typically defined as a perception of sound that is not related to an external sound source. Even though only a fraction of those who experience tinnitus are significantly disturbed [1–4], it is affecting about 17% of the general population around the world (44 million people in the USA) [5]. Being classified as a symptom and not a disease, tinnitus does require treatment as it can cause significant emotional and somatic distress and can significantly influence patients' quality of life, particularly when it becomes a chronic problem [6]. Tinnitus Retraining Therapy (TRT) [7], developed by Dr. Jastreboff since the mid '80s, is a treatment model with a high rate of success and is based on a novel neurophysical approach. This method uses a combination of sound masking with low level, broad-band noise, and counseling to achieve the habituation of tinnitus, i.e., the patient is no longer aware of their tinnitus, because even though the pitch and loudness of tinnitus might remain the same, tinnitus is no longer so intrusive.

Action rules learning explored the premium treatment operations to improve the efficiency of the therapy and potentially lower the treatment cost, where decision trees were

M. Ceci et al. (Eds.): ISMIS 2022, LNAI 13515, pp. 438–445, 2022.
https://doi.org/10.1007/978-3-031-16564-1_42

used to extract rules [8–12] and the results were promising. Decision trees have been widely used in data mining to efficiently extract rules for the purpose of classification; however, this type of learning algorithm is not flexible to learning potential interesting rules among all factors and features [13]. Association rules based on unsupervised learning have been successfully used in exploring the relationships among attributes without the limitation to preset any class labels to form the right-side part of the rules, the consequent. This characteristic will allow more potentially interesting rules to be uncovered [15]. The authors proposed to use unsupervised learning with association rules to explore more interesting rules to learn essential patterns in the data relative to tinnitus and its related symptoms.

2 Data Preprocessing

Tinnitus Retraining Therapy combines medical evaluation, counseling, and sound therapy to successfully treat a majority of patients. Based on a questionnaire from the patient as well as an audiological test, a preliminary medical evaluation of patients is required before beginning TRT. Sensitive information containing privacy from the medical evaluation were removed before the data analysis stage. However, some medical information, such as a list of medications the patient may take and other conditions that might be present, such as diabetes, was included in the tinnitus dataset since these information is relevant to the research of tinnitus.

Tinnitus is a symptom of many pathologies – even in one patient, several different types of tinnitus might coexist. For example, tinnitus is accompanied by hyperacusis in about 40% of the cases. Hyperacusis is a decreased tolerance of sound and can be a serious problem. Some patients experience hyperacusis without tinnitus. Tinnitus Retraining Therapy can restore totally or partially the normal level of sensitivity to sound. Patient categorization (see the table below) is performed after the completion of the medical evaluation, which is a structured interview guided by special forms and audiological evaluation. Patients were classified into the following categories in the initial visit based on their symptoms as shown in the following table [6].

Besides the above patient categories, a clustering based on clinic visit pattern had been performed. We postulate that the visit frequency and the number of total visits are important factors to the success of the therapy according to the clinic doctors. Therefore, we used a clustering method, where records were grouped by similar visiting patterns, while the visiting history of each patient is discretized into durations, starting from the date of his/her initial visit date in terms of weeks, and transformed as vectors of the same length for grouping. Assuming two patients denoted by p, q, patient p visits are represented by a vector $v_p = [v_1, v_2,..., v_n]$ whereas vector $w_q = [w_1, w_2,..., w_m]$ represents visits of patient q where the number of total visits of patient p is n and the number of total visits of patient q is m. The distance $\rho\ (p, q)$ between p, q and the distance $\rho\ (q, p)$ between q, p is defined as

$$\rho(q, p) = \frac{\sum_{i=1}^{k} \min(v_i - w_i, v_i - w_{i+1}, v_i - w_{i+2})}{k}, k = \min(m, n), \quad (1)$$

where w_{i+1} will be set to 0 when $i + 1 > m$ so on and so forth. Also, $v_i - 0$ would be bigger than any other distance differences and therefore will be ignored. This algorithm quickly

Table 1. Categories of patients with Tinnitus and Hyperacusis.

Type	Impact on life	Tinnitus	Subjective hearing loss	Hyperacusis	Prolonged sound induced exacerbation	Treatment
0	Low	Present	–	–	–	Abbreviated counseling
1	High	Present	–	–	–	Sound generators set at mixing point
2	High	Present	Present	–	–	Hearing aid with stress on enrichment of the auditory background
3	High	Irrelevant	Irrelevant	Present	–	Sound generators set above threshold of hearing
4	High	Irrelevant	Irrelevant	Present	Present	Sound generators set at the threshold

drops the less matching dates of a visit and selected pairs of visits even of different index number.

A threshold of an empirical value was applied to remove those records which have large distance values to form a tolerance class, so that all members of a patient cluster were ensured to have similar visiting patterns. Then visit-based temporal features were computed for all group members. Constructing tolerance classes helps to identify the right groups of patients for which useful temporal features can be built. By increasing the threshold value, larger classes would be produced for the process of knowledge extraction, but the information included in temporal features would be less accurate. On the other hand, if the threshold value is too small, the size of tolerance classes might be also too small to get any useful information through the knowledge extraction process. Since the original database was small, the authors only produced a small number of classes so that a good number of the patient records were kept in several classes. Additionally, visits except for those producing the minimum distances were kept. This means, after the data preprocessing, all the patients have the same total number of visits with similar visiting frequency in the dataset.

Important to the equation is the constraint that recovery of any patient with only one visit cannot be evaluated. Due to the intuition of the process, it is reasonable to assume that one visit would not be enough to fix the problem and that it was due to an interruption with some unknown cause. Therefore, such records have been removed from dataset for the rest of the experiments.

During a period of medical treatment, a doctor may change the treatment from one category to another based on the recovery specifics of the patients and the symptoms occurred during treatment. Additionally, the category of patient may change over time (e.g., the patient category may be changed from 3 to 1 if hyperacusis is disappearing). Other typical categorical features which may change over time in the dataset include sound-instrument types as well as visiting frequencies. At this stage, there were two types of data: one is numerical, such as scores for emotions, functions, and catastrophes related to the tinnitus problems; the other is categorical, such as instruments used in the therapy and patient categories. In terms of stability, we categorize the data into two types: Stable and Flexible [14]. Stable, compared with Flexible, is defined as an attribute that should maintain the same value over time throughout most of the records.

3 Temporal Feature Extraction

We employ temporal feature extraction method in order to study the effect of treatment duration on patients' recovery, which are evaluated based on three areas: functional, emotional, and catastrophic. We examine both the individual score and the total scores of them to evaluate whether the improvement has been made in a short or a long duration of treatment. Specifically, temporal feature extraction involves segmenting the visits into vectors based on the duration of patients' visit and use the recovery speed for each duration vector to evaluate the performance of different durations.

$$A_i(p) = \frac{A_k^i - A_{k-1}^i}{W_k - W_{k-1}}, k = 2, \ldots n, p \in m_j, j = 1, ..t \tag{2}$$

$$B_i(p) = \frac{A_n^i - A_1^i}{W_n - W_1}, p \in m_j, j = 1, ..t \tag{3}$$

$$C_i(p) = A_k^i - A_{k-1}^i, k = 2, \ldots n, p \in m_j, j = 1, ..t \tag{4}$$

$$D_i(p) = A_n^i - A_1^i, p \in m_j, j = 1, ..t, \tag{5}$$

where $A_i(p)$, representing the derivative of a segment of a visit vector, is the recovery speed of the i_{th} type for patient p of a group m_j, W is the number of weeks from the initial visit till the k_{th} visit, and n is the total visit; $B_i(p)$, representing the derivative of the whole visit vector, is the overall recovery speed of a patient; $C_i(p)$ represents the difference of the scores of a segment of a visit vector; $D_i(p)$ represents the difference of the scores of the whole vector.

The sounds pieces used in the treatment are not recorded completely as a full piece of acoustic recording; instead, only their loudness level, instrument type, and the main fundamental frequencies are documented in each visit. Therefore, complex sound descriptors for large sampling rate about sound vibrations are not applicable in this case. Moreover, to evaluate individual sound descriptors of all the visits for a patient, the median values were included in a larger dataset into different experiments in this research. The next section describes the details of the experiment design.

4 Experiment Design and Results

Two types of dataset were used in the experiments: the control group in which patients had similar visiting patterns, and the experimental group in which patients received sound masking treatment. The original dataset included 744 patients, whose total visits varies from 1 to over 15. We removed 203 patient records since they only have one visit. In the control group, each sub-dataset had 37 features, while the experiment group had 62 features. Altogether, nineteen clusters were generated, among which five clusters contained patients with four transformed visits, while fourteen clusters contained patients with three transformed visits. The temporal features showed that the majority of the patients had the best recovery rate in their first week.

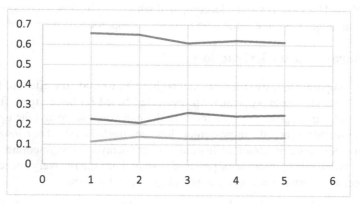

Fig. 1. Recovery rate comparison in the three-week durations of the five clusters with four transformed visits (blue – first week, red – second week, green – third week).

The above figure shows that, among those patients who took four visits in the three week-durations of TRT, about 60%–65% of them had the best recovery results in the first week (blue line). The red line shows the percentage of patients (about 20%–27%), whose best recovery occurred in the second week. The green line shows the percentage of those (about 11%–15%), whose best recovery occurred in the third week. Nevertheless, the results did not indicate that a patient should stop after the first week of treatment.

Figure 2 shows that among patients, who took three visits in a two-week duration of TRT, about 58%–65% of them have the best recovery in the first week for each of the fourteen clusters as shown in the blue line in the figure below.

We use five clinical categories as listed in Table 1. Among 212 patients, 33 were in category 0, 75 were in category 1, 50 were in category 2, 45 were in category 3, 9 were in category 4. The figure below shows that patients of the clinic category 0 had the best average recovery speed, while patients of category 4 had the least. The original dataset is very unbalanced, where patients of category 4 were less than 4% (26 out of 735 patients); therefore, patterns found related to this category may need more data to support. The blue marks represent individual patient recovery speed of each category, while the red line denotes the average speed of each category.

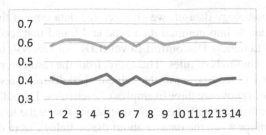

Fig. 2. Recovery rate comparison in two-week durations of the 14 clusters with three transformed visits (blue – first week, red – second week)

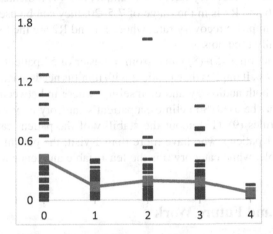

Fig. 3. Overall recovery speeds among the five categories of patients.

Weka 3.8.6 was used to generate association rules by the Apriori algorithm. Similar rules from multiple clusters were observed, while only representative rules were listed and compared in this paper, since the total number of association rules were very large.

$LR12=120 \wedge LL12=120== > class=y$ *(confidence 88%, lift 1.01, support 31%) (1)*

$LL8=120== > class=recovery$ *(confidence 88%, lift 1.01, support 35%) (2)*

$LR8=120== > class=recovery$ *(confidence 88%, lift 1.01, support 34%) (3)*

$R1=(3–16] \wedge R2=(7.5–20]== > class=y$ *(confidence 91%, lift 1.05, support 22%) (4)*

$p2=T \wedge ful=A== > class=y$ *(confidence 91%, lift 1.06, support 62%) (5)*

$p1=T \wedge ful=A== > class=y$ *(confidence 97%, lift 1.13, support 63%) (6)*

$ful=A== > class=y$ *(confidence 89%, lift 1.04, support 67%) (7)*

$fu2=T== > class=y$ *(confidence 90%, lift 1.04, support 37%) (8)*

$C1=0== > C2=0$ *(confidence 29%, lift 1.02, support 43%) (9)*

$C1=1== > C2=1$ *(confidence 94%, lift 3.40, support 34%) (10)*

$C1=2== > C2=2$ *(confidence 90%, lift 4.56, support 28%) (11)*

$C1=3== > C2=3$ *(confidence 86%, lift 4.24, support 17%) (12)*

$C1=4== > C2=4$ *(confidence 81%, lift 20.52, support 5%) (13)*

Analyzing the audiology dataset, we changed the data type of the sound loudness values of different levels into nominal data type and found that recovery was usually associated with the Loudness Discomfort Level (LDL) of levels 4, 6, 8, and 12 of the value of 120 dB by thousands of rules. Therefore, from the resultant rules, we observed that LDL 4, 6, 8, and 12 may be set to the value of 120 dB to achieve positive results. Rules (1)–(3) as shown above are three examples of such rules. The authors also observed that the rules were consistent with the clinic operations where the lower the LDL level, the lower its value range when the dataset about the audiology were discretized into 10 bins per sound loudness column. Because 85% of the records in this dataset were related to recovery, while only 15% were related to no recovery, rules of confidence over 85% are useful to improve the recovery rate. For example, rule (4) indicates that when R1 is in the range of 3–16 and R2 is in the range of 7.5–20, they can be used together in the same clinic visit to improve recovery rate, where R1 and R2 are the loudness values of two types of clinically used noises.

Another example, rules (5)–(8) came from a cluster of 52 patients, where the total recovery rate was 79%. It implies that a patient with tinnitus needs at least two following-up treatments with both audiology and counseling, where in between such treatments a telephone visit may be used. Sometimes, a patient's category may vary after a period of treatment. The rules (9)–(13) about the stability of the patient categories indicate that, among the 551 patients who have more than 1 visit, the patient categories 1 and 2 are relatively stable, while category 0 is the least stable and tends to change to other categories.

5   Conclusion and Future Work

Tinnitus is not a well-defined disease. The authors developed an unsupervised learning system to extract interesting rules among the medical records of the TRT process to uncover the essentials of tinnitus and its related symptoms. The authors discovered that the 1st week treatment achieves the best recovery rate for the majority of the patients in the dataset though the recovery continued through the rest of the treatment process. The system successfully identified treatment patterns to improve the recovery rate. Action rules may be uncovered among the vast amounts of the rules from the association rules learning to identify how changes in treatment affects the recovery rate.

References

1. McFadden, D.: Tinnitus: facts, theories, and treatments. National Academy Press, Washington (DC) (1982)
2. Coles, R.R.A.: Epidemiology, aetiology and classification. In: Reich, G.E, Vernon, J.A. (eds.) Proceedings of the Fifth International Tinnitus Seminar, 1995, Portland, Or., U.S.A. Portland, Or. :American Tinnitus Association, pp. 25–30 (1996)
3. Pilgramm, M., et al.: Tinnitus in the federal Republic of Germany: a representative epidemiological study. In: Hazell, J.W.P. (ed.) Proceedings of the Sixth International Tinnitus Seminar, 1999, Cambridge, UK, London. Tinnitus and Hyperacusis Centre, pp. 7–64 (1999)
4. Davis, A., El Refaie, A.: Epidemiology of tinnitus. In: Tyler, R. (ed.) Tinnitus handbook. San Diego, Singular, Thomson Learning, pp. 1–23 (2000)

5. Tinnitus and Hyperacusis Center Homepage. https://www.tinnitus-pjj.com/. Accessed 01 May 2021
6. Snow, J., Ballenger, J.: Ballenger's Otorhinolaryngology Head and Neck Surgery, 16th edn. BC Decker Inc, Hamilton, Ontario, Canada (2003)
7. Jastreboff, P.J.: Phantom auditory perception (tinnitus): mechanisms of generation and perception. Neurosci. Res. **8**(4), 221–254 (1990)
8. Zhang, X., Ras, Z.W., Jastreboff, P.J., Thompson, P.L.: From tinnitus data to action rules and tinnitus treatment. In: 2010 IEEE International Conference on Granular Computing (GrC), pp. 620–625. IEEE (2010)
9. Thompson, P.: Mining for knowledge to build decision support system for diagnosis and treatment of tinnitus. In: Unc Charlotte Electronic Theses And Dissertations (2011)
10. Tarnowska, K.A., Ras, Z.W., Jastreboff, P.J.: Decision Support System for Diagnosis and Treatment of Hearing Disorders. Springer International Publishing, Cham (2017). https://doi.org/10.1007/978-3-319-51463-5
11. Tarnowska, K.A., Ras, Z.W., Jastreboff, P.J.: Mining for actionable knowledge in Tinnitus datasets. In: Wang, G., Skowron, A., Yao, Y., Ślęzak, D., Polkowski, L. (eds.) Thriving Rough Sets. Studies in Computational Intelligence, vol. 708, pp. 367–395. Springer, Cham (2017). https://doi.org/10.1007/978-3-319-54966-8_18
12. Tarnowska, K.A.: Emotion-based music recommender system for Tinnitus patients (EMOTIN). In: Ras, Z.W., Wieczorkowska, A., Tsumoto, S. (eds.) Recommender Systems for Medicine and Music. Studies in Computational Intelligence, vol. 946, pp. 197–221. Springer, Cham (2021). https://doi.org/10.1007/978-3-030-66450-3_13
13. Tan, P., Steinbach, M., Karpatne, A., Kumar, V.: Introduction to Data Mining, 2nd edition, Pearson Education (2019)
14. Ras, Z.W., Wieczorkowska, A.: Action-rules: how to increase profit of a company. In: Zighed, D.A., Komorowski, J., Żytkow, J. (eds.) Principles of Data Mining and Knowledge Discovery. Lecture Notes in Computer Science (Lecture Notes in Artificial Intelligence), vol. 1910, pp. 587–592. Springer, Heidelberg (2000). https://doi.org/10.1007/3-540-45372-5_70
15. Martin, D., Rosete, A., Alcalá-Fdez, J., Herrera, F.: A new multi-objective evolutionary algorithm for mining a reduced set of interesting positive and negative quantitative association rules. IEEE Trans. Evol. Comput. **18** (1), 54–69 (2014)

Industrial Applications

TrueDetective 4.0: A Big Data Architecture for Real Time Anomaly Detection

Luciano Argento[✉], Erika De Francesco, Pasquale Lambardi,
Paolo Piantedosi, and Carlo Romeo

Relatech S.p.a., Via Anguissola, 23, 20146 Milano, MI, Italy
{luciano.argento,erika.francesco,pasquale.lambardi,paolo.piantedosi,
carlo.romeo}@relatech.com
https://www.relatech.com/

Abstract. Industry suffers from many machine-related problems, such as breakdown, failures, personnel safety, and management cost. Predictive maintenance is an industrial and research area that is permeating goods and services production systems, aimed at preventing critical issues in machinery and workplaces, and reducing the costs in terms of resources, time and money caused by incoming risk events that can slow or even stop the production. This paper presents TD4 a Big Data IoT architecture able to: (i) acquire huge amounts of data from real-time sensor streams, (ii) analyze and prepare the data, scaling over a network of distributed working nodes, (iii) perform real-time fault prediction. Experiments on well-known benchmarks show the applicability of the proposed architecture on different real scenarios.

Keywords: Predictive maintenance · Big data · IoT · Distributed systems · Scalable AI · Smart industry

1 Introduction

In manufacturing industries, machinery monitoring represents a crucial operation to allow the production chain to operate seamlessly. Monitoring usually consists of checking the operating parameters of machinery, which are then compared with thresholds; whenever excessive deviations are recorded, maintenance is carried out to bring the parameters back below the thresholds. In general, many industrial settings require the use of complex machinery and systems, which are sometimes subject to various types of malfunctions and failures. The occurrence of a machine failure could bring the production chain to a halt, with serious consequences on the company's safety and/or economy. In this context, the use of predictive maintenance approaches, in combination with Machine Learning technologies, can dramatically reduce the costs associated with machinery failures, maintenance personnel and their safety, since it could provide actionable knowledge that can trigger targeted maintenance interventions. Prediction can

© The Author(s), under exclusive license to Springer Nature Switzerland AG 2022
M. Ceci et al. (Eds.): ISMIS 2022, LNAI 13515, pp. 449–458, 2022.
https://doi.org/10.1007/978-3-031-16564-1_43

be achieved through the use of monitoring and forecasting approaches in order to conduct timely planning and thus avoid serious damage to the plant. This makes it possible to prolong the life of machines and their proper operation. This work proposes TrueDetective 4.0 (TD4) a Big Data compliant software architecture designed to capture, process and analyze an infinite stream of data in real time. The system ingests data from multiple sources (e.g. IoT sensors) and provides smart batch processing, aimed at building train sets for Machine Learning models used to discover anomalies or unexpected flow of information in the data streams. In addition, TD4 defines and exploit an explanation algorithm to detail the cause of the prediction. Once processed, the data is channeled into a data analytics store, ready to be analyzed and monitored.

The remainder of the paper is organized as follows: Sect. 2 describes the related work; Sect. 3 presents the architecture of TD4; Sect. 4 illustrates the experiments conducted and Sect. 5 concludes the paper.

2   Related Works

In [6], authors propose a cloud-based architecture operating on multiple sensors that can be applied in every industrial domain, especially in the Health Domain. The incoming data is stored in the cloud, making it able to run the same algorithm by many different devices at the same time while reducing development costs. The work [2] presents a flexible framework that combines: relevant environmental data, high-level enterprise systems, state-of-the-art indicator models, and forecasting techniques. The architecture results in a great variety of customizable techniques.

The authors of [4] realize a semi-supervised machine learning approach to improve the sensor feature selection in a transport system. After measuring the sensor contribution and potential sensor contribution metrics, their architecture dynamically updates the sensors by removing the old ones and adding newer candidates.

In [7], the authors present a new framework focused on maintenance cost reduction for a telecommunication network. This framework includes real-time production monitoring and business processes to create an innovative business process and notation meta-model.

A Big Data architecture for the medical domain is described in [10]. The health status of the biomedical devices is analyzed in real-time and sent to the architecture, which classifies incoming data and notifies the right device.

The authors of [8] propose a 4-layer transport-oriented architecture able to realize a policy to manage a great number of products, thanks to the implementation of a recurrent LSTM machine learning model, embedding all the preprocessing steps and providing analysis tool.

The solution we propose takes inspiration from all the most relevant achievements reported in the industrial literature. In particular, we focus on the design and development of a Big Data architecture capable of performing preprocessing and forecasting in real time, to meet the industrial sector needs. The solution can easily integrate with new data analysis components and machine learning algorithms to extend its capabilities. Our architecture benefits from a high degree of abstraction that makes it possible to easily adapt it to different application domains. At the time of this writing, the developed anomaly detection module is based on a recurrent Deep Learning architecture, using LSTM cells in the AutoEncoder [1] configuration (Fig. 1).

3 TD4 Architecture

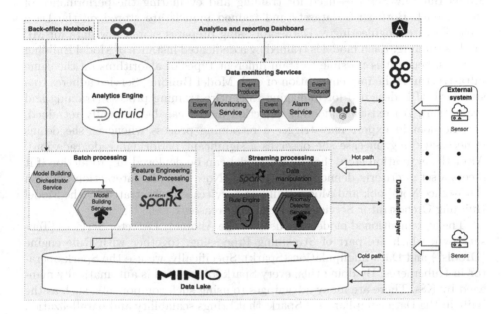

Fig. 1. Architecture of the proposed solution.

The architecture of TD4 comprises the following high-level constituent parts: Data Transfer Layer, Batch Processing, Streaming Processing, Data Lake, Analytics Engine, Data Monitoring Services, Analytics and Reporting Dashboard and Back-office Notebook. The architecture is managed and orchestrated through Kubernetes[1] (aka K8s), with great benefits for cloud deployment. The Data Transfer Layer (Kafka)[2] component is responsible for managing the data flow from the external to the internal systems and between the inner components of the architecture itself. Kafka was chosen because is a real-time, distributed messaging platform that supports horizontal scalability, low latency, fault tolerance, as well as built-in QoS (Quality of Service) and tools to develop custom QoS to meet specific data delivery requirements. Ingested data follow two paths: i) the cold path, through which they flow into the Data Lake where they will

---

[1] https://kubernetes.io/.

[2] https://kafka.apache.org/.

be processed in batch mode and ii) the hot path, for real time anomaly detection stream processing. The Data Lake (MinIO)[3] represents the distributed data store where data, produced by both the external system and other internal components, are channelled and preserved for long time storage. Batch processing is carried out by the subcomponents of Batch Processing, i.e. Model Building Services (TensorFlow)[4] and Feature Engineering & Data Processing (Spark)[5] These operations require human intervention; a domain expert performs data analysis through the Back-office Notebook component, that is a client that offers tools to interactively conduct data analysis and, ultimately, build the datasets that Model Building Services need for training and evaluating the performance of their model. Both such datasets and trained models are stored on the Data Lake. Feature Engineering & Data Processing prepares stored raw data for ML models training. Each model is trained by a dedicated instance of Model Building Services, hence it is possible to train different types of algorithms on the same infrastructure. The implementation of each Model Building Service adheres to a standard API, designed for managing the entire training process (training and evaluation). An existing anomaly detection model can be manually retrained, i.e. the domain expert may trigger a retraining process whenever she deems it necessary (e.g. in case she observes a significant performance degradation). After the evaluation step, the trained model can be deployed in production if it meets a set of criteria defined by the expert. Note that the interaction between Back-office Notebook and Model Building Services is orchestrated by the Model Building Orchestrator Service for decoupling reasons.

The aforementioned models are used by the Anomaly Detector Services (TensorFlow), which are part of Streaming Processing, together with Rule engine (Drools[6]) and Data manipulation (Spark). Specifically, we use the Spark operator in Kubernetes. By doing this, every Spark job created is automatically managed by K8s. There are many advantages to using such components, such as the built-in Big Data compliance of Spark, that brings scalability and parallelization to the architecture. But also, Spark is effortlessly integrated with Kafka, and can be easily connected to MinIO. The Data that flows through the hot path are first preprocessed by Data Manipulation. Data Manipulation treats the incoming data as time-series and aggregates them by means of moving average windows; furthermore it takes advantage of a watermarking mechanism to manage data arriving late. The output of this process is feed into a specific instance of Anomaly Detector Services for further analysis. Each Anomaly Detector Service is designed to detect anomalies in the behaviour of a target machine and execute an explanation algorithm to describe the result of the detection. The output produced by the latter component is processed by Rule Engine, whose goal is to determine whether or not there are anomalies, based on a set of rules, accurately created by a domain expert. Rule Engine encodes business logic in the form of rules represented as "if - then"

---

[3] https://min.io/.
[4] https://www.tensorflow.org/.
[5] https://spark.apache.org.
[6] https://www.drools.org/.

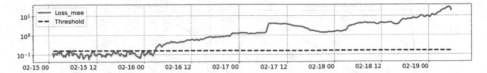

Fig. 2. Autoencoder reconstruction error over time for the second bearing time-series. The system failure occurs at the end of the time-series, Feb. 19, while the first anomaly signals are from Feb. 16.

conditions, which are used to determine the presence or absence of an anomaly in the functioning of the external system. The result of this operation is written to the Data Transfer Layer, through which it gets consumed by multiple components. The data associated with the result are stored on the Data Lake and, subsequently, they can be transferred to the Analytics Engine (Druid)[7], whose purpose is to represent such data in a way that facilitates the analysis and visualisation of Big Data. Druid was chosen because is a scalable distributed system designed as a real-time analytics database that offers low latency queries. Furthermore Druid has a cloud-native, fault tolerant architecture. On top of that it is fully integrated with the Apache ecosystem (Spark and Kafka). The output of Rule Engine is also consumed by Data Monitoring Services (Node.js[8]), which help the system provide real-time machine behaviour monitoring. Finally, real-time machine behaviour monitoring and data analysis and reporting functionalities are made available to the end user through the Analytics and Reporting Dashboard component (Angular.js[9], Superset[10]).

4   Experimentation

This section describes the experiments conducted to evaluate: i) the machine learning model performance and ii) the system performance in terms of Streaming Processing Delay (SPD). The SPD is defined as the time elapsed from the first data acquisition, needed for anomaly detection, and the output produced by an anomaly detector service. Two standard datasets were used for the experiments: the former is the *Bearing dataset*, which is composed of four time-series coming from accelerometers that measure the accelerations of some bearings installed on a shaft coupled to an AC motor [5]; the latter is the *Turbofan dataset*, it contains Run-to-Failure simulated data from turbo fan jet engines [9]. These datasets were used to simulate telemetry data produced by the external system.

Machine Learning Performance. With regard to the Bearing dataset, the ML performance were assessed via observation of the time-series graphs, depicting the reconstruction error, on a preserved test slice of the data. The reason

---

[7] https://druid.apache.org/.

[8] https://nodejs.org/en/.

[9] https://angularjs.org/.

[10] https://superset.apache.org/.

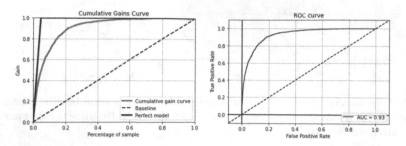

Fig. 3. Autoencoder ROC curve and cumulative gain chart for the Turbofan dataset. The ROC-AUC is 0.93.

behind this choice is that the dataset consists of only three different time-series, therefore a more in depth analysis wouldn't be statistically significant. In all cases the reconstruction error starts increasing around three days before the system's failure occurs, this gives the maintenance team enough time to investigate the behaviour and possibly address the problem. This is shown in Fig. 2. The ML performance for the Turbofan dataset was evaluated using the standard metrics used for binary classification problems: ROC-AUC and Cumulative Gain Chart. The results are shown in Fig. 3.

Streaming Processing Delay. The SPD was estimated as the elapsed time from the moment one anomaly prediction terminates to the next one. This estimate is correct only insofar the delay between messages delivered by the external system is minimized. In doing so it was necessary to set the spark structured streaming watermark in such a way that neither the input messages would be lost nor the output messages would ever be emitted. This can be done by setting the watermark smaller than the total time length of the simulation dataset, but bigger than the moving window grouping. A simulation was conducted to test the feasibility of parallelization with varying levels of problem hardness and hardware resources[11]. Using K8s, specifically Minikube, the maximum resources for a given run were fixed. Since the data produced by one specific external machine are not needed to evaluate the operations of a second, different, external machine, parallelization was achieved by separating the data streams into different pipelines. The system is able to execute multiple pipelines to process data related to different machines. A pipeline consists of a dedicated set of partitions on the Kafka topics used for communication, some dedicated Spark executors that read from those topics and a K8s pod for the Anomaly Detector service. Each pipeline is made of 2 spark executors with 1 core and 2 GB of dedicated RAM (for the Bearing dataset), 1 spark executors with 1 core and 2 GB of dedicated RAM (for the Turbofan dataset) and 1 anomaly detector service K8s pod with 1 core and 1 GB of dedicated RAM (for both). The spark executors are managed by a single spark driver with 1 core and 2 GB of RAM,

---

[11] The execution environment is a single computer node with 12 Intel Xeon Gold 6136 CPUs and 32 GB of RAM memory.

Table 1. Times are expressed in seconds. The confidence intervals (0.95) were estimated using bootstap.

Bearing dataset		
	Pipelines number	
Difficulty multiplier	1	2
1	3.78 (3.40, 4.20)	3.49 (3.24, 3.76)
2	8.83 (8.20, 9.46)	7.74 (7.38, 8.12)
10	43.69 (42.90, 44.48)	35.95 (35.51, 36.42)

Table 2. Times are expressed in seconds. The confidence intervals (0.95) were estimated using bootstap.

Turbofan dataset			
	Pipelines number		
Difficulty multiplier	1	2	4
1	6.74 (6.63, 6.85)	3.60 (3.54, 3.65)	2.36 (2.30, 2.42)
2	13.01 (12.86, 13.16)	6.68 (6.60, 6.75)	3.65 (3.60, 3.69)
5	31.21 (30.65, 31.79)	15.74 (15.46, 16.01)	8.27 (8.12, 8.42)

independently of the number of pipelines. With regard to the Bearing dataset, the testing was conducted by simulating two different external machines. Furthermore, the hardness was increased, specifically it was first doubled and then subjected to a tenfold increase, w.r.t. the external machine message rate. Two different hardware configurations were tested: i) a single sequential pipeline, where the data from both machines are processed by a single data pipeline, and ii) a parallelized configuration, where each machine has its own dedicated processing pipeline. The results are shown in Table 1. In contrast, the hardness on the Turbofan dataset was increased in terms of the number of simulated external machines. Tests with 20, 40 and 100 simulated machines were carried out. In this case three different hardware configurations were tested, a single sequential pipeline and two parallelized configurations used to run 2 and 4 pipelines. The data from a specific machine was always assigned to a single pipeline. The results are shown in Table 2. As can be seen, as problem hardness grows the speedups obtainable with parallelization become more significant.

To conclude we performed one further analysis. Our goal was to verify that the architecture is suitable for monitoring multiple machines. The evaluation was based on the Gustafson's law [3]. If the problem scales the right way as the hardness increases, then the percentage of time spent on parallelizable tasks will increase (Tables 4 and 3). Equation (1) shows how to calculate the time spent in inherently sequential and parallelizable task, assuming the overhead is negligible, due to communication between pipelines.

Table 3. $P/S\%$, a higher number implies better parallelizability. The fact that the $P/S\%$ implied by the $n = 2$ data are higher than the ones implied by $n = 4$ could be explained by the hypothesis $C_n = 0$ starting to be violated.

Turbofan dataset: $P/S\%$		
	n pipelines	
Difficulty multiplier	2	4
1	1429.99 (1379.07, 1402.59)	652.33 (626.87, 675.60)
2	3726.59 (3655.58, 3752.21)	2393.81 (2379.734, 2395.94)
5	12046.07 (11464.09, 13714.63)	4908.77 (4903.21, 4944.35)

Table 4. $P/S\%$, a higher number implies better parallelizability.

Bearing dataset: $P/S\%$	
	n pipelines
Difficulty multiplier	2
1	18.43 (10.81, 26.63)
2	32.72 (25.15, 39.61)
10	54.89 (52.58, 56.82)

$$\begin{cases} T_1 = S + P \\ T_n = S + \frac{P}{n} + C_n \end{cases} \quad if \ C_n = 0 \Rightarrow \quad \begin{cases} S = \frac{nT_n - T_1}{n-1} \\ P = \frac{n(T_1 - T_n)}{n-1} \end{cases} \tag{1}$$

Here T_1 is the SPD for a single sequential pipeline, T_n is the SPD for n pipelines, S is the inherently sequential time, P is the parallelizable time spent on a single sequential pipeline and C_n is the time spent for inter-process communication. In our case, due to the presence of a single spark diver, this assumption is not completely correct, but is still mostly true due to the low number of pipelines. To address a greater number of pipelines we could use more separated drivers.

5   Conclusions and Future Work

Predictive maintenance is a fundamental approach used in industry to prevent failures, reduce costs and increase the safety of human operators. Numerous implementations have been presented over the years to address the problem of fault prediction. In this paper, we presented TrueDetective 4.0, an IoT Big Data architecture capable of ingesting and processing an infinite stream of data produced by an industrial production system, in real time. The data stream is fed to a recurrent Deep Learning model that predicts anomalies within the data to alert the human operator of any critical issues. The system gives a human operator full control over data collected from machinery for data analysis, dataset building and ML model training. In the future, we aim at evaluating the system's scalability. We also plan to enhance our architecture with the introduction of a model service system that will automatically store and manage trained Deep Learning models in order to obtain the best prediction performance. In addition, we are considering the implementation of digital twins to visualize the machines being analysed in real-time via AR/VR, in order to empower the interaction with the human operator.

Acknowledgement. This paper has been partially supported by the project "True Detective 4.0: Strumenti e Servizi Intelligenti di Monitoraggio in Tempo Reale per la Manutenzione Predittiva di apparati, per l'Ottimizzazione dei Processi Produttivi e di Automazione Industriale e per la Gestione della Sicurezza Fisica in Ambito Aziendale" funded by the Ministry of Economic Development (MISE), project code number F/190105/01-03/X44. Terms and conditions enforced by the project regulation do not allow us to make public the source code of the software platform.

References

1. Alfeo, A.L., Cimino, M.G., Manco, G., Ritacco, E., Vaglini, G.: Using an autoencoder in the design of an anomaly detector for smart manufacturing. Pattern Recogn. Lett. **136**, 272–278 (2020)
2. Groba, C., Cech, S., Rosenthal, F., Gossling, A.: Architecture of a predictive maintenance framework. In: 6th International Conference on Computer Information Systems and Industrial Management Applications (CISIM 2007), pp. 59–64 (2007). https://doi.org/10.1109/CISIM.2007.14
3. Gustafson, J.L.: Gustafson's Law, pp. 819–825. Springer, Boston (2011). https://doi.org/10.1007/978-0-387-09766-4_78
4. Killeen, P., Ding, B., Kiringa, I., Yeap, T.: IoT-based predictive maintenance for fleet management. Procedia Comput. Sci. **151**, 607–613 (2019). https://doi.org/10.1016/j.procs.2019.04.184, www.sciencedirect.com/science/article/pii/S1877050919306519, the 10th International Conference on Ambient Systems, Networks and Technologies (ANT 2019)/The 2nd International Conference on Emerging Data and Industry 4.0 (EDI40 2019)/Affiliated Workshops
5. Lee, J., Qiu, J., Yu, G., Lin, J.: Rexnord technical services: bearing data set (2007). https://ti.arc.nasa.gov/project/prognostic-data-repository, NASA Ames Prognostics Data Repository
6. Motaghare, O., Pillai, A.S., Ramachandran, K.: Predictive maintenance architecture. In: 2018 IEEE International Conference on Computational Intelligence and Computing Research (ICCIC), pp. 1–4 (2018). https://doi.org/10.1109/ICCIC.2018.8782406
7. Sahba, R., Radfar, R., Rajabzadeh Ghatari, A., Pour Ebrahimi, A.: Development of industry 4.0 predictive maintenance architecture for broadcasting chain. Adv. Eng. Inform. **49**(C) (2021). https://doi.org/10.1016/j.aei.2021.101324
8. Salierno, G., Morvillo, S., Leonardi, L., Cabri, G.: An architecture for predictive maintenance of railway points based on big data analytics. In: Dupuy-Chessa, S., Proper, H.A. (eds.) CAiSE 2020. LNBIP, vol. 382, pp. 29–40. Springer, Cham (2020). https://doi.org/10.1007/978-3-030-49165-9_3

9. Saxena, A., Goebel, K.: Turbofan engine degradation simulation data set (2008). https://ti.arc.nasa.gov/project/prognostic-data-repository, NASA Ames Prognostics Data Repository
10. Çoban, S., Gökalp, M.O., Gökalp, E., Eren, P.E., Koçyiğit, A.: Predictive maintenance in healthcare services with big data technologies. In: 2018 IEEE 11th Conference on Service-Oriented Computing and Applications (SOCA), pp. 93–98 (2018). https://doi.org/10.1109/SOCA.2018.00021

Optimising the Machine Translation Workflow

Analysis, Development, Benchmarking, Testing and Maintenance

Nicola Poeta[1](✉), Enrico Giai[2](✉), and David Turnbull[3](✉)

[1] STAR7 S.p.A., Global Content, Service Line Leader, Alessandria, Italy
nicola.poeta@star-7.com
[2] STAR7 S.p.A., Language Technologies Expert, Alessandria, Italy
enrico.giai@star-7.com
[3] STAR7 S.p.A., Language Lead, Alessandria, Italy
david.turnbull@star-7.com

Abstract. Machine Translation (MT) has now become an essential part of the localisation industry. New roles connected with it have emerged, and new technologies have been adopted.

As Language Service Providers (LSPs) need to implement these systems in their workflows – because of client demand, to improve cost efficiency, or to meet the rising demand for translated content – the need for clear guidelines to be followed in the adoption process grows.

In this paper, we describe in detail eight steps to integrate an MT workflow into the translation process. These steps have been identified by analysing existing literature and thoroughly validated through real-world MT implementations by STAR7.

The first step is to identify appropriate use cases. Then we must select the most suitable MT engine. Testing and benchmarking using scoring systems to evaluate the performance of the system is essential, as is supplier engagement in order to involve and inform all stakeholders. As the operational stage begins, several different Post-Editing Machine Translation (PEMT) workflows can be adopted. Quality Assurance (QA) and Language Quality Assessment (LQA) steps can then take place. Finally, feedback from all stakeholders can be collected to improve both the workflow and/or the MT engine performance itself.

Keywords: LQA · MT · PEMT

1 Introduction

Among the general public and non-expert users, the deployment of Machine Translation (MT) in localisation workflows may be thought of as a linear process, a 'silver bullet' that will solve the problem of multilingual communication at high speed and low cost. The demise of translators and translation agencies (or Language Service Providers – LSPs) has long been presaged [12].

However, despite the wide availability of commercial machine translation, research shows that demand for paid translation services – performed by individual translators or

© The Author(s), under exclusive license to Springer Nature Switzerland AG 2022
M. Ceci et al. (Eds.): ISMIS 2022, LNAI 13515, pp. 459–466, 2022.
https://doi.org/10.1007/978-3-031-16564-1_44

LSPs – continues to grow [11]. MT has proven not to be a replacement for these services, but rather a facilitator and driver of it. Professionally optimised MT workflows are not, in fact, simple and linear, but instead demand a wide variety of emerging skillsets at every stage of a cyclical process of continuing evolution and improvement.

In this paper, we set out to show the various stages of the optimised MT workflow, which requires contributions from multiple stakeholders: translation users, translation buyers, LSPs and their employees, individual linguists and engineers.

2 Related Work

The implementation of Machine Translation in the language industry has already been investigated by scholars like Lommel & DePalma [7] and, more recently, by Álvarez, Badia & Oliver [1]. Experts seem to agree on the fact that Machine Translation should not be seen as a substitute for human translation [5, 9]; rather, it should be a component of a translation workflow, such as in the case of Machine Translation Post-Editing (MTPE).

MTPE has been addressed by major organisation such as the ISO, which developed the ISO 18587:2017 standard on post-editing of machine translation output [3]. This standard describes in detail the requirements for the process of full, human post-editing of machine translation output and post-editors' competences.

However, the implementation of MTPE and MT workflows in general by LSPs is not usually clearly disclosed. In this paper, we aim to fill this gap by providing useful insights on applying MT workflows in real-world scenarios.

3 Methodology

The workflow described in this paper is the result of the analysis of both existing literature and empirical, day-to-day work experience. By applying general processes and methods as described by Translation Studies scholars and adapting them to our specific needs, we have developed what we believe is an efficient method of implementing Machine Translation in translation workflows.

To further prove our point, in this paper we will also provide data reporting the performance of Machine Translation output and PE effort.

4 Results

4.1 Analysing the Use Case

Machine translation output is far from a homogenous product. The quality of 'raw' (unedited) machine-translation output may vary substantially between language combinations, subject matters and file formats. Since even neural MT engines depend on vast quantities of translated material for training purposes, it stands to reason that language combinations and subject matters with less reference material will perform less well than their more common counterparts, so for example it is easier to train an MT engine for English ↔ Spanish in the area of EU regulations, where there is a large corpus available, compared to an engine for Swedish ↔ Tagalog in the area of medical technology, where little relevant data already exists.

Furthermore, translations requiring a high degree of creativity – such as 'transcreation' for advertising slogans, etc. – may not be suitable for MT.

The first step, therefore, in the optimised MT workflow is to identify suitable projects and to pair it with the appropriate technology. This may be done by a blend of experience on the part of the LSP – knowing what is likely to work and what is not – and through objective testing and benchmarking.

4.2 Choosing the Right MT Technology

Machine Translation technology is not new but has evolved especially rapidly in recent years [2, 13].

With the exponential growth in computational power, Machine Translation systems have been able to perform tasks quicker, more efficiently and using models ranging from statistical algorithms fed by big data or even neural networks [6] (Table 1).

Table 1. Summary of available MT Technologies

RBMT Rule-Based MT	SMT Statistical MT	NMT Neural MT
1930s → translation at sentence level	1980s → translation at phrase level	2016–2017 → phrase-level translation
Uses rule sets to transfer the words (i.e. their meaning) and grammatical structure of the source content into the target language	It calculates possible translations of a source segment and finds the most probable translation. It works through statistical models based on the analysis of parallel corpora	NMT attempts to determine the meaning of the source content and replicate it in the target language by sending it through several layers of 'neurons' that work together to determine the most likely sentence in the target language
Creates the translation of a source segment through grammar parsers, lexical info and syntax generation algorithms	Combination of translation models based on n-grams of bilingual units	The system learns from human translations and calculates the probabilities of the most suitable translations, without applying linguistic rules as such

While NMT is now widely regarded as the state of the art – due also to its commercial and/or free availability from a range of well-known providers such as Google, DeepL and Amazon, the professional decision of whether to deploy proprietary or commercial MT, and the choice between SMT or NMT, depends on several factors including subject matter, document type and language pair.

4.3 Testing and Benchmarking

Testing and benchmarking using scoring systems to evaluate the performance of the system is essential. Among the most widespread assessment metrics are 'BLEU score' and 'Edit Distance.'

The 'BLEU score' measures the quality of an MT engine, comparing the MT translation of texts with existing human translations of the same texts. It then measures how close the MT translation is to the human translation.

The 'Edit Distance' measures the quality of machine-translated segments, comparing the segments before and after editing. It measures how much the machine translation has to be corrected (Fig. 1).

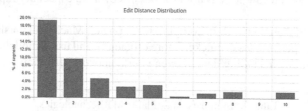

Fig. 1. Edit distance distribution on the post-editing of a 1,400-word English to Italian text

4.4 Training the Stakeholders

The optimised MT workflow requires stakeholders to have a firm grasp of the technologies in use, their strengths and limitations.

The principal stakeholders involved in the MT workflow include all those involved in setting up the project at the LSP and the linguists who actually perform post-editing work.

Besides the critical nature of the correct technology choice, at STAR7 we acknowledge the importance of engaging all stakeholders in the process, in this we build on a school of thought among Translation Studies scholars [1, 10], that the optimised MT workflow needs to be fair, realistic and sustainable – in the sense that it can be maintained over the long-term with no ill effects on any stakeholders [4].

To implement this, STAR7 uses a variety of approaches: generic training and information on machine translation; project-specific information setting out the use case, quality expectations, and project instructions, in accordance with the ISO standard.

4.5 Defining the PEMT Workflow

The issue of 'quality' is a persistent bugbear of the translation industry: who defines the parameters of a 'good translation'?

In most MT workflows, 'quality' can be defined simply as meeting the needs of the end users, however these may vary substantially.

By dint of this, the optimised MT workflow may also vary substantially: ranging from a near-fully automated 'human-out-of-the-loop' process to one in which translators simply use machine translation output as a way to enhance productivity (and potentially quality), as in the 'augmented translation' workflow [8].

Therefore, once a project has been identified as suitable for MT, and an MT type and engine has been selected, the next step is to define the quality expectations and therefore the amount of human work that should go into producing a final translation that meets the end user's needs.

Again, this can be achieved based on the LSP's experience with its client, the subject matter and the text type, and in direct consultation with the client.

Different levels of post-editing – ranging from 'light' to 'full' – have been defined and are widely accepted within the industry [3]. However at STAR7 we have adopted a more granular scale of five different approaches, each with detailed instructions of what the linguist should focus on in order to meet user requirements.

Once the PEMT workflow has been defined, the file is processed and handed off, where applicable, to linguists for translation, post-editing of MT and quality assurance.

4.6 Performing the Quality Assurance (QA)

MT engines – like human translators – are prone to a range of predictable and unpredictable mistakes. Historically the industry has sought to ensure 'quality' by means of the standardised Translate-Edit-Proofread (TEP) methodology (TEP), however a number of automated and semi-automated tools are now available.

In addition, customised regular expressions designed to mitigate the inherent weaknesses of current MT technology may be developed.

4.7 Performing the Language Quality Assessment (LQA)

The quality of raw and post-edited machine translation can be tracked using Language Quality Assessment tools, generating standardised, quality-profile adjusted 'scores' that can be harnessed to leverage data on the performance of MT engines and post-editors (Fig. 2).

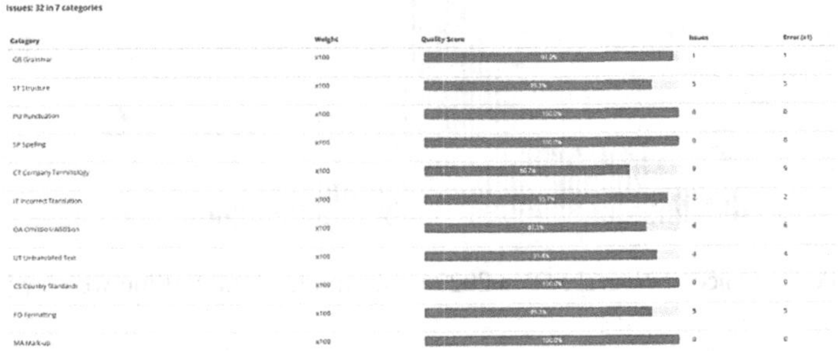

Fig. 2. Example of scores in the LQA of a sample of a 1,400-word English to Italian text

4.8 Implementing Feedback and Corrections

Feedback and corrections may arrive from all stakeholders and the optimised MT workflow can be tweaked as a result: the type of MT, MT engine, post-editing workflow and QA processes may all be adjusted as a result of stakeholder experience.

In addition, proprietary MT engines can be retrained and commercial MT engines can, in some cases, be customised using glossaries.

5 Conclusion

We have described how, in order to yield optimal results at the desired quality level, the Machine Translation workflow must be carefully calibrated and considered from multiple perspectives.

The different steps require a variety of high-level skills ranging from localisation engineering, MT engine creation and evaluation, 'soft' communication skills in dealing with end users and suppliers to manage expectations, and high-level language and technology skills to use the MT output appropriately and ensure its current and future quality. And yet, the power of modern MT is such that, despite this panoply of highly skilled human intervention, productivity can still routinely be improved to a material extent, with potential positive repercussions for all stakeholders.

Figure 3 displays the overall performance of NMT on a sample set of 2.8 million words (weighted).

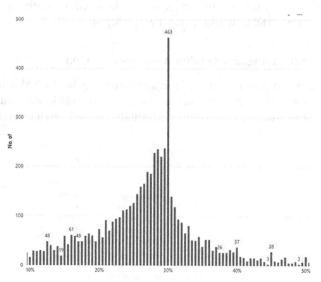

Fig. 3. Performance of Neural MT in 2022 on a sample of 2.8 million weighted words

On the other hand, Fig. 4 shows a breakdown of post-edited words in the case of SMT and displays the PE effort by target language (the source being UK English).

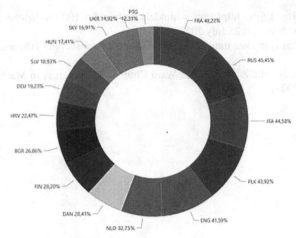

Fig. 4. Percentage of post-edited words in 2022 using statistical machine translation on a sample of 24 million words

Therefore, the optimised MT workflow has the potential to offer enormous benefits to all its stakeholders, but equally its imprudent and unscrupulous use can raise as many problems: unmet expectations, degraded translation quality and loss of skills and job satisfaction among linguists. Ongoing training and customisation of engines through reliable data (glossaries, human-translated and PEMT'd content, etc.) is critical in triggering the correct PEMT process.

References

1. Alvarez-Vidal, S., Badia, T., Oliver, A.: Implementing MTPE into a real industrial scenario: what do translators need for a fair MT workflow?. In: Fair MT Workflow. MT Summit XVII, Dublin (2019)
2. Hutchins, J.: Machine translation: a concise history. J. Transl. Stud. **13**(1), 29–70 (2010)
3. ISO 18587:2017. https://www.iso.org/standard/62970.html. Accessed 25 July 2022
4. Kenny, D., Moorkens, J., do Carmo, F.: Fair MT: Towards ethical, sustainable machine translation. Transl. Spaces **9**(1), 1–11 (2020)
5. Klein, J.: Use MT to simplify and speed up your alignment for TM creation. In: Proceedings of the 14th Conference of The Association for Machine Translation in the Americas, vol. 2: MT User Track, pp. 237–269. https://aclanthology.org/2020.amta-user.pdf. Accessed 05 July 2022
6. Liu, Q., Zhang, X.: Machine translation general. In: Chan, S. (ed.) The Routledge Encyclopedia of Translation Technology. Routledge, London-New York (2015)
7. Lommel, A., DePalma, D.: Europe's Leading Role in Machine Translation. Common Sense Advisory, Cambridge Mass (2016)
8. Lommel, A.: Augmented Translation: Are We There Yet? https://csa-research.com/Blogs-Events/Blog/augmented-translation-2020. Accessed 25 July 2022
9. Monti, J.: Dal sogno meccanico alla e-translation: la traduzione automatica è realtà?, Media Duemila – Mensile di Cultura informatica e ICT, 219, 60–67 (2004)
10. Moorkens, J.: "A tiny cog in a large machine": Digital Taylorism in the Translation Industry. Transl. Spaces **9**(1), 12–34 (2020)

11. Nimdzi 100 Top LSPs. https://www.nimdzi.com/nimdzi-100-top-lsp/#size-of-the-market-and-projection. Accessed 25 July 2022

12. Schaier, K.: Can computer translation replace human translation? Hispania **79**(1), 96–107 (1996)

13. Wang, H., Wu, H., He, Z., Huang, L., Ward Church, K.: Progress in Machine Translation, Engineering (2021)

A Comparative Study: Classification Vs. Matrix Factorization for Therapeutics Recommendation

Seda Polat Erdeniz[1,2(✉)], Michael Schrempf[2], Diether Kramer[2], and Alexander Felfernig[1]

[1] Graz University of Technology, Graz, Austria
{spolater,felfernig}@ist.tugraz.at
[2] Die Steiermärkische Krankenanstaltengesellschaft m. b. H. (KAGes), Graz, Austria
{seda.polaterdeniz,michael.schrempf,diether.kramer}@kages.at
https://www.ase.ist.tugraz.at
https://www.kages.at

Abstract. Hospital information systems (HIS) hold various healthcare information of patients. Most of them are held as structural data by a database table. This information include history of diagnoses, medications, applied procedures and laboratory results of patients which can be used by machine learning methods to predict some useful information about patients. These predictions can be the progress of a disease, which is called *prognosis*, or it can also be *therapeutics* which includes medications and procedures. In this paper, we explain how to recommend therapeutics using various machine learning approaches, especially by comparing classification methods with matrix factorization (a recommender systems approach). In order to evaluate the performance of compared methods, we applied experiments on real patients' electronic health records (EHR). We observed that matrix factorization outperforms the compared classification approaches in terms of accuracy. Therefore, it is feasible to employ matrix factorization in clinical decision support systems to provide therapeutics recommendations which improves the daily performance of physicians, so the life quality of the patients.

Keywords: Recommender systems · Matrix factorization · Multi-label classification · Healthcare

1 Introduction

Hospital information systems (HIS) hold various types of patient data and these data can be useful to extract new information using various approaches of machine learning (ML) [9]. The extracted information can be a prediction of a diagnosis, prediction of a prognosis (progress of a disease) [16], a recommendation of therapeutics [14], and so on.

Recommender systems have shown tremendous value for the prediction of personalized item recommendations for individuals in a variety of settings (e.g.,

M. Ceci et al. (Eds.): ISMIS 2022, LNAI 13515, pp. 467–476, 2022.
https://doi.org/10.1007/978-3-031-16564-1_45

marketing, e-commerce, etc.). User-based collaborative filtering is a popular recommender system approach, which leverages an individuals' prior satisfaction with items, as well as the satisfaction of individuals that are "similar". Recently, there have been applications of collaborative filtering based recommender systems for clinical risk prediction. In these applications, individuals represent patients, and items represent clinical data, which includes an outcome. Matrix factorization based collaborative filtering algorithms [10] introduce a rating matrix R (a.k.a., user-item matrix) which describes preferences of users for the individual items the users have rated.

In this paper, we focus on *recommendation of therapeutics*. *Therapeutics* [14] are treatments used to alleviate or prevent a particular disease after a diagnosis is made. Examples of therapeutics include drug therapy, medical devices, nutrition therapy and stem-cell therapies. For example, a medication or a procedure applied at the hospital are healing procedures. Therapeutics predictions (or recommendations) can be automated using machine learning algorithms and can be provided to physicians via Clinical Decision Support Systems (CDSS) [2]. In order to solve a *therapeutics recommendation* problem, there can be two AI approaches: applying recommender system models (e.g., matrix factorization) or predictive models (classification or regression models, such as Random Forest or Logistic Regression).

Hao and Blair [8] had a research about comparison of user-based collaborative filtering method and traditional classification methods for data imputation and clinical risk predictions (single target). They employed logistic regression and random forest models with imputation methods and compared it with collaborative filtering (a recommender systems algorithm). At the end, they show that traditional classification methods (logistic regression and random forest) outperform user-based collaborative filtering in terms of misclassification rate, sensitivity and specificity. However, in this paper, our objective is to compare the recommendation performance (in terms of accuracy) of Matrix Factorization with three traditional classification models. The target predictions are different from the work of Hao and Blair [8]. They compared the performance of models for risk prediction which includes one target variable, whereas in this paper, we aim to predict/recommend multiple target variables (medications or procedures) at the same time.

In the rest of this paper, first we provide a background section which explains relevant definitions. Then, we provide the details of the used HIS data and provide short explanations of compared methods. In experimental results section, we evaluate the compared methods based on real-world HIS patients data of the public healthcare provider Steiermärkische Krankenanstaltengesell- schaft m.b.H. (KAGes) [16].

2   Background

Classification models and recommender systems are important topics in this paper. Therefore, in this section we provide required definitions in these topics in the following paragraphs.

Classification is a type of predictive modeling which involves assigning a class label to input examples [12]. In classification models, a model can predict one target with two or more values or multiple targets with two or more values. Prediction of one target feature with two values are called Binary Classification. Prediction of one target with more than two values are called Multi-class Classification. Finally, as we use in the rest of the paper, there can be also classification models which predict more than one target features which is called Multi-label Classification.

Binary Classification refers to those classification tasks that have two class labels [12]. Examples include:

- Email spam detection (spam or not),
- Churn prediction (churn or not), and
- Conversion prediction (buy or not).

Multi-class Classification refers to those classification tasks that have more than two class labels [12]. Examples include:

- Face classification,
- Plant species classification, and
- Optical character recognition.

Multi-label Classification refers to those classification tasks that have two or more class labels, where one or more class labels may be predicted for each example [12]. Consider the example of photo classification, where a given photo may have multiple objects in the scene and a model may predict the presence of multiple known objects in the photo, such as "bicycle," "apple," "person," etc. This is unlike binary classification and multi-class classification, where a single class label is predicted for each example.

Multi-label Binary-Classification refers to those multi-label classification tasks that have labels all with binary values. Multi-label Binary-Classification is a supervised learning problem with several labels or outputs to predict, that is when the array of predictions is a 2d array. When there is no correlation between the labels, a very simple way to solve this kind of problem is to build n independent models, i.e. one for each output, and then to use those models to independently predict each one of the n labels. However, because it is likely that the output values related to the same input are themselves correlated, an often better way is to build a single model capable of predicting simultaneously all n labels. First, it requires lower training time since only a single estimator is built. Second, the generalization accuracy of the resulting estimator may often be increased.

Recommender systems [6] suggest items (alternatives, solutions) that are of potential interest for a user. In this paper, we use a multi-label binary classification problem as a recommendation problem and solve it also using a recommender systems approach (matrix factorization) [10]. Examples of questions to recommender system are:

- which movie should I watch?,

- which bike should I purchase?, and
- which destination should I travel? .

Recommendation systems also provide multi label predictions (as Multi-label Classification Models) with the predicted acceptance probabilities of all recommendable items. Then, the highest probability items are recommended to users. This makes it different from the classifiers which actually classify an item with a label.

In this paper we are evaluating methods for therapeutics recommendation based on a multi-label binary-classification problem. As a result, we do not actually make a classification, instead our target is to predict the acceptance probabilities of recommendable items. Therefore, in this paper we use the accuracy_probability formula (see Formula 2).

Accuracy measures the correctness of predictions. Usually in classification problems, it is calculated as in Formula 1 where the number of correct predictions are divided by the total number of predictions.

$$accuracy\_class = \frac{\#(correct\_predictions)}{\#(all\_predictions)} \tag{1}$$

However, in a binary classification problem, to measure the accuracy we can also use Formula 2 based on predicted probability by the classifier (e.g., using predict_proba function in scikit-learn in python). This accuracy formula actually subtracts mean absolute error from 1. In mean absolute error calculation, we use actual values of a binary classification problem (0 or 1) and predicted probabilities (between 0 and 1). Therefore, we obtain a normalized mean absolute error which is in between 0 and 1. When the mean absolute error is 0, the accuracy of the prediction becomes 1, which means the highest accuracy. This eliminated errors from accuracy because of the threshold defined for classes.

$$accuracy\_probability = 1 - \frac{\sum_{i=1}^{n} |actual\_i - predicted\_probability\_i|}{n} \tag{2}$$

3   Data

In the experimental results, we recommend procedures and medications for *major adverse cardiovascular events (MACE)* patients. Therefore, we used electronic health records (EHR) of more than 90,000 MACE patients' hospitalizations (each hospitalization is represented as one observation/row in the data) at KAGes [16]. Using the historical information (diagnoses, laboratories, medications, etc.) of patients, we aimed to recommend therapeutics for MACE diagnosed patients based on their similarities based on predictor variables with hospitalizations of other patients.

To prepare training and test data for classification methods and matrix factorization, we split the data slightly different for these methods. As shown in

Fig. 1, on the left hand side, we demonstrate how we split the data for classification models, which is the traditional way of splitting as it is done also by "train_test_split" method of "sklearn.model_selection" package of python. However, on the right hand side of this figure, we show how we use the same dataset for matrix factorization. We only exclude the labels of test data (colored in green), and use the rest of the data completely during matrix factorization. Afterwards, only the labels of test data part is taken from the dense matrix (the output of matrix factorization) generated by the matrix factorization method and compared with the original values of this part (which is shown as evaluation data).

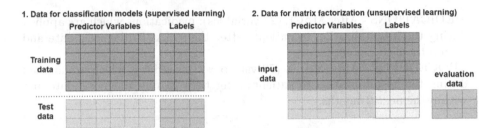

Fig. 1. Data split for classification models and matrix factorization

3.1   Predictor Variables

Predictor variables are used to train the models to find relations with the target variables (labels). Predictor variables include historical (maximum five years old) data before MACE diagnosis.

– DEMOGRAPHIC (24 variables): Being single or married, gender, age and academical degrees are some the example variables of these predictors.
– STATUS (3 variables): Sometimes unhealthy behaviour is documented in the medical reports and obesity or smoking could be noted in these documents.
– DIAGNOSES - Before MACE (473 variables): The ICD-10 diagnoses [13] of the patients were used as a source of features. This is because diagnoses are documented for the patient for each stay. One or more diagnoses can be saved by the physician. Example diagnoses with their ICD-10 codes are: I20 (Angina pectoris), I21 (Acute myocardial infarction), and I46 (Cardiac arrest).
– LABORATORIES - Before MACE (406 variables): Important data about the patient's condition is also provided by the laboratory data. For this purpose, we used abnormal laboratory information. Each of them are encoded into 6 levels of abnormalities as : Very low (—), low (–), slightly low (-), slightly high (+), high (++), and very high (+++).
– MEDICATIONS - Before MACE (27 variables): One of these sources is the medication, which was extracted from the medical reports. These documents contain the recommended drugs for medical treatment. These reports had been created before the discharge of the patient's stay.

- PROCEDURES - Before MACE (78 variables): The treatments are documented for the calculation of the costs of stay. For standardisation purposes, we used the catalogue of services in Austria[1] which contains the available procedures. Example procedures are: DD040 (Percutaneous transluminal coronary angioplasty), DD050 (Implantation of a stent in the coronary vessels), and DD060 (Implantation of a drug-eluting stent in the coronary vessels).

3.2 Labels (Recommendable Items)

Target variables (labels or recommendable items) include new (maximum two months old) data after MACE diagnosis.

- MEDICATIONS - After MACE (27 variables): medications which are applied during the hospitalization to patients after their MACE diagnosis date and within the next 2 months.
- PROCEDURES - After MACE (78 variables): procedures which are applied during the hospitalization to patients after their MACE diagnosis date and within the next 2 months.

4 Methods

Based on the defined data in the previous section, we evaluate and compare the performance of three multi-label classifiers: k-Nearest Neighbors, Random Forest, and Neural Networks and a recommendation algorithm: Matrix Factorization. In this section, we provide descriptions of these methods and mention used Python methods which clarifies how their parameters are defined and tuned during our evaluation.

4.1 k-Nearest Neighbors

The k-nearest neighbors algorithm (kNN) [11] is a supervised learning method. It can be used for classification and regression problems. In both cases, the input consists of the k closest training examples in a data set. The output is the average of the values of k nearest neighbors. In order to select a value for k (n_neighbors), we used 3-fold cross validation and in order to provide *multi target outputs* at the same time, we used "MultiOutputClassifier"[2] with "KNeighborsClassifier"[3] in Python.

---

[1] https://www.sozialministerium.at/dam/sozialministeriumat/Anlagen/Themen/Ge-sundheit/Krankenanstalten/LKF-Modell-2022/Kataloge-2022/LEISTUNGSKATALOG-BMSGPK-2022.pdf.

[2] https://scikit-learn.org/stable/modules/generated/sklearn.multioutput.MultiOutputClassifier.html.

[3] https://scikit-learn.org/stable/modules/generated/sklearn.neighbors.KNeighborsClassifier.html.

4.2   Random Forest

Random Forest (RF) [3,4] is also a supervised learning method which is an ensemble of tree-structured classifiers. Every tree of the forest gives a unit vote, assigning each input to the most probable class label. It is a fast method, robust to noise and it is a successful ensemble which can identify non-linear patterns in the data. An RF model usually has one output – the output/prediction variable. However it can also be trained to provide *multi target outputs* at the same time. For this purpose, we used "MultiOutputClassifier" with "RandomForestClassifier"[4] in Python.

4.3   Neural Networks

Neural networks [17] as another supervised learning method, more specifically artificial neural networks (ANNs), mimic the human brain through a set of algorithms. At a basic level, a neural network is comprised of four main components: inputs, weights, a bias or threshold, and an output. Deep learning neural networks are an example of an algorithm that natively supports multi-label classification problems. Neural network models for multi-label classification tasks can be easily defined and evaluated using the Sequential model of Keras deep learning library in Python[5] where we compile the model with parameters loss='binary_crossentropy', optimizer='adam', and metrics='accuracy' and two layers.

4.4   Matrix Factorization

Matrix factorization [5,10] is a method of collaborative filtering algorithms used in recommender systems. It can be used as supervised or unsupervised. Matrix factorization algorithms work by decomposing the user-item interaction matrix into the product of two rectangular matrices. This family of methods became widely known during the Netflix prize challenge due to its effectiveness as reported by Simon Funk [15]. The prediction results can be improved by assigning different regularization weights to the latent factors based on items' popularity and users' activeness.

The Singular Value Decomposition (SVD) [1], a method from linear algebra that has been generally used as a dimensionality reduction technique in machine learning. SVD is also a matrix factorisation technique, which reduces the number of features of a dataset by reducing the space dimension from n-dimension to k-dimension (where k<n). In the context of the recommender system, the SVD is used as a collaborative filtering technique. It uses a matrix structure where each row represents a user, and each column represents an item. The elements of this matrix are the ratings that are given to items by users. We used "svds" method of "scipy"[6] library in Python for matrix factorization with k=50.

---

[4] https://scikit-learn.org/stable/modules/generated/sklearn.ensemble.RandomForestClassifier.html.

[5] https://keras.io/guides/sequential_model/.

[6] https://docs.scipy.org/doc/scipy/reference/generated/scipy.sparse.linalg.svds.html.

5   Experimental Results

We used MACE patients data to predict therapeutics in terms of medications and procedures. We used it as a multi-label binary-classification problem and applied classification algorithms and matrix factorization. As output, we had predicted probabilities for therapeutics (recommendable items). We implemented all compared methods in Python as mentioned in Sect. 4 and tested them on more than 90,000 MACE patients' hospitalizations data which is explained in Sect. 3. Afterwards, we compared their accuracy (accuracy_probability) performance according to Formula 2. We evaluated the performances separately for procedures recommendation and medications recommendation.

As it is shown in Table 1, we observed that Matrix Factorization performs best for both two types of therapeutics recommendations (medications and procedures) compared to other three multi-label binary classifiers.

Table 1. Accuracy results of *k-Nearest Neighbors*, *Random Forest*, *Neural Networks*, and *Matrix Factorization* for therapeutics recommendations (78 procedures and 27 medications).

	Procedures	Medications
k-Nearest neighbors	0.52	0.87
Random forest	0.52	0.86
Neural networks	0.60	0.93
Matrix factorization	**0.94**	**0.95**

We observed that the accuracy in recommending medications is much higher than recommending procedures for classification methods (kNN, Random Forest and Neural Networks). One reason of this is the number of medications are lower than procedures, so it is easier or more straightforward to select an accurate medication out of them. However, deciding on procedures is more complicated then medications. There can be various procedures options for the same case. It is not easier to find out the correct procedure.

In recommending procedures, the best performance after the Matrix Factorization is achieved by Neural Networks. However, compared to the performance of the Matrix Factorization (0.94), Neural Networks provided a very low accuracy (0.60). The other two methods kNN and Random Forest could achieve even lower performances (both 0.52). This can be caused because of the data splitting difference for matrix factorization algorithm. Matrix factorization may have a little bit more advantage with having more data during prediction. This advantage brings also a disadvantage in terms of runtime since it calculates all predictions with complete training and test data at one time. However, classification methods are first trained on the training dataset then used for predicting the test data.

As we compared the results of recommending medications, we observed the best performing one is again Matrix Factorization with 0.95 accuracy and it is followed by Neural Networks with 0.93 accuracy. Therefore, we can say both methods are performing well for recommending medications where the number of options are relatively limited (27 medications). On the other hand, kNN (0.87) and Random Forest (0.86) perform very similar to each other but lower the other two since these algorithms are not designed for multi-label predictions indeed. They perform better in solving traditional single target prediction problems (e.g. "has this patient a high cancer risk or not?") but for multi-label predictions their performance were not as good as Neural Networks or Matrix Factorization.

These results actually show the power of Matrix Factorization for *multi-label classifications* which was also proven in Netflix Movie Recommendation Competition in 2006 when Netflix announced a prize money of $1 million to those who will improve its root mean square performance by 10%. Netflix provided a training data set of 100,480,507 ratings that 480,189 users gave to 17,770 movies [7].

6 Conclusion

In hospital information systems, there is a important amount of electronic healthcare records of patients. This data can be used to extract new information to improve the life quality of patients. In this paper, we focused on therapeutics recommendation by testing and comparing ML methods based on EHR data. We used real-world EHR data of patients at KAGes hospitals. We examined performance of matrix factorization in comparison with three classification models for multi-label classification.

As observed in the experimental results, Matrix Factorization provides the best accuracy among all the compared methods. This makes it feasible to be used in healthcare to recommend therapeutics which improves the quality of patients by applying them the accurate therapeutics. Moreover, it also increases the performance of physicians by reducing the time spent to decide on an accurate medication or procedure for each patient since they will not need to spend time to compare the past therapeutics applied to similar patients.

However, before deploying such a method in a production system CDSS, some other challenges must be also considered. In order to break the "resistance to AI", the usage principles should be well defined with the physicians. ML outputs should not bring them an overhead in their already busy daily schedules but help them to lower their stress by reminding some useful information via ML based recommendations. To overcome "need of trust", it should be provided with explanations which can be learned by explainable AI methods. Last but not least, to be able to provide an added value to the "highly knowledgeable users", the performance of the tool should be evaluated in a pilot application by physicians.

References

1. Abdi, H.: Singular value decomposition (svd) and generalized singular value decomposition. Encycl. of Measur. Stat. 907–912 (2007)
2. Castillo, R.S., Kelemen, A.: Considerations for a successful clinical decision support system. CIN: Comput. Inf. Nurs. **31**(7), 319–326 (2013)
3. Chaudhary, A., Kolhe, S., Kamal, R.: An improved random forest classifier for multi-class classification. Inf. Process. Agric. **3**(4), 215–222 (2016)
4. Devetyarov, D., Nouretdinov, I.: Prediction with confidence based on a random forest classifier. In: Papadopoulos, H., Andreou, A.S., Bramer, M. (eds.) AIAI 2010. IAICT, vol. 339, pp. 37–44. Springer, Heidelberg (2010). https://doi.org/10.1007/978-3-642-16239-8_8
5. Erdeniz, S.P., Felfernig, A., Samer, R., Atas, M.: Matrix factorization based heuristics for constraint-based recommenders. In: Proceedings of the 34th ACM/SIGAPP Symposium on Applied Computing, pp. 1655–1662 (2019)
6. Felfernig, A., et al.: An overview of recommender systems in the internet of things. J. Intell. Inf. Syst. **52**(2), 285–309 (2018). https://doi.org/10.1007/s10844-018-0530-7
7. Hallinan, B., Striphas, T.: recommended for you: the Netflix prize and the production of algorithmic culture. New Media Soc. **18**(1), 117–137 (2016)
8. Hao, F., Blair, R.H.: A comparative study: classification vs. user-based collaborative filtering for clinical prediction. BMC Med. Res. Methodol. **16**(1), 1–14 (2016)
9. Khanzode, K.C.A., Sarode, R.D.: Advantages and disadvantages of artificial intelligence and machine learning: a literature review. Int. J. Libr. Inf. Sci. (IJLIS) **9**(1), 3 (2020)
10. Koren, Y., Bell, R., Volinsky, C.: Matrix factorization techniques for recommender systems. Computer **42**(8), 30–37 (2009)
11. Mucherino, A., Papajorgji, P.J., Pardalos, P.M.: K-nearest neighbor classification. In: Data mining in agriculture, vol. 34, pp. 83–106. Springer, New York (2009). https://doi.org/10.1007/978-0-387-88615-2_4
12. Novaković, J.D., Veljović, A., Ilić, S.S., Papić, Ž, Milica, T.: Evaluation of classification models in machine learning. Theor Appl. Math. Comput. Sci. **7**(1), 39–46 (2017)
13. O'malley, K.J., Cook, K.F., Price, M.D., Wildes, K.R., Hurdle, J.F., Ashton, C.M.: Measuring diagnoses: ICD code accuracy. Health Serv. Res. **40**(5p2), 1620–1639 (2005)
14. Organization, W.H., et al.: Therapeutics and covid-19: living guideline, 31 March 2021. World Health Organization, Technical Report (2021)
15. Piatetsky, G.: Interview with simon funk. Acm Sigkdd Explor. Newsl. **9**(1), 38–40 (2007)
16. Schrempf, M., Kramer, D., Jauk, S., Veeranki, S.P., Leodolter, W., Rainer, P.P.: Machine learning based risk prediction for major adverse cardiovascular events. In: Navigating Healthcare Through Challenging Times, pp. 136–143. IOS Press (2021)
17. Zhang, G.P.: Neural networks for classification: a survey. IEEE Trans. Syst. Man Cybern. Part C (Appl. Rev.) **30**(4), 451–462 (2000)

Document Layout Analysis
with Variational Autoencoders:
An Industrial Application

Ali Youssef[iD], Gabriele Valvano[iD], and Giacomo Veneri[(✉)][iD]

TPS Artificial Intelligence - Baker Hughes, Florence, Italy
{giacomo.veneri,gabriele.valvano,ali.youssef}@bakerhughes.com

Abstract. We present a novel method for Document Layout Analysis that detects documents that are not compliant with a given template. The major challenge we solve is dealing with a highly unbalanced dataset with only a few, hard-to-distinguish, non-compliant documents. Our model learns to detect inadequate documents based on localised non-compliant characteristics, including stamps, handwritten text, and misplaced signatures. Nevertheless, the model must not report documents containing other artefacts such as amendments or notes, which we deem acceptable. We address these challenges via generative modelling, using anomaly detection techniques to validate document layout. In particular, we first let the model learn the compliant document distribution. Then, we detect and report out-of-distribution samples for their automated rejection. In the paper, we investigate and compare two major approaches to anomaly detection: 1) classifying anomalies as those samples that cannot be accurately generated by the model; and 2) detecting samples whose mapping to a known proxy distribution is not possible. Both methods can be trained without annotations and obtain a classification accuracy of ∼90% on real-world documents, outperforming alternative supervised solutions.

Keywords: Document validation · VAE · Anomaly detection · Unbalanced dataset · Document layout analysis · Deep learning

1 Introduction

Large conglomerate companies receive thousands of offers and requests from suppliers, internal functions, and customers every day. These documents must be compliant with a standard layout: for instance, a specific document must report one or more stamps, a certain number of signatures, a date, and no other artefacts such as external notes, recommendations or amendments. Nevertheless, as we show in Fig. 1, submitted forms and documents may still present undesired incongruities with standard layouts, making documents quality verification necessary. In this context, developing automated scanning systems is an appealing opportunity to reduce the expensive and time-consuming need for manual checks

© The Author(s), under exclusive license to Springer Nature Switzerland AG 2022
M. Ceci et al. (Eds.): ISMIS 2022, LNAI 13515, pp. 477–486, 2022.
https://doi.org/10.1007/978-3-031-16564-1_46

Fig. 1. Non-compliant documents (crops) with artefacts and missing information. (a, b, c) are examples of unacceptable amendments; (d) missing information on the role, and (e) missing signature.

performed by humans [2]. For example, it would be beneficial to have a Document Layout Analysis (DLA) system that, within a pool of documents provided as scanned images, can detect those non-compliant with standard templates.

Unfortunately, developing automated DLAs is not trivial due to the data scarcity. In fact, most of the documents stored on companies databases must be compliant by definition, making it hard to collect negative samples that we can use to train simple classifiers. In other terms, every company interested in developing automated DLAs will face the problem of having **highly unbalanced datasets** , with just a small percentage of non-compliant documents (e.g. <4%). This unbalance is problematic and makes the training of simple efficient classifiers harder [13]. In our work, we bypass the strong class imbalance issue by formulating the non-compliant document detection as an anomaly detection problem [16]. In particular, we investigate generative approaches based on Variational Autoencoders (VAE) and compare document-specific anomaly scores computed in pixel space vs feature space.

VAEs [7,11,14] are auto-encoding based deep learning models that learn to map an input distribution (i.e. the documents) onto a known prior distribution. After the mapping is complete, the VAE's decoder samples from the prior distribution (usually a Gaussian) and to accurately reconstruct the input image. Herein, we suggest training the VAE to learn the mapping from a distribution of compliant documents to a Gaussian distribution, and vice versa [15]. After training is performed on compliant documents, we can expect that non-compliant documents won't be correctly mapped onto the prior distribution and their reconstruction cost becomes higher than expected. We measure non-compliance either in feature space, using a learned classifier, or in pixel space, using the document self-reconstruction SSIM [18].

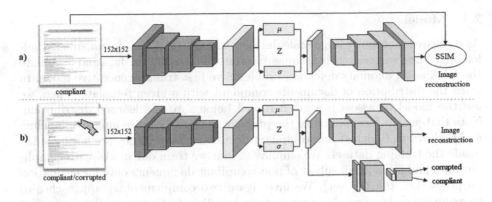

Fig. 2. VAE-based anomaly detection strategies. **(a)** A VAE learns to auto-encode compliant input documents. After training, we use the reconstruction error (based on the SSIM index) as anomaly score to detect non-compliant documents. **(b)** A VAE learns to auto-encode compliant input documents while a classifier learns to detect documents subjects to image corruption and classify them as non-compliant. At inference, we encode new documents and use the classifier to predict anomaly scores.

1.1 Related Work

Recently, there has been an increasing interest in Document Layout Analysis (DLA) [2,19]. Some authors propose to split DLA into several sub-task: separately detecting tables, texts, formulas, and figures [3,4,8]. For each sub-task, these methods apply different rules for DLA validation. In the last few years, several authors have investigated the use of Convolutional Neural Networks to analyse documents with holistic approaches [1]. Li et al. [10] have recently explored the opportunity to use generative models to build realistic layouts that it is possible to use to detect out-of-distribution documents. Other techniques are strongly specialised to recognise tables [12] or part of the text [9]. Unfortunately, outliers and pervasive noise are commonplace in the real world, and one may not have access to clean training data as required by standard deep denoising deep learning techniques. Among others, deep autoencoders [20] have demonstrated their effectiveness in discovering non-linear features across many problem domains. Moreover, these methods do not require a tedious labelling procedure to split documents into sub-parts. On the contrary, the model can directly learn from the data the most important features to analyse. In our work, we apply a Variational Autoencoder (VAE) [5,7,14] to validate the compliance of a document to a given layout. We show that our technique can outperform standard supervised methods.

2 Proposed Method

Notation. In the manuscript, we denote functions $\Phi(\cdot)$ using capital Greek letters, italic lowercase for scalars s, and bold lowercase for documents (i.e. 2D images) $\mathbf{x} \in \mathbb{R}^{h \times w}$, where $h, w \in \mathbb{N}$ are the image height and width.

2.1 Model

As discussed, obtaining large-scale datasets of non-compliant documents is challenging. Thus, we treat such documents as out-of-distribution data and formulate the DLA as an anomaly detection problem: we first train a generative model to learn the distribution of documents compliant with a given template. Then, we use the model to assess if a given record belongs to the learned distribution. Note that such a formulation of the problem does not require intensive labeling and it allows us to detect non-compliant documents even if they are not included inside the training dataset. We emphasize that we train our model with compliant data and use a small subset of non-compliant documents only for evaluation purposes (i.e., the *test* set). We investigate two complementary approaches to measure anomaly scores: 1) in *pixel space*; and 2) in *feature space*. Below, we first describe how we can learn a high-level representation of the compliant document distribution with a VAE; then, we detail how we detect anomalies.

Learning the Distribution of Compliant Data. To learn the distribution of compliant documents, we use a Variational Autoencoder, consisting of an encoder $E(\cdot)$ and a decoder $\Delta(\cdot)$. Given a document $\mathbf{x}$, the encoder learns to map the input sample $\mathbf{x}$ to a compressed representation $\underline{z} = E(\mathbf{x})$. More specifically, we use the KL divergence to encourage $\underline{z}$ to follow a multivariate Gaussian distribution [7]. The decoder $\Delta(\cdot)$ performs a stochastic sampling in the neighborhood of the encoded $\underline{z}$ and reconstructs the input image $\tilde{\mathbf{x}} = \Delta(\underline{z}) \approx \mathbf{x}$ subject to a self-reconstruction loss. Such a process ultimately forces the encoder to learn an efficient compression of the data into the lower-dimensional vector $\underline{z}$.

Detecting Non-Compliant Documents in Pixel Space. After training the VAE to correctly auto-encode compliant documents, we can use the self-reconstruction loss of a document as anomaly score. In particular, we can use the score $\mathcal{L} = 1 - SSIM$, where SSIM is the structural similarity index. Given a specific document template, we expect the reconstruction of compliant data (which is in-training distribution) to have a higher quality than the reconstruction of non compliant (out-of-distribution) data. Hence, we distinguish *compliant* vs *non-compliant* documents based on the heuristic h $\in \{0, 1\}$:

$$h_{px} = bool(\mathcal{L} > t) \tag{1}$$

where t is a threshold tuned on a validation set as the average reconstruction error, $h_{px} = 1$ detects anomalous (non-compliant) documents, and $h_{px} = 0$ compliant data. We summarize this first approach in Fig. 2a.

Detecting Non-Compliant Documents in Feature Space. Detecting non-compliant documents in pixel space may have the limitation of missing small anomalies, which incur in small reconstruction errors. In these cases, it would be hard to distinguish physiological reconstruction errors from reconstruction mistakes due to small anomalies. Besides, it would be interesting to learn to detect these documents learning an ad-hoc function. In particular, we could encourage the VAE encoder to map in-distribution documents onto a Gaussian space, and out-of-distribution samples far away from such a Gaussian. To this end, we introduce an anomaly detector in the form of a classifier Γ receiving input vectors $\underline{z}$ extracted by the VAE encoder. Given $\underline{z}$, the classifier must learn to say apart normal from anomalous documents. As we train the whole model using only normal, compliant documents, we synthetically generate non-compliant documents via data-augmentation[1] and train the classifier to recognise them using the sigmoid cross-entropy loss. Note that the model is still trained in a self-supervised fashion since no human labels are used for training.

At inference, we simply use the classifier to predict whether an input document is compliant or not, according to the heuristic h_f:

$$h_f = sigmoid(c) \qquad c = \Gamma(\mathrm{E}(\mathbf{x})) = \Gamma(\underline{z}) \qquad (2)$$

This model is illustrated in Fig. 2b.

Implementation. The VAE has 4 convolutional layers mapping input images onto a 4-dimensional latent space. The classifier, is a 3-layered fully connected neural network, trained with sigmoid cross-entropy loss. We used Adam optimizer and decayed the learning rate from 10^{-3} to 10^{-6}. Our learning rate scheduler first decreases the learning rate with a multiplicative factor of 0.1 after each epoch until epoch 10, then it exponentially decays it with an exponent of -0.1.

2.2 Data

For our experiments, we used two different datasets, containing documents appearing with radically different layouts. The first dataset consists of 2100 compliant and 115 non-compliant documents with layout L1. We show an example of such data in Fig. 3. The second group of data contains 260 compliant and 46 non-compliant documents. Out of the non-compliant documents, we randomly selected a subgroup of samples that we assume to be available during model optimisation. For our model, we used these samples to select the threshold discussed in Eq. 1, while we used them to supervise the training of the baselines that we will discuss in Sect. 2.3. We used the remaining non-compliant documents (i.e. 92 samples for L1, 28 samples for L2) as part of the test set. In the test set, we also included 206 randomly selected samples from the compliant data.

Table 1. Number of compliant and non-compliant documents for training and test set of dataset having layout L1 and L2.

Dataset	Compliant		Non-compliant	
	Train	Test	Train	Test
L1	2000	146	23	92
L2	200	60	18	28

---

[1] We corrupt compliant documents by: i) using random noise; ii) masking out parts of the documents with white boxes having side of 1–10% image size; and iii) applying image jigsaw with random patches of having 10–20% of the image size.

We summarize the specifics of the two datasets and their training splits in Table 1. In Fig. 1, we report a few examples of artefacts, amendments and missing information, which make documents non-compliant with internal guidelines. Note that the variety of possible artefacts and their limited number in the available datasets make vanilla supervised approaches inefficient. In both cases, we treat documents as 2D images that we feed to the model. We resize data to 152×152 pixel size and normalised them in $0 \div 1$ range. Then, we apply the data augmentation as: i) small random rotations in $0 \div 1$ degrees degree; ii) planar translations along the width and height axis, with a magnitude in $0 \div 2\%$ pixels; and iii) image zoom in between $99 \div 101\%$.

2.3 Baselines

We compare our method with standard supervised approaches. As baselines, we use a standard machine learning algorithm, i.e. Support Vector Classifier (SVC), and two popular deep learning models: VGG16 [17] and ResNet50 [6] pretrained with ImageNet weights. Since in our work we deal with two classes (i.e. compliant vs non-compliant documents) rather than the numerous classes inside ImageNet, we substituted the final fully connected layer of the deep learning models with a dense two-units softmax layer, which we train with a 10% dropout regularizer. To deal with unbalanced data we re-balance the minority class artificially generating new images (*upsampling* procedure) according to the data augmentation procedure described in Sect. 2.2.

3 Results

As illustrated in Table 2, our anomaly-detection based methods are the best across metrics and datasets. In particular, compared to SVC baseline, obtaining up to 13% accuracy increase on L1 data, 11% increase on L2 data.

We argue that such a performance increase is most probably due to the large class imbalance problem between compliant vs non-compliant documents. As anomaly detection-based approaches do not require any knowledge of the minority class, they are more flexible and lead to better results.

Comparing the detection of anomalous documents in pixel space (VAE+SSIM) vs in features-space (VAE+classifier), we notice that the latter is more consistent across metrics. Overall, using a classifier in the latent space seems more promising, probably because the model learns to amplify small anomalies that may be hidden by standard decoding reconstruction errors. However, the model working in pixels space (VAE+SSIM) seems more efficient with the small dataset.

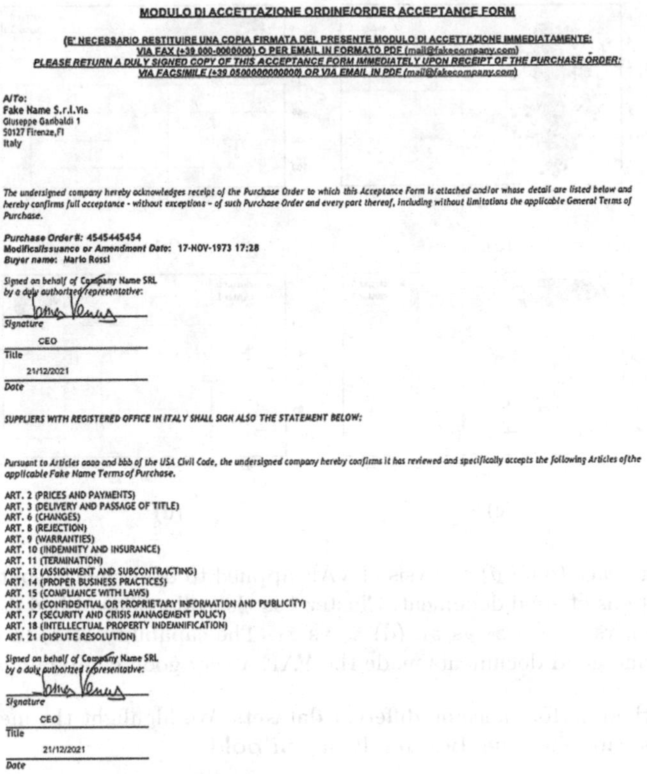

MODULO DI ACCETTAZIONE ORDINE/ORDER ACCEPTANCE FORM

(E' NECESSARIO RESTITUIRE UNA COPIA FIRMATA DEL PRESENTE MODULO DI ACCETTAZIONE IMMEDIATAMENTE:
VIA FAX (+39 000-0000000) O PER EMAIL IN FORMATO PDF (mail@fakecompany.com)
PLEASE RETURN A DULY SIGNED COPY OF THIS ACCEPTANCE FORM IMMEDIATELY UPON RECEIPT OF THE PURCHASE ORDER:
VIA FACSIMILE (+39 0500000000000) OR VIA EMAIL IN PDF (mail@fakecompany.com)

A/To:
Fake Name S.r.l.Via
Giuseppe Garibaldi 1
50127 Firenze,FI
Italy

The undersigned company hereby acknowledges receipt of the Purchase Order to which this Acceptance Form is attached and/or whose detail are listed below and hereby confirms full acceptance - without exceptions - of such Purchase Order and every part thereof, including without limitations the applicable General Terms of Purchase.

Purchase Order#: 4545445454
Modifica/issuance or Amendment Date: 17-NOV-1973 17:28
Buyer name: Mario Rossi

Signed on behalf of Company Name SRL
by a duly authorized representative:

Signature

CEO
Title

21/12/2021
Date

SUPPLIERS WITH REGISTERED OFFICE IN ITALY SHALL SIGN ALSO THE STATEMENT BELOW:

Pursuant to Articles aaaa and bbb of the USA Civil Code, the undersigned company hereby confirms it has reviewed and specifically accepts the following Articles of the applicable Fake Name Terms of Purchase.

ART. 2 (PRICES AND PAYMENTS)
ART. 3 (DELIVERY AND PASSAGE OF TITLE)
ART. 6 (CHANGES)
ART. 8 (REJECTION)
ART. 9 (WARRANTIES)
ART. 10 (INDEMNITY AND INSURANCE)
ART. 11 (TERMINATION)
ART. 13 (ASSIGNMENT AND SUBCONTRACTING)
ART. 14 (PROPER BUSINESS PRACTICES)
ART. 15 (COMPLIANCE WITH LAWS)
ART. 16 (CONFIDENTIAL OR PROPRIETARY INFORMATION AND PUBLICITY)
ART. 17 (SECURITY AND CRISIS MANAGEMENT POLICY)
ART. 18 (INTELLECTUAL PROPERTY INDEMNIFICATION)
ART. 21 (DISPUTE RESOLUTION)

Signed on behalf of Company Name SRL
by a duly authorized representative:

Signature

CEO
Title

21/12/2021
Date

Fig. 3. Sample from dataset L1. Compliant document of layout L1. Documents must be signed twice and should not contain external amendments. Dataset L2 is similar to dataset L1, signed once and with similar heading.

To better investigate the capability of the VAE+SSIM latent space to differentiate between compliant and non-compliant data, we performed a qualitative analysis of the 4×2 (μ ad σ) dimensions of $\underline{z}$. We find that the latent dimension z_0 can cluster non-compliant documents (Fig. 4(a)). However, not all the non-compliant documents are well detected (see red dots inside the green cluster), which supports our choice of training a separate model to detect anomalies from the feature space (VAE+classifier).

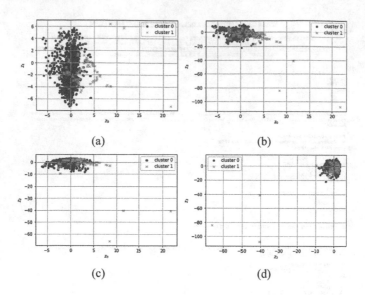

(a)   (b)

(c)   (d)

Fig. 4. Latent space (only μ) analysis of VAE applied to document layout L1. Cluster 0 is the collections of good document. Cluster 1 is the collections of bad document. (a) z_0 vs z_1. (b) z_0 vs z_3. (c) z_0 vs z_2. (d) z_2 vs z_3. The capability of z_0 to discriminate between bad and good documents made the VAE a very good candidate for the task.

Table 2. Method performance on different datasets. We highlight the methods investigated in this paper in   blue . Best results are in **bold**.

Dataset	Method	Accuracy	Precision	Recall	f1-score
	SVC	79%	81%	80%	79%
	VGG16	63%	47%	62%	52%
L1	ResNet50	67%	44%	67%	53%
	VAE+SSIM	**92%**	62%	**92%**	93%
	VAE+Classifier	84%	**85%**	83%	84%
	SVC	76%	70%	78%	74%
	VGG16	85%	68%	68%	67%
L2	ResNet50	88%	67%	71%	69%
	VAE+SSIM	**87%**	69%	**86%**	73%
	VAE+Classifier	85%	**70%**	81%	73%

4   Industrial Application

Given these promising results, we deployed the model in a production environments. In Fig. 5, we briefly describe the usage of the tool.

As illustrated, an external user can submit a document using an existing supplier portal. Then, a Robotic Process Automation (RPA) interacts with the

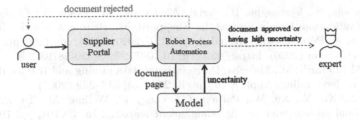

Fig. 5. Application of the tool in an industrial production environment.

described deep learning model. The model classifies the input as compliant vs non-compliant, providing also an uncertainty score (i.e. the $\mathcal{L}^{-1}$ for eq. 1, and $1 - h_f$ for Eq. 2). When the model detects non-compliant documents, the system rejects them showing an error message. Instead, the RPA forwards to a human expert the documents classified as compliant or having high prediction uncertainty. In this way, we consistently reduce the number of non-compliant documents in the system, we also avoid the false positives (i.e., documents detected as non-compliant while being compliant) from hampering the perceived quality of the system by external users. The model was successfully integrated into our internal workflow in December 2020, proving effectiveness and reliability.

5   Conclusion

We apply a VAE to recognize if a document was compliant with a predefined layout discarding documents with unwanted artefacts and amendments superposed by users. In particular, we formulate the Document Layout Validation as an anomaly detection problem, and we investigate two different approaches to detect non-compliant documents. These approaches learn the distribution of compliant documents, and analyse input data either in pixel space or in feature space. Both approaches are able to identify non-compliant documents with an accuracy of about 90%.

The investigated methods outperforms standard supervised approaches, showing that VAEs are a valuable and reliable method also for real industrial applications. Indeed, the model was successfully integrated in our internal workflow (as described in Sect. 4) providing reliable and effective results.

References

1. Ares Oliveira, S., Seguin, B., Kaplan, F.: dhsegment: a generic deep-learning approach for document segmentation. In: International Conference on Frontiers in Handwriting Recognition, pp. 7–12 (2018)
2. Binmakhashen, G.M., Mahmoud, S.A.: Document layout analysis: a comprehensive survey. ACM Comput. Surv. **52**(6), 1–36 (2019)

3. Bragagnolo, S., Verhaeghe, B., Seriai, A., Derras, M., Etien, A.: Challenges for layout validation: lessons learned. In: Shepperd, M., Brito e Abreu, F., Rodrigues da Silva, A., Pérez-Castillo, R. (eds.) QUATIC 2020. CCIS, vol. 1266, pp. 107–119. Springer, Cham (2020). https://doi.org/10.1007/978-3-030-58793-2_9

4. Ceci, M., Berardi, M., Malerba, D.: Relational data mining and ILP for document image understanding. Appl. Artif. Intell. **21**(4–5), 317–342 (2007)

5. Ding, Z., Xu, Y., Xu, W., Parmar, G., Yang, Y., Welling, M., Tu, Z.: Guided variational autoencoder for disentanglement learning. In: CVPR, pp. 7917–7926. IEEE (2020)

6. He, K., Zhang, X., Ren, S., Sun, J.: Deep residual learning for image recognition. In: CVPR, pp. 770–778. IEEE Computer Society (2016)

7. Kingma, D.P., Welling, M.: Auto-encoding variational bayes. In: Bengio, Y., LeCun, Y. (eds.) ICLR (2014)

8. Kosaraju, S.C., et al.: Dot-net: document layout classification using texture-based CNN. In: International Conference on Document Analysis and Recognition (ICDAR), pp. 1029–1034 (2019)

9. Kuang, Z., et al.: Mmocr: a comprehensive toolbox for text detection, recognition and understanding. ArXiv preprint abs/2108.06543 (2021)

10. Li, J., Yang, J., Hertzmann, A., Zhang, J., Xu, T.: Layoutgan: generating graphic layouts with wireframe discriminators. In: ICLR. OpenReview.net (2019)

11. van den Oord, A., Vinyals, O., Kavukcuoglu, K.: Neural discrete representation learning. In: Guyon, I., et al. (eds.) NeurIPS, pp. 6306–6315 (2017)

12. Paliwal, S.S., D, V., Rahul, R., Sharma, M., Vig, L.: Tablenet: deep learning model for end-to-end table detection and tabular data extraction from scanned document images. In: 2019 International Conference on Document Analysis and Recognition (ICDAR), pp. 128–133 (2019)

13. Prati, R., Batista, G., Monard, M.C.: Data mining with imbalanced class distributions: concepts and methods, pp. 359–376 (2009)

14. Razavi, A., van den Oord, A., Vinyals, O.: Generating diverse high-fidelity images with VQ-VAE-2. In: Wallach, H.M., Larochelle, H., Beygelzimer, A., d'Alché-Buc, F., Fox, E.B., Garnett, R. (eds.) NeurIPS, pp. 14837–14847 (2019)

15. Ruff, L., et al.: A unifying review of deep and shallow anomaly detection. In: Proceedings of the IEEE (2021)

16. Sabuhi, M., Zhou, M., Bezemer, C.P., Musilek, P.: Applications of generative adversarial networks in anomaly detection: a systematic literature review. IEEE Access **9**, 161003–161029 (2021)

17. Simonyan, K., Zisserman, A.: Very deep convolutional networks for large-scale image recognition. In: Bengio, Y., LeCun, Y. (eds.) ICLR (2015)

18. Wang, Z., Bovik, A., Sheikh, H., Simoncelli, E.: Image quality assessment: from error visibility to structural similarity. IEEE Trans. Image Process. **13**, 600–612 (2004)

19. Wu, X., Hu, Z., Du, X., Yang, J., He, L.: Document layout analysis via dynamic residual feature fusion. In: 2021 IEEE International Conference on Multimedia and Expo (ICME), pp. 1–6. IEEE (2021)

20. Zhou, C., Paffenroth, R.C.: Anomaly detection with robust deep autoencoders. In: International Conference on Knowledge Discovery and Data Mining, pp. 665–674. ACM (2017)

Author Index

Printed in the United States
by Baker & Taylor Publisher Services

Printed in the United States
by Baker & Taylor Publisher Services